NOW YOU'RE TALKING!

ALL YOU NEED TO GET YOUR HAM RADIO TECHNICIAN LICENSE

3rd Edition

Edited by **Larry D. Wolfgang**, WR1B
Joel P. Kleinman, N1BKE

Production Staff

Dan Wolfgang, Electronic Publications Assistant: Layout

Joe Shea, Production Assistant: Layout

David Pingree, N1NAS, Senior Technical Illustrator: Technical Illustrations and Computer Cover Art

Steffie Nelson, KA1IFB, Proofreader

Paul Lappen, Production Assistant: Layout

Jodi Morin, KA1JPA, Assistant Production Supervisor: Layout

Sue Fagan, Graphic Design Supervisor: Cover Design

Joe Costa, Technical Illustrator: Technical Illustrations and HT Cover Art

Michelle Bloom, WB1ENT, Production Supervisor: Layout

Published by:

The American Radio Relay League

Newington, CT 06111-1494

Printed in USA

Library of Congress Catalog Card Number: 93-70669

ISBN: 0-87259-597-8

Third Edition
Second Printing, 1997

This book may be used for Novice and Technician exams given beginning July 1, 1997. The book may be used for exams given until July 1, 2001. (This ending date assumes no FCC rules changes to the licensing structure or privileges for these two license classes, which would force the VEC Question Pool Committee to modify these question pools.)

FOREWORD

You are about to enter the wonderful world of Amateur Radio! This hobby is truly all about the very best things in life: great friends, devotion to public service, intellectual and scientific curiosity, exploration and just plain fun.

I was a member of the crew of STS-78 aboard Space Shuttle Columbia, which flew in June 1996. The other crew members and I enjoyed the tremendous honor and pleasure of talking from space with so many friends in Amateur Radio around the world. During that time I couldn't help but reflect on all the deeply personal and emotional moments Amateur Radio had brought to me. I have many fond memories of my own great mentor in the traditions and ethics of Amateur Radio, my "Elmer," Mr Bill Brown, W4NIW (now a Silent Key — but he will live forever in my heart). I remember the sheer exhilaration of listening to signals from around the world as a young boy in a darkened room with the glow of a tube radio my dad helped me build. As a teenager I talked with Senator Barry Goldwater and King Hussein of Jordan on the radio and didn't realize who they were until I looked up their call signs later. Sitting on St. Peter and St. Paul Rocks on a DXpedition in the middle of the Atlantic Ocean with my friends Al, K8CW; Stu, WA2MOE; Phillip, PY2CPU and Jacinto, PY2BZD, I was in awe of a place where Charles Darwin had visited on the voyage of the Beagle. I experienced these same wonderful feelings while talking from space to the great fraternity of ham operators around the world.

There is no greater feeling for a person in space than the warm personal contact with another human being on Earth. It will be so important for us to be able to communicate with one another in this way as we fulfill our planet's destiny in space. I am absolutely sure that wherever we travel in our universe, whether it is on the International Space Station, back to the lunar surface to live or even to Mars, there will be Amateur Radio and fellow hams reaching out like they always do.

There is truly something for everyone in this fantastically diverse world of Amateur Radio. I am always amazed with the many activities Amateur operators get involved in: building, testing, contests, clubs, computers, public service, satellites and so many more.

What a great thing it is that anywhere we go, be it on Earth or in space, we can always call CQ and reach out with our first names and call signs to all our fellow human beings of the Earth!

I hope I have my chance to talk with you!

73,

Chuck, N4BQW

Mission Specialist Charles E. Brady, Jr

PREFACE

Welcome to the exciting world of Amateur Radio! You are about to join the more than 650,000 licensed Amateur Radio operators in the United states and over two and a half million people from around the world who call themselves "hams." Hams are found in virtually every country in the world. They have earned the very special privilege of being able to communicate directly with one another, by radio, without regard to the geographic and political barriers that so often limit our understanding of the world.

Whether across town or across the sea, hams are always looking for new friends. So wherever you may happen to be, you are probably near someone—perhaps a whole club—who would be glad to help you get started. If you need help contacting hams, instructors, Volunteer Examiners or clubs in your area, contact us here at ARRL Headquarters. We'll help you get in touch with someone near you. (See the contact information listed at the bottom of this page.)

Whether you want to learn Morse code and enter Amateur Radio as a Novice licensee or choose to skip the code and start with the Technician license, you'll find plenty of Amateur Radio activity to keep you busy. You'll also find plenty of friendly folks who are anxious to help you get started. Amateur Radio has many interesting areas to explore. You may be interested in one particular aspect of the hobby now, but be willing to investigate new avenues occasionally. You'll discover a world of unlimited potential!

Most of the active radio amateurs in the United States are members of the American Radio Relay League. ARRL has been the hams' own organization for over 80 years. We provide training materials and other services, and represent our members nationally and internationally. *Now You're Talking!* is just one of the many ARRL publications for all levels and interests in Amateur Radio. You don't need a ham license to join. If you're interested in ham radio, we're interested in you. It's as simple as that!

David Sumner, K1ZZ

Executive Vice President

Newington, Connecticut

March 1997

New Ham Desk
ARRL Headquarters
225 Main Street
Newington, CT 06111-1494
(860) 594-0200

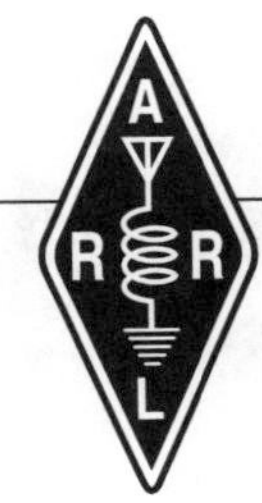

Prospective new amateurs call:
800-32-NEW-HAM (800-326-3942)
You can also contact us via e-mail: **newham@arrl.org**
or check out our World Wide Web site: **http://www.arrl.org/**

CONTENTS

TWO TRACKS TO YOUR FIRST HAM RADIO LICENSE

For many years, the Novice license was the entry route for most newcomers to Amateur Radio. The Novice license written exam covers basic theory and regulations. It also includes a 5-word-per-minute (wpm) Morse code test. Novices are allowed to operate code on four high-frequency (HF) bands, where world-wide communication is possible without satellites or repeater stations. Novices are also allowed to use voice on one HF band and on two higher-frequency bands used mostly for local communications. Other privileges include radioteletype and computer-to-computer communications on certain bands. The Technician class license was usually the next step up for many radio amateurs.

The Technician license has now become the favorite entry-level license for most newcomers. In fact, today there are more Technician licensees than any other license class. Effective February 14, 1991, FCC removed the Morse-code requirement for obtaining a Technician license.

The Technician exam includes the Novice written exam and a slightly more technical exam covering electronics and radio theory along with questions about the Federal Communications Commission (FCC) rules and regulations.

Technician licensees have full VHF and UHF privileges granted to higher-class licensees, but *not* the HF privileges granted to Novices. Those Novice-band privileges can be earned, however, by any Technician licensee who passes a 5-wpm code test. This results in a Technician Plus license.

Any holder of a new Technician license who passes the code test is issued a *Certificate of Successful Completion of Examination* (CSCE) and is immediately granted Novice operating privileges on the HF bands. (A Technician Plus license will arrive in the mail from the FCC after the upgrade application is processed.) Novices who upgrade to Technician (or who upgrade within two years of the expiration of their Novice licenses) are also automatically granted a Technician Plus license when they pass the Technician written exam.

When you are ready to take a Novice exam, contact your local ham radio club, class instructor or ARRL Headquarters if you have any questions about testing procedures.

WANT MORE INFORMATION?

Looking for more information about ham radio in your local area? Interested in taking a ham radio class? Ready for your license exam? Call 1-800-32 NEW HAM (1-800-326-3942). Do you need a list of ham radio clubs, instructors or examiners in your local area? Just let us know what you need! You can also contact us via e-mail:

newham@arrl.org

or check out our World Wide Web site:

http://www.arrl.org/

You can even write to us at:
New Ham Desk
ARRL Headquarters
225 Main Street
Newington, CT 06111-1494

How To Use This Book

You're about to begin an exciting adventure: a fun-filled journey into the world of Amateur Radio. *Now You're Talking!* is *the* study guide to help you reach your goal. This book introduces you to basic radio theory in an easily understood style. You'll also learn Federal Communications Commission (FCC) rules and regulations—just as the FCC requires. You will learn enough to pass your Novice or Technician written exam with ease. But we in the American Radio Relay League care much more about you and your future in Amateur Radio than to stop there! We won't abandon you once your new "ticket" is in hand.

Now You're Talking! is also a pathway to a successful beginning after you've passed your license exam. In this book you'll find the practical knowledge needed to become an effective communicator. We all hope you will take enough pride in the achievement of earning your license to be a considerate communicator as well.

But that's all a few weeks and a few pages down the line. To ensure your success and head you in the right direction toward your first on-the-air contact, here's how to use this book to your best advantage.

Now You're Talking! has been designed and written by a staff with a great deal of experience, backed by decades of Amateur Radio tradition. This book provides everything you'll need to learn—and understand—what you should know to operate an Amateur Radio station. We explain all the rules and regulations required of Novice and Technician candidates, not only to pass the test, but to operate properly (legally) once your license is hanging proudly in your shack.

Self Study or Classroom Use?

We designed *Now You're Talking!* both for self study and for classroom use. An interested student will find this book complete, readable and easy to understand. Read carefully, and test yourself often as you study. Before you know it, you'll be ready to pass that exam!

Why deprive yourself of the company of fellow beginners and the expertise of those "old-timers" in your hometown, though? *Now You're Talking!* goes hand-in-hand with a very effective ARRL-sponsored training program run by over 6000 volunteer instructors throughout the United States. If you would like to find out about a local class, just contact the New Ham Desk at ARRL Headquarters—we'll be happy to assist. (Our phone number and e-mail address were listed earlier.)

Hams are very social animals who derive a great deal of pleasure from helping a newcomer along the way. The most effective learning situation is often the one you share with others. There are knowledgeable people to turn to when you have a question or problem. You can practice the Morse code with fellow students, and quiz one another on basic electronics concepts. It doesn't matter if you're studying on your own or joining a class. Use *Now You're Talking!* to study for your Novice or Technician exam and you'll be on the air in no time at all.

Using This Book

Now You're Talking! leads you from one subject to the next in a logical sequence that builds on the knowledge learned in earlier sections. It presents the material in easily digested and well-defined "bite-sized" sections. You will be directed to turn to the question pool in Chapter 13 (and 14 if you're preparing for the Technician exam) as you complete a section of the material.

This review will help you determine if you're ready to move on. It will also highlight those areas where you need a little more study. In addition, this approach takes you through the entire question pool. By the time you complete the book, you will be familiar with all the questions used to make up your test. Please take the time to follow these instructions. Believe us, it's better to learn the material correctly the first time than to rush ahead, ignoring weak areas and unresolved questions.

Every page of Chapters 1 through 10 presents information you'll need to pass the exam and become an effective operator. Pay attention to diagrams, photographs, sketches and captions; they contain a wealth of information you should know. You'll also find a few anecdotes and "mini-articles" (called Sidebars) that will help put the tradition of Amateur Radio in perspective. Our roots go back to the beginning of the 20th Century, and our community service continues even as you read this. (Chapter 11 provides some information you will surely find helpful when you are ready to learn the Morse code and Chapter 12 will help guide you through the confusing process of selecting equipment for your station.)

As you study *Now You're Talking!*, you'll find that several chapters have sections of material with additional information for Technician licensees. These sections contain information needed for the Technician exam. If you are studying for a Novice license you may want to skip those sections. While you won't find any exam questions from this material on your Novice exam, you may find it interesting and helpful to read.

Start at the beginning. The Introduction summarizes the fun you will have when you earn your license and join the thousands of other active radio amateurs. Chapter 1 explains the need for international and national regulation. It describes the regulating bodies and the relevant sections of Part 97 of the FCC Rules and Regulations.

Chapters 2 through Chapter 10 break basic radio theory and amateur practices into well-defined sections. Study the material in each section carefully. It explains the theory you'll need to answer the questions in Chapters 13 and 14.

Study the material presented in this book and follow the instructions to review the exam questions. You'll cover small sections of the text and a few questions at a time. This is the best way to determine how well you understand the most important points. The text explains the theory in straightforward terms, so you shouldn't have any problems. Review the related sections if you have any difficulty, and then go over the questions again. In this way you will soon be ready for your exam. Before you know it, you'll be on the air!

Don't be afraid to ask for help if you don't understand something, though. If you're participating in an official ARRL-registered class, you'll have the chance to ask the experts for help. Ask your instructor about anything you are having difficulty with. You may also find it helpful to discuss the material with your fellow students. If you're not in a class and run into snags, don't despair! The folks at the New Ham Desk at ARRL Headquarters will be happy to put you in touch with an Amateur Radio operator in your area who can help answer your questions.

Most people learn more when they are actively involved in the learning process. Turning to the questions and answers when directed in the text helps you be actively involved. Fold the edge of the question-pool pages to cover the answer key and page-reference information. Check your answers after you study each group of questions, and review the appropriate text for any questions you get wrong. Paper clips make excellent place markers to help you keep your spot in the text and the question pool.

If You Decide to Learn the Code

Chapter 11 lets you in on a little secret: The Morse code is not only easy to learn; it can be so enjoyable that you may become "addicted." You should read this chapter before beginning to learn the Morse code. If you think you have decided *not* to learn the code you should read this chapter to learn about the fun some people have with the code. You may even decide it wouldn't be as difficult as you may have thought.

People try to learn the code by many different methods. Some methods make the process much easier. The cassette tapes (or audio CDs if you prefer the higher-quality sound and other advantages offered by CDs) in *Your Introduction to Morse Code* teach you the Morse code at your own pace. You should review any characters that give you difficulty. The technique used with *Your Introduction to Morse Code* also makes it easy to increase your speed once you've learned the code.

After reading Chapter 11, begin studying the code and theory in "parallel;" that is, split your study time between daily sessions with the code and study sessions with the text. Move ahead at a pace that allows you to master the material. You do have to keep moving, however. Don't allow yourself to stay on one section too long, even if you have to come back and review later.

Before You Take Your Test

If you need help in locating someone to administer a Novice or Technician test, or for a schedule of Volunteer Examiner test sessions in your area, write to the ARRL/VEC Office, ARRL Headquarters, 225 Main St, Newington, CT 06111-1494. If you need help locating a ham, instructor or club, contact the New Ham Desk at ARRL Headquarters as listed earlier. We can put you in touch with examiners, instructors and clubs in your area!

Give *Now You're Talking!* a chance to guide you the way it was intended—by following these instructions. You'll soon be joining us on the air. Each of us at the American Radio Relay League Headquarters and the entire ARRL membership wishes you the very best of success. We are all looking forward to that day in the not-too-distant future when we hear your signal on the ham bands. 73 (best regards) and good luck!

INTRODUCTION: Welcome to Amateur Radio

Join us for a fascinating journey through the wonderful world of Amateur Radio.

How did you first hear about Amateur Radio? That evening news report about the Amateur Radio operators who relayed messages after a devastating hurricane or terrible earthquake? The funny-looking antenna in your neighbor's yard? A birthday greeting via Amateur Radio from your uncle who lives clear across the country? Or maybe you heard about hams who talked to astronauts during a space-shuttle mission or on the Russian *Mir* space station.

Obviously, you know something about Amateur (ham) Radio; you have this book. But you want to know more. *Now You're Talking!* will introduce you to the wonderful hobby of Amateur Radio. It will answer your many questions on the subject, and lead you to your first license.

What Can I Do As a Ham?

Ham radio offers so much variety, it would be hard to describe all its activities in a book twice this size! Most of all, ham radio gives you a chance to meet other people who like to communicate. That's the one thing all hams have in common. You can communicate with other hams on a simple hand-held radio that fits in your pocket. Thanks to "repeater stations" (*repeaters*) operated by local ham clubs, your range with a tiny hand-held radio or "H-T" may be 50 miles or more.

Repeaters also make it possible for mobile ham stations to keep in touch. Mobile operating is very popular during commuting hours. There's almost always someone to talk to on a repeater. Most people would like to do something to help their communities. Someday, you may spot another motorist who needs help, and use the repeater telephone link to call the police. Or, your repeater club may provide communications for a parade, foot race, or even during an emergency like a flood. Best of all, you'll be meeting other people nearby who are as excited about Amateur Radio personal communications as you are.

Imagine talking to a missionary operating a battery-powered station deep in the Amazon jungle, or a sailor attempting an around-the-world solo journey! You can talk to hams all over the world in many ways. The most popular way is by bouncing your signal off the ionosphere, a layer in the upper atmosphere. Other hams like to use the OSCAR satellites. OSCAR means Orbiting Satellite Carrying Amateur Radio. Hams have designed and built over 30 OSCARs since 1961. If you like a challenge, you can even bounce your signals off the moon. It's possible to contact other hams in more than 100 countries by this method, as strange as it sounds!

How do you talk to these other hams? Well, you can use voice of course, but there are other ways as well. Maybe you've tried computer-to-computer conversations over the telephone lines. As a ham, you can have similar conversations with other hams around the world. The best part is, you don't have to pay for the call!

The earliest radio communication was done with Morse code. You don't have to know the code to become a ham anymore, but many hams still enjoy using this funny language. To them, the beeping of Morse code is like listening to a favorite song.

Maybe you're interested in photography or video. Many hams have television equipment, including color systems. Hams led the way in *slow-scan television*, which lets you send color photographs, slides and artwork to other hams thousands of miles away.

How about radio-controlled models? Yes, many hams enjoy using exclusive ham frequencies to fly gliders and powered planes, or pilot graceful sailboats. The Technician license (with no Morse code exam) is all you need to join them.

If you like to work on electronic circuits, ham radio gives you the chance to build your own transmitters and receivers, and actually use them to talk to other people. Even if you'd just like to plug in a radio and go on the air, you might enjoy building some part of your station, from scratch or from a kit.

Unlike shortwave or scanner listening, Amateur Radio doesn't make you sit on the sidelines. When East and West Germany became one country, hams didn't have to learn about it from the TV news: they talked directly to hams in both halves of the country! Instead of listening to police calls on Halloween, you can join your club's "Goblin Watch," and help make the streets safer for Trick-or-Treaters.

Ham radio operators are proud of their hobby. That's why you see so many license plates with ham call signs. Communicating on the air isn't all we do, though. We get together at club meetings, hamfests and conventions. To welcome newcomers we sponsor thousands of classes each year to help anyone who's interested join our great hobby.

Two Paths to Choose From

There are six classes of Amateur Radio license in the United States. All US amateur licenses have 10-year terms and are renewable. Licenses are issued by the Federal Communications Commission (FCC). There is no license fee. Although you can take the exams and start right out at the highest class, most beginners enter Amateur Radio with either the Novice or Technician license.

There are several differences between the two entry-level licenses, in exams and operating privileges. They are described in the following sections. To pass the Novice license exam, you're required to know Morse code, but there's no code exam for the Technician license. No matter which path you choose, this book has all the information you need to pass the written exams.

The Novice License

To receive a Novice license you have to pass a 35-question written exam (Element 2) in basic electronics theory, FCC regulations and operating practices. A passing grade is 74%, or at least 26 correct answers out of the 35 questions. You also have to copy the international Morse code at 5 words per minute (wpm) and either answer correctly 7 out of 10 questions about your copy or copy one minute without errors on a 5-minute exam.

Novice licensees are granted a wide range of privileges on several amateur bands. Some of these bands are mainly useful for local communications, but the "high-frequency" (HF) Novice bands offer the ability to talk to other radio amateurs around the world, without the use of satellites (we'll tell you more about satellite communication later). The Morse code and many abbreviations such as "Q signals" are understood internationally. This makes Morse code the ideal language for communicating with hams in foreign countries. Plenty of stations in other countries operate in the Novice bands. Quite a few Novices have even contacted hams in more than 100 different countries. Chapter 1 has more information about Novice-license privileges.

Novice exams are given by three Volunteer Examiners (VEs) with General, Advanced or Amateur Extra licenses.

Be On the Frontier of Technological Advances

For nearly 100 years hams have carried on a tradition of learning by doing. From the earliest days of radio, hams have built their transmitters from scratch, wrapping strands of copper wire salvaged from Model T automobiles around oatmeal boxes. Through experimenting with building their own equipment, hams have pioneered advances in technology such as the techniques for single sideband voice. Hams were the first to bounce signals off the moon to extend signal range. Amateurs' practical experience has led to many technical refinements and cost reductions beneficial to the commercial radio industry. The photo at the right shows a complete amateur television (ATV) transmitter constructed in a small package. This photo is from a project described in the November 1996 issue of *QST*.

The examiners must be accredited by a Volunteer Examiner Coordinator (VEC).

For help learning the code we recommend *Your Introduction to Morse Code*, a package of two audio CDs or two audio cassettes that teach you the code, letter by letter. When you know the code, this package gives you practice at speeds a little faster than the 5 wpm speed required for the Novice exam. By practicing code at 6 to 7 wpm you will have a bit of additional confidence when you take your exam. Chapter 11 of this book has more information about learning Morse code.

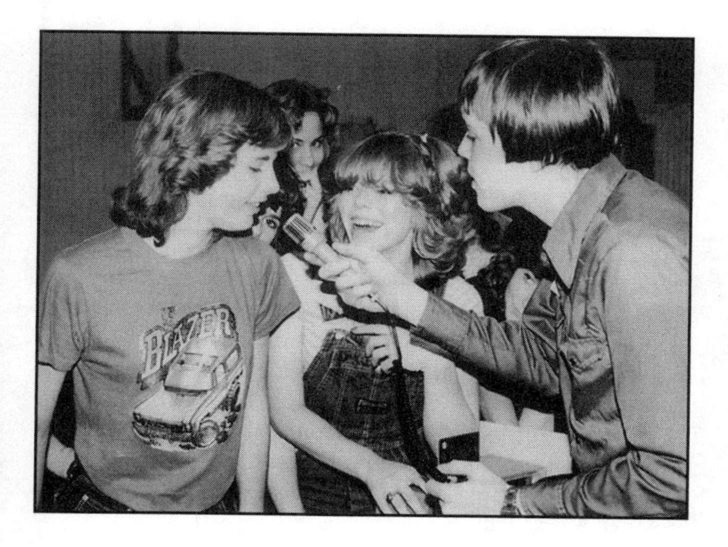

The Technician License

If you aren't interested in learning the Morse code right now, the Technician license is for you. To receive a Technician license you have to pass the same 35-question written exam as for the Novice license (Element 2), plus an additional 30-question Technician (Element 3A) exam. The additional exam covers the same subjects, but in more detail. You must get at least 22 out of 30 answers correct on the Element 3A written exam to pass that one.

Technician licensees are granted full amateur privileges on the extensive VHF, UHF and microwave amateur bands, including Amateur Radio satellite bands. Yes, with a Technician license you can talk to other amateurs around the world via satellite! Technician licensees can use the popular 2-meter band, where thousands of amateur-owned repeater stations provide solid communication over large areas.

If you start out as a Technician and then decide you'd also like to use the high-frequency Novice bands, all you have to do is pass a 5-wpm Morse code exam. Your new privileges will begin the day you pass the exam. There is enough Morse code activity on the Technician bands to help you if you decide to learn the code. Chapter 1 has more information about Technician-license privileges.

The Technician license exam is given (as are all amateur license exams) by teams of Volunteer Examiners (VEs), who are accredited by a Volunteer Examiner Coordinator (VEC). Thousands of exam sessions are held each year, so you shouldn't have to travel far or wait long to take one. Contact the ARRL/VEC Office for information on exam sessions near you. Call 860-594-0300, send e-mail to **vec@arrl.org** or write to ARRL/VEC, 225 Main Street, Newington, CT 06111-1494.

We Come From All Walks of Life

Having fun communicating with other hams and experimenting with antennas and radio circuits — that's what Amateur Radio is all about. That's why people from all walks of life become hams. Young or old, we all enjoy the thrill of meeting and exchanging ideas with people from across town or from the other side of the Earth. The excitement of building a new project or getting a circuit to work properly and then using that project to talk with someone "over the air" is almost beyond description.

Carm Prestia was a diamond-in-the-rough police sergeant, patrolling a bustling university town tucked away in the mountains of Pennsylvania when he became interested in Amateur Radio. By night, he packed a .38-caliber Police Special to protect thousands of his fellow townspeople. By day, he plugged in a soldering iron in pursuit of the world's greatest hobby: Amateur Radio.

"I love to talk to people on the radio," Carm explains as he unstraps his portable radio from his uniform belt. "I talk all night at work, but I still go home and fire up the ham gear."

Carm's radio room (hams all over the world affectionately call this room their "shack") is in a corner of his basement. His equipment table holds a transmitter for sending and a receiver for listening. His radio gear works with an antenna outside, above his backyard. Carm can talk with a friend in the next town one minute and with a ham halfway around the world in Australia the next.

Each Amateur Radio station has its own distinctive call sign. The Federal Communications Commission (FCC) issued Carm his call sign, WB3ADI.

Ham radio operators are so proud of their call signs that the two often become inseparable in the minds of friends. Barry, K7UGA, of Arizona has worked (talked with) thousands of hams on the air. Many of them didn't know that his last name is Goldwater or that he was a United States senator.

Age Is No Barrier

Age is no barrier to getting a ham license and joining in the fun. There are hams of all ages, from five years to more than 80 years. Samantha Fisher received her Technician license at age 9. Known to her ham friends as KA1VBQ, she enjoys talking on 10 meters.

Then there's Luke Ward, KO4IQ, of Alexandria, Virginia, who was licensed at age 7. Luke upgraded to General at age 8, and became an Amateur Extra licensee when he turned 11. Luke hopes to experiment with "moonbounce"—reflecting VHF/UHF signals off the moon to communicate over long distances.

Like many hams, Dan Broniatowski, N8QWI, enjoys chasing DX from his Cleveland Heights, Ohio, station. Dan was licensed at the age of 10, and received his General license less than a year later.

Of course, young people aren't the only ones who are active radio amateurs. Involved in many hobbies, Evelyn Fox of Merrimac, Wisconsin, played contract bridge with her AARP (American Association of Retired Persons) group on the 40-meter band.

Evelyn was over 75 years old when she became interested in Amateur Radio. That didn't stop her from taking on the job of learning radio theory and the international Morse code. She found the theory a pleasant challenge. She joined a club, attended its classes, and became WB9QZA. Not bad for someone who knew nothing about electronics when she first got started.

Bill Bliss, KD6AHF received his Novice license about 3 weeks before his birthday. So what, you say? He was 82 years old on that birthday! Bill says, "Some think that a bit old to start ham radio, but I have the rest of my life to enjoy it."

Herb Kline is further proof that it's never too late to become a ham. Herb, KB2QGD, of Adams, New York passed his Novice exam just before his 83rd birthday and upgraded to Technician Plus a few weeks later. With plans to upgrade to General, Herb likes to operate CW. He is also learning about computers.

"Brewing It" At Home

Ham radio operators pop up in some of the least expected places. Dr. Peter Pehem, 5Z4JJ, is one of Africa's flying doctors. He works out of a small village on the north slope of Mount Kilimanjaro in Kenya. Pete has been bitten by an OSCAR bug, but can't do anything about it while on medical duty. But he can and does attack it with great pleasure when he's off duty.

Somebody gave Pete an old radiotelephone, a vacuum tube and some coaxial cable. The doctor added empty aspirin tins and a quartz crystal from his airplane radio. Right out there in the African bush, he fired up a homemade transmitter, built on the aspirin tins. Then he talked to the world through OSCAR, an Orbiting Satellite Carrying Amateur Radio.

Pete proved something with his home-made gear: You don't need a shack full of the latest commercial equipment to have fun on the air. New hams find this out every day. Tom Giugliano, WA2GOQ, of Brooklyn, New York, contacted 26 states using a pre-World War II transmitter and receiver. He used a simple home-made wire antenna. Other hams have bridged the oceans to contact hams in Europe and Japan using simple equipment running less than 1 watt of power. As a Novice, you'll be permitted to run 200-watts output on the HF bands. That's more than enough to contact other hams around the world. Technician Plus licensees, who have passed a 5-wpm code test, have the same privileges. On the VHF, UHF and microwave bands, Technicians are allowed to use 1500-watts output, although that much power is rarely needed or used.

There was a time, many years ago, when no commercial equipment was available. The earliest hams, beginning more than 75 years ago, tried to find more efficient ways of communicating with each other. All early radio sets were "home brew" (home built) and were capable only of communication over several miles. Some transmitters were nothing more than a length of copper wire wrapped around an oatmeal box, attached to a few other basic parts and a wire antenna. Often the transmissions were one-way, with one transmitting station broadcasting to several receive-only stations. Over the years, hams have continually looked for ways to transmit farther and better. They are constantly developing and advancing the state of the art in their quest for more effective ways of communicating.

Peering Back Through Time

It all started on a raw December day in 1901. Italian inventor and experimenter Guglielmo Marconi launched the Age of Wireless from an abandoned barracks at St John's, Newfoundland. He listened intently for a crackling series of buzzes, the letter S in international Morse code, traversing the 2000 miles from Cornwall, England. That signal was the culmination of years of experimentation. News of Marconi's feat stimulated hundreds of electrical hobbyists to build their own "wireless" equipment. They became the first hams.

Later, Marconi set up a huge station at Cape Cod that was unlike anything today's ham has experienced. Marconi's 3-foot-diameter spark-gap rotor fed 30,000 watts of power to a huge antenna array suspended from four 200-foot towers on the dunes at South Wellfleet, Massachusetts.

By 1914, Marconi had set up a station and antennas for daily transmission across the Atlantic. At the same time, Amateur Radio operators all over America were firing up their own homebuilt transmitters. Soon, several hundred amateurs across the country joined Hiram Percy Maxim in forming the American Radio Relay League (ARRL), based in Hartford, Connecticut. These amateurs set up a series of "airborne trunk lines" through which they could relay messages from coast to coast. If you're interested in learning

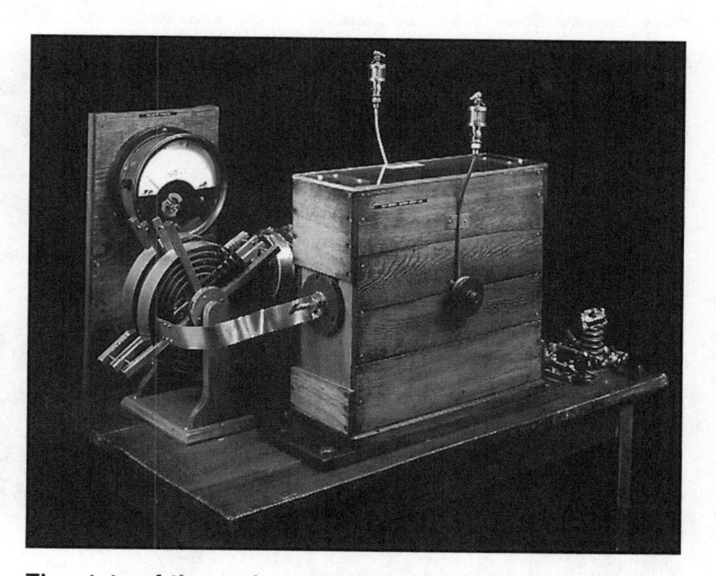

The state-of-the-art in amateur gear has come a long way since the days of "Old Betsy," Hiram Percy Maxim's own spark-gap transmitter. Spark-gap transmitters were the very first type of radio transmitters, used in the early 1900s.

Some Early Amateur Ingenuity

One young lad of seventeen, known to possess an especially efficient spark, CW and radio telephone station, was discovered to be the son of a laboring man in extremely reduced circumstances. The son had attended grammar school until he was able to work, and then he assisted in the support of his family. They were poor indeed. Yet despite this the young chap had a marvelously complete and effective station, installed in a miserable small closet in his mother's kitchen. How had he done it? The answer was that he had constructed every last detail of the station himself. Even such complicated and intricate structures as head telephones and vacuum tubes were homemade! Asked how he managed to make these products of specialists, he showed the most ingenious construction of headphones from bits of wood and wire. To build vacuum tubes he had found where a wholesale drug company dumped its broken test tubes, and where the electric light company dumped its burned-out bulbs, and had picked up enough glass to build his own tubes and enough bits of tungsten wire to make his own filaments. To exhaust the tubes he built his own mercury vacuum pump from scraps of glass. His greatest difficulty was in securing the mercury for his pump. He finally begged enough of this from another amateur. And the tubes were good ones — better than many commercially manufactured and sold. The greatest financial investment that this lad had made in building his amateur station was 25 cents for a pair of combination cutting pliers. His was the spirit that has made Amateur Radio. — Clinton B. Desoto, in *200 Meters and Down*

more about the history of Amateur Radio, you'll enjoy *200 Meters and Down* by Clinton B. DeSoto. This book is available directly from the ARRL. *Fifty Years of ARRL* is also an interesting account of the ARRL's first half century, but that book is out of print. Perhaps you can borrow a copy from a local library or a ham in your area.

There were more hams experimenting all the time. Commercial broadcasting stations began to spring up after World War I. This brought a great deal of confusion to the airwaves. Congress created the Federal Radio Commission in 1927 to unravel the confusion and assign specific frequencies for specific uses. Soon amateurs found themselves with their own frequency bands.

Continued experimentation over the years has brought us tubes, transistors and integrated circuits. Equipment has grown smaller and more sophisticated. In the early days of radio communications, equipment was large and heavy. Sometimes it would take a roomful of equipment to accomplish what can now be done with the circuitry in a tiny box.

WE PITCH IN WHEN NEEDED

Traditionally, amateurs have served their countries in times of need. During wartime, amateurs have patriotically taken their communications skills and technical ability into the field. During natural disasters, when normal channels of communication are interrupted, hams provide an emergency communications system.

Public service has been a ham tradition since the very beginning. Whether it's a walk-a-thon, the Olympic Torch Run or the aftermath of a hurricane, hams are always there to help with communications, which they provide at no cost to the group involved.

Practically all radio transmitters operating in the United States provide some public service at one time or another. Hams don't wait for someone to ask for their help; they pitch in when needed. They provide communications on March of Dimes walk-a-thons, help plug the dikes when floods threaten and warn of approaching hurricanes. Hams bring assistance to sinking ships, direct medical supplies into disaster areas, and search for downed aircraft and lost children.

Amateur Radio operators recognize their responsibility to provide these public-service communications. They train in various ways to be effective communicators in times of trouble. Every day, amateurs relay thousands of routine messages across the country. They send many of these messages through "traffic nets" devoted to developing the skill of sending and receiving messages efficiently. (A net is a gathering of hams on a single frequency for some specific purpose. In this case the net's purpose is to "pass traffic" — relay messages.) This daily operation helps prepare hams for real emergencies. Also, each June thousands of hams across the country participate in the ARRL Field Day. They set up portable stations, including antennas, and use emergency power. Relaying messages and operating on Field Day help hams to test their emergency communications capabilities and to help them identify problems that could arise in the event of a disaster.

MAYDAY — We're Going Down

In September 1991, a Pacific storm caught two boats, the *Molly Sue* and the *Dauntless*, eastbound from Hawaii to California. The *Molly Sue* and the *Dauntless* had intended to travel together, but the *Dauntless*, a 65-foot schooner, outran the other boat by 500 miles in the storm, out of maritime radio range.

The storm's winds ripped the *Molly Sue*'s sails and broke her halyard, injuring a crewman in the process — injuries that appeared serious. Alerted by hams operating on the California-Hawaii net, with whom the *Molly Sue*'s skipper was in contact, the Coast Guard diverted a freighter to help. Unfortunately, the *Molly Sue* needed new sails. A local ham picked up a set of sails from the skipper's home in San Diego, and delivered them to the Coast Guard. The Coast Guard parachuted the sails to the *Molly Sue*.

Meanwhile, the *Dauntless*, back in radio range, reported she was taking on water and running out of fuel. The net relayed this word to the Coast Guard, and the *Dauntless* was towed to San Francisco. The *Molly Sue*, fitted with new sails, arrived safely in San Diego.

An FCC official commended the hams on the net for their "shining example of the communications skills, dedication and public spirit of the amateur service."

In Quakes, Hurricanes and Floods: Hams Are There

In October 1989, a major earthquake struck the San Francisco area. Thousands of inquiries came in to ARRL Headquarters and to individual hams across the country. Friends, relatives and business associates worried about people in the affected area.

Why did they choose to ask Amateur Radio operators to help them? Over the decades, hams have volunteered their services in times of emergency to relay vital information to and from stricken areas. With local power and telephone communications knocked out, San Francisco hams went to work. Using amateur stations that fit in briefcases, they relayed messages between residents and their worried relatives and friends outside the stricken area. These tiny stations included a hand-held Amateur Radio transceiver, a lap-top computer and a radio modem called a terminal-node controller, or TNC. Local hams went back to their regular routine of working and spending time with their families only after workers restored regular communications channels.

In 1992, Hurricane Andrew chewed across southern Florida, then took aim at Louisiana. The city of Homestead, Florida, and the adjacent air base were almost totally destroyed. Despite the damage to their own homes, local hams provided essential communications in the first hours after Andrew passed. In Andrew's aftermath, amateurs from other states volunteered their time and equipment to aid relief efforts. A few weeks later, Hurricane Iniki struck Hawaii. Again, radio amateurs played key communications roles when commercial communications systems were destroyed.

The summer of 1993 is one that won't soon be forgotten across the middle of the US. In early April heavy rain caused rivers and streams to rise. Crews were laying sandbags along some stretches of the mighty Mississippi to fend off the rising water. In June, many Field Day plans turned to real emergency communications events as flooding occurred along the Minnesota River in the St. Peter, Minnesota area. The rain continued through July, and people in Kansas, Kentucky, Iowa, Illinois, Missouri, Minnesota and Wisconsin were affected. It was September before all the flood waters had retreated. Through it all, Amateur Radio operators were there, providing vital emergency communications and helping coordinate sand bagging and rescue efforts. In Saint Charles county Missouri, hams operated from the emergency operations center (EOC) 24 hours per day from July 2 through August 20. Volunteers came from across the country to relieve weary radio operators and help wherever needed.

Radio amateurs help out even when the emergency is outside the US. In September 1989, Hurricane Hugo did terrible damage to island nations in the Caribbean, and Puerto Rico. Most of the islands were completely cut off from the outside world; all normal means of communication were destroyed. Through volunteer Amateur Radio networks, shipments of relief and medical supplies were coordinated. Island residents were able to contact loved ones on the mainland and other islands. Amateur Radio "jump teams" even sped to the affected areas to help restore communications.

No Barriers To Enjoying Ham Radio

Amateur Radio holds no roadblocks for people with disabilities. Many people who are unable to walk, see or talk are able to enjoy their Amateur Radio hobby, conversing with friends in their home town or across the world. Some local ham clubs even take classes to a home to help a person with a disability discover ham radio.

Although unable to leave his bed, Otho Jarman was able to earn his ham license. Bill Haney, KE3CO, (then WA6CMZ) helped Otho learn the code and pass his Novice exam through such a club program. Bill spent one hour each week teaching Otho the code, and in seven weeks he passed his Novice exam. Other members of the Barstow (California) Amateur Radio Club helped Otho put his station together. They obtained equipment for him and constructed antennas.

Soon, Otho, WB6KYM, was on the air talking with hams in Mozambique, Nicaragua and Puerto Rico. He could communicate with the world from his bed. Sixteen years earlier, at age 22, Otho had broken his spine when he dove into a reservoir to rescue a drowning child. Although interested in Amateur Radio for years, he had not had an opportunity to learn about it until the club came along. Now he monitors local frequencies from 7 AM to 10 PM, often just chatting or giving directions to motorists.

"Amateur Radio can take a disabled person out of his living room, out of his bed or out of his wheelchair, and put him in the real world," Otho says.

The Courage Center, of Golden Valley, Minnesota, sponsors the HANDI-HAM system to help people with physical disabilities obtain amateur licenses. The system provides materials and instruction to persons with disabilities interested in obtaining ham licenses. The Center also provides information to other hams, "verticals," who wish to help people with disabilities earn a license. Write to:

Courage HANDI-HAM System
Courage Center
3915 Golden Valley Road
Golden Valley, MN 55422

NOW YOU'RE TALKING!

Once you've earned your license, you can join the thousands of other hams on the air. Then you'll begin to experience the thrill of Amateur Radio firsthand. As a ham, you can contact other hams in your town and around the world. You can talk into a microphone to use either single sideband (SSB) or frequency modulation (FM) to communicate. If you want, you can experience the thrill of using the international Morse code (called "CW" by hams). You can even communicate by packet radio or radioteletype using a computer. You can do all this on parts of special frequency bands set aside by the FCC for Amateur Radio operators.

For your first contact, you'll probably tune around looking for a station calling "CQ" (calling for any station to make contact). Perhaps you'll even try calling a CQ on your own. Suddenly you'll hear your own call coming back! It's hard to describe the excitement. Someone else is sending your call sign back to let you know that they hear you and want to make contact.

Each time you send a CQ, you'll wonder who will answer. It could be a ham in the next town, the next state or clear across the country. The whole world is full of hams to talk to.

There are many wonderful things you can do as a ham. After you have been on the air for a while, you'll be known as one of the regulars on the band. It's surprising how many people from all over you'll recognize and who will recognize you. Many a fast friendship has developed through repeated on-the-air contacts.

Soon you'll be collecting contacts with different states and exchanging QSL cards (postcards) with other hams you've talked to. These special cards commemorate each contact. They also serve as proof of the contact as you begin working toward some of the awards issued by the American Radio Relay League. [The Worked All States (WAS) award is one popular example.] You'll learn more about the special language and abbreviations that hams use as you study. Chapter 2 has a list of common "Q signals" used by most hams.

There is a certain intrigue about DX (long-distance communication) that catches many hams. Talking to hams from other lands can be quite an experience. After all, foreign hams are people just like you who enjoy finding out about other people and places! Also, hams in other countries often speak enough English to carry on a limited conversation, so you'll have little problem there. Whether you start as a Novice or a Technician, Amateur Radio offers many opportunities to contact DX stations.

If you enjoy a little competition, perhaps you'll like on-the-air contesting. The object of a contest is to work as many people in as many different areas as possible in a certain time. Each year, the American Radio Relay League sponsors a number of contests and "operating events." Several of these events have provisions to encourage Novice and Technician operators. For example, the Field Day rules (always the fourth full weekend in June) include a provision for a Novice/Technician station that encourages clubs to set up such a station. Other contests offer extra points for contacts with Novice/Technician stations. All these events give you the chance to contact old friends and make new ones. You might work some new states or countries, and if you use Morse code you will increase your code speed. You are certain to improve your general operating skills and ability. Most of all, though, you will have fun.

Other Modes

Novice and Technician licensees may use just about every type of communication available to Amateur Radio operators. You should become familiar with these modes. In addition to voice and Morse code, you may want to investigate some of the less traditional modes.

With slow-scan television (SSTV), hams send still photos to each other, one frame at a time. It takes about 8 seconds for the bright band of light to creep down the screen to make a complete black and white picture. (Your home TV makes 30 complete pictures per second.) But why restrict yourself to black and white pictures. Several methods have been devised to send full-color pictures by SSTV! Each of these requires a longer transmission time, and the picture quality and resolution varies between these formats. SSTV pictures may be transmitted around the world via shortwave ham transmitters. Many hams use computer software and adapters to enjoy the fun of SSTV even without a video camera.

Facsimile (fax) is a means of sending drawings, charts, maps, graphs and pictures. You can even play games over the air by transmitting fax pictures of each move.

Using the "digital" modes, a ham can type out a message and send it over the air to a friend's station. Even if the friend is away, her radioteletype system can receive and hold the message until she returns. Early digital systems used mechanical machines cast off by news services. Today, many hams use personal computer systems. These display the message silently on a TV screen rather than using roll after roll of paper with noisy, clacking typewriter keys. There is a growing number of modern computer-controlled systems capable of relaying messages and storing them for later reception by the intended ham.

Ham Satellites

Hams use satellites to communicate in voice, code, radioteletype and packet radio around the world. Long-distance telephone, radio and TV broadcasters also use communications satellites to relay signals around the world.

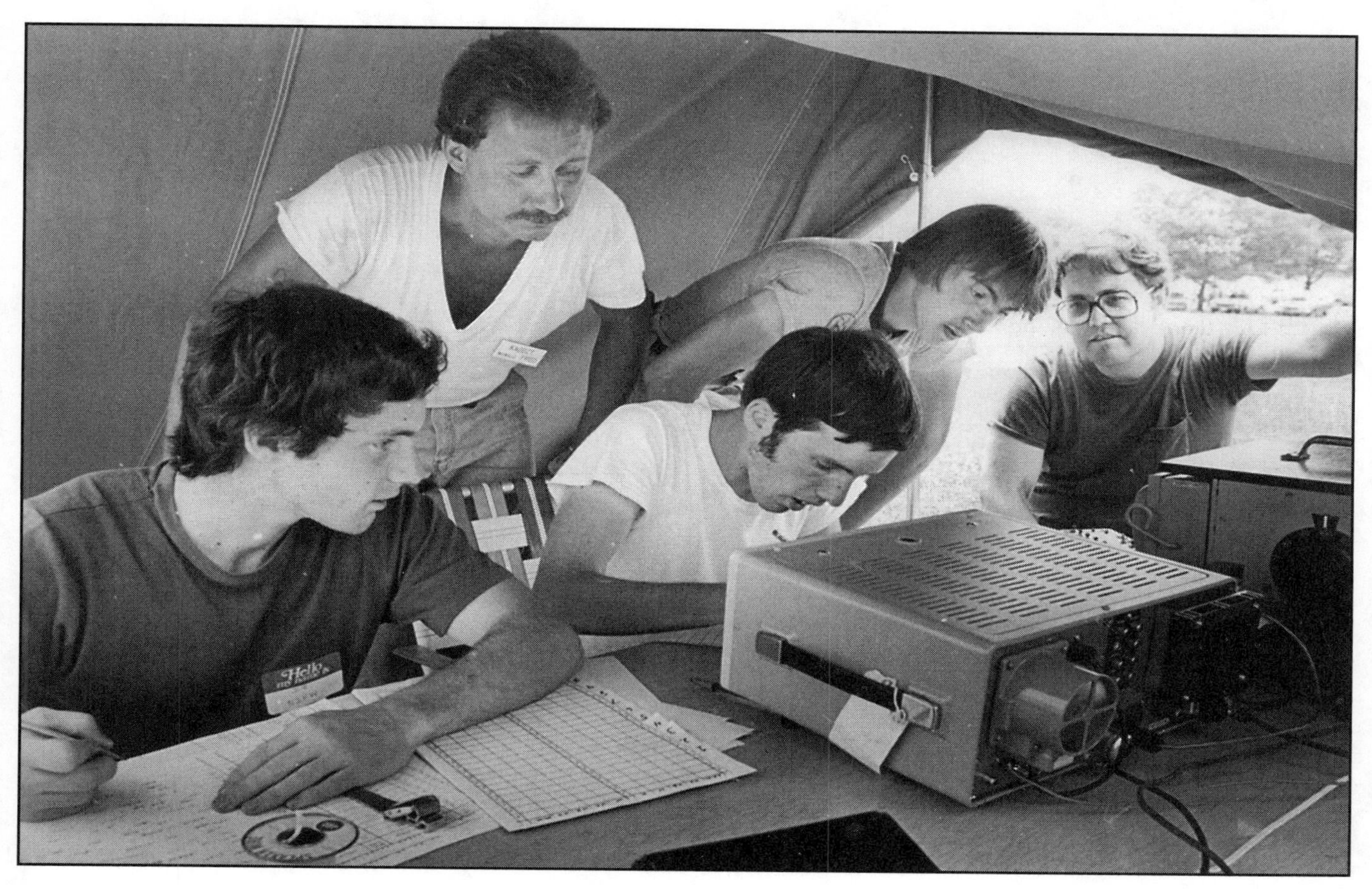

Hams are special, though, because their satellites are known as OSCARs (Orbiting Satellites Carrying Amateur Radio), which were designed and built by amateurs!

Some schools use the OSCARs to instruct students in science and math. No license is needed to listen, so many students across the country have eavesdropped on Amateur Radio transmissions through an OSCAR. A receiver and an antenna are all you need to introduce students to the exciting world of space technology.

Hams from many countries worked in a joint effort tc build several OSCARs. Weighing less than your living-room TV and powered by sun-charged batteries, OSCARs retransmit hams' "uplink" signals down to other earthbound stations.

There are many hobbies within the hobby of Amateur Radio. Some hams collect QSL cards and awards. Others love the thrill of talking to rare DX stations in foreign lands, while still others enjoy exchanging pictures by slow-scan television. You'll find one or more of these pastimes enjoyable, and in time you'll probably try most of them.

WHAT ARE YOU WAITING FOR?

You can operate in any of these exciting modes when you become an Amateur Radio operator. Take the time to explore the adventure of ham radio! As a new or prospective amateur operator, you will begin to discover the wide horizons of your new pastime. You will learn more about the rich heritage we all share. Take this sense of pride with you, and follow the Amateur's Code.

The Amateur's Code

The Radio Amateur is:

CONSIDERATE . . . never knowingly operates in such a way as to lessen the pleasure of others.

LOYAL . . . offers loyalty, encouragement and support to other amateurs, local clubs, and the American Radio Relay League, through which Amateur Radio in the United States is represented nationally and internationally.

PROGRESSIVE . . . with knowledge abreast of science, a well-built and efficient station and operation above reproach.

FRIENDLY . . . slow and patient operating when requested; friendly advice and counsel to the beginner; kindly assistance, cooperation and consideration for the interests of others. These are the hallmarks of the amateur spirit.

BALANCED . . . radio is an avocation, never interfering with duties owed to family, job, school, or community.

PATRIOTIC . . . station and skill always ready for service to country and community.

The original Amateur's Code was written by Paul M. Segal, W9EEA, in 1928.

TABLE 1—ELEMENT 2 (NOVICE) SYLLABUS

(Required for all operator licenses)

SUBELEMENT N1 — COMMISSION'S RULES [10 Exam Questions — 10 Groups]

N1A Basis and purpose of amateur service and definitions

N1B Station/Operator license; classes of US amateur licenses, including basic differences and privileges of the various license classes

N1C Novice control operator frequency privileges

N1D Novice eligibility, exam elements, mailing addresses, US call-sign assignment and life of license

N1E Novice control operator emission privileges

N1F Transmitter power on Novice sub-bands and digital communications [limited to concepts only]

N1G Responsibility of licensee, control operator requirements

N1H Station identification, points of communication and operation, and business communications

N1I International and space communications, authorized and prohibited transmissions

N1J False signals or unidentified communications and malicious interference

SUBELEMENT N2 — OPERATING PROCEDURES [2 Exam Questions — 2 Groups]

N2A Preparing to transmit; choosing a frequency for tune-up, operating or emergencies; Morse code; RST signal reports; Q signals; voice communications and phonetics

N2B Radio teleprinting; packet; repeater operating procedures; special operations

SUBELEMENT N3 — RADIO-WAVE PROPAGATION [1 Exam Question — 1 Group]

N3A Line of sight, ground wave, HF propagation characteristics; sunspots and the sunspot cycle; and reflection of VHF/UHF signals

SUBELEMENT N4 — AMATEUR RADIO PRACTICES [4 Exam Questions — 4 Groups]

N4A Preventing unauthorized use; lightning protection and station grounding

N4B Safety interlocks, antenna installation safety procedures

N4C SWR meaning and measurements

N4D RFI and its complications, resolution and responsibility

SUBELEMENT N5 — ELECTRICAL PRINCIPLES [4 Exam Questions — 4 Groups]

N5A Metric prefixes, i.e. pico, nano, micro, milli, centi, kilo, mega, giga

N5B Concepts and measurement of current, voltage, resistance; concept of conductor and insulator

N5C Ohm's Law (any calculations will be kept to a very low level—no fractions or decimals) and the concepts of energy and power, and open and short circuits

N5D Concepts of frequency, including AC vs. DC, frequency units, AF vs. RF and wavelength

SUBELEMENT N6 — CIRCUIT COMPONENTS [2 Exam Questions — 2 Groups]

N6A Electrical function and/or schematic representation of resistor, switch, fuse, or battery

N6B Electrical function and/or schematic representation of a ground, antenna, transistor, integrated circuit or vacuum tube

SUBELEMENT N7 — PRACTICAL CIRCUITS [2 Exam Questions — 2 Groups]

N7A Functional layout of station components including transmitter, transceiver, receiver, power supply, antenna, antenna switch, antenna feed line, impedance-matching device and SWR meter

N7B Station layout and accessories for telegraphy, radiotelephone, radioteleprinter (RTTY) or packet

SUBELEMENT N8 — SIGNALS AND EMISSIONS [2 Exam Questions — 2 Groups]

N8A CW, phone, RTTY and data emission types

N8B Harmonics and unwanted signals; chirp; superimposed hum; equipment and adjustments to help reduce interference to others

SUBELEMENT N9 — ANTENNAS AND FEED LINES [3 Exam Questions — 3 Groups]

N9A Wavelength vs. antenna length; multiband antenna advantages and disadvantages

N9B Yagi parts, concept of directional antennas

N9C Feed lines, baluns and polarization via element orientation

SUBELEMENT N0 — RF SAFETY [5 Exam Questions — 5 Groups]

N0A RF safety fundamentals

N0B RF safety terms and definitions

N0C RF safety rules and guidelines

N0D Routine station evaluation

N0E Practical applications

TABLE 2—ELEMENT 3A (TECHNICIAN) SYLLABUS

(Required for all operatorlicenses except Novice)

SUBELEMENT T1—COMMISSION'S RULES
[5 Exam Questions — 5 Groups]

T1A Station control; frequency privileges authorized to the Technician and Technician Plus class control operator; term of licenses, grace periods and modifications of licenses

T1B Emission privileges for Technician and Technician Plus class control operator; frequency selection and sharing; transmitter power

T1C Digital communications, station identification, ID with authorization of Certificate of Successful Completion of Examination

T1D Correct language, phonetics, beacons and radio control of model craft and vehicles

T1E Emergency communications; broadcasting; permissible one-way, satellite and third-party communication; indecent and obscene language

SUBELEMENT T2 —OPERATING PROCEDURES
[3 Exam Questions — 3 Groups]

T2A Repeater operation; autopatch, definition and proper use; courteous operation; repeater frequency coordination

T2B Simplex operations; RST signal reporting; choice of equipment for desired communications; communications modes including amateur television (ATV), packet radio and SSB/CW weak signal operations

T2C Distress calling and emergency drills and communications—operations and equipment, Radio Amateur Civil Emergency Service (RACES)

SUBELEMENT T3 — RADIO-WAVE PROPAGATION
[3 Exam Questions — 3 Groups]

T3A VHF/UHF/Microwave Propagation

T3B Ionospheric absorption, causes and variation, maximum usable frequency

T3C Amateur satellite and EME operations

SUBELEMENT T4 — AMATEUR RADIO PRACTICES
[4 Exam Questions — 4 Groups]

T4A Electrical wiring, including switch location, dangerous voltages and currents

T4B Meters and their placement in circuits, including volt, amp, multi, peak-reading and RF watt; ratings of fuses and switches

T4C Marker generator, crystal calibrator, signal generators and impedance-match indicator

T4D Dummy antennas and S meters

SUBELEMENT T5 —ELECTRICAL PRINCIPLES
[2 Exam Questions — 2 Groups]

T5A Definition and unit of measurement of resistance, inductance and capacitance

T5B Concepts and calculation of resistance, inductance and capacitance values in series and parallel circuits

SUBELEMENT T6 —CIRCUIT COMPONENTS
[2 Exam Questions — 2 Groups]

T6A Resistors, construction types, variable and fixed, color code, power ratings, schematic symbols

T6B Inductor and capacitor schematic symbols; construction of variable and fixed inductors and capacitors; factors affecting inductance and capacitance

SUBELEMENT T7 — PRACTICAL CIRCUITS
[1 Exam Question — 1 Group]

T7A Transmitter and receiver block diagrams; purpose and operation of low-pass, high-pass and band-pass filters

SUBELEMENT T8 — SIGNALS AND EMISSIONS
[2 Exam Questions — 2 Groups]

T8A Concepts and types of modulation

T8B RF carrier, definition and typical bandwidths and FM deviation

SUBELEMENT T9 — ANTENNAS AND FEED LINES
[3 Exam Questions — 3 Groups]

T9A Parasitic beam and non-directional antennas

T9B Polarization, impedance matching and SWR, feed lines, balanced vs. unbalanced (including baluns)

T9C Line losses by line type, length and frequency

SUBELEMENT T0 — RF SAFETY
[5 Exam Questions — 5 Groups]

T0A RF safety fundamentals

T0B RF safety terms and definitions

T0C RF safety rules and guidelines

T0D Routine station evaluation

T0E Practical applications for VHF/UHF and above operations

1 KEY WORDS

Amateur operator — A person holding a written authorization to be the control operator of an amateur station.

Amateur service — A radiocommunication service for the purpose of self-training, intercommunication and technical investigations carried out by amateurs, that is, duly authorized persons interested in radio technique solely with a personal aim and without **pecuniary** interest.

Amateur station — A station licensed in the amateur service, including necessary equipment, used for amateur communication.

Broadcasting — Transmissions intended to be received by the general public, either direct or relayed.

Control operator — An amateur operator designated by the licensee of a station to be responsible for the transmissions of an amateur station.

Control point — The locations at which the control operator function is performed.

Earth station — An amateur station located on, or within 50 km of, the Earth's surface intended for communications with space stations or with other Earth stations by means of one or more other objects in space.

Emission — The transmitted signal from an amateur station.

Emission privilege — Permission to use a particular emission type (such as Morse code or voice).

False or deceptive signals — Transmissions that are intended to mislead or confuse those who may receive the transmissions. For example, distress calls transmitted when there is no actual emergency are false or deceptive signals.

Frequency bands — A group of frequencies where amateur communications are authorized.

Frequency privilege — Permission to use a particular group of frequencies.

Grace period — The time the FCC allows following the expiration of an amateur license to renew that license without having to retake an examination. Those who hold an expired license may not operate an amateur station until the license is reinstated.

Malicious (harmful) interference — Intentional, deliberate obstruction of radio transmissions.

MAYDAY — From the French *m'aidez* (help me), MAYDAY is used when calling for emergency assistance in voice modes.

Operator/primary station license — An amateur license actually includes two licenses in one. The operator license is that portion of an Amateur Radio license that gives permission to operate an amateur station. The primary station license is that portion of an Amateur Radio license that authorizes an amateur station at a specific location. The station license also lists the call sign of that station.

Peak envelope power (PEP) — The average power of a signal at its largest amplitude peak.

Pecuniary — Payment of any type, whether money or other goods. Amateurs may not operate their stations in return for any type of payment.

SOS — A Morse code call for emergency assistance.

Space station — An amateur station located more than 50 km above the Earth's surface.

Third-party communications — Messages passed from one amateur to another on behalf of a third person.

Third-party communications agreement — An official understanding between the United States and another country that allows amateurs in both countries to participate in third-party communications.

Third-party participation — The way an unlicensed person can participate in amateur communications. A control operator must ensure compliance with FCC Rules.

Unidentified communications or signals — Signals or radio communications in which the transmitting station's call sign is not transmitted.

Volunteer Examiner (VE) — A licensed amateur who is accredited by a **Volunteer Examiner Coordinator (VEC)** to administer amateur license exams.

Volunteer Examiner Coordinator (VEC) — An organization that has entered into an agreement with the FCC to coordinate amateur license exams.

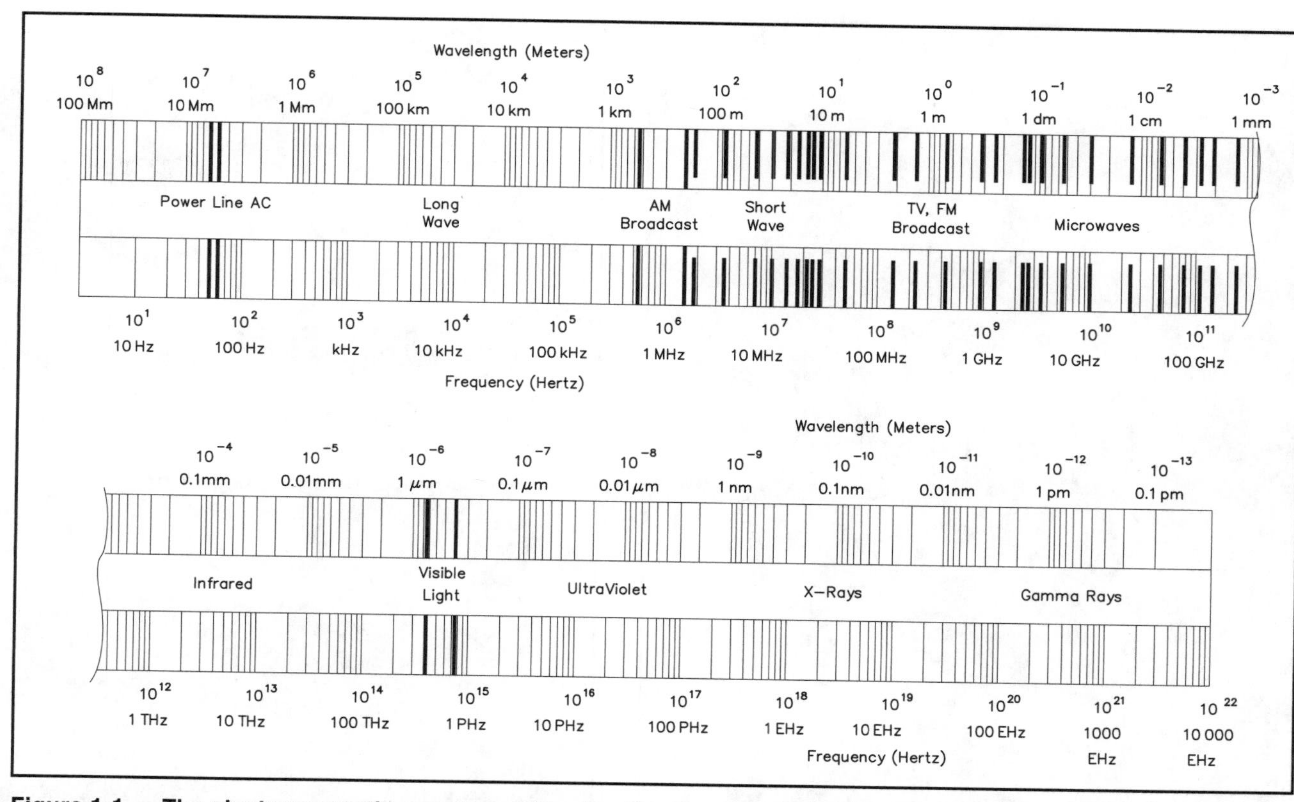

Figure 1-1 — The electromagnetic spectrum, showing the range from the 60-hertz ac power line in your home, through the radio frequencies, X-rays and gamma rays. Amateur Radio operations use a small, but significant, portion of these frequency bands. The amateur bands are represented as short bars on the graph.

1 Federal Communications Commission's Rules

The FCC Rules, Part 97, establish station operating standards, Technical standards and emergency communication standards. Amateur Radio operators must know—and follow— these rules.

When you tune an AM or FM broadcast radio to your favorite station, you select a specific spot on the tuning dial. There are many stations spread across that dial. Each radio station occupies a small part of the entire range of "electromagnetic waves." Other parts of this range, or spectrum, include microwaves, X-rays and even infrared, ultraviolet and visible light waves.

Figure 1-1 shows the electromagnetic spectrum from below the radio range all the way through the X-rays. Amateur Radio occupies only a small part of the total available space; countless users must share the electromagnetic spectrum.

You may be thinking: "Who decides where Amateur Radio frequencies will be, and where my favorite FM broadcast station will be?" That's a good question, and the answer has several parts.

Radio signals travel to distant corners of the globe, so there must be a way to prevent total chaos on the bands. The International Telecommunication Union (ITU) has the important role of dividing the entire range of communications frequencies between those who use them. Many radio services have a need for communications frequencies. These services include commercial broadcast, land mobile and private radio (including Amateur Radio). ITU member nations decide which radio services will be given certain bands of frequencies, based on the needs of the different services. This process takes place at ITU-sponsored World Radio Conferences (WRCs).

An avid outdoorsman, Luigi, IC8GVV, always carries his hand-held radio when he climbs the cliffs near his home on the Isle of Capri. He used his radio to call for help when his climbing partner fell and broke a wrist.

Kenya's first Novice, 14-year-old Anand, 5Z4RAN, was licensed in June 1996. When he upgrades, the final N in Anand's call will be dropped.

Iris, ZS2AA, age 93, and Elaine, ZR2ELF, age 12, are reported to be the oldest and youngest female hams in South Africa.

Nearly every country on Earth has its share of active Amateur Radio operators. Here, distinguished visitors observe the operation of demonstation station A43AT at the Arabian Telecom '96 Exhibition and Forum in Oman. The station was organized and set up by the Royal Omani Amateur Radio Society.

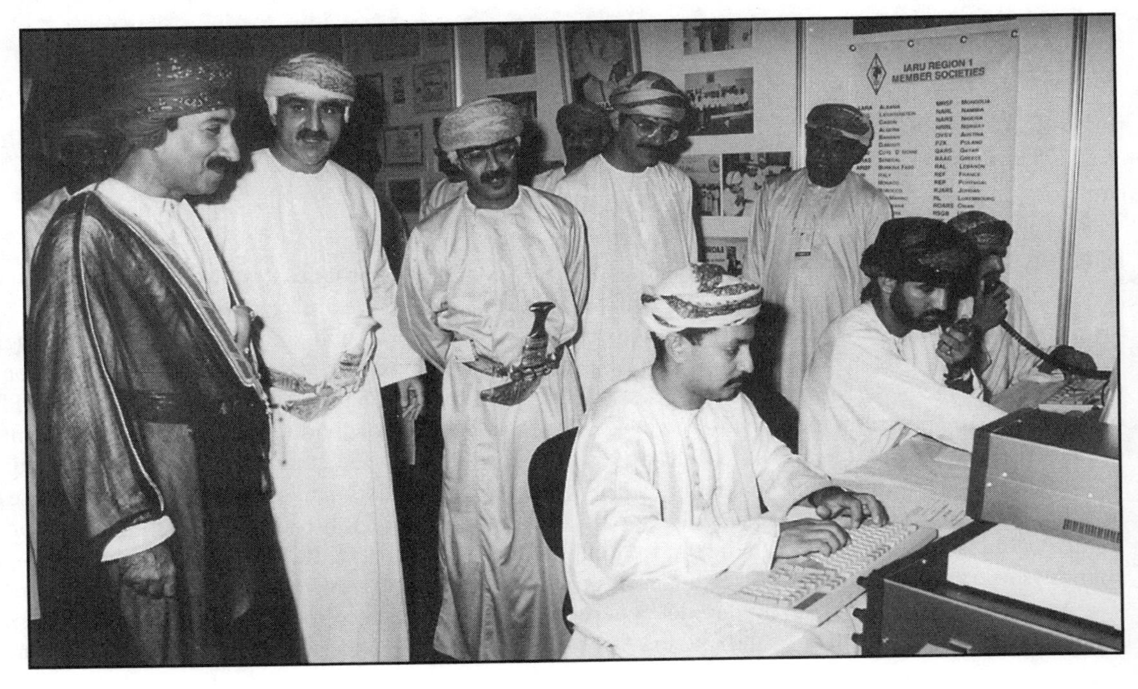

In the case of Amateur Radio, the ITU has long recognized that hams are invaluable in times of emergency or disaster. At an ITU World Administrative Radio Conference held in 1979, Amateur Radio gained three new HF bands as well as some new UHF and microwave frequencies. Mighty praise indeed!

The ITU makes these allocations on an international basis. The Federal Communications Commission (FCC) decides the best way to allocate frequency bands to those services using them in the US. The FCC is the governing body in the United States when it comes to Amateur Radio. In fact, an entire part of the FCC Rules is devoted to the amateur service — Part 97. The amateur service rules in Part 97 describe station operation standards, technical standards and emergency communications. As your amateur career develops, you'll become more familiar with Part 97.

To gain the privilege of sending a radio signal over the airwaves in the US, you must pass a license exam. To earn a Novice license, you will need to pass a 35-question exam covering basic radio theory and FCC regulations, as well as a Morse code exam. To earn a Technician license, you will need to pass two written exams — the 35-question Novice written test plus a more comprehensive 30-question written exam. There is no Morse code requirement for the Technician exam. Whether you choose to learn the code and try for the Novice license or work toward the code-free Technician license, this book will prepare you to earn your first Amateur Radio license.

The Five Principles

In Section 97.1 of its Rules, the FCC describes the basis and purpose of the amateur service. It consists of five principles:

97.1(a) Recognition and enhancement of the value of the amateur service to the public as a voluntary noncommercial communication service, particularly with respect to providing emergency communications.

Probably the best-known aspect of Amateur Radio is our ability to provide life-saving emergency communications. Normal communications channels often break down during hurricanes, earthquakes, tornadoes, airplane crashes and other disasters. Amateur Radio is frequently the first available means of contact with the outside world from the affected area. Red Cross and other civil-defense agencies rely heavily on the services of volunteer radio amateurs.

One of the more noteworthy aspects of Amateur Radio is its noncommercial nature. In fact, amateurs may not accept any form of payment for operating their radio stations. (There is one limited exception to this rule, which we will explain

In the days before the giant eruption that demolished vast areas around the Mount St. Helens volcano in Washington, volunteer amateurs monitored its behavior. Hams are ready to do whatever they can when someone needs their communications services. Because of the suddenness and extent of that eruption, two hams lost their lives while helping to monitor the volcano.

later.) This means that hams make their services available free of charge. This is true whether they are assisting a search-and-rescue operation in the Sierra Nevada, relaying Health-and-Welfare messages from a disaster-stricken Caribbean island or providing communications assistance at the New York City Marathon. Talk about value!

Why do hams work so hard if they can't be paid? It gives them an immense feeling of personal satisfaction! It's like the good feeling you get when you lend a hand to an elderly neighbor, only on a much grander scale. Hams operate their stations only for personal satisfaction and enjoyment. They don't talk about business matters on the air. (FCC Rules don't allow business communications over Amateur Radio.)

97.1(b) Continuation and extension of the amateur's proven ability to contribute to the advancement of the radio art.

In the early days of radio there were no rules, but in 1912 Congress passed a law to regulate the airwaves. Amateurs had to keep to a small range of frequencies (known as "short waves"). There they would remain "out of the way"—everyone knew that radio waves couldn't travel very far at those frequencies. Ha! Amateurs soon overcame the restrictions. They were among the first to experiment with radio propagation, the study of how radio waves travel through the atmosphere. When vacuum tubes became available, amateurs began to develop much-improved radio communication circuits.

Today, the traditions and spirit in Amateur Radio remain. Amateurs continue to experiment with state-of-the-art technologies. Advancement of the radio art takes a major portion of an amateur's energies. The FCC promotes this amateur experimentation and technical development by establishing rules that are consistent with amateur techniques.

97.1(c) Encouragement and improvement of the amateur service through rules which provide for advancing skills in both the communications and technical phases of the art.

Along with some of the technical aspects of the service, amateurs also hold special training exercises in preparation for communications emergencies. Simulated Emergency Tests and Field Days, where amateurs practice communicating under emergency conditions, are just two ways amateurs sharpen their operating skills.

The Commission's rules even specify a "service within a service" called the Radio Amateur Civil Emergency Service, or RACES. RACES provides amateur communications assistance to federal, state and local civil defense organizations in times of need.

97.1(d) Expansion of the existing reservoir within the amateur radio service of trained operators, technicians and electronics experts.

Self-training, station operating standards, intercommunication and technical investigation are all important parts of the amateur service. We need more amateurs who are experienced in communications methods, because they are a national resource to the public.

97.1(e) Continuation and extension of the amateur's unique ability to enhance international goodwill.

Hams are unique, even in this time of worldwide jet travel. They journey to the far reaches of the earth and talk with amateurs in other countries every day. They do this simply by walking into their ham shacks. International peace and coexistence are very important today. Amateurs represent their countries as ambassadors of goodwill. Amateur-to-amateur communications often cross the cultural boundaries between societies. Amateur Radio is a teacher in Lincoln, Nebraska, trading stories with the headmaster of a boarding school in a London suburb. It is a tropical-fish hobbyist learning about fish in the Amazon from a missionary stationed in Brazil. Amateur Radio is a way to make friends with other people everywhere.

To summarize, the five principles on which the amateur service is based are to recognize the value of emergency communication skills, advance the radio art, improve communication and technical skills, to increase the number of trained radio operators and electronics experts, and to improve international goodwill. These five principles provide the basis and purpose for the amateur service, as set down by the FCC. These principles place a large responsibility on the amateur community — a responsibility you will share. It is the Commission's duty to ensure that amateurs are able to operate their stations properly, without interfering with other radio services. All amateurs must pass an examination before the FCC will issue a license authorizing amateur station operation. It's a very serious matter.

While the FCC expects you to operate your station properly, the construction of your station is up to you. There are no FCC requirements for amateur station construction.

[At this point you should turn to Chapter 13 and study the questions that cover this material in the Novice question pool. Before you turn to Chapter 13, though, it may be helpful to understand a bit about the numbering system used for these questions. The numbers match the study guide or syllabus printed at the end of the Introduction. The syllabus forms a type of outline, and the numbering system follows this outline format. There are 10 subelements, labeled N1 through N0. This chapter covers subelement N1 for the Novice written exam and subelement T1 for the additional Technician written exam, so all the question references will be to questions with numbers that begin "N1" or "T1."

Each subelement is further divided into question groups. These groups represent the number of questions to be selected from each subelement. There are 10 questions on the exam from subelement N1, Commission's Rules. So there are 10 question groups labeled N1A through N1J. Individual questions end with a two-digit number, such as N1A01, N1B11, N1E13 and so on.

You should study questions N1A01 through N1A06 in Chapter 13 now. If you have difficulty with any of these questions, review the material in this section.]

The FCC defines some important terms in Section 97.3 of the amateur rules. The **amateur service** is "A radio communication service for the purpose of self-training, intercommunication and technical investigations carried out by amateurs, that is, duly authorized persons interested in radio technique solely with a personal aim and without pecuniary interest." **Pecuniary** means related to money or other payment. In other words, you can't be paid for operating your station or providing a communications service for some individual or group. (There is one exception, which we will explain later in this chapter.) Amateurs learn various communication skills on their own, and carry out technical experiments in electronics and radio principles.

An **amateur operator** is a person to whom the FCC has granted a license in the amateur service. In the United States, the Federal Communications Commission issues these licenses. A US amateur license allows you to operate wherever the amateur service is regulated by the FCC. An amateur operator performs communications in the amateur service.

How does the FCC define an **amateur station**? "A station licensed in the amateur service, including the apparatus necessary for carrying on radiocommunications." The person operating an amateur station has an interest in self-training, intercommunication and technical investigations or experiments.

An Amateur Radio license is really two licenses in one — an *operator license* and a *station license*. The operator license is one that lets you operate a station within your authorized privileges on amateur-service frequencies. You must have an amateur license to operate a transmitter on amateur service frequencies. This license is your permission to control the transmissions of an amateur station.

The station license authorizes you to have an amateur station and its associated equipment. It also lists the call sign that identifies that station. The FCC calls this license an amateur **operator/primary station license**. Figure 1-2 shows an actual Amateur Radio license document. One piece of paper includes both the operator and the station license.

The operator license portion lists your license class and gives you the authority to operate an amateur station. The station license portion includes the address of your primary, or main, Amateur Radio station. The station license also lists the call sign of your station. It is your written authorization for an amateur station. You can operate an amateur station at locations other than the one listed for your primary location, however. For example, you might operate a portable station while you are away from home or a mobile station while you are riding in a vehicle. In fact, your amateur license authorizes you to operate a station anywhere that the FCC regulates the amateur service. In addition to the 50 states, this also means US possessions such as certain Pacific islands and onboard ships registered in the US even while sailing in international waters.

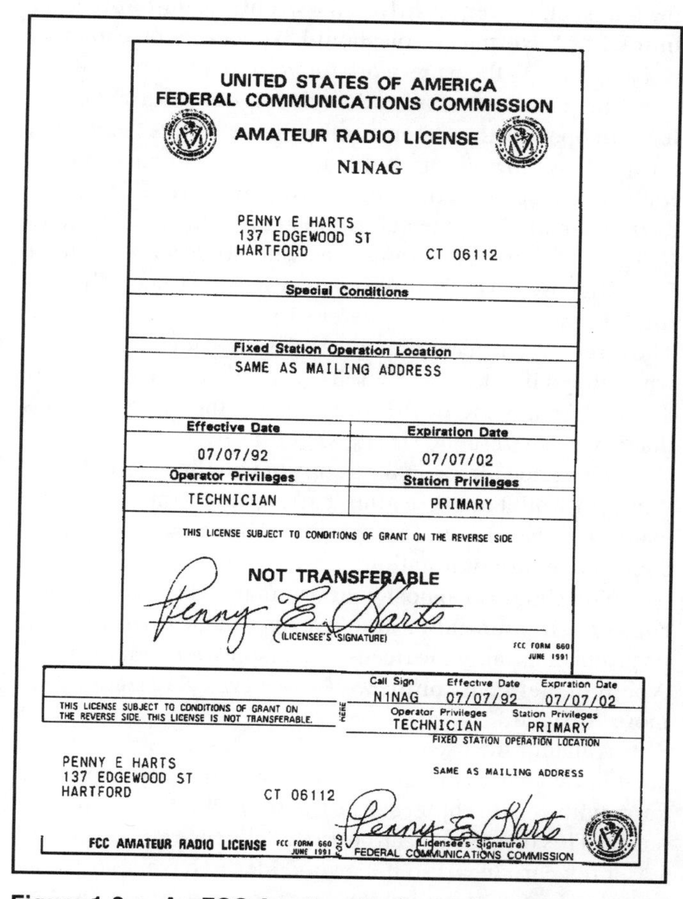
UNITED STATES OF AMERICA
FEDERAL COMMUNICATIONS COMMISSION
AMATEUR RADIO LICENSE
N1NAG

PENNY E HARTS
137 EDGEWOOD ST
HARTFORD CT 06112

Special Conditions

Fixed Station Operation Location
SAME AS MAILING ADDRESS

Effective Date	Expiration Date
07/07/92	07/07/02
Operator Privileges	Station Privileges
TECHNICIAN	PRIMARY

THIS LICENSE SUBJECT TO CONDITIONS OF GRANT ON THE REVERSE SIDE

NOT TRANSFERABLE

(LICENSEE'S SIGNATURE)

FCC FORM 660
JUNE 1991

THIS LICENSE SUBJECT TO CONDITIONS OF GRANT ON THE REVERSE SIDE. THIS LICENSE IS NOT TRANSFERABLE.

Call Sign	Effective Date	Expiration Date
N1NAG	07/07/92	07/07/02
Operator Privileges	Station Privileges	
TECHNICIAN	PRIMARY	

FIXED STATION OPERATION LOCATION
SAME AS MAILING ADDRESS

PENNY E HARTS
137 EDGEWOOD ST
HARTFORD CT 06112

FCC AMATEUR RADIO LICENSE FCC FORM 660 JUNE 1991
(Licensee's Signature)
FEDERAL COMMUNICATIONS COMMISSION

Figure 1-2 — An FCC Amateur Radio license consists of two parts — the station license and the operator's license.

Now this doesn't mean that the owners of private property can't restrict you from operating while you are on their property, however. For example, you may find that when you go to enter an amusement park or other public attraction, the security people won't allow you to carry your VHF or UHF hand-held radio along. They may not understand Amateur Radio, and may have other security concerns.

You might also wonder if you would be allowed to operate your radio while on board a commercial airplane or even on board a ship. The answer is a qualified yes. The FCC Rules specify that your station must be approved by the pilot or master of the ship. Section 97.11 lists some other qualifications for such an operation. For example, you may not operate your station while the airplane is operating under "Instrument Flight Rules." The practical answer to this question is that you will probably not be allowed to use your radio on a commercial airplane.

Your Amateur Radio license is valid as soon as the FCC posts the information about your license in their electronic data base. You don't have to wait for the actual license document to arrive in the mail before you begin to transmit. You can check the FCC database on one of the internet license "servers" (or someone else can check it for you). The address for one such server is: **http://www.ualr.edu/~hamradio/index.html**. Normally you should expect it to take about a week to process the paperwork for your license application.

There are several other documents that can grant privileges to operate an amateur station in the US. For example, a Canadian citizen who holds a valid Canadian Amateur Radio license can use that license to operate in the US. (A US citizen can also use his or her amateur license in Canada.) Citizens of some other countries who hold valid Amateur Radio licenses in their home country may be issued "Reciprocal Permits for Alien Amateur Licensees" by the FCC. Those reciprocal permits then serve as their amateur licenses while they visit the US. There is currently no form of international Amateur Radio license issued by the United Nations that is valid within the US, however.

It is a good idea to make a copy of your license and carry it in your wallet or purse after it arrives in the mail. You are using the *operator license* portion when you serve as the control operator of a station.

You should also post your original license, or a photocopy of it, in your station after it arrives in the mail. You will be proud of earning the license, so display it in your station. A copy of the license on the wall also makes your station look more "official."

Amateur licenses are printed on a laser printer, and issued in two parts (Figure 1-2). One part is small enough to carry with you; the other can be framed and displayed in your shack. This means you can carry your license *and* display it! Laser ink can smear, so it's a good idea to have your wallet copy laminated. If the small part is too large for your wallet, you can make a reduced-size copy on a photocopier. If you carry a copy, you can leave your original safely at home. Although you can legally carry a copy, your original license must be available for inspection by any US government official or FCC representative, however. Don't lose the original license!

The FCC issues all licenses for a 10-year term. You should always renew your license for another 10 years, about 60 to 90 days before the present one expires. If your license is issued in September 1997, it expires in September 2007. It's a good idea to form the habit of looking at the expiration date on your license every now and then. You'll be less likely to forget to renew the license in time. Use an FCC Form 610 to apply for a license renewal. Always attach a photocopy of your current license. You can also ask one of the Volunteer Examiner Coordinators' offices to file your renewal application electronically if you don't want to mail the form to the FCC. You must still mail the form to the VEC, however. The ARRL/VEC Office will electronically file application forms for any ARRL member free of charge.

If you do forget to renew your license, you have up to two years to apply for a new license. After the two-year **grace period**, you will have to take the exam again. Your license is not valid during this two-year grace period, however. You may not operate an amateur station with an expired license. All the grace period means is that the FCC will renew the license if you apply during that time.

If your license is lost, mutilated or destroyed, request a new one from the FCC. A Form 610 isn't required; a letter will do. Be sure to explain why you are requesting a new license. (You don't have to explain that your dog ate your license — just that it was destroyed.) If you move or change your name, you will need to modify your license. Notify the FCC of the new information on an FCC Form 610, and attach a photocopy of your license. Then FCC will be able to write to you if necessary.

To obtain a new Form 610, call 800-418-3676. You can also write to: Federal Communications Commission, Forms Distribution Center, 2803 52nd Avenue, Hyattsville, MD 20781 (specify "Form 610" on the envelope), or : Form 610, ARRL, 225 Main Street, Newington, CT 06111-1494. (Please include an SASE with your request.) The Form 610 also is available from the FCC's fax on demand service. Call 202-418-0177 and ask for form number 000610. Form 610 also is available via the Internet. The World Wide Web location is: **http://www.fcc.gov/Forms/Form610** or you can receive the form via ftp to: **ftp.fcc.gov/pub/Forms/Form610**. Many clubs and Amateur Radio equipment dealers may also have a supply of these forms. If you obtain one from a source other than the FCC or ARRL, be sure it is dated November 1993 or later in the lower right-hand corner. Versions of the form dated earlier than November 1993 are obsolete and no longer acceptable. (As of this writing — March 1997 — the latest Form 610 is dated March 1995.)

[Turn to Chapter 13 and study questions N1A07, N1A08, N1A09, N1B01, N1B02, N1B03, N1B04, N1B06, N1B07, N1B08 and N1B09. Also study questions N1D05, N1D06, N1D10, N1D11 and N1D12. You should also be able to answer questions N1H09 and N1H12 after reading this text. Review this section if you have difficulty with any of these questions.]

[**If you are studying for a Technician license**, you should also turn to Chapter 14 and study questions T1A10, T1A11 and T1A12. Review this section if you have difficulty with any of these questions.]

The Control Operator

A **control operator** is an "amateur operator designated by the station licensee to be responsible for the transmissions from that station to assure compliance with the FCC Rules." In effect, the control operator operates the Amateur Radio station. Only a licensed ham may be the control operator of an amateur station. If another licensed radio amateur operates your station with your permission, he or she assumes the role of control operator.

If you let another amateur with a higher class license than yours control your station, he or she may use any operating privileges allowed by the higher class license. On the other hand, if you are the control operator at the station of another amateur who has a higher class of license than yours, you can use only the privileges allowed by your license.

Any amateur operator may designate another licensed operator as the control operator, to share the responsibility of station operation. The FCC holds both the control operator and the station licensee responsible for proper operation of the station. If you are operating your own Amateur Radio station, then you are the control operator at that time.

Every Amateur Radio station has a **control point**. The control point is where the station operating controls are located. This may be at the actual radio controls, or it may be at some other location, such as when the radio is controlled by a computer link, telephone connection or some other method.

A control operator must be present at the station control point whenever a transmitter is operating. This means that you may not allow an unlicensed person to operate your radio transmitter while you are not present. There is one time when a transmitter may be operated without a control operator being present, however. Some types of stations, such as repeater stations, may be operated by automatic control. In this case there is no control operator at the transmitter control point.

Your Novice or Technician license authorizes you to be the control operator of an amateur station in the Novice or Technician frequency bands. This means you can be the control operator of your own station or someone else's station. In either case, you are responsible to the FCC for the proper operation of the station. It is interesting to note that your amateur license places no restrictions on the number of transmitters that you can control at the same time. A Novice licensee can control any number of transmitters at one time.

A Canadian citizen who holds a Canadian amateur license can be a control operator of a US amateur station. So can a ham from another country who has a reciprocal operating permit issued by the FCC. Otherwise, the control operator must hold a valid US amateur license.

If you allow another licensed ham to operate your station, you are still responsible for its proper operation. You are always responsible for the proper operation of your station. Your primary responsibility as the station licensee is to ensure the proper operation of your station. Unless your station records show that another amateur was the control operator of your station at a certain time, the FCC will assume that you were the only control operator.

What kind of station records should you keep? The FCC does not *require* you to keep any particular information about the operation of your station. Many amateurs find it helpful to keep a station logbook with certain information about the operation of their station. Logbooks are useful for recording dates, calls, names and locations of those stations you contact. When you confirm contacts by sending "QSL cards" (see Chapter 2) a logbook is a convenient way to keep track of these exchanges. Your log will provide a useful and interesting history, if you choose to keep one. Figure 1-3 shows an example of the information many amateurs keep in a logbook.

There are commercially prepared logbooks, such as the ARRL Log Book, or you can use a notebook or other form that you prepare for your own use. Many amateurs find it helpful to use one side of a page to log amateur contacts and the back of the page to record information about station equipment, changes, and other information.

It isn't necessary to use a station logbook. If you don't, though, it will not be possible to show when someone else has

DATE	FREQ.	MODE	POWER	TIME	STATION WORKED	REPORT SENT	REPORT REC'D	TIME OFF	COMMENTS (QTH / NAME / QSL VIA)	QSL S	QSL R
16 Nov	3715	CW	100	1800	KA1EBV	589	479	1835	MANOMET, MA JOHN	✓	✓
	"	CW	100	1840	WB7TPY	599	599	1855	TUFTS UNIVERSITY DAVE	✓	
17 Nov	28.125	CW	100	2002	KC4AAA	469	559	2003	AT SOUTH POLE! BIG PILEUP BURO	✓	
	28.380	SSB	100	2010	KAØHJD	59	58	2016	DES MOINES, IA KRISTEN	✓	
	"	SSB	100	2010	WØSH	58	57	2016	PALM BAY, FL GARY COLLECTS OLD TELEGRAPH KEYS	✓	
	7.130	CW	100	2110	KA1GQT	579	589	2055	LINDA IS A NURSE	✓	✓
20 NOV	21.126	CW	100	1645	NA1L	599	599	1705	BRISTOL, CT DALE IS A LAWYER		
21 NOV	28.380	RTTY	60	1018	WB8IMY	589	579	1052	CT STEVE WORKS AT ARRL		
	28.125	CW	100	1804	XZ1A	599	479	1804	MYANMAR! BILL Huge Pileup K5FUV	✓	✓

Figure 1-3 — Most amateurs keep a station logbook to record information about each contact.

been the control operator of your station. Whenever you give someone else permission to be the control operator at your station, enter the other person's name and call sign, and the time, date and frequencies used, in your logbook.

[Before you go on to the next section, you should turn to Chapter 13 and study questions N1A10, N1B05 and N1G01 through N1G11. If you have difficulty with any of these questions, review the material in this section.]

[**If you are studying for a Technician license,** you should also turn to Chapter 14 and study questions T1A01 and T1A02. Review this section if you are uncertain of the answer to these questions.]

Amateur License Classes

There are six kinds, or levels, of amateur license. They vary in degree of knowledge required and **frequency privileges** granted. Higher-class licenses have more comprehensive examinations. In return for passing a more difficult exam you earn more frequency privileges (frequency space and modes of operation).

The first step is either the Novice or Technician license. The FCC issues these "beginner's" licenses to those who demonstrate the ability to operate an Amateur Radio transmitter safely and properly.

An applicant for a Novice license must show a basic proficiency in Morse code by passing a test at 5 words per minute (wpm). The written part of the exam covers some very basic radio fundamentals and knowledge of a few rules and regulations. With a little study you'll soon be ready to pass the Novice exam.

An applicant for a Technician license must pass a slightly more difficult written test, but there is no Morse code requirement. Because each step up the Amateur Radio license ladder requires the applicant to pass the lower exams, a Technician must also pass the Novice written exam.

Anyone (except an agent or representative of a foreign government) is eligible to qualify for an Amateur Radio operator license. There is no age requirement. To hold an Amateur Radio station license, you must have a valid operator license. (Remember, both licenses are printed on the same piece of paper.)

A Novice or Technician license gives you the freedom to develop operating and technical skills through on-the-air experience. These skills will help you upgrade to a higher class of license, with additional privileges.

As a Novice, you can communicate with other amateur stations in the exotic reaches of the world. You can provide public service through emergency communications and message handling. You can enhance international goodwill just like operators with higher license classes. You also can use repeater stations on two popular amateur bands.

As a Technician, you can use a wide range of frequency bands—all amateur bands above 30 MHz, in fact. You'll use repeaters, packet radio and orbiting satellites to relay your signals over a wider area. By passing the 5-wpm Morse code test you will upgrade to a Technician Plus license and gain Novice privileges below 30 MHz.

Later, you will probably want to earn greater operating privileges. As you gain operating experience, you will prepare for a higher-class license — the gateway to more frequencies and modes. Don't get the idea that a Novice or Technician license is so limited that you can't enjoy the full range of amateur activities, though. You will be allowed to use all of the major operating modes on some band, and you can sample the full range of operating frequencies including the high frequency (HF), very high frequency (VHF) and ultra high frequency (UHF) ranges. You can even send messages for your unlicensed friends (called **third-party communications**) or allow them to talk over your radio (more on this later). As you upgrade, however, you will gain access to more operating frequencies, higher transmitter power and you will even be able to help prepare and administer Amateur Radio license exams!

There is a real thrill to having more frequencies available to talk with other amateurs on the far side of the globe. That is a powerful incentive to upgrade to the General license (the next step up from Technician Plus). You will have to pass a 13-wpm Morse code exam and another theory and rules exam to earn the General class license. The General license gives voice privileges on eight high-frequency (HF) bands. These bands typically carry signals over great distances. In addition there is one HF band exclusively for Morse code, radioteletype and other "digital" (computer) modes. For this reason, the General license is very popular.

As you progress and mature in Amateur Radio, you will develop specialized interests in such exciting modes as amateur television and satellite communication. You'll also want even more privileges, and that is where the Advanced and Amateur Extra licenses come in. To obtain one of these, you must be prepared to face exams on the more technical aspects of the hobby. For the Amateur Extra license, the top-of-the-line, there is also a 20-word-per-minute Morse code test.

To qualify for a higher-class license, you must pass the theory exam for each level up to that point. For example, suppose you want to go from Novice directly to Amateur Extra. You will have to pass the Technician, General and Advanced theory exams as well as the Amateur Extra test. (You can just take the 20-wpm code test, however, without passing the 13-wpm General-class code test.) In fact, you could take the Novice through Amateur Extra exams all at once.

[You should turn to Chapter 13 now and study questions N1B10 through N1B13, N1D01 and N1D02. Review the material in this section if you have difficulty with any of these questions.]

The Novice License

A Novice license allows you to operate on portions of six Amateur Radio bands. We normally identify these bands by specifying the frequency range they cover or by listing the wavelength in meters. (You will learn more about the relationship between frequency and wavelength later in this

Table 1-1

Amateur Operator Licenses†

Class	*Code Test*	*Written Examination*	*Privileges*
Novice	5 wpm (Element 1A)	Novice theory and regulations (Element 2)	Telegraphy on 3675-3725, 7100-7150 and 21,100-21,200 kHz with 200 W PEP output maximum; telegraphy RTTY and data on 28,100-28,300 kHz and telegraphy and SSB voice on 28,300-28,500 kHz with 200 W PEP max; all amateur modes authorized on 222.1-223.91 MHz, 25 W PEP max; all amateur modes authorized on 1270-1295 MHz, 5 W PEP max.
Technician		Novice theory and regulations; Technician-level theory and regulations. (Elements 2 and 3A)*	All amateur privileges above 50.0 MHz.
Technician Plus	5 wpm (Element 1A)	Novice theory and regulations; Technician-level theory and regulations. (Elements 2 and 3A)*	All Novice HF privileges in addition to all Technician privileges.
General	13 wpm (Element 1B)	Novice theory and regulations; Technician and General theory and regulations. (Elements 2, 3A and 3B)	All amateur privileges except those reserved for Advanced and Amateur Extra; see Table 1-2.
Advanced	13 wpm (Element 1B)	All lower exam elements, plus Advanced theory. (Elements 2, 3A, 3B and 4A)	All amateur privileges except those reserved for Amateur Extra class; see Table 1-2.
Amateur Extra	20 wpm (Element 1C)	All lower exam elements, plus Extra-class theory (Elements 2, 3A, 3B, 4A and 4B)	All amateur privileges.

†A licensed radio amateur will be required to pass only those elements that are not included in the examination for the amateur license currently held.

*If you hold a valid Technician license issued before March 21, 1987, you also have credit for Element 3B. You must be able to prove your Technician license was issued before March 21, 1987 to claim this credit.

book.) Table 1-1 lists the frequency range for each of the Novice bands. This table also serves as a comparison between Novice and Technician privileges and those given to higher class licensees.

Frequency Privileges

When operating, you must stay within your assigned **frequency bands**. Novice operators may transmit on portions of six bands in the radio spectrum. Amateurs usually refer to these frequency bands by their wavelength. Novices may operate in the 80, 40, 15 and 10-meter bands, and in the 1.25-meter (222-MHz) and 23-centimeter (1270-MHz) bands. These bands are further divided into subbands for the different classes of license. Novice operators have **frequency privileges** (or permission to operate) on these subbands:

3675-3725 kHz in the 80-meter band
(5167.5 kHz—Alaska only, emergency communications, single-sideband voice emissions only)
7100-7150 kHz in the 40-meter band*
21,100-21,200 kHz in the 15-meter band
28,100-28,500 kHz in the 10-meter band
222.0-225.0 MHz in the 1.25-meter band*
1270-1295 MHz in the 23-centimeter band

*In ITU Region 2 only. ITU Region 2 comprises all of North America, the Caribbean and the Eastern Pacific, including Hawaii. Contact ARRL Headquarters for more information.

We use the metric system of measurement in electronics. Chapter 5 includes a full explanation of the metric system and terms like *kilo*, *mega* and *centi* that we have used here. For now it is important that you know that *kilohertz* and *megahertz* are measures of the frequency of a radio signal. The *hertz* (Hz) is the basic unit of frequency. Kilo means thousand and mega means million. We can list any of the amateur frequency bands in either kilohertz or megahertz. For example, the 15-meter Novice band can be written either as 21.100 to 21.200 MHz, or 21,100 to 21,200 kHz and the 10-meter band can be written either as 28.100 to 28.500 MHz or 28,100 to 28,500 kHz.

[You must memorize these Novice frequency privileges. Before you go on to the next section, turn to Chapter 13 and study questions N1C01 through N1C11. Review these frequencies often as you study the remaining material in this book.]

Emission Privileges

Amateur operators transmit a wide variety of signals. These include Morse code, radioteletype, several types of voice communications and even television pictures. An **emission** is any radio-frequency (RF) signal from a transmitter.

There is a system for describing the various types of signals (or emissions) found on the amateur bands. Different modes are given identifiers, called emission types.

You should be familiar with the various emission types. The FCC lists the **emission privileges** for each license class by giving the emission types each may use. An emission privilege is FCC permission to use a particular emission type, such as Morse code or single-sideband phone. As a Novice licensee, you will be permitted to use all of the emission types (except pulse) on at least one frequency band. The emission types defined by the FCC are:

- CW — Morse code telegraphy.
- Data — Computer communications modes, often called *digital communications* because digital computers are used.
- Image — Television and facsimile communications.
- MCW — (Tone-modulated CW) Morse code telegraphy using a keyed audio tone.
- Phone — Speech (voice) communications.
- Pulse — Communications using a sequence of controlled signal variations.
- RTTY — Narrow-band direct-printing telegraphy communications (received by automatic techniques). Since digital computers are often used on radioteletype (RTTY), these signals are also often called *digital communications.*
- SS — Spread-spectrum communications in which the signal energy is spread across a wide bandwidth.
- Test — Transmissions containing no information.

On 80, 40 and 15 meters, Novices may transmit only Morse code (CW). The transmitter produces this Morse code signal by *keying* (switching on and off) the signal from a continuous-wave (CW) transmitter. On 10 meters, Novices may use CW from 28.1 to 28.5 MHz. Novice operators may also transmit radioteletype (RTTY) and data from 28.1 to 28.3 MHz, and single-sideband phone from 28.3 to 28.5 MHz. On the frequencies 222.0-225.0 MHz and 1270-1295 MHz, Novices may use all emissions authorized to higher-class licensees on these bands, including FM phone and digital modes, such as packet radio. Table 1-2 summarizes the amateur band limits and operating modes for each license class.

Hams often describe radioteletype (RTTY) and packet radio or other data emissions as *digital communications.* These are signals intended to be received and printed or displayed on a computer screen automatically. Information transferred directly from one computer to another is an example of digital communications. Data emissions also include telemetry and telecommand communications. Telemetry refers to signals sent from a remote location, such as a satellite, to provide information about the station and operating conditions. Telecommand communications are signals sent to a remote station, such as a satellite, to control the station.

[You have to memorize these Novice emission privileges. You should turn to Chapter 13 now and study questions N1E01 through N1E14. Also study questions N1F13 and N1F14. Review the material in this section if you have difficulty with any of the questions.]

Novice Transmitter Power

The FCC has issued rules explaining how transmitter power should be measured at the output of the transmitter. Novice licensees may use a maximum of 5 W **peak envelope power (PEP)** output on the 1270-MHz band, 25 W PEP on the 222-MHz band, and 200 W PEP on the 80, 40, 15 and 10-meter bands. The 200 W limitation also applies to any other licensed radio amateur who operates in the 80, 40 and 15-meter Novice bands. For example, an Amateur Extra licensee may only use 200 W PEP when operating on 21.150 MHz. Remember that Novices may use up to 200 W PEP on any Novice frequency in the 80, 40, 15 and 10-meter bands.

In addition to these Novice bands, higher-class licensees may use a maximum of 200 W PEP on the 30-meter band. On all other bands, the rules limit the maximum transmitter output power in the amateur service to 1500 W PEP output. Higher-class licensees may use 1500 W PEP in the Novice sections of 10 meters, 222 MHz and 1270 MHz. In practice, unless you're attempting something as challenging as "moonbounce" — bouncing a signal off the moon — amateurs rarely use more than a couple of hundred watts on the VHF and UHF bands.

So far, we have been talking about the *maximum* transmitter power the rules allow. There is another rule to consider, though. According to FCC Rules, an amateur station must use the *minimum* transmitter power necessary to maintain reliable communication. What this means is simple — if you don't need 200 W to contact someone, don't use it! For example, suppose you contact another amateur station and learn that your signals are extremely strong and loud, and perfectly readable. You should turn down your transmitter output power in that case.

[Before proceeding to the next section turn to Chapter 13 and study questions N1F01 through N1F12. Review this section if you have any difficulties.]

Table 1-2

US AMATEUR BANDS

December 20, 1994

160 METERS

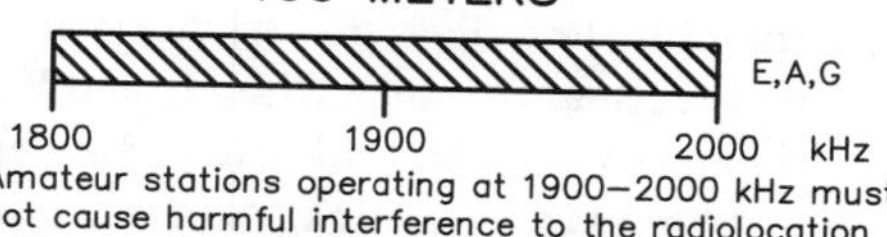

Amateur stations operating at 1900–2000 kHz must not cause harmful interference to the radiolocation service and are afforded no protection from radiolocation operations.

80 METERS

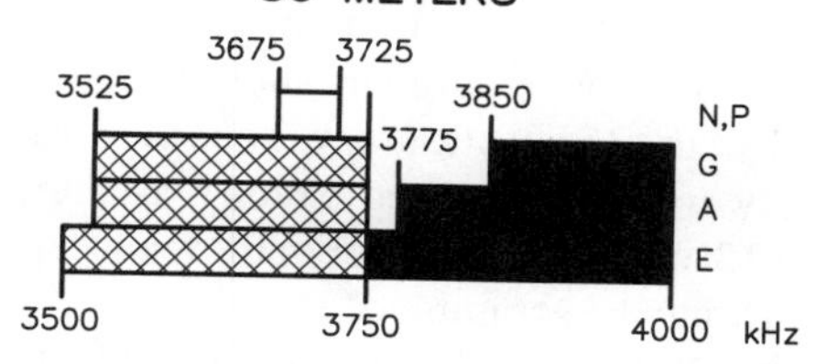

5167.5 kHz (SSB only): Alaska emergency use only.

40 METERS

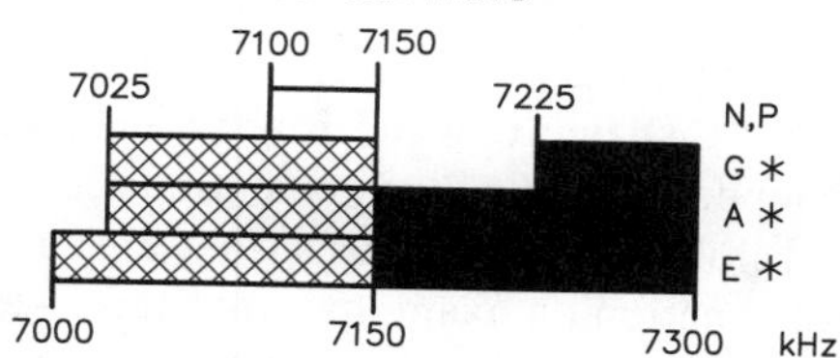

* Phone operation is allowed on 7075–7100 kHz in Puerto Rico, US Virgin islands and areas of the Caribbean south of 20 degrees north latitude; and in Hawaii and areas near ITU Region 3, including Alaska.

30 METERS

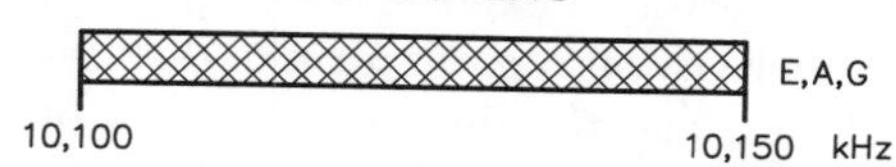

Maximum power on 30 meters is 200 watts PEP output. Amateurs must avoid interference to the fixed service outside the US.

20 METERS

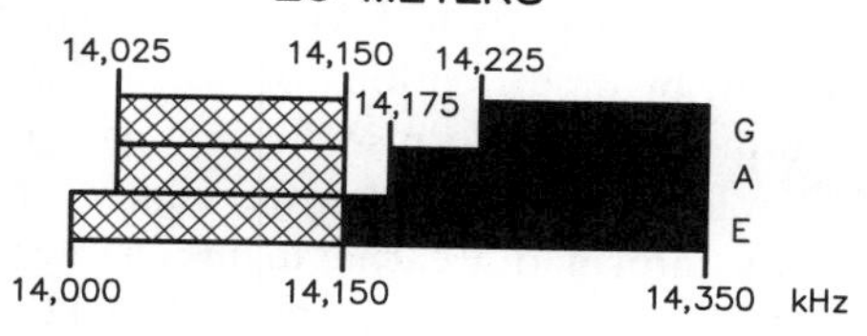

17 METERS

15 METERS

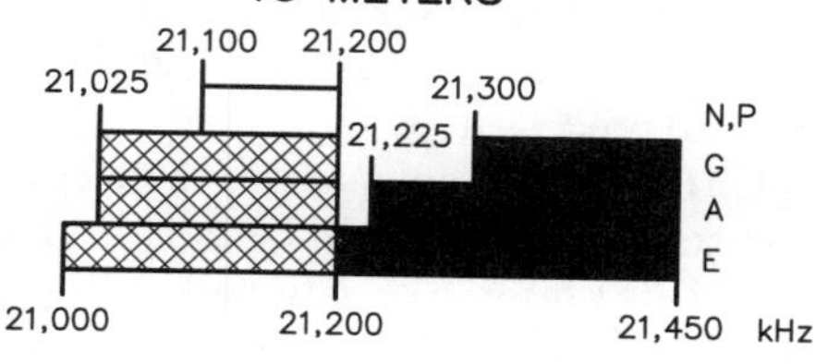

12 METERS

10 METERS

28,100 28,500

N,P

E,A,G

28,000 28,300 29,700 kHz

Novices and Technician Plus Licensees are limited to 200 watts PEP output on 10 meters.

6 METERS

2 METERS

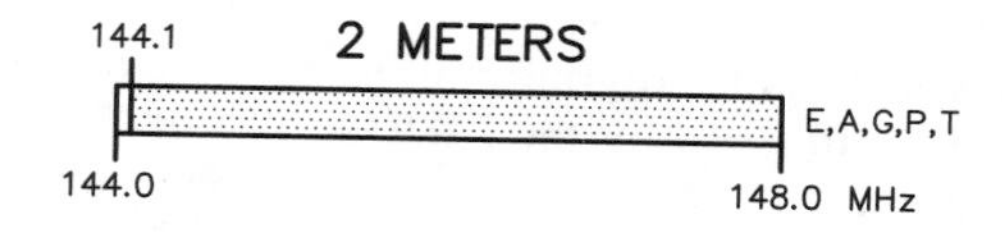

1.25 METERS

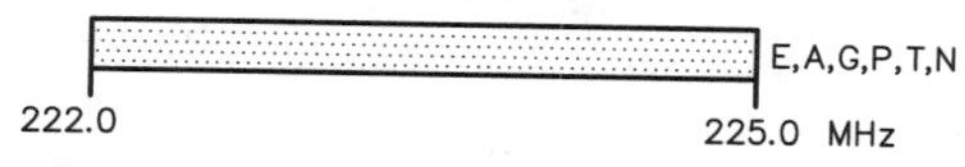

Novices are limited to 25 watts PEP output from 222 to 225 MHz.

70 CENTIMETERS**

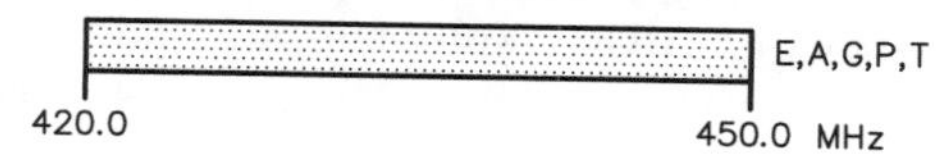

33 CENTIMETERS**

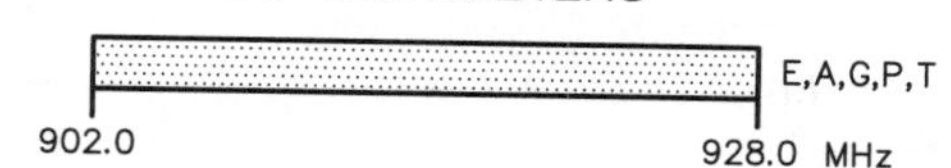

23 CENTIMETERS**

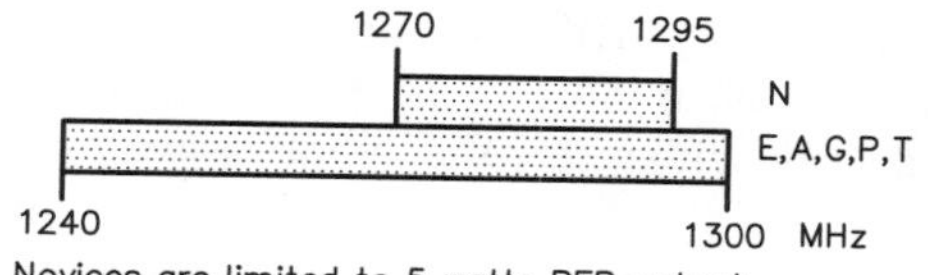

Novices are limited to 5 watts PEP output from 1270 to 1295 MHz.

US AMATEUR POWER LIMITS

At all times, transmitter power should be kept down to that necessary to carry out the desired communications. Power is rated in watts PEP output. Unless otherwise stated, the maximum power output is 1500 W. Power for all license classes is limited to 200 W in the 10,100–10,150 kHz band and in all Novice subbands below 28,100 kHz. Novices and Technicians are restricted to 200 W in the 28,100–28,500 kHz subbands. In addition, Novices are restricted to 25 W in the 222–225 MHz band and 5 W in the 1270–1295 MHz subband.

Operators with Technician class licenses and above may operate on all bands above 50 MHz. For more detailed information see The FCC Rule Book.

KEY

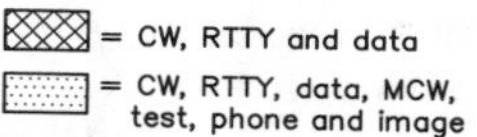

- = CW, RTTY and data
- = CW, RTTY, data, MCW, test, phone and image
- = CW, phone and image
- = CW and phone
- = CW, RTTY, data, phone, and image
- = CW only

E = EXTRA CLASS
A = ADVANCED
G = GENERAL
P = TECHNICIAN PLUS
T = TECHNICIAN
N = NOVICE

** Geographical and power restrictions apply to these bands. See The FCC Rule Book for more information about your area.

Above 23 Centimeters:

All licensees except Novices are authorized all modes on the following frequencies:
2300–2310 MHz
2390–2450 MHz
3300–3500 MHz
5650–5925 MHz
10.0–10.5 GHz
24.0–24.25 GHz
47.0–47.2 GHz
75.5–81.0 GHz
119.98–120.02 GHz
142–149 GHz
241–250 GHz
All above 300 GHz

For band plans and sharing arrangements, see The ARRL Operating Manual or The FCC Rule Book.

The Novice Exam

The FCC refers to the various exams for Amateur Radio licenses as exam *Elements*. For example, exam Elements 1A, 1B and 1C are the 5, 13 and 20 word-per-minute (wpm) code exams. Element 2 is the Novice written exam. Table 1-1 summarizes the exams and privileges that go with each license class.

The table shows that you must pass exam Element 1A and Element 2 for the Novice license. The purpose of Element 1A is to prove your ability to send and receive messages using the international Morse code at a speed of 5 words per minute. The Element 2 written exam covers basic FCC Rules and Novice operating procedures, along with basic electronics theory.

[Now turn to Chapter 13 and study question N1D03. Review this section if you have difficulty with this question.]

What Will My Novice Exam Be Like?

Three amateurs with General licenses or higher, who are accredited by a **Volunteer Examiner Coordinator (VEC)** can give the test for a Novice license. The examiners must be at least 18 years old and must not be relatives of anyone taking the exam. The exams can be given at the convenience of the candidates and the examiners, at any location they agree to. (All rules of the VEC program must be followed.) *The FCC Rule Book*, published by the ARRL, contains details on the VE program.

A **Volunteer Examiner (VE)** is someone who volunteers to test others for amateur licenses. There is probably an active Volunteer Examining Team somewhere in your area. These teams conduct tests for all classes of Amateur Radio licenses. The Volunteer Examiners give Novice exams at all of their regular exam sessions.

If you have any trouble locating examiners, write to: ARRL/VEC Office, ARRL Headquarters, 225 Main Street, Newington, CT 06111. We will refer you to someone near you who can arrange your test. We will also supply a list of ARRL/VEC Volunteer Exam sessions in your area. The ARRL/VEC oversees the thousands of ARRL Volunteer Examiners, arranges and publishes exam schedules and inspects application forms before they are submitted to the FCC.

The code test is normally given first. The examiners will usually send five minutes of 5-words-per-minute code, and then test your copy. The test usually takes one of two forms. The examiners may ask you 10 questions based on the contents of the transmission, and you must answer 7 of the 10 questions correctly. (These questions may have a fill-in-the-blank format or a multiple-choice format.) Or, the examiners may check your answer sheet for one minute of perfect ("solid") copy (25 characters in a row with no errors) out of the five-minute transmission.

The written test consists of 35 questions on general operating practices, rules and regulations, and basic radio theory. To pass, you must correctly answer 26 of the 35 questions on your exam. The FCC specifies 10 subelements, or divisions, for each exam element. These are the Commission's Rules, Operating Procedures, Radio-Wave Propagation, Amateur Radio Practices, Electrical Principles, Circuit Components, Practical Circuits, Signals and Emissions, Antennas and Feed Lines, and Radiofrequency environmental safety practices at an amateur station (RF Radiation Safety).

Your exam must include 35 questions taken from the pool of questions as released by the Volunteer Examiner Coordinators' Question Pool Committee, and printed in Chapter 13 of this book. Your examiners should choose one question from each of the 35 question-pool subsections. The questions and answers must be used exactly as they are printed in Chapter 13, although the order of the answer positions A, B, C and D may be scrambled.

[Turn to Chapter 13 and study question N1A11. If you have trouble answering this question review this section.]

Finding an Exam Opportunity

To determine where and when an exam will be given, contact the ARRL/VEC Office, or watch for announcements in the Hamfest Calendar and Coming Conventions columns in *QST*. Many local clubs sponsor exams, so they are another good source of information on exam opportunities. ARRL officials such as Directors, Vice Directors and Section Managers receive notices about test sessions in their area. See pages 10 and 12 in the latest issue of *QST* for names and addresses.

To register for an exam, send a completed Form 610 to the VE Team responsible for the exam session if preregistration is required. Otherwise, bring the form to the session. Registration deadlines, and the time and location of the exams, are mentioned prominently in publicity releases about upcoming sessions.

Filling Out Your FCC Form 610

What's next? A bit of paperwork! All applications for new amateur licenses, or modifications or renewals, are made on an FCC Form 610. Only forms with a revision date of November 1993 or later (found in the lower right corner) are acceptable. See the information earlier in this chapter for detailed information about getting a copy of the latest Form 610. Figure 1-4 shows a Form 610 completed for a successful Technician applicant.

You should complete the top half of the form, Section 1, as shown. Be sure to write your birth year in box 2. Probably the most common error is to write the current year in this space! Check box 4A, "EXAMINATION for new license" if you don't already have a license. Skip items 5 and 7, unless this is not your first license or you have another Form 610 waiting for the FCC to process it.

Unless you plan to install an antenna over 200 feet high, or your permanent station location will be in a designated wilderness area, wildlife preserve or nationally recognized scenic and recreational area, check "NO" on line 6. Be sure to sign and date your application on lines 8 and 9. A daytime telephone number is not required, but is helpful in case there are any questions about your application.

Your Volunteer Examiners will fill out Section 2, the Administering VE's report, on the bottom half of Form 610. That is where the examiners certify that you have passed the exam. They will check the appropriate box in Section A and fill out Sections C and D. The three Volunteer Examiners will print their names, call signs and the date of the exam, and sign the form on the lines at the bottom.

The back of Form 610 is for a physician's certification that you have a handicap that makes it impossible for you to pass a 13 or 20-wpm Morse code exam. Detailed instructions to the physician are included on the bottom of the form.

The Form 610 serves as the application for your license. You will use the same form to renew your license before it expires. This form also serves another very important purpose. Line 3 provides the FCC with a mailing address where they can contact you. If you move, or your address changes for any reason, use an FCC Form 610 to notify them of your new address. It's important that you receive any mail sent to you by the FCC. If you don't respond to an FCC letter concerning a rules violation, you may be fined or your license may be suspended or revoked!

[Turn to Chapter 13 at this time and review question N1D04. Review this section if you are uncertain of the answer.]

Taking the Exam

By the time examination day rolls around, you should have already prepared yourself. This means getting your schedule, supplies and mental attitude ready. Plan your schedule so you'll get to the examination site with plenty of time to spare. There's no harm in being early. In fact, you might have time to discuss hamming with another applicant, which is a great way to calm pretest nerves. Try not to discuss the material that will be on the examination as this may make you even more nervous. By this time, it's too late to study anyway!

What supplies will you need? First, be sure you bring your current *original* Amateur Radio license, if you have one. Also bring the *original* Certificate of Successful Completion of Examination (CSCE) if you have one of those for some of the exam elements. Also be sure to bring a photocopy of your license and CSCE. Bring along several sharpened number 2 pencils and two pens (blue or black ink). Be sure to have a good eraser. A pocket calculator may also come in handy. You may use a programmable calculator if that is the kind you have, but take it into your exam "empty" (cleared of all programs and constants in memory). Don't program equations ahead of time because you may be asked to demonstrate that there is nothing in the calculator's memory.

The VE Team is required to check two forms of identification before you enter the test room. This includes your *original* Amateur Radio license (if you have one). A photo ID of some type is best for the second form, but is not required by the FCC. Other acceptable forms of ID include a driver's license, a piece of mail addressed to you, a birth certificate or some other such document.

The following description of the testing procedure applies to exams coordinated by the ARRL/VEC, although many other VECs use a similar procedure.

Code Tests

If you are planning to take the Novice exam or are interested in using Morse code and gaining the additional privileges granted to Technicians who have passed a 5-wpm code test, this information will be of interest to you.

The code tests are usually given before the written exams. The 20-wpm exam is usually given first, then the 13-wpm exam and finally the 5-wpm test. There is no harm in trying the 20 or 13 wpm exams even though you are testing for the Novice or Technician exams.

Before you take the code test, you'll be handed a piece of paper to copy the code as it's sent. The test will begin with about a minute of practice copy. Then comes the actual test: five minutes of Morse code. You are responsible for knowing the 26 letters of the alphabet, the numerals 0 through 9, the period, comma, question mark, $\overline{\text{AR}}$, $\overline{\text{SK}}$, $\overline{\text{BT}}$ (or double dash =) and $\overline{\text{DN}}$ (fraction bar /). See Chapter 11 for explanations of what these characters mean. You may copy the entire text word for word, or just take notes on the content. At the end of the transmission, the examiner will hand you 10 questions about the text. Simply fill in the blanks with your answers. (You must spell each answer exactly as it was sent.) Some examining teams may choose to use a multiple-choice format for the code-test answers. If you get at least 7 correct, you pass! Alternatively, the exam team has the option to look at your copy sheet. If you have one minute of solid copy, they can certify that you passed the test on that basis. You must copy 25 characters without an error for the 5-wpm exam. The format of the test transmission is similar to one side of a normal on-the-air amateur conversation.

A sending test may not be required. The Commission has decided that if applicants can demonstrate receiving ability, they most likely can also send at that speed. But be prepared for a sending test, just in case! Subpart 97.503(a) of the FCC Rules says, "A telegraphy examination must be sufficient to prove that the examinee has the ability to send correctly by hand and to receive correctly by ear texts in the international Morse code at not less than the prescribed speed — "

Written Tests

The examiner will give each applicant a test booklet, an answer sheet and scratch paper. After that, you're on your own. The first thing to do is read the instructions. Be sure to sign your name every place it's called for. Do all of this at the beginning to get it out of the way.

Next, check the examination to see that all pages and questions are there. If not, report this to the examiner immediately. When filling in your answer sheet, make sure your answers are marked next to the numbers that correspond to each question.

FEDERAL COMMUNICATIONS COMMISSION
GETTYSBURG, PENNSYLVANIA

Approved by OMB 3060-0003 Expires 8/31/96 See instructions for information regarding public burden estimate.

APPLICATION FORM 610 FOR AMATEUR OPERATOR/PRIMARY STATION LICENSE

SECTION 1 - TO BE COMPLETED BY APPLICANT (See instructions)

1. Print or type last name: Gagne | Suffix: | First name: Jennifer | Middle initial: S
2. Date of birth: 11-28-69 (month, day, year)
3. Mailing address (Number and street): 225 Main St. | City: Newington | State code: CT | ZIP code: 06111

4. I HEREBY APPLY FOR (make an X in the appropriate box(es)):

- 4A. [X] **EXAMINATION** for a new license
- 4B. [ ] **EXAMINATION** for upgrade of my operator license class
- 4C. [ ] **CHANGE** my name on my license to my new name in Item 1. My former name was: (Last name) (Suffix) (First name) (MI)
- 4D. [ ] **CHANGE** my mailing address on my license to my new address in Item 3
- 4E. [ ] **CHANGE** my station call sign systematically (See instructions) Applicant's Initials ______
- 4F. [ ] **RENEWAL** of my license

5. Unless you are requesting a new license, attach the original or a photocopy of your license to the back of this Form 610 and complete Items 5A and 5B.
5A. Call sign shown on license
5B. Operator class shown on license

6. Would an FCC grant of your request be an action that may have a significant environmental effect? [X] NO [ ] YES (Attach required statement)

7. If you have filed another Form 610 that we have not acted upon, complete Items 7A and 7B.
7A. Purpose of other form
7B. Date filed (month, day, year)

WILLFUL FALSE STATEMENTS MADE ON THIS FORM ARE PUNISHABLE BY FINE AND/OR IMPRISONMENT, (U.S. CODE, TITLE 18, SECTION 1001), AND/OR REVOCATION OF ANY STATION LICENSE OR CONSTRUCTION PERMIT (U.S. CODE, TITLE 47, SECTION 312(A)(1)) AND/OR FORFEITURE (U.S. CODE, TITLE 47, SECTION 503).

I CERTIFY THAT ALL STATEMENTS AND ATTACHMENTS ARE TRUE, COMPLETE, AND CORRECT TO THE BEST OF MY KNOWLEDGE AND BELIEF AND ARE MADE IN GOOD FAITH; THAT I AM NOT A REPRESENTATIVE OF A FOREIGN GOVERNMENT; THAT I WAIVE ANY CLAIM TO THE USE OF ANY PARTICULAR FREQUENCY REGARDLESS OF PRIOR USE BY LICENSE OR OTHERWISE; AND THAT THE STATION TO BE LICENSED WILL BE INACCESSIBLE TO UNAUTHORIZED PERSONS.

8. Signature of applicant (Do not print, type, or stamp. Must match name in Item 1.)
× Jennifer Gagne
(860) 594-0328
Daytime Telephone Number

9. Date signed: 01-17-97 (month, day, year)

SECTION 2 - TO BE COMPLETED BY ALL ADMINISTERING VE's

A. Applicant is qualified for operator license class:

[ ]	NOVICE	(Elements 1(A), 1(B), or 1(C) and 2)
[X]	TECHNICIAN	(Elements 2 and 3(A))
[ ]	TECHNICIAN PLUS	(Elements 1(A), 1(B), or 1(C), 2 and 3(A))
[ ]	GENERAL	(Elements 1(B) or 1(C), 2, 3(A) and 3(B))
[ ]	ADVANCED	(Elements 1(B) or 1(C), 2, 3(A), 3(B) and 4(A))
[ ]	AMATEUR EXTRA	(Elements 1(C), 2, 3(A), 3(B), 4(A) and 4(B))

B. VEC receipt date:

C. Name of Volunteer-Examiner Coordinator (VEC): ARRL

D. Date of VEC coordinated examination session: 1/17/97

E. Examination session location: 225 Main St. Newington, CT 06111

I CERTIFY THAT I HAVE COMPLIED WITH THE ADMINISTERING VE REQUIREMENTS IN PART 97 OF THE COMMISSION'S RULES AND WITH THE INSTRUCTIONS PROVIDED BY THE COORDINATING VEC AND THE FCC

VE's name (Print First, MI, Last, Suffix)	VE's station call sign	VE's signature (must match name)	Date signed
1st: Larry D. Wolfgang	WR1B	Larry D. Wolfgang	1/17/97
2nd: Wayne K. Irwin	W1KI	Wayne K. Irwin	1/17/97
3rd: MARTIN G. COOK	N1FOC	Martin G Cook	1/17/97

FCC Form 610
March 1995

Figure 1-4 — A completed FCC Form 610. All applications for new Amateur Radio licenses must be submitted using this form.

ATTACH ORIGINAL OR A PHOTOCOPY OF YOUR LICENSE HERE:

SECTION 3 - TO BE COMPLETED BY PHYSICIAN

PHYSICIAN'S CERTIFICATION OF DISABILITY

Please see notice below

Print, type, or stamp physician's name: __________

Street address: __________

City, State, ZIP code: __________

Office telephone number: ( ) __________

I CERTIFY THAT I have read the Notice to Physician Certifying to a Disability, and that the person named in Item 1 on the reverse is severely handicapped, the duration of which will extend for more than 365 days beyond this date. Because of this severe handicap, this person is unable to pass a 13 or 20 words per minute telegraphy examination. I am licensed to practice in the United States or its Territories as a doctor of medicine (M.D.) or doctor of osteopathy (D.O.). I have considered the accommodations that could be made for this person's disability and have determined that, even with accommodations, this person would be unable to pass a 13 or 20 words per minute telegraphy examination.

WILLFUL FALSE STATEMENT IS PUNISHABLE BY FINE AND IMPRISONMENT (U.S. CODE TITLE 18, SECTION 1001)

➻ __________

PHYSICIAN'S SIGNATURE (DO NOT PRINT, TYPE, OR STAMP) M.D. or D.O. DATE SIGNED

PATIENT'S RELEASE

Authorization is hereby given to the physician named above, who participated in my care, to release to the Federal Communications Commission any medical information deemed necessary to process my application for an amateur operator/primary station license.

➻ __________

APPLICANT'S SIGNATURE (DO NOT PRINT, TYPE, OR STAMP) DATE SIGNED

NOTICE TO PHYSICIAN CERTIFYING TO A DISABILITY

You are being asked by a person who has already passed a 5 words per minute telegraphy examination to certify that, because of a severe handicap, he/she is unable to pass a 13 or 20 words per minute telegraphy examination. If you sign the certification, the person will be exempt from the examination. Before you sign the certification, please consider the following:

THE REASON FOR THE EXAMINATION - Telegraphy is a method of electrical communication that the Amateur Radio Service community strongly desires to preserve. We support their objective by authorizing additional operating privileges to amateur operators who increase their skill to 13 and 20 words per minute. Normally, to attain these levels of skill, intense practice is required. Annually, thousands of amateur operators prove by passing examinations that they have acquired the skill. These examinations are prepared and administered by amateur operators in the local community who volunteer their time and effort.

THE EXAMINATION PROCEDURE - The volunteer examiners (VEs) send a short message in the Morse code. The examinee must decipher a series of audible dots and dashes into 43 different alphabetic, numeric and punctuation characters used in the message. To pass, the examinee must correctly answer questions about the content of the message. Usually, a fill-in-the-blanks format is used. With your certification, they will give the person credit for passing the examination, even though they do not administer it.

MUST A PERSON WITH A HANDICAP SEEK EXEMPTION?

No handicapped person is required to request exemption from the higher speed telegraphy examinations, nor is anyone denied the opportunity to take the examinations because of a handicap. There is available to all otherwise qualified persons, handicapped or not, the Technician Class operator license that does not require passing a telegraphy examination. Because of international regulations, however, any handicapped applicant requesting exemption from the 13 or 20 words per minute examination must have passed the 5 words per minute examination.

ACCOMMODATING A HANDICAPPED PERSON - Many handicapped persons accept and benefit from the personal challenge of passing the examination in spite of their hardships. For handicapped persons without an exemption who have difficulty in proving that they can decipher messages sent in the Morse code, the VEs make exceptionally accommodative arrangements. They will adjust the tone in frequency and volume to suit the examinee. They will administer the examination at a place convenient and comfortable to the examinee, even at bedside. For a deaf person, they will send the dots and dashes to a vibrating surface or flashing light. They will write the examinee's dictation. Where warranted, they will pause in sending the message after each sentence, each phrase, each word, or each character to allow the examinee additional time to absorb and interpret what was sent. They will even allow the examinee to send the message, rather than receive it.

YOUR DECISION - The VEs rely upon you to make the necessary medical determination for them using your professional judgement. You are being asked to decide if the person's handicap is so severe that he/she cannot pass the examination even when the VEs employ their accommodative procedures. The impairment, moreover, will last more than one year. This procedure is not intended to exempt a person who simply wants to avoid expending the effort necessary to acquire greater skill in telegraphy. The person requesting that you sign the certification will give you names and addresses of VEs and other amateur operators in your community who can provide you with more information on this matter.

DETAILED INSTRUCTIONS - If you decide to execute the certification, you should complete and sign the Physician's Certification of Disability on the person's FCC Form 610. You must be an M.D. or D.O. licensed to practice in the United States or its Territories. The person must sign a release permitting disclosure to the FCC of the medical information pertaining to the disability.

FCC Form 610
March 1995

Go through the entire exam, and answer the easy questions first. Next, go back to the beginning and try the harder questions. The really tough questions should be left for last. Guessing can only help, as there is no additional penalty for answering incorrectly.

If you have to guess, do it intelligently: At first glance, you may find that you can eliminate one or more "distracters." Of the remaining responses, more than one may seem correct; only one is the best answer, however. To the applicant who is fully prepared, incorrect distracters to each question are obvious. Nothing beats preparation!

After you've finished, check the examination thoroughly. You may have read a question wrong or goofed in your arithmetic. Don't be overconfident. There's no rush, so take your time. Think, and check your answer sheet. When you feel you've done your best and can do no more, return the test booklet, answer sheet and scratch pad to the examiner.

The Volunteer-Examiner team will grade the exam right away. The passing mark is 74%. (That means no more than 9 incorrect answers on the Element 2 exam and no more than 8 incorrect answers on an Element 3A exam.) You will receive a Certificate of Successful Completion of Examination (CSCE) showing all exam elements that you pass. If you are already licensed, and you pass the exam elements required to earn a higher class of license, the CSCE authorizes you to operate with your new privileges. When you use these new privileges, you must sign your call sign, followed by the slant mark ("/"; on voice, say "stroke" or "slant") and the letters KT (for Novices upgrading to Technician Plus). There is one exception to this rule: If you previously held a Technician license and later pass the 5-wpm code test, you do not have to use the "KT" indicator when you use your new privileges.

If you pass only some of the exam elements required for a license, you will still receive a CSCE. That certificate shows what exam elements you passed, and is valid for 365 days. Use it as proof that you passed those exam elements so you won't have to take them over again next time you try for the license.

NOW THAT YOU'RE A HAM

Well, the big day has finally arrived — the FCC has granted your license and you found your new call sign listed on one of the Internet call sign servers (or your license arrived in the mail)! You have permission to operate an amateur station. Now you are ready to put your very own amateur station on the air!

As the proud owner of a new Amateur Radio ticket, you'll soon be an "on-the-air" person instead of an "off-the-air" person. You probably can't wait to make your first contact! You'll be putting all the information you had to learn for the exam to good use.

When you get your new ticket, "hot off the press," make a few photocopies of it. Then put the original license in a safe place. It's a good idea to put one of those copies in your wallet. That magic piece of paper is your **operator license**, which lets you operate a station within your Novice, Technician or Technician Plus privileges. It is also your **station license**, and lists the call sign that identifies your station.

Consider laminating your original license, as the ink sometimes lifts off the paper, even if you place the license behind glass in a frame. Clear plastic Con-Tac, or similar material is inexpensive and available in most department stores.

Call Signs

The FCC issues call signs on a systematic basis. When they process your application, you get the next call sign to come out of the computer. The FCC issues call signs from four "groups," depending on the class of license.

Novice licensees are given "Group D" call signs. These have a "two-by-three" format — two letters followed by a number, followed by three letters. The letters before the number make up the call sign *prefix*, and the letters after the number are the *suffix*. An example of a Novice call sign is KA1WWP.

Technician licensees have "Group C" call signs. These have a "one-by-three" format — a one-letter prefix, a number and a three-letter suffix. An example: N1NAG.

The call-sign formats of other Amateur Radio license classes follow. Group C calls are also given to General amateurs. Advanced licensees get calls from Group B — the "two-by-two" group. KB9NM is an example of a call from this group. Group A calls are for Amateur Extra class operators only. These calls are of the "one-by-two" or "two-by-one" format. (K8CH is a one-by-two call; WR1B is a two-by-one call.) In some parts of the country, all one-by-two, two-by-one and one-by-three call signs have been issued. If that is the case, your call sign may be assigned from a different group. For example, new Technician licensees may receive a call sign from Group D, rather than Group C.

Once you have a call sign, you may keep it as long as you want to (unless your license expires or is revoked). In other words, there's no requirement to change your call sign when you upgrade to a higher-license class. The FCC gives amateurs their choice in this matter.

The first letter of a US call will always be A, N, K or W.

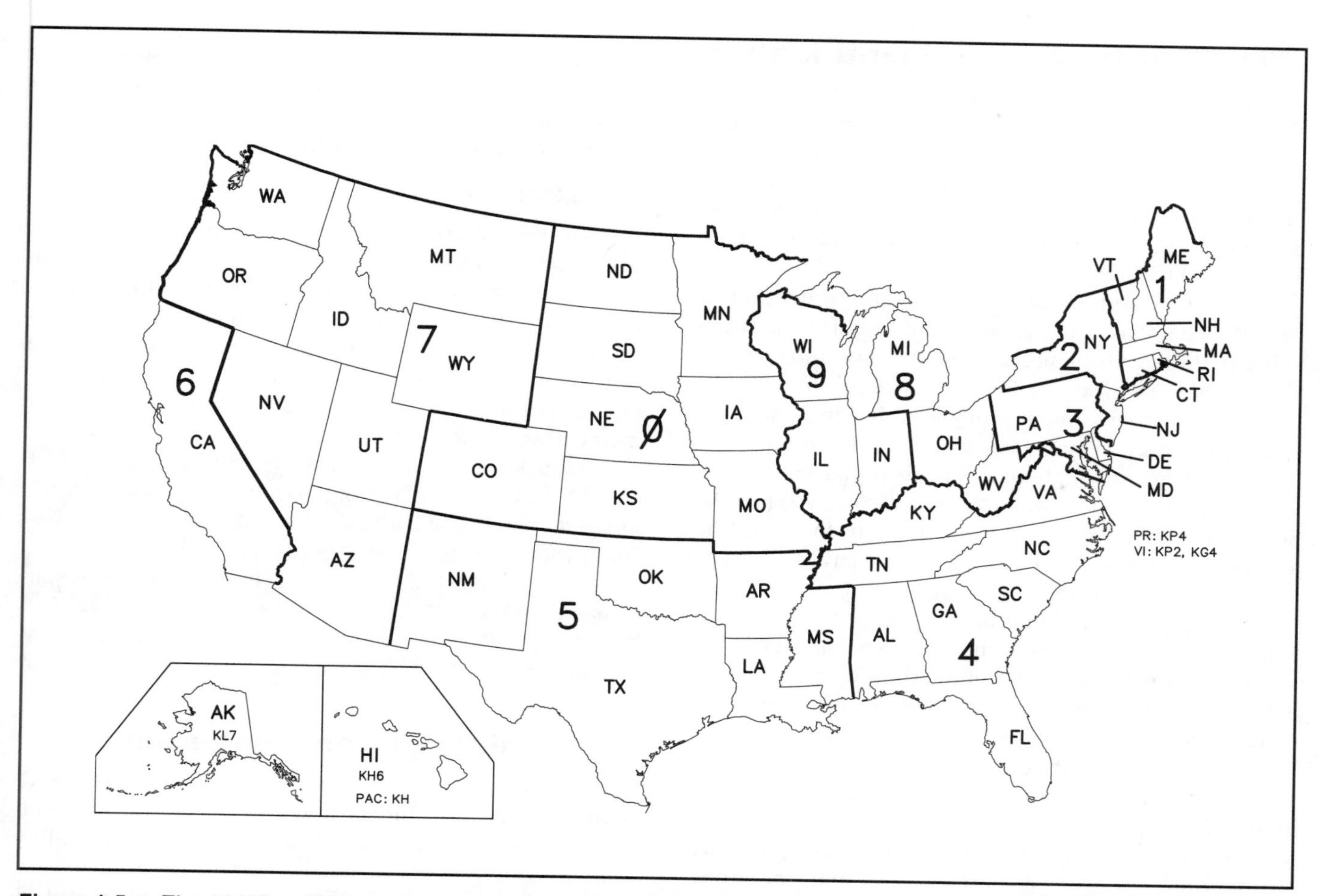

Figure 1-5 — The 10 US call districts. An amateur holding the call sign WA1STO lived in the first district when the FCC assigned her that call. Alaska is part of the seventh call district, but has its own set of prefixes: AL7, KL7, NL7 and WL7. Hawaii, part of the sixth district, has the AH6, KH6, NH6 and WH6 prefixes.

These letters are assigned to the United States as amateur call-sign prefixes. Other countries use other prefixes — LA2UA is a Norwegian call sign, and VU2HO is from India.

The number in a US call sign shows the district where the call was first issued. Figure 1-5 shows the 10 US call districts. Every US amateur call sign includes a single-digit number, 0 through 9 corresponding to these call districts. Amateurs may keep their calls when they move from one district to another. This means the number is not always an indication of where an amateur is. It tells only where he or she was living when the license was first issued. For example, WB3IOS received her license in Pennsylvania, in the third call district, but she now lives in Connecticut, in the first district.

[Turn to Chapter 13 now and study questions N1D07, N1D08 and N1D09. If you have any difficulty, review this section.]

OPERATING GUIDELINES

Keep your operation "above board" — set an operating example that will make you proud. You should be familiar with the basic operating and technical rules covered in Part 97. You should also know the standard operating practices used for the various modes you operate. Other chapters in this book include detailed information about operating your station and the various modes you will want to sample. In this section we will study the FCC Rules of operating and cover some very simple guidelines.

Time Out for Station Identification

FCC regulations are very specific about station identification. You must identify your station every 10 minutes or less during a contact and at the end of the contact. You do not have to identify with every transmission during a conversation with another ham or group of hams, though. Identify your station by transmitting the station call sign listed on your license at the required intervals.

The rules prohibit **unidentified communications or signals** (where the transmitting station's call sign is not transmitted). Be sure you understand the proper station identification procedures, so you don't violate this rule.

You don't *have* to transmit your call sign at the beginning of a contact. In most cases it will help you establish the communications, though. Unless you give your call sign, the other stations won't know who you are! When you take part in a net or talk with a group of hams regularly, they may learn to recognize your voice. Then you may not have to give your call sign for them to know who you are.

You do not have to transmit both call signs when you are talking with another ham — only your own. The exception is in cases of **third-party communications** with a station in a foreign country. (You'll learn more about third-party communications later in this chapter.) Then you do have to transmit both call signs at the end of the contact.

Let's look at an example of how to identify an Amateur Radio station properly. WB8IMY and WR1B have been in communication for 45 minutes and are now signing off. Each operator has already transmitted his call four times (once after each 10-minute interval). Each should now transmit his call one more time as they sign off, for a total of five times during the QSO. (*QSO* means a two-way contact or communication with another ham.)

Suppose the QSO had lasted only 8 minutes. Each station would be required to transmit their call sign only once (at the end of the communication). You may identify more often than this to make it easier to communicate on a crowded band. But the rules specify only that you identify every 10 minutes, and at the end of a contact. On voice, identification is simply "this is WB8IMY." In Morse code, use DE WR1B ("DE" means *from* in French).

There are two exceptions to the station identification rules. The first is when you are transmitting signals to control a model craft, such as a plane or boat.

Remote Control of Model Craft

Amateurs are permitted to use radio links to control model craft. This is called "telecommand" (control) of model craft. Most model control activity takes place on the 6-meter band, so you would need a Technician license to participate in this activity. Remote control operation is permitted with these restrictions:

- station identification is not required for transmissions directed only to the model craft. The control transmitter must have a label indicating the station's call sign and the licensee's name and address.
- control signals are not considered codes and ciphers. (Amateurs may not generally make up "secret codes" that will hide the meaning of their communications.)
- transmitter power cannot exceed 1 watt.

Space Stations

The second exception to station identification is for a **space station**. Space stations are stations located more than 50 km above the Earth's surface! Orbiting Satellites Carrying Amateur Radio (OSCARs) relay signals between amateur operators on the Earth (**Earth stations**), and as such the satellite itself does not have to transmit a station call sign. Operators transmitting to the satellite do have to transmit their call signs, however!

[You should turn to Chapter 13 now, and study questions N1H01 through N1H06. Also study questions N1J07 through N1J10. Review this section if you have any difficulty with these questions.]

[If you are studying for a Technician license, you should also turn to Chapter 14 and study questions T1D09 through T1D11. Review this portion of the text if you have any difficulty with these questions.]

Points of Communication

Who can you talk to with your new license? The FCC defines "points of communication" to specify what kinds of radio stations you may talk with. It's pretty simple, actually: *You may converse with all amateur stations at any time*. This includes amateurs in foreign countries, unless either amateur's government prohibits the communications. (There are a few countries in the world that do not allow Amateur Radio. There are also times when a government will not allow its amateurs to talk with people in other countries.)

The FCC must authorize any communication with any stations not licensed in the amateur service. An example of such authorization is when amateur stations communicate with military communications stations on Armed Forces Day each year.

Another example of amateurs communicating with non-amateur stations is during Radio Amateur Civil Emergency Service (RACES) operation. During an emergency, a registered RACES station may conduct civil-defense communications with stations in the Disaster Communications Service. During such emergencies, RACES stations may also communicate with other US Government stations authorized to conduct civil-defense communications.

[Turn to Chapter 13 now and study questions N1H07 and N1H08. Also study question N1I01. Review this section if you have difficulty with any of these questions.]

Broadcasting

Amateur Radio is a two-way communications service. Amateur Radio stations may not engage in **broadcasting**: the transmission of information intended for reception by the general public. There are also restrictions regulating one-way transmissions of information of general interest to other amateurs. Amateur stations may transmit one-way signals while in beacon operation or radio-control operation. Novice control operators don't have these privileges, however.

[Now turn to Chapter 13 and study question N1I05. If you are uncertain of the answer to that question, review this section.]

[If you are studying for a Technician license, you should also turn to Chapter 14 and study question T1E03. Review this section if you are uncertain of the definition of the term **broadcasting**.]

Third-Party Communications

Communication on behalf of anyone other than the operators of the two stations in contact is **third-party communications**. (Many hams call it *third-party traffic*.) For example, sending a message from your mother-in-law to her relatives in Scarsdale on Valentine's Day is third-party communications. The control operator of one station (the first party) sends communications to the control operator of another station (the second party) for someone else (the third party). Third-party messages include those that are spoken, written, keystroked, keyed, photographed or otherwise originated by or for a third party, and transmitted by an amateur station either live or delayed.

Third-party communications of a personal nature is okay, but passing messages involving business matters is not. The Amateur Service is not the place to conduct business. This also means that you may not receive any type of payment in return for transmitting or receiving third-party communication.

You can pass third-party messages to other stations in the United States. Outside the US, FCC Rules strictly limit this type of communication to those countries that have **third-party communications agreements** with the US. In general, you should consider that international third-

Table 1-3

International Third-Party Traffic — Proceed With Caution

Occasionally, DX stations may ask you to pass a third-party message to a friend or relative in the States. This is all right as long as the US has signed an official third-party traffic agreement with that particular country, or the third party is a licensed amateur. The traffic must be noncommercial and of a personal, unimportant nature. During an emergency, the US State Department will often work out a special temporary agreement with the country involved. But in normal times, never handle traffic without first making sure it is legally permitted.

US Amateurs May Handle Third-Party Traffic With:

Prefix	Country
C5	The Gambia
CE	Chile
CO	Cuba
CP	Bolivia
CX	Uruguay
D6	Federal Islamic Republic of the Comoros
DU	Philippines
EL	Liberia
GB*	United Kingdom
HC	Ecuador
HH	Haiti
HI	Dominican Republic
HK	Colombia
HP	Panama
HR	Honduras
J3	Grenada
J6	St Lucia
J7	Dominica
J8	St Vincent and the Grenadines
JY	Jordan
LU	Argentina
OA	Peru
PY	Brazil
TG	Guatemala
TI	Costa Rica
T9	Bosnia-Herzegovina
V2	Antigua and Barbuda
V3	Belize
V4	St Christopher and Nevis
V6	Federated States of Micronesia
V7	Marshall Islands
VE	Canada
VK	Australia
VR6**	Pitcairn Island
XE	Mexico
YN	Nicaragua
YS	El Salvador
YV	Venezuela
ZP	Paraguay
3DAØ	Swaziland
4U1ITU	ITU Geneva
4U1VIC	VIC, Vienna
4X	Israel
6Y	Jamaica
8R	Guyana
9G	Ghana
9L	Sierra Leone
9Y	Trinidad and Tobago

Notes:

*Third-party traffic permitted between US amateurs and special-events stations in the United Kingdom having the prefix GB only, with the exception that GB3 stations are not included in this agreement.

**Since 1970, there has been an informal agreement between the United Kingdom and the US, permitting Pitcairn and US amateurs to exchange messages concerning medical emergencies, urgent need for equipment or supplies, and private or personal matters of island residents.

Please note that the Region 2 Division of the International Amateur Radio Union (IARU) has recommended that international traffic on the 20 and 15-meter bands be conducted on the following frequencies:

14.100-14.150 MHz
14.250-14.350 MHz
21.150-21.200 MHz
21.300-21.450 MHz

The IARU is the alliance of Amateur Radio societies from around the world; Region 2 comprises member-societies in North, South and Central America, and the Caribbean.

Note: At the end of an exchange of third-party traffic with a station located in a foreign country, an FCC-licensed amateur must also transmit the call sign of the foreign station as well as his own call sign.

From October 1955 *QST*

party communication is prohibited, except when:

- communicating with a person in a country with which the US shares a third-party agreement, or
- in cases of emergency where there is an immediate threat to lives or property, or
- the third party is eligible to be a control operator of the station.

In many countries, the government operates the telephone system and other communications lines. You can imagine why these governments are reluctant to allow their amateur operators to pass messages. They would be in direct competition with the government-operated communications system.

The ARRL monthly journal, *QST*, periodically publishes a list of countries with which the US has a third-party communications agreement. The list does change, and sometimes there are temporary agreements to handle special events. Check such a list, or ask someone who knows, before you try to pass a message to another country. Table 1-3 is a list of countries with which the US had third-party communications agreements at the time of printing for this book.

You may allow an unlicensed person to participate in Amateur Radio from your station. This is **third-party participation**. It is another form of third-party communications.

You (as control operator) must always be present to make sure the unlicensed person follows all the rules. You can allow your family members and friends to enjoy some of the excitement of Amateur Radio in this way. They can speak into the microphone or even send Morse code messages on a keyboard, as long as you are present to control the radio. They can also type messages on your computer keyboard to talk with someone using packet radio or radioteletype.

If you are allowing a non-amateur friend to use your station to talk to someone in the US, and a foreign station breaks in to talk to your friend, you should have your friend wait while you find out if the US has a third-party agreement with the foreign station's government.

There is another important rule to keep in mind about third-party participation. If the unlicensed person was an amateur operator whose license was suspended or revoked by the FCC, that person may not participate in any amateur communication. You can't allow that person to talk into the microphone of your transmitter or operate your Morse code key or computer keyboard.

[Now turn to Chapter 13 and study questions N1I08 through N1I11. Review this section if any of these questions give you difficulty.]

[**If you are studying for a Technician license**, you should also turn to Chapter 14 and study questions T1E07, T1E08 and T1E09. Review this section if needed.]

Business Communications

Amateur communication is noncommercial radio communication between amateur stations, solely with a personal aim and without **pecuniary interest** or business reasons. *Pecuniary* refers to payment of any type.

This definition tells us that amateur operators should not conduct any type of business communications. You may not conduct communications for your own business, or for your employer. You can, however, use your Amateur Radio station to conduct your own personal communications. This includes using the autopatch on your local repeater for such personal calls as making an appointment or ordering food. (In 1993, the FCC amended the business communications rules in Section 97.113 to allow such communication. Prior to this rule change, such calls would have been considered "illegal business communications.")

Of course you can use the autopatch to call a garage for help when your car breaks down in the middle of a busy expressway. This is an emergency, because your property (car), and possibly your life, are in immediate danger. In an emergency, you can use Amateur Radio in any way possible to call for help.

There are many examples of times when personal property or lives are in immediate danger. You may use Amateur Radio to call for help in such a situation, even though you seem to be violating other rules. Just be sure you really have an emergency situation first. For example, you can call for help on frequencies outside of your license privileges, and even on frequencies outside of the amateur bands if that is what it takes to establish the communications you need.

No one can use an amateur station for monetary gain. You must not accept payment in any form for the use of your station at any time. This also means you may not accept payment for transmitting a message for anyone, such as in third-party communication. Payment means more than just money. It refers to any type of compensation, which would include materials or services of any type. That is why the FCC uses the words *pecuniary interest* when stating that you can't be paid for operating your station.

There are two exceptions to this rule. A club station intended primarily for transmitting Morse code practice and information bulletins of interest to all amateurs may employ a paid control operator as long as certain conditions are met. That person can be paid to serve as the control operator only when the station is actually transmitting code practice or bulletins. The code practice and bulletins must last at least 40 hours per week. The station must transmit on at least six medium and high-frequency bands, and the schedule of transmissions must be published 30 days in advance. This exception allows the ARRL to pay a control operator for W1AW, the station at League Headquarters in Newington, Connecticut, for example. The station is dedicated to Hiram Percy Maxim, the League's first president. The W1AW operator can't make general contacts with other hams after the code-practice sessions, however.

The second exception is that teachers may use Amateur Radio stations as part of their classroom instruction at an educational institution. This permits a science teacher to use an amateur station to demonstrate satellite communications or a geography teacher to establish contact with an amateur in another part of the country to describe local terrain or weather conditions.

Another common question about the "no business" rule is whether it is okay to buy and sell Amateur Radio equipment over the air. The best way to answer this question is by taking a look at what the Rules say. Section 97.113(a) says, "No amateur station shall transmit: (3) Communications in which the station licensee or control operator has a pecuniary interest, including communications on behalf of an employer. Amateur operators may, however, notify other amateur operators of the availability for sale or trade of apparatus normally used in an amateur station, provided such activity is not conducted on a regular basis."

So if you decide to upgrade your station equipment and want to sell your old rig to another ham, go right ahead and mention it on the air. (Many hams prefer to "close" the deal off the air, in person or over the telephone.) Just when does such activity become "regular?" Well, if you find that you are buying or selling gear every week or so, and it is almost always passed on at a profit, you have probably crossed that line. If the local ham dealer starts referring to you as "the competition" you have definitely gone too far!

Many clubs hold "swap nets" on local repeaters and you will find lots of "For Sale" notices posted on your local packet radio bulletin board. Keep in mind, however, that this is for Amateur Radio equipment and not the rest of your household belongings, including the "kitchen sink!"

[Before you go on to the next section, turn to Chapter 13 and study questions N1H10, N1H11 and N1I04. If you have any difficulty with either of them, review this section.]

Other Assorted Rules

Under FCC Rules, amateurs may not transmit music of any form. This means you can't transmit your band's practice session or play the piano for transmission over the air. It also means you can't play a song from your favorite tape or CD for your friend to hear. You should take care not to transmit *unintentional* music, either. For example, this could happen if you have a broadcast radio playing in the background when you pick up the microphone to talk with someone. There is one exception to the "No Music" rule. If you obtain special permission from NASA to retransmit the audio from a space shuttle mission for other amateurs to hear, and during that retransmission NASA or the astronauts play some music over the air, you won't get in trouble.

Amateurs may not use obscene or indecent language. Remember that anyone, of any age, can hear your transmission if they happen to be tuned to your transmitting frequency. Depending on your operating frequency and other conditions, your signals can be heard around the world. While there is no list of "banned words" or other specific list, you should avoid any questionable language.

You can't use codes or ciphers to obscure the meaning of transmissions. This means you can't make up a "secret" code to send messages over the air to a friend. (Control signals transmitted for remote control of model craft are not considered codes or ciphers. Neither are telemetry signals, such as a satellite might transmit to tell about its condition. A space station —satellite — control operator can use specially coded signals to control the satellite.

Amateurs may not cause **malicious (harmful) interference** to other communications of any type, amateur or non-amateur. You may not like the other operator's practices, or you may believe he or she is violating the rules. You have no

Shuttle astronaut Kathy Sullivan holds the amateur call N5YYV. She put it to good use when she flew aboard shuttle mission STS-45 in March and April 1992, making 2-meter voice and packet contacts with hams and schoolchildren across the country. One FCC Rule applies specifically to shuttle communications: Amateurs can retransmit US Government shuttle communications to a ham audience — if they get NASA's permission beforehand. *(photo courtesy NASA)*

right to interfere with their communications, however. You may never deliberately interfere with another station's communications. Repeatedly transmitting on a frequency already occupied by other amateurs, such as in a net operation, is a form of harmful or malicious interference.

Many of our amateur bands are shared with other radio services. For example the 23-cm band is shared with radiolocation stations outside the US. If you discover that your operation is interfering with one of these other stations, you should stop operating on that frequency and take steps to eliminate this form of harmful interference.

Some Amateur Radio transceivers can transmit on frequencies outside of the amateur bands. You should not use such a radio to transmit signals on other frequencies. Such transmissions, even if sent as a "joke" can cause serious harmful interference. For example, if someone were to transmit on police frequencies, that transmission might block real police calls that could involve an emergency.

Amateurs may not transmit **false or deceptive signals**, such as a distress call when no emergency exists. You must not, for example, start calling **MAYDAY** (an international distress signal) unless you are in a life-threatening situation.

Distress Calls

If you should require immediate emergency help, and you're using a voice (telephony) mode, call **MAYDAY**. Use whatever frequency offers the best chance of getting a useful answer. "**MAYDAY**" is from the French *m'aidez* (help me). On CW (telegraphy), use $\overline{\text{SOS}}$ to call for help. Repeat this call a few times, and pause for any station to answer. Identify the transmission with your call sign. Stations that hear the call sign will realize the $\overline{\text{SOS}}$ is legitimate. Repeat this procedure for as long as possible, or until you receive an answer. In a life or property-threatening emergency, you may send a distress call on *any* frequency, even outside the amateur bands, if you think doing so will bring help faster.

Be ready to supply the following information to the stations responding to an $\overline{\text{SOS}}$ or **MAYDAY**:

- *The location of the emergency*, with enough detail to permit rescuers to locate it without difficulty.
- *The nature of the distress.*
- *The type of assistance required* (medical, evacuation, food, clothing or other aid).
- *Any other information* to help locate the emergency area.

If you receive a distress signal, you are also allowed to transmit on *any* frequency to provide assistance.

Space Stations and Earth Stations

The FCC defines a **space station** as "An amateur station located more than 50 km (about 30 miles) above the Earth's surface." This obviously includes amateur satellites, and also includes any operation from the Space Shuttle, the Russian *Mir* space station and any future operations by astronauts in space. Any licensed ham can be the licensee or control operator of a space station. Likewise, any licensed amateur may operate through or communicate with a space station, as long as their transmissions take place on frequencies available for that license class.

Earth stations are stations located on the Earth's surface, or within 50 km of it. They are intended for communications with **space stations** or with other Earth stations by means of one or more objects (satellites) in space. Any amateur can be the control operator of an Earth station, subject to the limitations of his license class. Some specific transmitting frequencies are authorized for Earth stations. These are located in the 40, 20, 17, 15, 12, 10 and 2-meter bands, in the 23 and 70-cm bands, and in all other amateur bands that are higher in frequency. There are no frequencies available for Earth stations in the 6-meter band, however. A complete list appears in *The FCC Rule Book*.

[Now it's time to turn to Chapter 13 again, and study a few questions. You should be able to answer questions N1I02, N1I03, N1I06, N1I07, N1I12, N1I13 and N1I14. Also study questions N1J01 through N1J06 and N1J11. Review this section if you have difficulty with any of these questions.]

[**If you are studying for a Technician license**, you should also turn to Chapter 14 and study questions T1C11, T1E05, T1E06, T1E10 and T1E11.]

Official Notices of Violation

The FCC is the agency in charge of maintaining law and order in radio operation in the US. The International Telecommunication Union (ITU) sets up international rules that the government telecommunication agencies of each country follow. Both sets of rules provide the basic structure for Amateur Radio in the United States.

Suppose you receive an official notice from the FCC informing you that you have violated a regulation. Now what should you do? Simple: Whatever the notice tells you to do. Usually this will involve some station modification and making a response to the FCC about what corrective steps you have taken.

There exists a philosophy in Amateur Radio that is deeply rooted in our history. This philosophy is as strong now as it was in the days of the radio pioneers. We are talking about the *self-policing* of our bands. Over the years, amateurs have become known for their ability to maintain high operating standards and technical skills. We do this without excessive regulation by the FCC. The Commission itself has praised the amateur service for its tradition of self-policing. Perhaps the underlying reason for this is the amateur's sense of pride, accomplishment, fellowship, loyalty and concern. Amateur Radio is far more than just a hobby to most amateurs.

As a new or prospective amateur operator, you will begin to discover the wide horizons of your new pastime. You will learn more about the rich heritage we all share. Take this sense of pride with you, and follow the Amateur's Code.

[This completes your study of the first part of the chapter. You have studied all of the FCC Rules material required for the Novice exam. The remainder covers the rules and regulations you'll need to know to pass the Element 3A exam for the Technician license. By now you should have no difficulty with the questions in subelement N1 of the Novice question pool.]

KEY WORDS

Bandwidth — The width of a frequency band outside of which the mean power is attenuated at least 26 dB below the mean power of the total emission, including allowances for transmitter drift or Doppler shift.

Beacon station — An amateur station transmitting communications for the purposes of observation of propagation and reception or other related experimental activities.

Frequency coordination — Allocating repeater input and output frequencies to minimize interference between repeaters and to other users of the band.

One-way communications — Transmissions that are not intended to be answered. The FCC strictly limits the types of one-way communications allowed on the amateur bands.

Peak envelope power (PEP) — The average power of a signal at its largest amplitude peak.

Repeater station — An amateur station that automatically retransmits the signals of other stations.

Temporary state of communications emergency — When a disaster disrupts normal communications in a particular area, the FCC can declare this type of emergency. Certain rules may apply for the duration of the emergency.

1 Commission's Rules For The Technician Exam

THE TECHNICIAN LICENSE

With Technician license in hand, you'll be able to operate on all the VHF, UHF and microwave bands allocated to the amateur service — and there are a lot of them! Table 1-2 lists the frequency ranges for the most popular amateur bands. This table also shows the frequency and emission privileges for each license class on these bands. For a complete list of the amateur bands, including those in the largely experimental microwave region, see *The FCC Rule Book*, published by the ARRL.

If you decide to learn the international Morse code and are able to pass a 5 word-per-minute code test, you will have earned a Technician Plus license. A Certificate of Successful Completion of Examination from a VE session, showing you passed a Morse code exam (or your new Technician Plus license) allows you to operate on the four HF Novice subbands that normally provide direct worldwide communication.

The Technician Exam

The Technician exam consists of two written tests, Elements 2 and 3A. Chapters 13 and 14 of this book contain every question in the question pools from which your exam will be composed. If you study this book carefully, you will be able to pass the Technician exam. If you wish to qualify for the Technician Plus license, you'll also have to pass the Element 1A 5-wpm Morse code exam. In that case you'll need a computer program, cassette tapes or audio CDs that teach international Morse code. The ARRL offers a set of two cassettes or audio CDs called *Your Introduction to Morse Code*. The ARRL also offers *Morse Tutor* for IBM-PC and compatible computers. If you already hold a valid Novice license, then you have credit for passing Elements 1A and 2, and will not have to retake them. In that case when you upgrade to Technician Plus, you will retain your Novice HF privileges.

The Element 3A exam consists of 30 questions taken from a pool of more than 300. The FCC specifies 10 subelements, or divisions, for each exam element. These are the Commission's Rules, Operating Procedures, Radio-Wave Propagation, Amateur Radio Practices, Electrical Principles, Circuit Components, Practical Circuits, Signals and Emissions, Antennas and Feed Lines and Radio frequency environmental safety practices at an amateur station (RF Radiation Safety).

The question pools for all amateur exams are maintained by a Question Pool Committee selected by the Volunteer Examiner Coordinators. The FCC allows Volunteer Examiners to select the questions for an amateur exam, but they must use the questions exactly as they are released by the VEC Question Pool Committee. If you attend a test session coordinated by the ARRL/VEC, your test most likely will be designed by the ARRL/VEC, or by a computer program designed by the ARRL/VEC. The questions and answers will be exactly as they are printed in Chapters 13 and 14. The answer positions (A, B, C, D) may change, however.

If you hold a Certificate of Successful Completion of Examination (CSCE) for the 5-wpm code test, make sure you keep it in a safe place. Along with your actual Technician license, it serves as authorization to operate on the HF Novice bands until the FCC grants your new Technician Plus license.

[You should turn to Chapter 14 now and study questions T1A08 and T1A09. Review this section if either of these questions gives you difficulty.]

TECHNICIAN EMISSION PRIVILEGES

As you can see from Tables 1-1 and 1-2, Technician class operators enjoy *all frequency privileges* allocated to the amateur service *above 30 MHz*. Technician licensees can operate on all authorized frequencies, using all authorized modes, on the VHF, UHF and microwave amateur bands, as can General, Advanced and Amateur Extra licensees. They may use up to 1500-watts peak-envelope power (PEP) output on these bands. In addition, Technicians who pass a 5-wpm code test are granted a Technician Plus license, and gain the same emission privileges that Novices have on the four HF Novice bands. Novices and Technician Plus class operators are limited to 200 watts PEP output on the HF bands. These bands are:

- On the 80-meter band, 3.675 to 3.725 MHz, CW only.
- On the 40-meter band, 7.1 to 7.15 MHz, CW only.
- On the 15-meter band, 21.1 to 21.2 MHz, CW only.
- On the 10-meter band, 28.1 to 28.3 MHz, CW, RTTY and data; and 28.3 to 28.5 MHz, CW and SSB phone.

This gent brought his 10-GHz microwave station to the summit of Mount Monadnock in New Hampshire to take part in an ARRL-sponsored contest. Many hams enjoy operating away from the comforts of home.

Technician licensees may operate on the VHF and UHF bands listed below, plus several others that are higher in frequency. For a complete list of amateur frequency privileges, including the microwave bands where frequency is measured in gigahertz, see *The FCC Rule Book*, published by the ARRL.

On the 50-MHz (6-meter) band: all emission types except pulse — CW, RTTY, data, MCW, test, phone and image — are authorized on 50.1 MHz to 54.0 MHz. On 50.0 to 50.1 MHz, only CW is allowed. All Technician class and higher licensees have these privileges.

On the popular 144-MHz (2-meter) band: all emission types (except pulse, but including image transmissions) from 144.1 to 148 MHz, and CW only from 144.0 to 144.1 MHz. Although the 2-meter band is best known for repeater and packet operation, hams also use it for such *weak-signal activities* as meteor scatter, moonbounce and aurora. All Technician class and higher licensees have these privileges.

On the 222-MHz (1¼-meter) band: all emission types (except pulse) from 222.0 to 225.0 MHz. All amateur licensees have these privileges, including Novices, except Novices are limited to no more than 25 W PEP.

On the 420-MHz (70-cm) band: all emission types (except pulse) from 420 to 450 MHz. All Technician class and higher licensees have these privileges.

On the 902-MHz (33-cm) band: all emission types from 902 to 928 MHz. All Technician class and higher licensees have these privileges.

On the 1240-MHz (23-cm) band: all emission types (except pulse) from 1240 to 1300 MHz. All Technician class and higher licensees have these privileges. Novices have privileges on 1270 to 1295 MHz, limited to no more than 5 W PEP.

On the 2300-MHz (13-cm) band: all emission types from 2300 to 2310 and 2390 to 2450 MHz. All Technician class and higher licensees have these privileges.

Transmitter Power

In general, outside the HF Novice bands, Technicians may operate with a *maximum* **peak envelope power (PEP)** output of 1500 W. We've emphasized the word *maximum* because the FCC Rules also stipulate: "An amateur station must use the minimum power necessary to carry out the desired communications." Transmitter power is measured in watts of PEP output at the antenna terminals of the transmitter or amplifier.

The FCC defines **peak envelope power** as "the average power supplied to the antenna transmission line by a transmitter during one RF cycle at the crest of the modulation envelope." This sounds pretty technical, but it isn't too difficult to understand the basic principle. The *modulation envelope* refers to the way the information signal varies the transmitter output. Think of it as increasing and decreasing the transmitted signal. All we have to do is find the highest point, or maximum output-signal level. Then we look at one cycle of the radio-frequency (RF) signal and measure the average power during that time. Figure 1-6 illustrates this measurement on a radio-signal waveform. You might use a *wattmeter* to make that measurement for you, and the meter

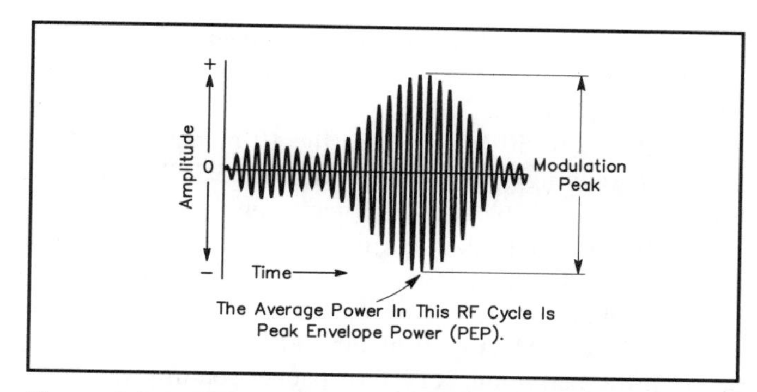

Figure 1-6 — This drawing represents a modulated RF waveform. The peak-envelope power (PEP) of this signal is calculated or measured by finding the average power of one RF cycle that occurs at the modulation peak.

will do all the work. So you just have to read a meter scale or read the numbers off a digital display.

There are some exceptions to the maximum power rules:

- All who are authorized to operate in the 80, 40 and 15-meter Novice subbands may use a maximum of 200 watts PEP.
- Technicians who are authorized to operate on the Novice bands (those who have passed a 5-wpm code test) are also limited to 200 W PEP on the 10-meter Novice sub band, 28.1 to 28.5 MHz.
- All amateurs are limited to 200 W PEP on the 30-meter band (10.1-10.15 MHz). You will be able to use this band when you upgrade to a General license.
- Stations in beacon operation are limited to 100 W PEP output.
- Stations operating near certain military installations may use a maximum of 50 W PEP on the 450-MHz band.

[You will have to memorize these operating frequency and mode privileges. After you have studied them, turn to Chapter 14 and review questions T1A03 through T1A07. Also answer questions T1B01 through T1B05, T1B12 and T1B13. Review this section if you are unsure of the answer to any of these questions.]

Emission Standards

FCC Rules stipulate that no amateur station transmission shall occupy more **bandwidth** than necessary for the information rate and emission type being transmitted, in accordance with good amateur practice. Bandwidth is a measure of how much space your signal takes up. A CW signal has a smaller bandwidth than does a single-sideband phone signal, and an SSB signal in turn has a smaller bandwidth than an AM or FM voice signal.

The Rules specify certain characteristics of RTTY and other emissions, and you should be aware of them to ensure that you're not in violation of the Rules.

RTTY Sending Speed

When transmitting one of the specified digital codes

listed in Section 97.309 of the Rules, there are certain sending-speed limitations. The specified codes are Baudot, AMTOR and ASCII.

In the 28 to 50-MHz range (the 10-meter band), the maximum sending speed (symbol rate) for a RTTY or packet transmission is 1200 bauds (bits per second).

Between 50 and 222 MHz (the 6 and 2-meter bands), the maximum sending speed for a RTTY or packet transmission is 19.6 kilobauds.

Above 222 MHz (1¼-meter and shorter wavelength bands), the maximum RTTY or packet sending speed is 56 kilobauds.

RTTY Frequency Shift

Below 50 MHz (the 10-meter band), the maximum frequency shift permitted for RTTY is 1000 Hz.

Above 50 MHz (the 6-meter and shorter wavelength bands), there is no maximum frequency shift.

Authorized Bandwidth

When you are transmitting a digital code other than one of those specified in Section 97.309 of the Rules, there is a bandwidth limitation imposed.

Between 50 and 222 MHz (the 6 and 2-meter bands), the authorized bandwidth of a RTTY, data or multiplexed emission using a specified digital code is 20 kHz.

Between 222 and 450 MHz (the 1¼-meter and 70-cm bands), the authorized bandwidth of a RTTY, data or multiplexed emission using an unspecified digital code is 100 kHz.

[Before continuing, take a look at questions T1C01 through T1C08 in Chapter 14. You may find it helpful to refer to Table 1-2 while studying these questions.]

FREQUENCY SHARING

If we are to make the best use of the limited amount of available spectrum, there must be ways to ensure that harmful interference is kept to a minimum. If two amateur stations want to use the same frequency, both stations have an equal right to do so. The FCC encourages efficient, interference-free sharing of the ham bands by limiting transmitter output power, by assigning services either primary or secondary status on a frequency band and by encouraging repeaters to be *coordinated* (operating frequencies are recommended by a regional group organized for this purpose). We've already discussed transmitter power and power limitations; now we'll take a look at primary and secondary allocations, and repeater coordination.

Keep in mind that the FCC has the authority to modify the terms of your amateur license any time they determine that such a modification will promote the public interest, convenience and necessity. This might occur, for example, if your station causes interference to another station with a primary allocation and you are unable to resolve the interference. In that case they might place time, power or frequency restrictions on your station. (Don't panic, though! Such action is very rare.)

Primary and Secondary Allocations

The basic frequency-sharing principle is straightforward: Where a band of frequencies is allocated to different services of the same category, the basic principle is the *equality of right to operate*. "Category" refers to whether a service operates on a primary or secondary basis. The amateur service is allocated many different frequency bands, but some of them are allocated on a primary basis and some are secondary. A station in a secondary service must not cause harmful interference to, and must accept interference from, stations in a primary service.

By sharing our frequencies with several other services, including the US military, hams can have the use of a greater amount of spectrum than would otherwise be the case. If you are operating on a band on which the amateur service has a secondary allocation, and a station in the primary service causes interference, you should change frequency immediately. You may also be interfering with the other station, and that is prohibited by Part 97 of the FCC Rules. *The FCC Rule Book* shows which amateur bands are primary allocations and which are secondary.

One special sharing arrangement is worth additional comment. Amateurs have a secondary allocation at 219 to 220 MHz. Amateur use of this band segment is limited to stations acting as packet radio network relay stations in point-to-point fixed digital message systems. Stations are limited to 50 watts PEP when operating on this band. Before you can operate such a station in this band you must meet several requirements. The Rules in Section 97.303(e) give the details of these requirements. You must give written notice to the ARRL about any such operation at least 30 days prior to making any transmissions on the band. You must make sure you are not within certain distances of Automated Maritime Telecommunications Systems (AMTS) stations unless you obtain their permission.

Amateur stations using the 219 to 220-MHz band must not cause harmful interference to stations of any radio service that has a primary allocation on this band or adjacent frequencies. Such services include the AMTS stations, television stations broadcasting on channels 11 or 13 and Interactive Video and Data Service systems.

Repeater Frequency Coordination

A **repeater station** is an amateur station that automatically retransmits the signals of other stations. Repeaters in the same area that use the same or similar frequencies can interfere with each other. As more and more repeater stations have been set up, there have been more and more cases of repeater interference. One effective way of dealing with the problem is **frequency coordination**. Volunteer frequency coordinators have been appointed to ensure that new repeaters use frequencies that will tend not to interfere with existing repeaters in the same area. The FCC encourages frequency coordination, but the process is organized and run by hams and groups of hams who use repeaters.

The FCC has ruled in favor of coordinated repeaters if there is harmful interference between two repeaters. In such a case, if a frequency coordinator has coordinated one but not the other, the licensee of the *uncoordinated* repeater is responsible for solving the interference problem. If both repeaters are coordinated, or if neither is, then both licensees are equally responsible for resolving the interference.

[Before moving along, turn to Chapter 14 and review questions T1A13 and T1B06 through T1B11. Also review questions T1C09 and T1C10. In addition, turn to questions T2A19 and T2A20 at this time. Study this section again if you have difficulty with any of these questions.]

COMMUNICATIONS

The FCC not only permits, but encourages, licensed hams to assist in emergencies, notwithstanding the FCC Rules that apply at all other times. Section 97.401(a) of the Rules says:

> "When normal communication systems are overloaded, damaged or disrupted because a disaster has occurred, or is likely to occur . . . an amateur station may make transmissions necessary to meet essential communication needs and facilitate relief actions."

The FCC recognizes that "amateurs may provide essential communications in connection with the immediate safety of human life and immediate protection of property when normal communication systems are not available." Section 97.405 of the Rules goes on to say:

> 97.405 Station in distress.
> (a) No provision of these rules prevents the use by an amateur station in distress of any means at its disposal to attract attention, make known its condition and location, and obtain assistance.
> (b) No provision of these rules prevents the use by a station, in the exceptional circumstances described in paragraph (a), of any means of radiocommunication at its disposal to assist a station in distress.

If you're in the middle of a hurricane, forest fire or blizzard, and you offer your communications services to the local authorities, you can do whatever you need to do to help deal with the emergency. This includes allowing a physician to operate your radio or helping the Red Cross to assess damages.

Hurricane debris on St Maarten. Note the "Emergency Dial 911" sign on the downed telephone pole.

Other cases may not be as clear cut. Suppose you hear a ship at sea sending a distress call on a frequency outside of your license privileges or even outside of a ham band. Can you call the ship and offer assistance? If it appears that no other station is able to provide the required communications, yes, you can. If you came across a serious auto accident (or were involved in one) and you were unable to obtain help on a ham band could you transmit outside of the ham band where you hear another station operating? Again, the answer is yes. Just be sure you have a real emergency and that you have no other way to obtain the necessary help.

In the wake of a major disaster, the FCC may suspend or change its Rules to help deal with the immediate problem. Part 97 says that when a disaster disrupts normal communications systems in a particular area, the FCC may declare a **temporary state of communication emergency**. The declaration will set forth any special conditions or rules to be observed during the emergency. Amateurs who want to request that such a declaration be made should contact the FCC Engineer in Charge in the area concerned.

[Before you go on to the next section, review these questions from Chapter 14: T1E01 and T1E02. Review this section as needed.]

ONE-WAY TRANSMISSIONS

Certain types of transmissions are designated as one-way. Only certain kinds of **one-way communications** or transmissions are permitted on the amateur bands.

Emergencies

Since normal restrictions are suspended when life or property is in immediate danger, as discussed above, one-way transmissions are not considered broadcasting (which is not allowed) under emergency conditions.

Beacon Stations

A **beacon station** is simply a transmitter that alerts listeners to its presence. In the amateur service, beacons are used primarily for the study of radio-wave propagation — to allow amateurs to tell when a band is open to different parts of the country or world. The FCC defines a beacon station as an amateur station transmitting communications for the purposes of observation of propagation and reception or other related experimental activities.

The FCC Rules address beacon operation this way:

- Automatically controlled beacon stations are limited to certain parts of the 28, 50, 144, 222 and 432-MHz amateur bands, and all amateur bands above 450 MHz.
- The transmitter power of a beacon must not exceed 100 W.
- Any license class, except Novice, can operate a beacon station.

Other Permitted One-Way Transmissions

Aside from the examples already listed (emergency communications, remote control of a model craft and beacon operation), Part 97 allows the following kinds of one-way amateur communications:

- brief transmissions necessary to make adjustments to a station;
- brief transmissions necessary to establishing two-way communications with other stations;
- transmissions necessary to assisting persons learning, or improving proficiency in, the international Morse code; and
- transmissions necessary to disseminate information bulletins (these are to be directed only to amateurs and must consist solely of subject matter of direct interest to the amateur service).

Amateurs are not allowed to transmit one-way communications intended for reception by the general public. This would be considered broadcasting, which is prohibited.

[Before continuing, turn to Chapter 14 and review questions T1D06, T1D07, T1D08 and T1E04. Review this section if necessary.]

STATION IDENTIFICATION

Under the FCC Rules, you must clearly make known the source of your transmissions to anyone receiving them. No station may transmit unidentified communications or signals, or transmit as the station call any call not authorized to the station.

You must identify your station at the end of each contact, and every 10 minutes during the contact. It doesn't matter if you are operating from your home station, from a portable location or in a vehicle as a mobile station. It doesn't matter what mode you are operating. You must always identify your station every 10 minutes when you are operating.

You can use any language you want to communicate with other amateurs. Amateur Radio gives you a great opportunity to practice your "foreign" language skills with other hams who speak the language you are learning. They will be very helpful with questions you may have, and you will find they are quite pleased that you would try to speak with them in their native language. When you give your station identification, however, you must use English.

The FCC recommends that you use a phonetic alphabet as an aid to station identification. The International Telecommunication Union phonetic alphabet uses words that are internationally recognized substitutes for letters, to make it easier to understand the letters. You should avoid using "cute" words or phrases to identify your station, especially when you are talking with amateurs who have a limited understanding of the English language. There is more information about using phonetics in Chapter 2.

Many repeaters now feature voice identification, but it's okay to identify a repeater with Morse code, too. If a repeater identifies with automatically sent Morse code, the code speed must be 20 wpm or less.

If you're a Novice who has passed the Technician exam and hold a Certificate of Successful Completion of Examination (CSCE), special rules apply. To identify when operating on a voice mode with your new Technician privileges, give your call sign followed by a word that describes the slant mark and the identifier "KT." An example: "KA9GND slant KT." Another: "KA2FCC stroke KT." When operating CW or on digital modes, use the fraction bar (/) followed by KT, as in KA1WWP/KT.

Technicians who pass the 5-wpm code test and thus qualify for the Technician Plus license (including HF Novice privileges) need not use the "KT" identifier when using their new privileges or when operating on the Technician bands (50 MHz and above). You must retain your Certificate of Successful Completion of Examination (CSCE) that shows you passed the code test until the FCC grants your new Technician Plus license.

You should know one more thing about station identification: There is only one type of emission that can always be used when you identify your station, regardless of frequency or operating mode: Morse code (CW).

[Congratulations! This completes your study of the FCC Rules for the Novice and Technician written exams. Be sure to turn to Chapter 14 and review questions T1C12 through T1C16 now. Also study questions T1D01 through T1D05. Review the material in this section as needed.]

Table 1-4

Standard ITU Phonetics

A—Alfa (**AL** FAH)
B—Bravo (**BRAH** VOH)
C—Charlie (**CHAR** LEE) or (**SHAR** LEE)
D—Delta (**DELL** TAH)
E—Echo (**ECK** OH)
F—Foxtrot (**FOKS** TROT)
G—Golf (GOLF)
H—Hotel (HOH **TELL**)
I—India (**IN** DEE AH)
J—Juliett (**JEW** LEE ETT)
K—Kilo (**KEY** LOH)
L—Lima (**LEE** MAH)
M—Mike (MIKE)
N—November (NO **VEM** BER)
O—Oscar (**OSS** CAH)
P—Papa (PAH **PAH**)
Q—Quebec (KEH **BECK**)
R—Romeo (**ROW** ME OH)
S—Sierra (SEE **AIR** RAH)
T—Tango (**TANG** GO)
U—Uniform (**YOU** NEE FORM) or (OO NEE FORM)
V—Victor (**VIK** TAH)
W—Whiskey (**WISS** KEY)
X—X-RAY (**ECKS** RAY)
Y—Yankee (**YANG** KEY)
Z—Zulu (**ZOO** LOO)

Note: The **boldfaced** syllables are emphasized. The pronunciations shown in this table were designed for those who speak any of the international languages. The pronunciations given for "Oscar" and "Victor" may seem awkward to English-speaking people in the US.

2 KEY WORDS

Audio-frequency shift keying (AFSK) — A method of transmitting radioteletype information by switching between two audio tones fed into an FM transmitter Most often used on VHF. Also see **frequency-shift keying (FSK)**.

Autopatch — A device that allows repeater users to make telephone calls through a repeater.

Baudot — A code used in radioteletype communications. Each character is represented with a string of five bits of digital information. Each character has a different combination of bits.

Closed — A repeater that restricts access to those who know a special code. See **CTCSS**.

Connected — The condition in which two packet-radio stations are sending information to each other. Each is acknowledging when the data has been received correctly.

CQ — "Calling any station": the general call when requesting a conversation with anyone.

CTCSS — A sub-audible tone system used on some repeaters. When added to a carrier, a CTCSS tone allows a receiver to accept a signal. Also called **PL**.

CW — A communications mode transmitted by on/off keying of a radio-frequency signal. Another name for international Morse code.

Data — Computer-based modes, such as RTTY and packet.

DE — The Morse code abbreviation for "from" or "this is."

Digipeater — A packet-radio station used to retransmit signals that are specifically addressed to be retransmitted by that station.

Digital communications — Computer-based communications modes. Also see **Data**.

Dummy load — A station accessory that allows you to test or adjust transmitting equipment without sending a signal out over the air. Also called dummy antenna.

Duplex operation — Receiving and transmitting on two different frequencies. Also see **simplex operation**.

DX — Distance, foreign countries.

Emergency — A situation where there is a danger to lives or property.

Emission types — Term for the different modes authorized for use on the Amateur Radio bands. Examples are CW, SSB and FM.

FM — Mode of voice (phone) communications used on repeaters. Abbreviation for frequency modulation.

Frequency coordinators — Individuals or groups that recommend repeater frequencies.

Frequency-shift keying (FSK) — A method of transmitting radioteletype information by switching an RF carrier between two separate frequencies. FSK RTTY is most often used on HF. Also see **audio-frequency shift keying**.

Input frequency — A repeater's receiving frequency. To use a repeater, transmit on the input frequency and receive on the output frequency.

K — The Morse code abbreviation for "any station respond."

Monitor mode — One type of packet radio receiving mode. In monitor mode, everything transmitted on a packet frequency is displayed by the monitoring TNC. This occurs whether or not the transmissions are addressed to the monitoring station.

Morse code (see **CW**).

Narrow-band direct-printing telegraphy — The technical term for **radioteletype (RTTY)**.

Network — A term used to describe several packet stations linked together to transmit data over long distances.

Open — A repeater that can be used by all hams who have a license that authorizes operation on the repeater frequencies.

Output frequency — A repeater's transmitting frequency. To use a repeater, transmit on the input frequency and receive on the output frequency.

Packet radio — A system of digital communication whereby information is broken into short bursts. The bursts ("packets") also contain addressing and error-detection information.

Phone — Another name for **voice communications**.

Phonetic alphabet — Standard words used on voice modes to make it easier to understand letters of the alphabet, such as those in call signs. The call sign KA6LMN stated phonetically is Kilo Alfa Six Lima Mike November.

PL (see **CTCSS**)

Procedural signal (prosign) — One or two letters sent as a single character. Amateurs use prosigns in CW contacts as a short way to indicate the operator's intention. Some examples are K for "Go Ahead," or $\overline{\text{AR}}$ for "End of Message." (The bar over the letters indicates that we send the prosign as one character.)

Q signals — Three-letter symbols beginning with Q. Used on CW to save time and to improve communication. Some examples are QRS (send slower), QTH (location), QSO (ham conversation) and QSL (acknowledgment of receipt).

QRL? — Ham radio Q signal meaning "Is this frequency in use?"

QSL card — A postcard that serves as a confirmation of communication between two hams.

QSO — A conversation between two radio amateurs.

Radioteletype (RTTY) — Radio signals sent from one teleprinter machine to another machine. Anything that one operator types on his teleprinter will be printed on the other machine. Also known as **narrow-band direct-printing telegraphy**.

Ragchew — A lengthy conversation between two radio amateurs.

Reflection — Signals that travel by line-of-sight propagation are reflected by large objects like buildings.

Repeater — An amateur station that receives a signal and retransmits it for greater range.

RST — A system of numbers used for signal reports: R is readability, S is strength and T is tone. (On single-sideband phone, only R and S reports are used.)

73 — Ham lingo for "best regards." Used on both phone and CW toward the end of a contact.

Sidebands — The sum or difference frequencies generated when an RF carrier is mixed with an audio signal. Single-sideband phone (SSB) signals have an upper sideband (USB — that part of the signal above the carrier) and a lower sideband (LSB — the part of the signal below the carrier). SSB transceivers allow operation on either USB or LSB.

Simplex operation — Receiving and transmitting on the same frequency. See **duplex operation**.

SSB — Emission mode that describes the type of voice emission used on the HF bands. Abbreviation for single sideband.

Time-out timer — A device that limits the amount of time any one person can talk through a repeater.

Voice communications — Hams can use several voice modes, including FM and SSB.

2 Operating Your Amateur Station

Amateur Radio is about communicating. You must know some basic operating procedures to be an effective communicator.

If you're like most people, as soon as your ticket arrives you'll want to get on the air and begin "working" other Amateur Radio operators. You'll be joining more than 2 million other radio enthusiasts around the world.

The operating world is rich and varied. With your ham license, you can relax and enjoy a conversation (a ragchew) with another ham. You can send and receive messages anywhere in North America via the National Traffic System (handle third-party messages, or *traffic*). You can contact large numbers of stations in a very short time in competition with other hams (enter *contests*). You can try to contact hams in as many different countries or distant places as possible (chase **DX**).

These are just a few of the activities that Amateur Radio operators enjoy. The more you operate, the more proficient you'll be on the air.

In this chapter, we'll give you some tips for operating with your new Novice or Technician license — on CW, phone, packet and through FM repeaters. Then we'll take a look at some other operating modes you may want to try.

OPERATING SKILL

Poor operating procedure is ham radio's version of original sin. It is a curse that will not go away. Hams call these poor operators *lids*. No one wants to be a lid or to have a conversation with a lid. Calling someone a lid is the ultimate insult. (Don't do it on the air. Just be sure your operating practices are such that no one *wants* to call *you* a lid.)

It is actually easier to be a good operator than to fall prey to sloppy habits. You can transmit a message without needless repetition and without unnecessary identification. On **CW (Morse code)**, you don't have to spell out each and every word. Good operating makes hamming more fun for everyone.

Clearly, the initial glamour of Amateur Radio is the opportunity to talk to people who share a common bond, the hobby of ham radio. It doesn't matter if they're across town or in another country. Novices and Technician Plus licensees (with Novice privileges) have plenty of chances to work **DX** (contact distant stations — usually in other countries) on the 10 and 15-meter HF ham bands (28.1 to 28.5 MHz and 21.1 to 21.2 MHz). The 40 and 80-meter HF bands (7.1 to 7.15 MHz and 3.675 to 3.725 MHz) also provide good DX opportunities at times, as well as long-distance communications across the US. The VHF and UHF bands are useful primarily for shorter-range contacts.

Chapter 9 presents information on antennas for all types of locations. Once you choose the frequency bands you want to operate on, you'll want to install an antenna that will get you on the air. The success of your antenna installation will be evident once you actually begin making contacts: Either they'll hear you or they won't. You won't need a beam antenna to work DX. At times of high sunspot activity, many Novices report outstanding results using dipoles or converted CB antennas. Your goal is to put the best possible signal on the air. Of course, this will vary depending on your location and your budget. Don't worry about competing with the ham down the street who has a huge antenna atop a 70-foot tower.

In Amateur Radio, it's important to enjoy and take pride in your own accomplishment. You've heard it said "It's what you do with what you've got that really counts." That is especially true in Amateur Radio. To make up for any real or imagined lack of equipment, concentrate on improving your own operating ability.

Good operating skills can be like adding 10 dB to your

Table 2-1

Q Signals

These Q signals are the ones used most often on the air. (Q abbreviations take the form of questions only when they are sent followed by a question mark.)

QRG	Your exact frequency (or that of ___) is ____kHz. Will you tell me my exact frequency (or that of ______)?
QRL	I am busy (or I am busy with _____). Are you busy?
QRM	Your transmission is being interfered with _____ (1. Nil; 2. Slightly; 3. Moderately; 4. Severely; 5. Extremely.) Is my transmission being interfered with?
QRN	I am troubled by static _____. (1 to 5 as under QRM.) Are you troubled by static?
QRO	Increase power. Shall I increase power?
QRP	Decrease power. Shall I decrease power?
QRQ	Send faster (____wpm). Shall I send faster?
QRS	Send more slowly (_____wpm). Shall I send more slowly?
QRT	Stop sending. Shall I stop sending?
QRU	I have nothing for you. Have you anything for me?
QRV	I am ready. Are you ready?
QRX	I will call you again at ___hours (on ___kHz). When will you call me again?
QRZ	You are being called by ____ (on ___kHz). Who is calling me?
QSB	Your signals are fading. Are my signals fading?
QSK	I can hear you between signals; break in on my transmission. Can you hear me between your signals and if so may I break in on your transmission?
QSL	I am acknowledging receipt. Can you acknowledge receipt (of a message or transmission)?
QSN	I did hear you (or ___) on ___kHz. Did you hear me (or ___) on ____kHz?
QSO	I can communicate with ____ direct (or relay through ___). Can you communicate with ___ direct or by relay?
QSP	I will relay to ___. Will you relay to ___?
QST	General call preceding a message addressed to all amateurs and ARRL members. This is in effect "CQ ARRL."
QSX	I am listening to ___ on ___kHz. Will you listen to ___on ___kHz?
QSY	Change to transmission on another frequency (or on ___kHz). Shall I change to transmission on another frequency (or on ___kHz)?
QTB	I do not agree with your counting of words. I will repeat the first letter or digit of each word or group. Do you agree with my counting of words?
QTC	I have ___messages for you (or for ___). How many messages have you to send?
QTH	My location is ____. What is your location?
QTR	The time is ____. What is the correct time?

signal. That's like increasing your power by 10 times! (More detailed operating information can be found in *The ARRL Operating Manual*. This book is available from your local radio dealer or directly from ARRL.)

A good rule for Amateur Radio contacts (**QSOs**) is to talk as you would during a face-to-face conversation. When you meet someone for the first time, you introduce yourself once; you don't repeat your name over and over again. An exception would be if the other person is hard of hearing or you're meeting in a noisy place. Even then you would only repeat if the person couldn't hear you. When you are on the air, the other operator will tell you if she can't hear you clearly. Problem situations on HF include static (QRN), interference (QRM) or signal fading (QSB). Table 2-1 lists many common **Q signals**. Although hams use some Q signals in face-to-face conversations, when you're on the air, use them on CW only. On voice, say what you mean.

Now, let's take a look at some guidelines that will help you get accustomed to operating your new radio with your new privileges. We'll start with CW.

Operating CW

Even if you're anxious to key the mike, don't pass up the information in this section. There are many similarities between CW and voice operation. Non-CW modes such as voice (SSB and FM) and specialized digital modes are covered later in the chapter. The FCC refers to these types of radio signals as **emission types**.

An unmodulated carrier wave (a steady signal with no information included) is called a **test emission**. **CW**, also called international **Morse code**, is transmitted by on/off keying of a radio-frequency signal. This is demonstrated whenever you use a key to send CW. To send code, you press a lever. To stop sending, you let up on the lever. Code is either on or off. What could be simpler?

Calling CQ

There is no point in wasting words, and that includes when you are trying to make a radio contact. Hams establish a contact when one station calls CQ and another replies. **CQ** literally means "Seek you: Calling any station." You can usually tell a good ham by the length of the CQ call. A good operator sends short calls separated by concentrated listening periods. Long CQs drive away more contacts than they attract! Generally, a "3 × 3" call is more than sufficient. Here's an example:

CQ CQ CQ DE KA7XYZ KA7XYZ KA7XYZ K

As you may have figured out, "3 × 3" refers to calling CQ three times, then **DE** ("from" or "this is") then your call sign three times. Perhaps the best way to get started is to listen for someone else's CQ. The Novice bands are usually alive with signals. You may choose to call CQ yourself, however, and that's okay. Always listen before you transmit, even if the frequency appears clear. To start, send **QRL?** ("Is this frequency in use?"). If you hear a **C** ("yes") in reply, then try another frequency. It is not uncommon to hear only one of the stations in a QSO. A frequency may seem clear even though there is a contact in progress. It's the worst of bad manners to jump on a frequency that's already busy. Admittedly, the Novice bands, especially 40 and 80 meters, are often crowded. Even so, if you listen first you'll avoid interfering with an in-progress QSO.

Whatever you do, be sure to *send at a speed you can reliably copy*. Sending too fast will surely invite disaster! This is especially true when you call CQ. If you answer a CQ, answer at a speed no faster than that of the sending station. Don't be ashamed to send PSE QRS (please send more slowly).

Correct CW Procedures

Establishing a contact — The best way to do this, especially at first, is to listen. When you hear someone calling CQ, answer them. If you hear a CQ, wait until the ham indicates she is listening, then call her. For example:

WA1STO WA1STO DE WL7AGA WL7AGA $\overline{\text{AR}}$

($\overline{\text{AR}}$ is equivalent to over).

In answer to your call, the called station will reply WL7AGA DE WA1STO R. . . . That R (roger) means she has received your call. That's all it means — *received*. It does *not* mean (a) correct, (b) I agree, (c) I will comply, or anything else. It is not sent unless everything from the previous transmission was received correctly. Perhaps WA1STO heard someone calling her but didn't quite catch the call because of interference (QRM) or static (QRN). In this case, she might come back with:

QRZ? DE WA1STO $\overline{\text{AR}}$ ("Who is calling me?").

Calling CQ — CQ means "I wish to contact any amateur station." Avoid calling CQ endlessly. It clutters up the air and drives off potential new friends. The typical CQ goes like this:

CQ CQ CQ DE WB3IOS WB3IOS WB3IOS K.

The letter K is an invitation for any station to go ahead.

The QSO — During a contact, it is necessary to identify your station only once every 10 minutes. Keep the contact friendly and cordial. Remember, the conversation is not private. Many others, including nonamateurs, may be listening. Both on CW and phone, it is possible to be informal, friendly and conversational. This is what makes the Amateur Radio QSO enjoyable. During the contact, when you stand by, use K ("go") at the end of your transmission. If you don't want someone else to join the QSO, use $\overline{\text{KN}}$. That says you want only the contacted station to come back to you. Most of the time K is sufficient (and shorter!).

Ending the QSO — When it's time to end the contact, don't keep talking. Briefly express your pleasure at having worked the other operator. Then, sign out:

SK WL7AGA DE WA1STO.

If you are leaving the air, add CL to the end, right after your call sign.

These procedures help hams communicate better on CW. They have no legal standing, however. FCC regulations say little about our internal procedures. They show we're not just hobbyists, but that we are an established, self-regulating communications service. Hams take pride in our distinctive procedures, and understandably so.

You will learn the meaning of many other **Q signals** as you gain operating experience. QTH means "location." Followed by a question mark, it asks, "What is your location?" QSL means "I acknowledge receipt" (of message or information). A **QSL card** is a written confirmation of an amateur contact. QSY means "change frequency."

As you gain experience, you will develop the skill of sorting out the signal you want from those of other stations. Space in the ham bands is limited. Make sure your QSO takes up as little of it as possible to avoid interfering with other QSOs. When two stations are communicating, their transmitters should be on exactly the same frequency. We call this procedure *operating zero beat*. Both transmissions will have the same pitch when you hear them in a receiver. When you tune in the other station, adjust your receiver for the strongest signal and the "proper" tone for your radio. This will usually be between 500 and 1000 Hz.

Zero-beat operating helps the other operator know where to listen for your signal. This is easy with a transceiver. The transmitter will automatically be close to zero beat when you tune in the received signal to a comfortable pitch. With a separate receiver and transmitter, you must be sure your transmitter is zero beat with that of the other station.

Working "Split"

There is an exception to the zero-beat rule. Sometimes a DX station has a "pileup" of stations trying to make contact. If they are all calling on the DX station's frequency, no one (including the DX station!) can copy anything. Under these conditions, the DX station may choose to work *split*. When you hear a DX station, but you never hear anyone else calling or working him, chances are he's working split. Listen to the directions given by the DX operator. Chances are he or she will say something like "U 5" or "U 10." This is telling you to call the DX station 5 or 10 kHz higher (Up) in frequency. Tune higher in the band, and look for the pileup. It should be easy to find. DX stations work split when operating SSB, too. (For more helpful hints on work-ing DX, see ARRL's *The DXCC Companion*.)

For amateurs to understand each other, we must standardize our communications. You'll find, for instance, that most hams use abbreviations on CW. Why? It's faster to

Table 2-2

Some Common Abbreviations Used on CW

Although abbreviations help to cut down unnecessary transmission, it's best not to abbreviate unnecessarily when working an operator of unknown experience.

AA	All after	GN	Good night	SASE	Self-addressed, stamped envelope
AB	All before	GND	Ground	SED	Said
ABT	About	GUD	Good	SIG	Signature; signal
ADR	Address	HI	The telegraphic laugh; high	SINE	Operator's personal initials or nickname
AGN	Again	HR	Here, hear	SKED	Schedule
ANT	Antenna	HV	Have	SRI	Sorry
BCI	Broadcast interference	HW	How	SSB	Single sideband
BCL	Broadcast listener	LID	A poor operator	SVC	Service; prefix to service message
BK	Break; break me; break in	MA, MILS	Milliamperes	T	Zero
BN	All between; been	MSG	Message; prefix to radiogram	TFC	Traffic
BUG	Semi-automatic key	N	No	TMW	Tomorrow
B4	Before	NCS	Net control station	TNX-TKS	Thanks
C	Yes	ND	Nothing doing	TT	That
CFM	Confirm; I confirm	NIL	Nothing; I have nothing for you	TU	Thank you
CK	Check	NM	No more	TVI	Television interference
CL	I am closing my station; call	NR	Number	TX	Transmitter
CLD-CLG	Called; calling	NW	Now; I resume transmission	TXT	Text
CQ	Calling any station	OB	Old boy	UR-URS	Your; you're; yours
CUD	Could	OC	Old chap	VFO	Variable-frequency oscillator
CUL	See you later	OM	Old man	VY	Very
CW	Continuous wave (that is, radiotelegraphy)	OP-OPR	Operator	WA	Word after
DE	From, this is	OT	Old-timer; old top	WB	Word before
DLD-DLVD	Delivered	PBL	Preamble	WD-WDS	Word; words
DR	Dear	PSE	Please	WKD-WKG	Worked; working
DX	Distance, foreign countries	PWR	Power	WL	Well; will
ES	And, &	PX	Press	WUD	Would
FB	Fine business, excellent	R	Received as transmitted; are	WX	Weather
FM	Frequency modulation	RCD	Received	XCVR	Transceiver
GA	Go ahead (or resume sending)	RCVR (RX)	Receiver	XMTR (TX)	Transmitter
GB	Good-by	REF	Refer to; referring to; reference	XTAL	Crystal
GBA	Give better address	RFI	Radio frequency interference	XYL (YF)	Wife
GE	Good evening	RIG	Station equipment	YL	Young lady
GG	Going	RPT	Repeat; I repeat	73	Best regards
GM	Good morning	RTTY	Radioteletype	88	Love and kisses
		RX	Receiver		

Shortly after you begin your on-the-air operation, you will experience the excitement of collecting QSL cards from all over the world.

Table 2-3

The RST System

READABILITY

1—Unreadable.
2—Barely readable, occasional words distinguishable.
3—Readable with considerable difficulty.
4—Readable with practically no difficulty.
5—Perfectly readable.

SIGNAL STRENGTH

1—Faint signals barely perceptible.
2—Very weak signals.
3—Weak signals.
4—Fair signals.
5—Fairly good signals.
6—Good signals.
7—Moderately strong signals.
8—Strong signals.
9—Extremely strong signals.

TONE

1—Sixty-cycle ac or less, very rough and broad.
2—Very rough ac, very harsh and broad.
3—Rough ac tone, rectified but not filtered.
4—Rough note, some trace of filtering.
5—Filtered rectified ac but strongly ripple-modulated.
6—Filtered tone, definite trace of ripple modulation.
7—Near pure tone, trace of ripple modulation.
8—Near perfect tone, slight trace of modulation.
9—Perfect tone, no trace of ripple or modulation of any kind.

The "tone" report refers only to the purity of the signal. It has no connection with its stability or freedom from clicks or chirps. Most of the signals you hear will be a T-9. Other tone reports occur mainly if the power supply filter capacitors are not doing a thorough job. If so, some trace of ac ripple finds its way onto the transmitted signal. If the signal has the characteristic steadiness of crystal control, add X to the report (for example, RST 469X). If it has a chirp or "tail" (either on "make" or "break") add C (for example, 469C). If it has clicks or noticeable other keying transients, add K (for example, 469K). Of course a signal could have both chirps and clicks, in which case both C and K could be used (for example, RST 469CK).

send a couple of letters than it is to spell out a word. But there's no point in using an abbreviation if no one else understands you. Over the years, amateurs have developed a set of standard abbreviations (Table 2-2). If you use these abbreviations, you'll find that everyone will understand you, and you'll understand them. In addition to these standards, we use a set of **procedural signals**, or *prosigns*, to help control a contact. The "Correct CW Procedures" sidebar explains these prosigns in more detail.

Answering a CQ

What about our friend KA7XYZ? In our last example he was calling CQ. What happened? Another ham, N2SN, heard him and answered the CQ:

KA7XYZ KA7XYZ DE N2SN N2SN $\overline{\text{AR}}$

Notice the "2 × 2" format. N2SN sent the call sign of the station he was calling twice, then sent DE, then sent his call sign twice. If KA7XYZ thought he heard someone calling him, but wasn't quite sure, he would send:

QRZ? DE KA7XYZ $\overline{\text{AR}}$

The Q signal QRZ means "who is calling me?" In that case, N2SN will send his call sign again, usually two or three times.

Use the prosign $\overline{\text{AR}}$ (the letters A and R run together with no separating space) in an initial call to a specific station before officially establishing contact. When calling CQ, use the prosign **K**, because you are inviting any station to reply. ("K" means "Any station go ahead and transmit.") KA7XYZ comes back to N2SN in a to-the-point manner (one you should copy).

N2SN DE KA7XYZ R GE UR RST 599 DENVER CO $\overline{\text{BT}}$ NAME BOB HW BK

In this transmission, **RST** refers to the standard *readability*, *strength* and *tone* system of reporting signal reception. You'll exchange a signal report in nearly every Amateur Radio QSO. Don't spend a lot of time worrying about what signal report to give to a station you're in contact with. The scales are simply a general indication of how you are receiving the other station. As you gain experience with the descriptions given in Table 2-3, you'll be more comfortable estimating the proper signal report.

A report of RST 368 would be interpreted as "Your signal is readable with considerable difficulty, good strength, with a slight trace of modulation." The tone report is a useful indication of transmitter performance. When the RST system was developed, the tone of amateur transmitters varied widely. Today, a tone report of less than 9 is cause to ask a few other amateurs for their opinion of the transmitted signal. Consistently poor tone reports means your transmitter has problems.

A further word about signal reports, one that applies to all types of operating: If your signal report is *too* good, *reduce your power*. FCC Rules say that amateurs must use the minimum power necessary to maintain communications. Whether you're operating through an FM repeater, on VHF or UHF simplex or on HF, if you don't have to use an amplifier or the highest power your transceiver is capable of, decrease your transmitting power. You'll be less likely to interfere with other stations (and if you're using a hand-held transceiver or other portable radio you'll conserve your battery).

The basic information is only transmitted once. The other station will request repeats if necessary. Notice too that BT (B and T run together) is used to separate portions of the text. This character is really the double dash (=), and is usually written as a long dash or hyphen on your copy paper. HW? means "how do you copy?" BK signifies that KA7XYZ is turning it over (back) to N2SN for his basic info. KA7XYZ does not sign both calls all over again. FCC Rules require identification only at the end of a QSO, and once every 10 minutes.

At the end of the contact send 73 and sign off. **73** is a common abbreviation that means "Best regards."

Conversation is a two-way phenomenon. There is no reason that an Amateur Radio QSO can't be a back-and-forth process. No one wants to listen to a long, unnecessary monologue. Propagation conditions might change, hampering the QSO. KA7XYZ and N2SN will relate better if each contributes equally. Sometimes there is interference or marginal copy because of weak signals. It may help to transmit your call sign when you are turning the conversation back to the other station.

Tips for Better CW Operating

In summary, here are the points to keep in mind:

Listen before transmitting. Send QRL? ("Is this frequency in use?") before transmitting. Listen again! It's worth repeating: Listen!

Send short CQs and listen between each.

Send no faster than you can ***reliably copy***.

Use standard abbreviations whenever possible — become familiar with them.

Use prosigns and Q signals properly.

Identify at the end of a QSO (the entire contact, not each turnover) and ***every 10 minutes***.

Use R only if you've received 100 percent of what the other station sent.

My First On-The-Air Contact

I slowly reach for the key. All I have to do is push down on the key. Then I'll be on the air and able to talk to the world. But I can't. My hand doesn't move, except to tremble slightly. It's not the rig, not the key. It's me! I can't find the nerve to do it. I can't send a signal to the world.

That's how it was for me when I tried to make my first contact. It had been weeks since I passed the Novice exam. Each day that passed without the arrival of my new ticket seemed endless. I knew that if that day would just end, the next would surely see the arrival of my license. Days, weeks passed. Then a month had gone by. The station was there waiting for me. All the equipment was in place and ready for operation. If only the license would arrive.

Finally, after what seemed to be several years, it did arrive. Can you believe it? It arrived on my birthday. What a present! The letter carrier had hardly made it back to the sidewalk before I was at the rig. I turned on the transmitter and receiver. Nothing! What could be wrong? All the wires connected? Oh no! The power cords were just hanging there. I'd forgotten to plug them into the wall socket. The few seconds it took seemed like hours. Then I heard the sound of *radio*. I was ready to enter the world of ham radio. Or so I thought.

After several minutes of total failure at my attempt to send CQ, I gave up. I needed help. But where to get it? Of course, the instructor from my Novice training class. He could, and would help me.

The next day I got in touch with him and explained my problem. Would he help me? Yes, he would be glad to stand behind me during that first contact. But, he was busy with something else. I'd have to wait a day or so. Another day without a contact? It might as well have been another year.

Although it didn't seem possible, the next day rolled around, and with it, my instructor. A few minutes of preparation and I was ready. I would give it my best, to sink or swim. Tuning on my radio, I found another station calling CQ. His sending was slow and steady. It had to be for *me* to copy it. "Go ahead, give him a try," advised my instructor. "What have you got to lose?"

. . . CQ CQ CQ CQ. His call seemed endless. Finally, he signed . . . CQ DE W3AOH. I reached confidently for the key. W2AOH DE WN2VDN AR.

"That was a pretty short call you gave him. You should have sent your call sign more than once. He may have missed it." But he didn't. There it was, slow and steady. "WN2VDN DE W3AOH BT TNX CALL BT UR . . . "He had answered my call . . . my call. I'd done it! I made a contact!"

"Pay attention and copy the code!" my instructor yelled at me. "He's still sending." Dahdididit dahdidah dah dahdahdah didídah dahdahdah . . . dahdididit . . . What on earth did that mean? I'd forgotten the code. Those strange sounds coming out of the speaker didn't make any sense at all. What was I going to do?

Fortunately, my instructor was copying it all. "He's just signed it over to you. Your turn now."

I hit the key again. This time there was no fear. The key was part of my hand, part of my mind. W3AOH DE WN2VDN . . . Until, that is, after I sent my call. I didn't know his name, his location, anything. So I sent the only thing I could think of. "Thanks for coming back to me OM. You're my first contact. I'm a bit nervous." Behind me, the instructor is saying, "You should have sent him your name and location, and his signal report." Too late, I was already signing it over to W3AOH.

Slow and steady came the reply. "Welcome aboard. Hope you enjoy ham radio as much as I do. . . ." This time I had no trouble copying him. After the contact was completed, I breathed a very large sigh of relief. But, I had done it. I had actually "talked" to another person via Morse code. I was a bona fide ham.

My instructor asked if he should hang around a little longer. But I didn't hear him. I had already begun looking around for another CQ to answer. I didn't need his help any longer; I was a ham.

Be courteous.

Zero beat the other station's frequency before calling (except when working "split").

Tuning Up

What is the most exciting, most memorable and perhaps most terrifying moment in your entire ham experience? Your first on-the-air contact with another station! Before you have that experience, you'll want to know how to operate your station equipment.

The best place to start is the instruction manual. Before you even turn on your radio, read the instructions carefully so you'll be familiar with each control. Without turning on the equipment, try adjusting the controls. Nothing will happen, but you'll learn the location and feel of each important control.

After studying the manual and finding the important controls, you'll be ready to tune up. You must have your FCC license in hand before you can transmit or even tune up on the air!

If your transmitter requires tuning when you change bands, connect your transmitter output to a **dummy load** while you tune. This avoids on-the-air interference. Never tune up on the air because you could interfere with other hams.

Once you have tuned up the transceiver according to the instruction manual, disconnect the dummy load and connect the antenna (an antenna switch makes this easy). Now, you're ready to operate! If you use a Transmatch, you may have to transmit a brief low-power signal to adjust the circuit.

[Before you go on to the next section, turn to Chapter 13 and study questions N2A01, N2A02, N2A04 through N2A16 and N2A19 through N2A22. You should also study questions N8A01 through N8A03. Review this section if you have any difficulty with these questions.]

[**If you are studying for a Technician license**, you should also turn to Chapter 14 and study question T2A14 and questions T8A01 and T8A02. Review this section as needed.]

SSB Voice Operating

To make the most of your voice privileges, you'll need to learn a different set of operating procedures. Those described here apply to voice operation on 10-meter **single sideband (SSB)** and on the "weak-signal" SSB frequencies on the VHF/UHF bands. Operating through FM repeaters is covered later in this chapter.

Any **voice communication** mode is known as **phone** under FCC Rules. AM, **SSB** and **FM voice** are all phone **emission types**.

What does single-sideband (SSB) mean? Begin with a steady radio frequency (RF) signal such as you would get by pressing the key of a Morse code transmitter and just holding it down. This is called the *RF carrier*. Then combine this signal with a voice signal from a microphone. The process of combining such an RF carrier with any information signal is called *modulation*. (Actually, keying the RF carrier to form Morse code is a form of modulation.) If we use *amplitude* modulation, the resulting signal has two *sidebands*, one higher in frequency than the carrier frequency and one lower in frequency than the carrier frequency. These are called the *upper sideband* and the *lower sideband*. For a single-sideband voice signal, the carrier and one of the sidebands is removed, and only one sideband is transmitted. See Figure 2-1. The RF carrier is the signal that we modulate to produce a radiotelephone signal.

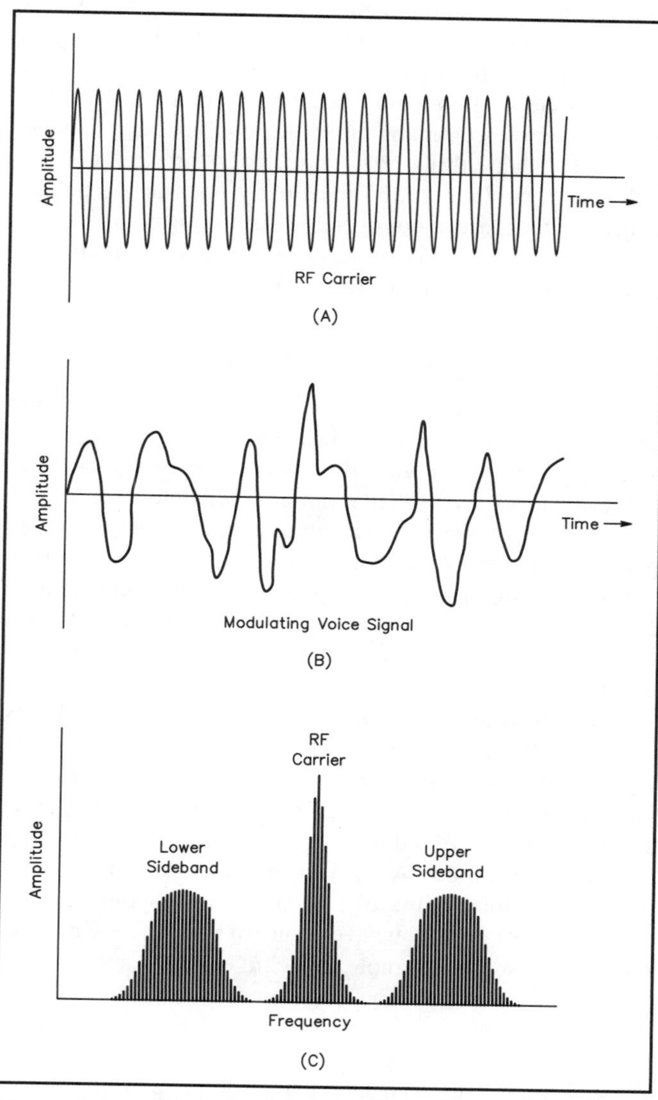

Figure 2-1 – A single-sideband (SSB) signal is formed by combining an RF carrier signal (A) with a voice or other information signal (B). The resulting signal includes the original RF carrier as well as two sidebands, one on either side of the carrier. The upper and lower sidebands are shown at C. To transmit an SSB signal one of the sidebands and the RF carrier are removed in the transmitter, and the remaining sideband is transmitted.

SSB is the most common voice mode on HF. You can use either the *lower* **sideband** or the *upper* **sideband** to transmit an SSB signal. Amateurs normally use the upper sideband for 10-meter phone operation.

Operating techniques and procedures vary for different modes, and even a little from band to band. If you're accustomed to 80-meter CW, you may be uncertain about how to make a 10-meter SSB contact. A great way to become familiar with new techniques is to spend time listening. Be discriminating, though. Take a few moments to understand the techniques used by the proficient operators — the ones who are the most understandable and who sound the best. Don't simply mimic whatever you hear. This is especially true if you're going on the air for the first time.

Whatever band or mode you are using, there are three fundamental things to remember. These apply for any type of voice operating you might try. The first is that courtesy costs very little. It is often rewarded by bringing out the best in others. Second, the aim of each radio contact should be 100% effective communication. A good operator is never satisfied with anything less. Third, your "private" conversation with another station is actually open to the public. Many amateurs are uncomfortable discussing controversial subjects over the air. Also, never give any confidential information on the air. You never know who may be listening.

Table 2-4

Voice Equivalents to Code Procedure

Voice	*Code*	*Meaning*
over	$\overline{\text{AR}}$	after call to a specific station
end of message	$\overline{\text{AR}}$	end of message
wait, stand by	$\overline{\text{AS}}$	please stand by
received	R	all received correctly (not a promise to take any specific action.)
go, go ahead	K	any station transmit
go only	$\overline{\text{KN}}$	addressed station only
clear	$\overline{\text{SK}}$	end of contact
closing station	CL	going off the air
break or back to you	BK	the receiving station's turn to transmit
stroke, portable	$\overline{\text{DN}}$	operation away from primary station location

Keep it Plain and Simple

Although it does not require the use of code or special abbreviations, proper voice procedure is very important. Voice operators *say* what they want to have understood. CW operators have to spell it out or abbreviate. One advantage of voice operation is speed: a typical voice QSO takes place at between 150 and 200 words per minute. Whether you're working a DX operator who may not fully understand our language, or talking to your friend down the street, speak slowly and clearly. That way, you'll have fewer requests to repeat information.

Avoid using CW abbreviations and prosigns such as "HI" and "K" for voice communications. Also, Q signals (QRX, QRV and so forth) are for CW, not voice, operation. You may hear operators using Q signals such as QSL, QSO and QRZ on voice, but you should generally avoid using them on voice modes. Abbreviations are used on CW to say more in less time, to improve the efficiency of your communications. On voice, you have plenty of time to say what you mean. On CW, for example, it's convenient to send "K" at the end of a transmission. On voice, it takes less than a second to say "go ahead." Table 2-4 shows the voice equivalents for common CW prosigns.

Use plain language and keep jargon to a minimum. In particular, avoid the use of "we" when you mean "I" and "handle" or "personal" when you mean "name." Also, don't say "that's a roger" when you mean "that's correct." Taken individually, any of these sayings is almost harmless. Combined in a conversation, however, they give a false-sounding "radioese" that is actually less effective than plain language. "Roger" for example, means "I have received what you sent." It doesn't necessarily mean that what was sent was "correct."

If the other operator is having difficulty copying your signals you should use the standard International Telecommunication Union (ITU) phonetic alphabet (See Chapter 1) to spell out the letters in your call sign, your name or any other piece of information that might be confused if the letters are not received correctly. This phonetic alphabet is generally understood by hams in all countries.

Initiating a Contact

There are two ways to initiate an SSB voice contact: call **CQ** (a general call to any station) or answer a CQ. At first, you may want to tune around and find another station to answer. If activity on a band seems low, a CQ call may be worthwhile.

Before calling CQ, it is important to find a frequency that appears unoccupied by any other station. This may not be easy, particularly during crowded band conditions. Listen carefully — perhaps a weak DX station is on frequency. If you're using a beam antenna, rotate it to make sure the frequency is clear. If, after a reasonable time, the frequency seems clear, ask if the frequency is in use, then sign your call. "Is the frequency in use? This is KA1IFB." If, as far as you can determine, no one responds, you are ready to make your call.

As in CW operation, keep voice CQ calls short. Long calls are considered poor operating technique. You may interfere with stations already on frequency who didn't hear your initial frequency check. Also, stations intending to reply to the call may become impatient and move to another frequency. Call CQ three times, followed by "this is," followed by your call sign three times and then listen. If no one comes back, try again. If two or three calls produce no answer, it may be that there is too much noise or interference, or that atmospheric conditions are not favorable. At that point, change frequency and try again. If you still get no answer, try looking around and answer someone else's CQ.

An example of a good CQ call is:

"CQ CQ Calling CQ. This is KB1AFE, Kilo Bravo One Alfa Foxtrot Echo, Kilo Bravo One Alfa Foxtrot Echo, calling CQ and standing by."

Notice that the operator used the International Telecommunication Union standard phonetic alphabet to spell out her call sign. This ensures that her call sign will be understood and letters won't be confused for other letters. There is no need to say what band is being used, and certainly no need to add "tuning for any possible calls, dah-di-dah!" or "K someone please!" and the like.

When replying to a CQ, say both call signs clearly. It's not necessary to sign the other station's call phonetically. You should always sign yours with standard phonetics, however. Remember to keep calls short. Say the call sign of the station you are calling only once. Follow with your call sign, repeated phonetically, several times. For example, "W1AW, this is KA3HAM, Kilo Alfa Three Hotel Alfa Mike, KA3HAM, over." Depending on conditions, you may need to give your call phonetically several times. Repeat this calling procedure as required until you receive a reply or until the station you are calling has come back to someone else.

Listening is very important. If you're using *PTT* (push-to-talk), be sure to let up on the transmit button between calls so you can hear what is going on. With *VOX* (voice operated switch), you key the transmitter simply by talking into the microphone. VOX operation is helpful because, when properly adjusted, it enables you to listen between words. Remember: It is extremely poor practice to make a long call without listening. Also, don't continue to call after the station you are trying to contact replies to someone else. Wait for the contact to end before trying again, or try another spot on the band.

Conducting the QSO

Once you've established contact, it is no longer necessary to use the phonetic alphabet for your call sign or to give the other station's call. FCC regulations say that you need to give your call only every 10 minutes and at the conclusion of the contact. (The exception is when handling international third-party traffic. Then, you must sign both calls.) This allows you to enjoy a normal two-way con–versation without the need for continual identification. Use "over" or "go ahead" at the end of a transmission to indicate that it's the other station's turn to transmit. (During FM repeater operation, it is obvious when you or the other station stops transmitting because you will hear the repeater carrier drop. In this case it may not be necessary to say "over.")

Signal reports on SSB are two-digit numbers using the RS portion of the RST system. No tone report is required. The maximum signal report would be "five nine" — that is, readability 5, strength 9. (See Table 2-3.) A signal report of "five seven" means your signal is perfectly readable and moderately strong. On the other hand, a signal report of "three three" would mean that the other operator is only able to understand your signal with considerable difficulty and your signals are weak in strength.

If you get a report of, say, "59 plus 20 dB," it means your received signal reads 20 decibels higher than signal strength 9 on the transceiver S meter. Decibels provide a convenient way to compare the power of signals. If one signal is twice as strong as another, there will be a 3-dB difference between them. When one signal is 10-dB stronger than another it means it is 10 times stronger. A difference of 20 dB means one signal is 100 times stronger and 30 dB means one signal is 1000 times stronger. If you receive a signal report of 59 plus 20 dB, and your transmitter is operating at 100 watts, you could reduce power to 1 watt and still have a "strength 9" signal!

On FM repeaters, RS reports are not used. FM signal reports are generally given in terms of signal quieting. Full quieting means the received signal is strong enough to block all receiver noise.

Voice contacts are often similar in content to CW QSOs. Aside from signal strength, most hams exchange name, location and equipment information (especially antennas!). Once these routine details are out of the way, you can talk about your families, the weather, your occupation or any appropriate subject.

Working DX

During the years of maximum sunspot activity, 10-meter worldwide communication on a daily basis is commonplace. (The propagation section of this chapter discusses the effects of sunspots in detail.) Ten meters is an outstanding DX band when conditions are right. A particular advantage of 10 meters for DX work is that effective beam-type antennas tend to be small and light, making for relatively easy installation.

There are a few things to keep in mind when you contact amateurs from outside the United States. While many overseas amateurs have an exceptional command of English (which is especially remarkable since few US amateurs understand foreign languages), they may not be familiar with many of our local sayings. Because of the language differences, some DX stations are more comfortable with the "bare-bones" type contact, and you should be sensitive to their preferences.

During unsettled band conditions it may be necessary to keep the contact short in case fading or interference occurs. Take these factors into account when expanding on a basic contact. Also, during a band opening on 10 meters or on VHF, it is crucial to keep contacts brief. This allows many stations to work whatever DX is coming through.

When the time comes to end the contact, end it. Thank the other operator (once) for the pleasure of the contact and say good-bye: "This is WB3IOS, clear." This is all that is required. Unless the other amateur is a good friend, there is no need to start sending best wishes to everyone in the household including the family dog! Nor is this the time to start digging up extra comments on the contact which will require a "final final" from the other station (there may be other stations waiting to call in).

Tips for Better SSB Operating

Listen with care. It is natural to answer the loudest station that calls, but sometimes a weaker signal may provide a more interesting contact. Not all amateurs can run high power, and sometimes the weaker signal may be from a more-distant station. Don't reward an operator who has cranked up the transmitter gain to the point of being difficult to understand, especially if a station with a nice sounding signal is also calling.

Use VOX or PTT. If you use VOX, don't defeat its purpose by saying "aah" to keep the transmitter on the air. If you use PTT, let go of the mike button every so often to make sure you are not "doubling" with the other station. A QSO should be an interactive conversation. Don't do all the talking.

Take your time. The speed of voice transmission (with perfect accuracy) depends almost entirely on the skill of the two operators concerned. Use a rate of speech that allows perfect understanding. The operator on the other end should have time to record important details of the contact. If you go too fast, you'll end up repeating a lot of information.

Use standard phonetics to make your call sign easier to understand.

[Before going on to the next section, turn to Chapter 13 and study questions N2A17, N2A18 and N2A23. Also study questions N8A04 through N8A06. Review this section if you have difficulty with any of these questions.]

[**If you are studying for the Technician exam**, also review questions T2A09, T2B05 through T2B07 in Chapter 14. You should also study questions T8A03, T8A07, T8A08, T8B01 and T8B02. Reread this section if you have any difficulty.]

VHF AND UHF FM REPEATER OPERATING

More hams use **frequency-modulated (FM) voice** than any other communications mode. Most hams have an FM rig of some type. They use it to keep in touch with their local friends. Hams often pass the time during their morning and evening commute talking on the air. In most communities, amateurs interested in a specialized topic (such as chasing DX) have an FM frequency where they meet regularly to exchange information. At flea markets and conventions, hand-held FM units are in abundance as hams compare notes on the latest bargain.

Generally, it's a good idea to use VHF or UHF for all local communications. The HF bands should be reserved for longer-distance contacts to reduce interference on the HF bands.

VHF and UHF FM voice operation takes two forms: simplex and repeater. **Simplex operation** means the stations are talking to each other directly, on the same frequency. This is similar to making a contact on the HF bands.

FM voice operation is well-suited to local VHF/UHF radio communication because the audio signal from an FM receiver is not affected by static-type electrical noise. Car engines and ignition systems produce quite a bit of static electrical noise, and many hams like to operate their FM radios while they are driving or riding in a car. (This is called *mobile* operation.) An AM or SSB receiver is affected much more by static-type electrical noise.

The communications range for VHF and UHF FM simplex is usually limited to your local area. If you live high on a mountain and use a high-gain directional antenna, you may be able to extend your range considerably. Unfortunately, most of us do not have the luxury of ideal VHF/UHF operating conditions. Often, we want to make contacts even though we live in a valley, are driving in a car or are using a low-power, hand-held transceiver.

Enter repeaters. A **repeater** receives a signal and retransmits it, usually with higher power and from a better location, to provide a greater communications range. Often located atop a tall building or high mountain, VHF and UHF repeaters greatly extend the operating range of amateurs using mobile and hand-held transceivers. See Figure 2-2. If a repeater serves an area, it's not necessary for everyone to live on a hilltop. You only have to be able to hear the repeater's transmitter and reach the repeater's receiver with your transmitted signal.

A repeater receives a signal on one frequency and simultaneously retransmits (repeats) it on another frequency. The frequency it receives on is called the **input frequency**, and the frequency it transmits on is called the **output frequency**.

To use a repeater, you must have a transceiver that can

transmit on the repeater's input frequency and receive on the repeater's output frequency. The input and output frequencies are separated by a predetermined amount that is different for each band. This separation is called the *offset*. For example, the offset on 1.25 meters is 1.6 MHz. A repeater on 1.25 meters might have its input frequency on 222.32 MHz and its output on 223.92 MHz. Repeater frequencies are often specified in terms of the output frequency (the frequency you set your receiver to listen on) and the offset. Your transmitter operates on a frequency that is different from the receive frequency by the offset amount.

Most transceivers designed for FM repeater operation are set up for the correct offset. They usually have a switch to change between **simplex operation** (transmit and receive on the same frequency) and **duplex operation** (transmit and receive on different frequencies). So, if you wanted to use the repeater in the preceding example, you would switch your transceiver to the duplex mode and dial up 223.92 to listen to the repeater. When you transmit, your rig will automatically switch to 222.32 MHz (1.6 MHz lower in frequency), the repeater input frequency.

When you have the correct frequency dialed in, just key your microphone button to transmit through ("access") the repeater. Most repeaters are **open** — that is, available for use by anyone in range. Some repeaters, however, have limited access. Their use is restricted to exclusive groups, such as members of a club. Such **closed** repeaters require the transmission of a continuous subaudible tone or a short "burst" of tones for access. These are called **CTCSS** (continuous tone-coded squelch system) or **PL** (Private Line — PL is a Motorola trademark) tones. There are also some repeaters available for use by everyone that require the use of special codes or subaudible tones to gain access. The reason for requiring access tones for "open" repeaters is to prevent interference from extraneous transmissions that might accidentally key the repeater. If you wish to join a group that sponsors a closed repeater, contact the repeater control operator.

Finding a Repeater

Most communities in the United States are served by repeaters. While the majority of repeaters (over 6000) are on 2 meters, there are more than 1600 repeaters on 222 MHz, more than 5000 on 440 MHz, over 70 on 902 MHz and more than 200 on 1270 MHz. More repeaters are being put into service all the time. Repeater frequencies are selected through consultation with **frequency coordinators** — individuals or groups that recommend repeater frequencies based on potential interference and other factors.

There are several ways to find the local repeater(s). Ask local amateurs or contact the nearest radio club. Each year, the ARRL publishes *The ARRL Repeater Directory*, a comprehensive listing of repeaters throughout the United States, Canada, Central and South America and the Caribbean. Besides finding out about local repeater

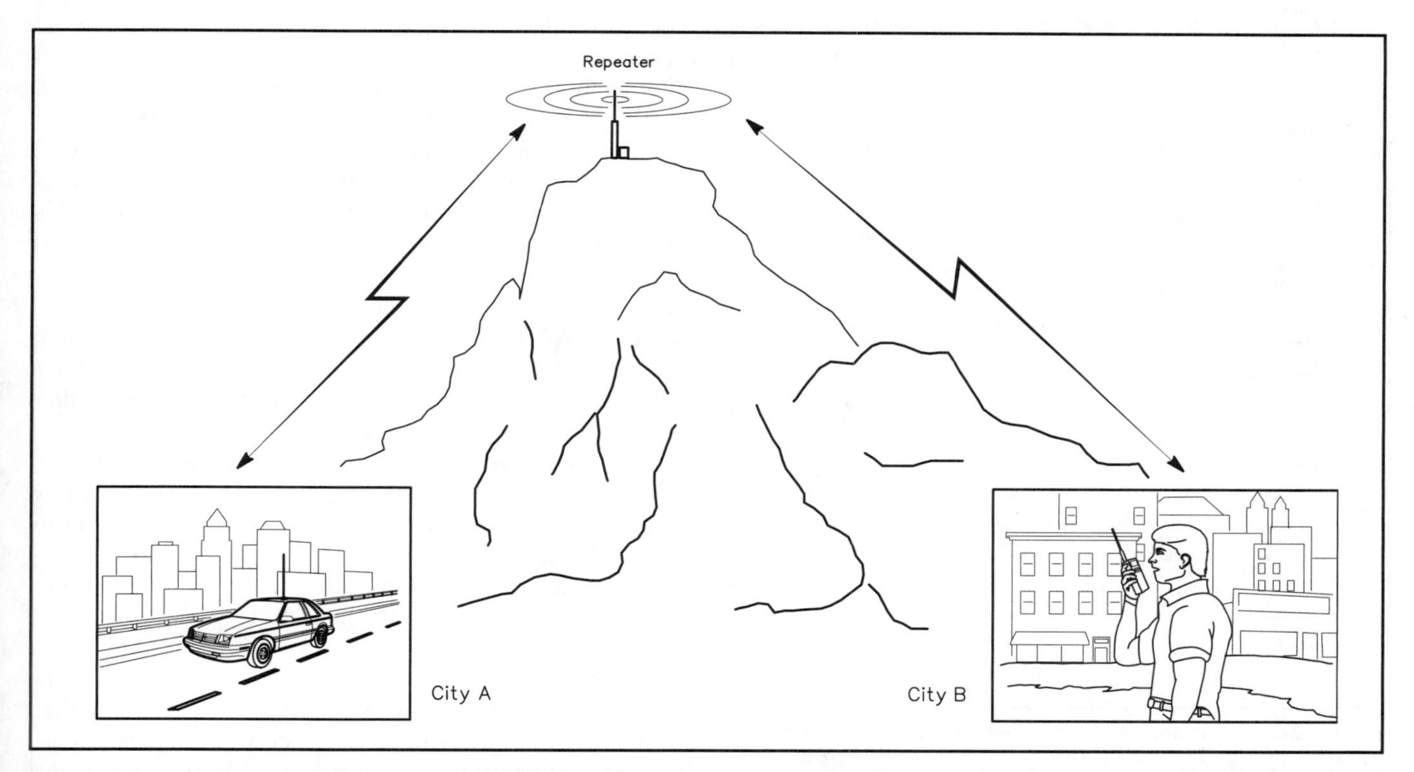

Figure 2-2 — Stations in city A can easily communicate with each other, but the hill blocks their communications with city B. The hilltop repeater enables the groups to communicate with each other.

activity, the *Directory* is handy for finding repeaters to use during vacations and business trips. See Figure 2-3.

Certain segments of each band are set aside for FM operation. For example, on 1.25 meters, repeater inputs are found between 222.32 and 223.28 MHz. The corresponding outputs are between 223.92 and 224.98 MHz. Frequencies between 223.42 and 223.9 MHz are set aside for simplex operation. On 23 cm, repeater inputs run between 1270 and 1276 MHz, with corresponding outputs between 1282 and 1288 MHz. Simplex operation is between 1294 and 1295 MHz.

Repeater Operating

Before you make your first FM repeater contact, you should learn some repeater operating techniques. It's worth a few minutes to listen and familiarize yourself with the procedures used by other hams in your area. Accepted procedures can vary slightly from repeater to repeater.

Your First Transmission

Making your first transmission on a repeater is as simple as signing your call. If the repeater is quiet, just say "N1GZO" or "N1GZO listening" — to attract someone's attention. After you stop transmitting, you will usually hear the unmodulated repeater carrier for a second or two. This *squelch tail* lets you know that the repeater is working. Someone interested in talking to you will call you after your initial transmission. Some repeaters have specific rules for making yourself heard. In general, however, your call sign is all you need.

222-225 MHz 367
PENNSYLVANIA

Location	*Output*	*Input*	*Call*	*Notes*	*Sponsor*
PENNSYLVANIA - TPARC					
EPA SECTION					
Voice FM Links	223.560	223.560			TPARC/ARCC
Voice FM Links	223.580	223.580			TPARC/ARCC
Voice FM Links	223.600	223.600			TPARC/ARCC
Voice FM Nat Smplx	223.500	223.500			TPARC/ARCC
Voice FM Smplx	223.520	223.520			TPARC/ARCC
Voice FM Smplx	223.540	223.540			TPARC/ARCC
BERKS					
Reading	224.640	223.340	WB3FYL	e l	Daub Dx As
BUCKS					
Churchville	224.580	–	W3CCX		MtAryVHFRC
Dublin	224.900	–	WB3KRW		WB3KRW
Feasterville	224.980	146.370	WB3BLG	e l	WB3BLG
Feasterville	224.980	–	WB3BLG	e l	WB3BLG
Langhorne	224.740	–	N3DFV		N3DFU
CARBON					
Palmerton	224.260	–	N3DVF	(ca) e l	EPA VHF
CHESTER					
Glen Mills	224.980	–	K3ADS		K3ADS
Parkesburg	223.940	–	KJ6AL	A e	KJ6AL
Parkesburg	224.040	–	N3JCN		N3JCN
Thorndale	224.360	–	N3NAB	(ca) e	ChesCtyBC
West Chester	224.680	–	AA3CH	(ca) t e	Metro-Com
Westtown	224.900	–	K3ADS		K3ADS
DELAWARE					
Chester	224.960	–	AA3CH		Metro-Com
Chester	224.960	448.050	AA3CH		Metro-Com
Chester	224.960	222.380	AA3CH		Metro-Com
Chester	224.960	1240.000	WB3JVX		Metro-Com
Darby	224.100	–	N3FCX	(ca) e	N3FCX
Darby	224.500	–	KM3N	(ca) e	DelCoARA
Newtown Square	223.820	–	AK3M	(ca) e	AK3M
LANCASTER					
Columbia	224.560	–	WB3FQY		Columbia
Cornwall	224.840	–	KA3CNT	l	SusqVy RA
Lititz	224.440	–	KA3CNT	A l	RabbitHill
LUZERNE					
Wilkes Barre	224.420	–	N3DAP	A e x	N3DAP
LYCOMING					
Williamsport	224.480	–	W3AVK		WestBrARA
MONTGOMERY					
Bryn Mawr	224.420	–	WB3JOE	(ca) e l	MARC
Norristown	223.860	–	K3HLN	t	PARA
Pottstown	224.020	–	K3ZMC		PART
Springfield Twp	224.820	–	K3VIT		K3VIT
West Conshocken	224.200	–	N3CVJ		N3CVJ
Wyncote	224.380	–	N3FSC		N3FSC
NORTHAMPTON					
Bangor	224.140	–	KA3ODJ		220ers
PHILADELPHIA					
Phila/North	224.180	–	K3QFP		K3QFP
Phila/Roxborough	224.060	–	WB3EHB		CCRG

Figure 2-3 — Anyone who operates on VHF FM should have a copy of *The ARRL Repeater Directory*. Directory information includes the repeater call sign, location, frequency and sponsor.

Don't call CQ to initiate a conversation on a repeater. It takes longer to complete a CQ than to transmit your call sign. (In some areas, a solitary "CQ" is permissible.) Efficient communication is the goal. You are not on HF, trying to attract the attention of someone who is casually tuning across the band. In the FM mode, stations are either monitoring their favorite frequency or not. Except for scanner operation, there is not much tuning across the repeater bands.

To join a conversation in progress, transmit your call sign during a break between transmissions. The station that transmits next will usually acknowledge you. Don't use the word "break" to join a conversation — unless you want to use the repeater to help in an **emergency**. To make a distress call over a repeater, say "*break break*" and then your call sign to alert all stations to stand by while you deal with the emergency.

A further word about emergencies: Regardless of the band, mode or your class of license, FCC Rules specify that, in case of emergency, the normal rules can be suspended. If you hear an emergency call for help, you should do whatever you can to establish contact with the station needing assistance, and immediately pass the information on to the proper authorities. If you are talking with another station and you hear an emergency call for help, stop your QSO immediately and take the emergency call.

To call another station when the repeater is not in use, just give both calls. For example, "N1II, this is N1BKE." If the repeater is in use, but the conversation sounds like it is about to end, wait before calling another station. If the conversation sounds like it is going to continue for a while, however, transmit only your call sign between their transmissions. After you are acknowledged, ask to make a quick call. Usually, the other stations will stand by. Make your call short. If your friend responds, try to meet on another repeater or on a simplex frequency. Otherwise, ask your friend to stand by until the present conversation ends.

Use plain language on a repeater. If you want to know someone's location, say "Where are you?" If you want to know whether someone you're talking with is using a mobile rig or a hand-held radio, just ask: "What kind of radio are you using?" You get the idea.

Courtesy Counts

If you are in the midst of a conversation and another station transmits his or her call sign between transmissions, the next station in line to transmit should acknowledge the new station and permit the new arrival to make a call or join the conversation. It is impolite not to acknowledge new stations, or to acknowledge them but not let them speak. The calling station may need to use the repeater immediately. He or she may have an emergency to handle, so let him or her make a transmission promptly.

A brief pause before you begin each transmission allows other stations to break in — there could be an emergency. Don't key your microphone as soon as someone else releases theirs. If your exchanges are too quick, you can prevent other stations from getting in.

The *courtesy tones* found on some repeaters prompt users to leave a space between transmissions. The beeper sounds a second or two after each transmission to permit new stations to transmit their call signs in the intervening time. The conversation may continue only after the beeper sounds. If a station is too quick and begins transmitting before the beeper sounds, the repeater may indicate the violation, sometimes by shutting down!

Keep transmissions as short as possible, so more people can use the repeater. Again, long transmissions could prevent someone with an emergency from getting the chance to call for help through the repeater. All repeaters encourage short transmissions by "timing out" (shutting down for a few minutes) when someone gets longwinded. The **time-out timer** also prevents the repeater from transmitting continuously, due to distant signals or interference. Because it has such a wide coverage area, a continuously transmitting repeater could cause unnecessary interference. Continuous operation can also damage the repeater.

You must transmit your call sign at the end of a contact and at least every 10 minutes during the course of any communication. You do not have to transmit the call sign of the station to whom you are transmitting.

Never transmit without identifying. For example, keying your microphone to turn on the repeater without saying your station call sign is illegal. If you do not want to engage in conversation, but simply want to check if you are able to access a particular repeater, simply say "N1KB testing."

Fixed Stations and Prime Time

Repeaters were originally intended to enhance mobile communications. During commuter rush hours, mobile stations still have preference over fixed stations on some repeaters. During mobile prime time, fixed stations should generally yield to mobile stations. When you're operating as a fixed station, don't abandon the repeater completely, though. Monitor the mobiles: your assistance may be needed in an emergency. Use good judgment: Rush hours are not the time to test your radio extensively or to join a net that doesn't deal with the weather, highway conditions or other subjects related to commuting. Third-party communications nets probably should not be conducted on a repeater during prime commuting hours.

Simplex Operation

After you have made a contact on a repeater, move the conversation to a *simplex* frequency if possible. The repeater is not a soapbox. You may like to listen to yourself, but others, who may need to use the repeater, will not appreciate your tying up the repeater unnecessarily. The easiest way to determine if you are able to communicate with the other station on simplex is to listen to the *repeater input frequency*. Since this is the frequency the other station uses to transmit to the repeater, if you can hear his signals there, you should be able to use simplex. If you want to perform an on-the-air test of a pair of hand-held radios, you should select an unoccupied simplex frequency.

Table 2-5

Common VHF/UHF FM Simplex Frequencies

2-Meter Band	***1.25-Meter Band***	***70-cm Band***
146.52*	223.42	446.0*
146.535	223.44	
146.55	223.46	***33-cm Band***
146.565	223.48	906.5*
146.58	223.50*	
146.595	223.52	***23-cm Band***
147.42		1294.5*
147.435		1294.000
147.45		1294.025
147.465		Every 25 kHz to
147.48		1295
147.495		
147.51		
147.525		
147.54		
147.555		
147.57		
147.585		

*National simplex frequency

The function of a repeater is to provide communications between stations that can't otherwise communicate because of terrain, equipment limitations or both. It follows that stations able to communicate without a repeater should not use one. That way, the repeater is available for stations that need it. (Besides, communication on simplex offers a degree of privacy impossible to achieve on a repeater. On simplex you can usually have extensive conversations without interruption.)

Select a frequency designated for FM simplex operation. Otherwise, you may interfere with stations operating in other modes without realizing it. (The reason for this is simple: Changing to a simplex frequency is far easier than changing the frequencies a repeater uses.) Table 2-5 lists the common simplex frequencies for the Novice bands on the $1^{1}/_{4}$ meter band and the 23 cm band. Each band has a designated national FM simplex calling frequency, which is the center for most simplex operation. On the $1^{1}/_{4}$ meter band this is at 223.50 MHz. On the 23 cm band the national simplex calling frequency is 1294.50 MHz, conveniently inside the Novice portion of the band. To see if you and the other station can communicate on a simplex frequency, listen on the repeater input frequency. If you can clearly hear what's going into the repeater, you don't need the repeater to communicate.

Autopatch: Use it Wisely

An **autopatch** allows repeater users to make telephone calls through the repeater. To use most repeater autopatches, you generate the standard telephone company tones to access and dial through the system. The tones are usually generated with a telephone-type tone pad connected to the transceiver. Tone pads are available from equipment manufacturers as standard or optional equipment. They are often mounted on the front of a portable transceiver or on the back of a fixed or mobile transceiver's microphone. Whatever equipment you use, the same autopatch operating procedures apply.

There are strict guidelines for autopatch use. The first question you should ask is "Is the call necessary?" If it is an **emergency**, there is no problem — just do it! Calling for an ambulance or a tow truck is okay. Other reasons may fall into a gray area. As a result, some repeater groups expressly forbid autopatch use, except for emergencies.

Don't use an autopatch where regular telephone service is available. One example of poor operating practice can be heard most evenings in any metropolitan area. Someone will call home to announce departure from the office. Why not make that call from work before leaving?

Never use the autopatch for anything that could be considered business communications. The FCC strictly forbids you to conduct communications in Amateur Radio for your business or for your employer. You may, however, use Amateur Radio to conduct your own personal communications. The rules no longer forbid you to use the autopatch to call your doctor or dentist to make an appointment, or to order food, for example.

Don't use an autopatch just to avoid a toll call. Autopatch operation is a privilege granted by the FCC. Abuses of autopatch privileges may lead to their loss for everyone.

You have a legitimate reason to use the autopatch? Here's how most systems operate. First, you must access (turn on) the autopatch, usually by pressing a designated key on the tone pad. Ask the other hams on a repeater how to learn the access code. Many clubs provide this information only to club members. When you hear a dial tone, you know that you have successfully accessed the autopatch. Now, simply punch in the telephone number you wish to call.

Once a call is established, remember that you are still on the air. Unlike a normal telephone call, only one party at a time may speak. Both you and the other person should use the word "over" to indicate that you are finished talking and expect a reply. Keep the call short. Many repeaters shut off the autopatch after a certain time.

Turning off the autopatch is similar to accessing it. A key or combination of keys must be punched to return the repeater to normal operation. Ask the repeater group sponsoring the autopatch for specific information about access and turn-off codes, as well as timer specifics. Don't forget to identify your station. Most groups expect you to give your call sign, the date and time just before accessing the autopatch and just after turning it off.

[Before you go on to the next section, turn to Chapter 13 and study the questions with numbers N2A03 and N2B08 through N2B15. Review this section if you have difficulty with any of these questions.]

[**If you are studying for the Technician exam**, turn to Chapter 14 and look over questions T2A04 through T2A08 and T2A10 through T2A13, T2A15 through T2A18, T2B01 through T2B04, and T2C03. You should also study questions T8A06, and T8B05. Read this section again if you have any difficulty.]

RADIOTELETYPE AND PACKET COMMUNICATIONS

Radioteletype and packet are terms used to describe amateur communications designed to be received and printed automatically. They are sometimes called **digital communications**, because they often involve direct transfer of information between computers. The FCC refers to these modes as **data emissions**. With the help of accessory equipment, the computer processes the signal and sends it over the air from your transceiver. The station on the other end receives the signal, processes it and prints it out on a computer screen or printer. In this section, you will learn about two exciting forms of communication: **radioteletype (RTTY)** and **packet radio**. Two of the main advantages of these modern data-transmission methods are that they provide high-speed data transmission and good communications reliability.

Emission Modes

RTTY and packet communications are similar to CW (Morse code). They both use two states to convey information. Instead of switching a carrier on and off as in CW operation, the carrier is left on continuously and switched between two frequencies. The transmitter carrier frequency "shifts" between the two frequencies, called MARK and SPACE. MARK is the ON state; SPACE is the OFF state. This is called **frequency-shift keying (FSK)**. On 10-meter RTTY, the MARK and SPACE frequencies are normally 170 Hz apart. This is known as 170-Hz shift RTTY.

On VHF, **audio-frequency shift keying (AFSK)** is used. AFSK is similar to FSK, except that an FM transmitter is used. Audio tones corresponding to MARK and SPACE are fed into the microphone jack and used to modulate the carrier. The most common VHF shift is 170 Hz, but you may hear 850-Hz shift as well.

It is interesting to note that you can also operate Morse code with an FM transceiver. You can feed an audio tone into the microphone jack, and then key the audio tone on and off to send Morse code. This type of signal is a *modulated CW* or *MCW* signal.

Sending Speeds

Just as you can send CW at a variety of speeds, RTTY and packet transmissions are sent at a variety of speeds. The signaling speed depends on the type of transmission and on the frequency band in use. A *baud* is the unit used to describe transmission speeds for digital signals.

For a single-channel transmission, a baud is equivalent to one digital bit of information transmitted per second. A 300-baud signaling rate represents a transmission rate of 300 bits of digital information per second in a single-channel transmission. A digital bit of information has two possible conditions, ON or OFF. We often represent a bit with a 1 for ON or a 0 for OFF. The MARK and SPACE tones of a radioteletype signal also represent the two possible bit conditions.

On HF, we use signaling rates of up to 300 bauds. Sending speeds on VHF are generally faster than on HF. A common VHF sending speed is 1200 bauds. The most important thing to remember about sending speeds is that both stations must use the same speed during a contact.

RTTY Communications

Radioteletype (RTTY) is a popular form of communications. Its technical name is **narrow-band direct-printing telegraphy**. Novices can operate RTTY on 10 meters; Novices and Technicians can operate RTTY on the 222-MHz and 1270-MHz bands. (Technicians can also operate RTTY on the other VHF and UHF amateur bands.) On 10 meters, Novice RTTY operation is allowed from 28.1 to 28.3 MHz. On 222 and 1270 MHz, you are allowed to operate RTTY on all of the frequencies that the FCC authorizes you to operate on. You should, however, follow the band plans, and operate RTTY only on those portions of the bands designated for that type of operation. The 222-MHz band plan calls for RTTY and other digital communications modes on 223.52 to 223.64 MHz when using simplex. On VHF, most RTTY activity is on repeaters. Check with local amateurs to find out where the 222-MHz RTTY activity is in your area.

There are three popular types of RTTY: *Baudot, AMTOR* and *ASCII*. Baudot and AMTOR are the most popular RTTY modes on HF. Until recently, high-speed ASCII was most often heard on VHF, but this mode has been almost completely replaced by packet radio. We will describe Baudot RTTY operation specifically here, but most of the operating techniques apply to AMTOR and ASCII as well.

Baudot Radioteletype

Radioteletype communication using the **Baudot code** (also known as the International Telegraph Alphabet

RTTY Equipment

Years ago, all RTTY operation was done with mechanical teleprinters. You may still see some of these machines in old movies — they are big, slow and noisy. Today, amateurs use RTTY *terminal units (TUs)* or *multimode communications processors (MCPs)* for RTTY operation. The TU connects between a transceiver and a computer and decodes the RTTY signals for display by the computer. MCPs let you operate nearly any mode, so you won't have to buy new equipment if you decide to give a new mode a try. Newer versions use digital signal processing for added flexibility: if a new digital mode comes along, you won't need to replace the MCP — just the software!

When you shop for a communications terminal, there are a few things to keep in mind. Some are designed to work with only one type of computer. Some require additional software in the computer, and others require only a simple terminal program. As always, the best way to find out about a particular unit is to ask someone who has one. The next best way is to read the Product Review column in *QST*.

RTTY equipment and MCPs are constantly being updated. To learn about the latest equipment, check out the ads in *QST* and other amateur magazines. Some older used units may be available at hamfests and flea markets. If you plan to buy a used communications terminal, take an experienced RTTY operator along to the hamfest with you. Remember, most hams love to talk about their favorite operating mode — ask around at your local club.

The trend in RTTY equipment today is toward the "all-mode" communications terminal. The *MFJ-1278B* (above) provides CW, Baudot RTTY, ASCII, AMTOR, PACTOR, fax, SSTV and packet-radio capability. The Kantronics *KAM Plus* (below) operates on AMTOR, ASCII, fax, packet, PACTOR, G-TOR and CW as well as RTTY. PACTOR and G-TOR combine the best features of AMTOR and packet, and are commonly used on HF. They are used with any of a number of popular personal computers.

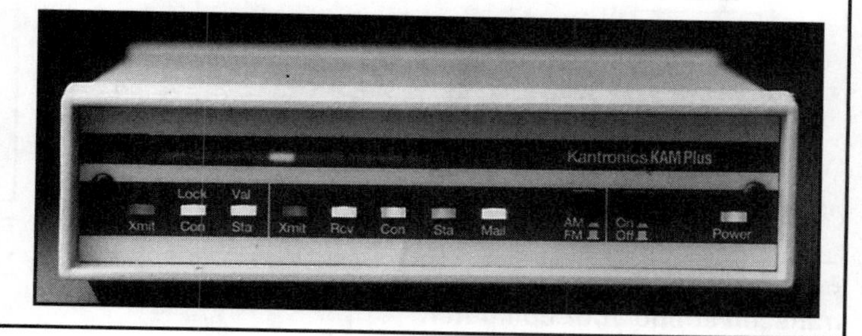

number 2, or ITA2) is used on the Amateur Radio HF bands in most areas of the world.

The Baudot code represents each character with a string of five bits of digital information. Each character has a different combination of bits. There are only 32 possible Baudot code combinations. This limits the number of possible characters. All text is in upper-case characters. To provide numbers and punctuation, you shift between the letters case (LTRS) and the figures case (FIGS).

There are three common speeds for 10-meter Baudot RTTY communications: 60 wpm (45 bauds), 75 wpm (56 bauds) and 100 wpm (75 bauds). You'll often hear 60, 75 and 100 wpm RTTY referred to as "60 speed," "75 speed," and "100 speed," respectively. Both stations must use the same sending speed to make a RTTY contact.

Setting Up

To set up a RTTY station using Baudot, the first thing you will need is a *multimode communications processor (MCP)*, a RTTY *terminal*, software to use the digital-signal processing capabilities of your computer sound card or other computer interface, or a mechanical *teleprinter*. (Noisy and heavy, mechanical teleprinters are rarely used nowadays.) See Figure 2-4.

If you use an MCP, you'll also need a personal computer and the appropriate software (computer program). As its name implies, the MCP allows operation on several modes, often including AMTOR, CW and packet. Dedicated RTTY terminals (also called RTTY modems or terminal units) operate only RTTY. No personal computer is required. In choosing a RTTY system, look for a modem with shift capabilities of 170 and 850 Hz. Check the *QST* Product Review column, and articles and ads in *QST* and other amateur magazines for information about these products.

Modem is a contraction of *mo*dulator-*dem*odulator. A modem takes electrical signals from the computer or terminal and turns them into audio tones to *modulate* the RF signal. It also takes audio tones from the receiver and *demodulates* them into electrical signals for the computer. It is an interface between the computer or terminal and the radio.

What about a rig for RTTY? You can use almost any HF transceiver. Ideally, the receiver used for 170-Hz-shift RTTY should have the minimum practical bandwidth, preferably between 270 and 340 Hz. Many CW filters have bandwidths around 500 Hz and are good for 170-Hz RTTY reception. Some transceivers can use only the SSB filter when in the SSB or RTTY mode. Some RTTY operators have modified their transceivers to use the CW filter when the RTTY mode is selected. A switch can be added so it is possible to select the CW filter with the mode switch set for SSB operation. A transceiver with a frequency display of 10-Hz resolution is also helpful, although not necessary.

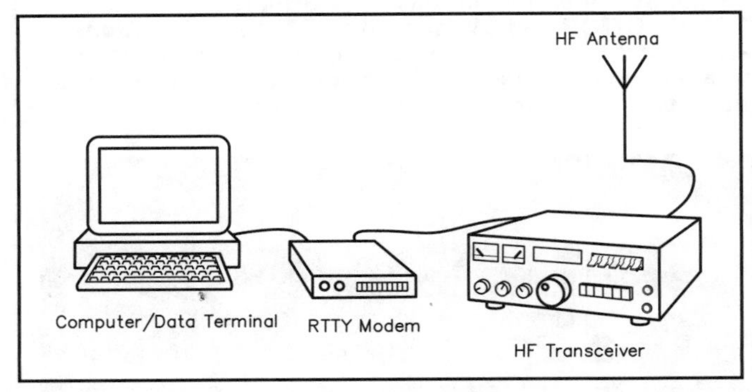

Figure 2-4 — A RTTY modem connects between your transceiver and your computer.

Many modern transceivers include an FSK mode. If yours does not, you can operate RTTY with the transceiver in the SSB mode. It is normal to use the lower-sideband mode for RTTY on HF SSB radio equipment. Most transceivers have two positions on the sideband-selection knob, labeled NORMAL and REVERSE. This labeling may cause some uncertainty when you use the rig for RTTY. On 10 meters, normal SSB operation uses upper sideband, so you will have to set the switch to the REVERSE position to select lower sideband for 10-meter RTTY.

If you are not using the correct sideband, your signal is "upside down," and other operators will have to change their normal operating setup to copy your signals. Be sure you select the lower sideband and transmit signals that are "right-side up." Consult your radio's manual for details.

On VHF, the most common practice is to use **AFSK**. Since most VHF RTTY work is done on local FM repeaters, almost any FM transceiver will work.

Your transmitter must be able to operate at full power for extended periods (called *100% duty cycle*) when transmitting radioteletype. Some transceivers, designed for SSB and CW operation (which do not require constant transmission at full power), may overheat and possibly fail if subjected to a long RTTY transmission. Many operators reduce to half power during long RTTY transmissions to avoid overheating problems.

Receiver-tuning accuracy is important. Thus, a tuning aid can be a great help in proper receiver adjustment. Most modems have some type of tuning indicator, possibly just flashing LEDs. Some have oscilloscopes that produce patterns such as those shown in Figure 2-5. The MARK signal is displayed as a horizontal line and the SPACE signal as a vertical line. Theoretically this should appear as a "+" sign on the 'scope screen, but in practice it may look more like a pair of crossed bananas! To avoid the high cost

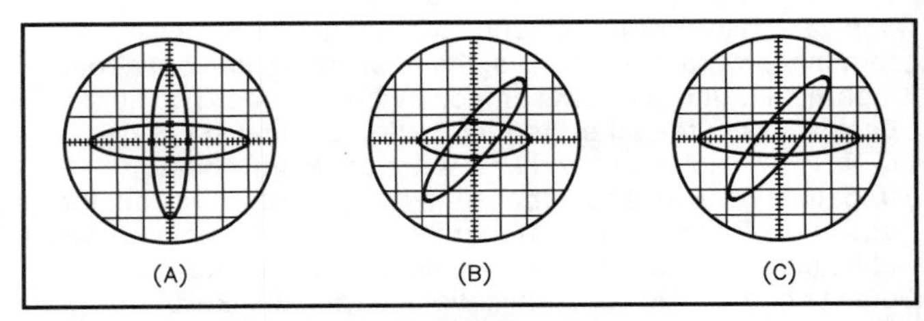

Figure 2-5 — Oscilloscope RTTY tuning patterns. A shows the pattern produced by a properly tuned signal. The displays at B and C indicate improperly tuned signals.

of an oscilloscope, some manufacturers offer LED displays that imitate the 'scope display.

Typing the Message

No one can disagree that touch typing (typing without looking at the keys) is the best way to go. But, you can have enjoyable contacts even if you can only type with two fingers! It is a good practice to keep the number of characters per line to a maximum of 69. That line length can be handled by virtually any teleprinter designed for Baudot. Many US teleprinters will print 72 or 74 characters, but some foreign printers take only 69.

After a mistake, the "oops" signal is XXXXX (some operators may use EEEEEE). There is no need to overdo this, though, because generally the other operator will figure out what you meant to say anyway, and a page filled with XXXXX can be confusing.

Station Identification

When you are transmitting RTTY, simply send your call sign at least once every 10 minutes, just as you would on any other mode. FCC Rules permit a station engaged in digital communications using Baudot, AMTOR or ASCII to identify in the digital code used for the communication. On the Novice 222 and 1270-MHz bands, identification may be given using one of the methods just mentioned or by voice. (Radiotelephone operation is permitted on the same frequency being used for the RTTY communication on these bands.) Identifying the station you are communicating with, in addition to your own station, is not legally required, but it will help those monitoring your QSO to determine band conditions.

Test Messages

A message consisting of repetitive RYs is useful for local and on-the-air testing of Baudot RTTY equipment:

RYRYRYRYRYRYRYRYRYRYRYRYRYRY DE W1AW

The use of the RY sequence goes back to the early Teletype-machine days. To set the mechanical machines, it was necessary to send a string of Rs and Ys while the receiving operator adjusted a knob called the "rangefinder." The military and commercial services also used RYRYRYRYRY as a channel-holding signal during idle periods. The letters R and Y both alternate between MARK and SPACE tones in the Baudot code, so it was natural to use these letters for the range-setting operation. (If we represent the SPACE tone with a 0 and the MARK tone with a 1, then R is 01010 and Y is 10101. So you can see that a string of RY characters sends alternating SPACE and MARK tones.)

Use the RY sequence sparingly. RYs have a musical sound that is easily recognizable, and is useful for tuning a signal in properly. This makes it good for rousing attention when you are sending a CQ call, or when calling another station on schedule, but it is unnecessary during routine contacts. Don't use RY when you are working DX.

Making Contact

There are two ways of establishing contact with other amateur stations using radioteletype. You can answer someone else's call, or you can try calling CQ yourself. When you call CQ, you depend on someone else to tune your signal properly, and if you are a newcomer to the mode, that may be an advantage. By listening for other calls, you have a good chance of finding stations that you might not contact otherwise. You will want to use both methods at various times, but whichever one you choose, learn the proper procedure and follow it.

Calling CQ

Since it is a little easier to explain, we will start by describing the procedure for calling CQ on RTTY. You must first locate a clear frequency on which to make your call. Courtesy should be the rule. Ask if the frequency is busy by sending QRL?, or by asking in clear text. There may be a QSO on frequency, but you may not hear both sides of it.

The general CQ call, like RY, has a recognizable, musical sound. Take advantage of this fact, and send CQ in a pattern something like this:

CQ CQ CQ CQ CQ DE WR1B WR1B WR1B K

Transmit "CQ" three to six times, followed by "DE," followed by your call sign three times. Many operators add their name and QTH (location) to the last line of the CQ call. This is a good idea, but keep it simple and short. As on voice, several short calls with pauses to listen are much better than one long call.

Calling Another Station

You have just heard 9H1EL in Malta calling CQ. He is booming in with an S9 signal, and this would be a new country for you. Make sure that he is tuned in properly, wait until he stops calling CQ and give him a call. Keep your calls short. Forget the RYs. Send 9H1EL's call sign only once, and your call sign no more than five times:

9H1EL DE WB8IMY WB8IMY WB8IMY K

Then stand by and listen. If he doesn't answer you or anyone else, try another short call.

Line Feeds and Other Things

When another station answers your call, gives you a report and then turns it back to you, what should you do? Always send a line feed (L/F) and a carriage return (C/R) first. This will put his computer and/or printer into the "letters" mode, and will eliminate the possibility of part of your message being garbled by numbers and/or punctuation marks.

Next, send his call and your call once each. On your first exchange, send your name and QTH message from the computer memory, if your system has such a message. If you have a type-ahead buffer on your system, you can compose some of your reply before it is actually transmitted. This saves time, especially if you are not a fast typist.

If you give another station a signal report of RST 599, it means you're copying solid. If you receive a similar report, it is unnecessary to repeat everything twice. Give honest reports — if the copy is solid, then the readability report is 5. Adapt your operating technique to the reported conditions. If copy is marginal, repeats or extra spaces are in order. But with solid copy, you can zip right along.

Don't send a string of carriage returns to clear the screen. The other operator may be copying on a printer, and the carriage returns will waste a lot of paper. The last two characters of each transmission should be carriage returns, however.

Signing Off

You have now completed your QSO and are about to sign off. What prosign do you use? Well, the standard prosigns mean the same thing on RTTY as they do on CW, and should be adequate. Table 2-6 summarizes the meaning of the most commonly used prosigns. For more on operating RTTY, see the ARRL's *Your HF Digital Companion*.

[Time to turn to Chapter 13 again. You should be able to answer questions N2B01, N2B02 and N2B03. Also study questions N8A07, N8A08, N8A10 and N8A11. Review this section if you have any difficulty.]

[**If you are studying for a Technician license**, you should also turn to Chapter 14 and study questions T8A05 and T8A11. Review this section as needed.]

Table 2-6
Common Prosigns Used On RTTY

K—Invitation to transmit.

KN—Invitation to the addressed station only to transmit.

SK—Signing off. End of contact.

CL or CLEAR—I am shutting my station down.

SK QRZ—Signing off and listening on this frequency for any other calls. The idea is to indicate which station is remaining on the frequency.

Some operators use various other combinations of prosigns:

SK KN—Signing off, but listening for one last transmission from the other station.

SK SZ—Signing off and listening on this frequency for any other calls.

PACKET RADIO

Packet radio is communications for the computer age. More and more hams are adding computers to their ham shacks.

Early ham-oriented computer programs allowed computers to send and receive CW and RTTY. Some farsighted hams, however, developed a new mode of communication that unleashes the power of the computer. That mode is **packet radio**.

Being a child of the computer age, packet radio has the computer-age features that you would expect.

- It is ***data*** communication — high speed and error-free packet-radio communication lends itself to the transfer of large amounts of data.
- It is *fast*, much faster than the highest speed CW or RTTY.
- It is *error-free* — no "hits" or "misses" caused by propagation variations or electrical interference.
- It is *spectrum efficient* — several stations can share one frequency at the same time.
- It is a ***network*** — packet stations can be linked together to send data over long distances.
- It is *message storage* — packet-radio bulletin boards (PBBS) provide storage of messages for later retrieval.

How Does Packet Work?

Packet radio uses a *terminal node controller (TNC)* as the interface between computer and transceiver. A TNC is an enhanced modem. We discussed what a modem is and what it does in the section on RTTY. The TNC accepts information from your computer or ASCII terminal and breaks the data into small pieces called *packets*. In addition to the information from your computer, each packet contains addressing, error-checking and control information. The addressing information includes the call sign of the station that sent the packet, and the call sign of the station the packet is being sent to. The address may also include call signs of stations that are being used to relay the packet. The error-checking information allows the receiving station to determine whether the received packet contains any errors. If the received packet contains errors, the receiving station waits until the transmitting station sends it again.

Breaking up the data into small parts allows several users to share the channel. Packets from one user are transmitted in the spaces between packets from other users. The address section allows each user's TNC to separate packets intended for him from packets intended for other users. The addresses also allow packets to be relayed through several stations before they reach their ultimate destination. Having information in the packet that tells the receiving station if the packet has been received correctly assures perfect copy.

What Do I Need to Get on Packet Radio?

All you need to set up a VHF packet-radio station is a VHF or UHF FM transceiver (with an antenna), a computer or ASCII terminal and a TNC or multimode controller, which allows operation on several modes. (We'll use "TNC" to refer to both.)

The TNC connects to the transceiver microphone input. It also connects to a computer or terminal. See Figure 2-6. For operation on 10 meters you'll need a 10-meter SSB transceiver in addition to the TNC and computer. The sidebar entitled "Packet Radio Equipment" lists some available equipment.

Your TNC manual should contain detailed instructions for wiring the TNC, radio and computer together. So many hams are on packet now that someone in the local radio club will probably be able to help you if you have problems, or ask around on the local voice repeater.

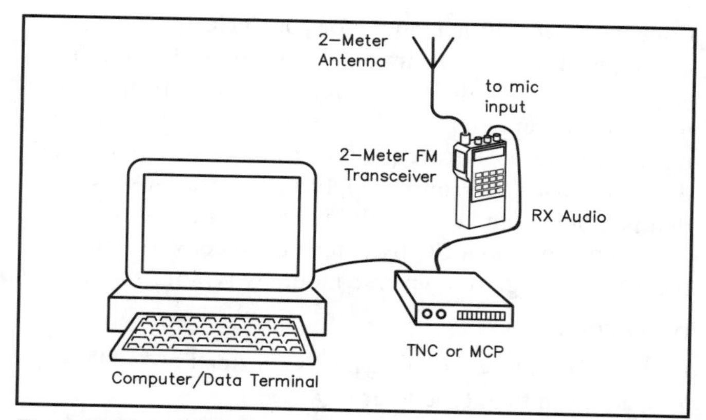

Figure 2-6 — A terminal node controller (TNC) connects between your transceiver and your computer, like a RTTY modem.

What Does It Look Like?

So far we've talked about the equipment you need and how packet works. But what does a packet contact look like? In the following examples, we'll look at the procedures used by most of the TNCs on the market today. Some TNCs may use different command formats — consult your operating manual if you're not sure.

First, you must tell the TNC your call sign. Most TNCs allow you to change your call sign at any time, so several family members could use one TNC with each of their call signs. Most TNCs also have a way to remember the call sign when the power is switched off. Before you can enter commands into the TNC, it must be in *command mode*. When the TNC is in command mode, you will see a prompt:

cmd:

This indicates that the TNC is waiting for input. To tell the TNC his call sign, K1RO types:

MYCALL K1RO <CR>

<CR> means "carriage return." On some computers this key may be labeled "ENTER" or have an arrow (←).

As in all other modes of Amateur Radio, packet allows you to "read the mail" or monitor channel activity. This is called the **monitor mode**, and looks like this:

WA6JPR>WB6YMH: HELLO SKIP, WHEN IS THE

NEXT OSCAR 30 PASS? K

WB6YMH>WA6JPR: HANG ON WALLY

I'LL TAKE A LOOK.

To enable monitor mode, simply type: MON ON at the cmd: prompt. You may also need to type MFROM ALL.

W1AW-4>K1CE: You have new mail, please kill after reading:
W1AW-4>K1CE: Msg# TR Size To From @BBS Date Title
W1AW-4>K1CE: 5807 N 420 K1CE K1MON W1AW 970308 HI AGAIN
W1AW-4>K1CE: K1CE de W1AW: at 2107z on 970308, 142 active msgs, last msg #5839
W1AW-4>K1CE: (A, B, D, H, I, J, K, L, R, S, T, U, W, X) >
W1AW-4>K1CE: Enter title for message:
W1AW-4>K1CE: Enter message, ^z (CTL-Z) to end, it will be message 5840
WR1B>N1OKK: Hi Mike, haven't heard from you in a long time. KK
W1AW-4>K1CE: K1CE de W1AW: at 2101z on 970308, 143 active msgs, last msg #5840
W1AW-4>K1CE: (A, B, D, H, I, J, K, L, R, S, T, U, W, X) >
W1AW-4>KE3Z: KE3Z de W1AW: at 2123z on 970308, 143 active msgs, last msg #5840
N1OKK>WR1B: No, Larry, I have been busy with other things. New business takes lots of time.
W1AW-4>KE3Z: (A, B, D, H, I, J, K, L, R, S, T, U, W, X) >
W1AW-4>KE3Z: Enter title for message:
N1OKK>WR1B: I'm having lots of fun, though. Hope to see you at the club meeting next week.
W1AW-4>KE3Z: Enter message, ^z (CTL-Z) to end, it will be message 5841

Figure 2-7 — In monitor mode, your TNC displays all the packet activity that your station hears on the frequency, whether or not the packets are addressed to your station. Notice that you may only copy one side of some exchanges.

Consult the operating manual for your TNC to be sure. The call signs of the stations involved appear as "FROM>TO:" and the contents of the packet follow. When it is in monitor mode, your computer will display everything that is transmitted on the packet frequency, whether or not it is addressed to you (see Figure 2-7). In addition, a packet station that is monitoring is not replying to any messages.

You can send a CQ by entering the converse mode of the TNC. You go to converse mode by typing:

CONV <CR>

(some TNCs allow you to type "K" instead of "CONV")

You can then type your CQ:

MIKE IN SAN DIEGO LOOKING FOR ANYONE IN SIMI

VALLEY

Your TNC adds your call sign as the FROM address and CQ as the TO address. The receiving station's TNC adds these addresses to the front of the text when it is displayed.

Making a Connection

You answer a CQ or establish a contact by using the CONNECT command. When two packet stations are connected, each station sends data packets specifically addressed to the other station. When a station receives an error-free packet, it transmits a reply (acknowledgment) packet to let the sender know the packet has been received correctly. The two stations take turns sending data or messages typed on the keyboard.

To connect to another station, you type:

Connect WA1LOU <CR>

where WA1LOU is the call sign of the station you wish to contact. (Most TNCs let you use "C" as an abbreviation for "connect.")

If WA1LOU's packet-radio station is on the air and receives your connect request, your station and his will exchange packets to set up a connection. When the connection is completed, your terminal displays:

***** CONNECTED to WA1LOU**

and your TNC automatically switches to the converse mode.

Now, everything you type into the terminal keyboard is sent to the other station. A packet is sent whenever you enter a carriage return. It's a good idea to use K, BK, O or > at the end of a thought to say "Okay, I'm done. It's your turn to transmit."

When you are finished conversing with the other station, return to the command mode by typing <CTRL-C> (hold down the CONTROL key and press the C key). When the command prompt (cmd:) is displayed, type:

Disconne <CR>

and your station will exchange packets with the other station to break the connection. (Most TNCs let you use "D" as an abbreviation.) When the connection is broken, your terminal displays:

***** DISCONNECTED**

If, for some reason, the other station does not respond to your initial connect request, your TNC will send the request again until the number of attempts equals the internal *retry* counter. When the number of attempts exceeds the retry counter, your TNC will stop sending connect requests and your terminal displays:

***** retry count exceeded**

***** DISCONNECTED**

A TNC can reject a connect request if it is busy or if the operator has set CONOK (short for CONnect OK) off. If this happens when you try to connect, your TNC displays:

***** WA1LOU busy**

***** DISCONNECTED**

Packet Radio Repeating

Sometimes terrain or propagation prevent your signal from being received by the other station. Packet radio gets around this problem by using other packet-radio stations to relay your signal to the intended station. All you need to know is which on-the-air packet-radio stations can relay signals between your station and the station you want to contact. Once you know of a station that can relay your signals, type:

Connect WA1LOU Via W1AW-5 <CR>

where WA1LOU is the call sign of the station you want to connect to and W1AW-5 is the call sign of the station that will relay your packets. The "-5" following W1AW is a secondary station identifier (SSID). The SSID permits up to 16 packet stations to operate with one call sign. For example, W1AW-5 is a 2-meter packet repeater and W1AW-6 is a 222-MHz packet repeater.

When W1AW-5 receives your connect request, it stores your request in memory until the frequency is quiet. It then retransmits your request to WA1LOU on the same frequency. This action is called digipeating, a contraction of "digital repeating." If WA1LOU's packet-radio station is on the air and receives the relayed connect request, your station and his will exchange packets through W1AW-5 to set up a connection. Once the connection is established, your terminal will display:

***** CONNECTED to WA1LOU VIA W1AW-5**

W1AW-5 will continue to relay your packets until the connection is broken (see Figure 2-8).

Digital and voice repeaters both repeat, but the similarity ends there. Notice that digital repeaters differ from typical voice repeaters in a number of ways. A digital repeater (**digipeater**) usually receives and transmits on the same frequency (whereas a voice repeater receives and transmits on different frequencies). A digipeater does not receive and transmit at the same time (as compared to a voice repeater, which immediately transmits whatever it receives). Rather,

cmd:c WU1I

*** CONNECTED to SNAKE

C WA1WYN

*** CONNECTED to WA1WYN If I don't answer leave a message on BBSWYN, 145.01

Hi Darryl. Are you in the shack? KK

Hi Larry. Yes, I'm here. What's up? KK

What time should I show up for class tomorrow? KK

Well, your presentation is near the end of class, so around 8 should be fine. KK

Okay, I'll be there by 8. How are the students doing? KK

The class seems to be going very well. The students are very interested, and have a lot of good questions. KK

Great. I'll see you tomorrow night. 73 SK

See you then.

cmd: D

***DISCONNECTED

Figure 2-8 — Two packet stations *connect* to have a QSO. Everything you type is sent to the other station. Packets not addressed to your station are ignored by your TNC. Text shown in bold type here is typed by the station operator and text not in bold is received by his station.

a digipeater receives a packet, stores it temporarily until the frequency is clear, and then retransmits the packet. Also, a digipeater repeats only packets that are specifically marked to be repeated by that station (the address in the packet contains the call sign of the digipeater). A voice repeater repeats everything it receives on its input frequency.

If one digipeater is insufficient to establish a connection, you can specify as many as eight stations in your connect request. Additional digipeaters are added to the connect command separated by commas. For example, typing:

Connect WA1LOU Via W1AW-5, WA2FTC-1 <CR>

after the command prompt (cmd:), causes your TNC to send the WA1LOU connect request to W1AW-5 which relays it to WA2FTC-1. Then, WA2FTC-1 relays it to WA1LOU.

Don't use more than one or two digipeaters at any one time, especially during the prime time operating hours (evenings and weekends). Each time you use a digipeater, you are competing with other stations attempting to use the same digipeater. Each station that you compete with has the potential of generating a packet that may collide with your packet (which causes your TNC to resend the packet). The more digipeaters you use, the more stations you compete with, greatly increasing the chance of a packet collision. As a result, it may be difficult to get one packet through multiple digipeaters, and your TNC will quickly reach its retry limit and disconnect the link.

Any packet-radio station can act as a digipeater. Most TNCs are set up to digipeat automatically without any intervention by the operator of the station being used as a digipeater. You do not need his permission, only his cooperation, because he can disable his station's digipeater function. (In the spirit of Amateur Radio, most packet-radio operators leave the digipeater function on, disabling it only under special circumstances.)

Similar to VHF/UHF voice repeaters, some stations are set up as dedicated digipeaters. They are usually set up in good radio locations by packet-radio clubs. Besides location, the other advantage of a dedicated digipeater is that it is always there (barring a calamity). Stations do not have to depend on the whims of other packet-radio operators, who may or may not be on the air when their stations' digipeater functions are most needed.

Although you are not allowed to be the control operator of a voice repeater until you upgrade from the Novice class, you may leave your TNC's digipeater function enabled. The FCC recognizes a distinction between digipeaters and voice repeaters in this case, and everyone realizes that an effective packet-radio system depends on having Novice digipeaters available.

The most common form of digipeater is the *node*. To reach a distant station, first connect to the node. Then, instruct the node to connect you to the distant station. The node acknowledges packets sent from either station, then relays them to the other station. This has a number of advantages over a simple digipeater.

Amateurs have set up nodes to connect with other nodes, so packet messages (data) can be sent over longer distances. Such a system of interconnected nodes forms a **network** of packet stations.

Amateurs have also established another type of network that operates in a manner very similar to the Internet. This system is known as the *TCP/IP network*. TCP/IP is short for *transmission control protocol/Internet protocol*, which describes the set of rules or commands that control the network operation. Here your station is assigned a net-

Packet Radio Equipment

The photos show several popular terminal node controllers (TNCs). The most popular units are really multimode communications processors (MCPs) that can be used for packet radio and a number of RTTY modes such as Baudot RTTY, AMTOR, ASCII and even CW.

The ads in *QST* and other ham magazines as well as *QST* Product Reviews are your best source of information about the latest equipment. Local hams will be happy to talk with you about what they are using, and even give you a demonstration.

Most of the packet controllers on the market today can trace their roots to two terminal node controller designs by the Tucson Amateur Packet Radio Corporation (TAPR); the TAPR TNC 1 and TNC 2. Equipment based on the TNC 2 is still available.

The MFJ TNC-2 Packet Controller is an inexpensive TNC 2 clone for packet operation. It is also compatible with global positioning system (GPS) receivers for automatic packet reporting system (APRS) operation.

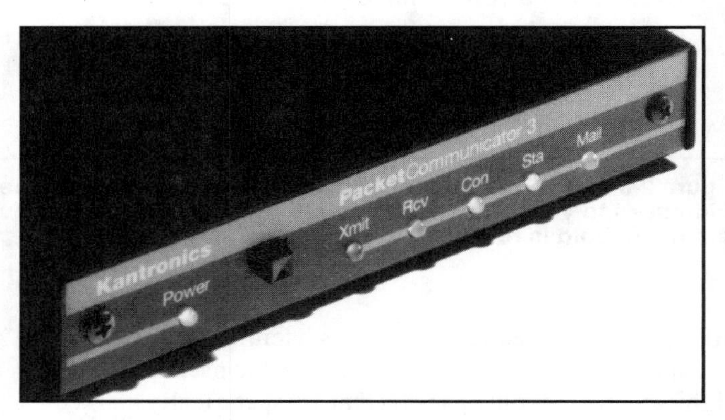

The Kantronics KPC-3 features a 1200-baud modem.

In addition to the usual digital modes, this MCP operates on PACTOR, a mode that combines the best features of packet and AMTOR.

The MFJ Data Radio is especially designed for packet operation. Available in 2-meter and 1.25-meter models, it will operate at up to 9600 baud. Just connect your TNC and an antenna, and you are on packet!

Radios for Packet Operation

VHF

Just about any VHF FM transceiver can be used on VHF packet. Even a small hand-held transceiver can be used, although an outside antenna is a good idea. If there is very little packet activity in your area, you may need a bit more power to reach the nearest dedicated digipeater. Many manufacturers sell amplifiers that increase the power from a hand-held transceiver to 10 or 25 watts. Most mobile FM transceivers are rated at 10 to 50 watts, and these radios make great packet rigs.

Older crystal-controlled transceivers may be used on packet, but a few words of caution are in order. Many older rigs cannot switch from receive to transmit quickly enough for packet operation. Most TNCs have an adjustable "transmit delay" that causes the TNC to key the radio and then wait some fraction of a second before actually sending the packet. Older radios may require this transmit delay. In addition, an older rig may have passed through the hands of several amateurs, and some of the internal settings may have been "adjusted." While this may not be a problem when you use the rig for voice communications, the rig may not work on packet. If you plan to use an older rig on packet, try to locate an experienced packet operator who can help you set up your packet station.

HF

Most SSB transceivers will work on HF packet. Again, your transceiver may require a bit of transmit delay from the TNC. Your TNC manual should cover wiring and operation guidelines in detail. On HF, packet operation uses lower sideband. PACTOR, G-TOR and some of the other "new" modes that combine the best properties of packet and AMTOR are the most popular form of digital communications on the HF bands.

Antennas

Since many packet stations are put on the air by amateurs who have already been active on VHF FM, the same antennas are often used for both modes. As a result, vertically polarized antennas are the standard for VHF packet. Yagi antennas are usually not required, unless there is very little packet activity in your area or you have a particularly bad location.

work address, and the stations automatically know how to route your messages and data to other stations on the network. If you want to participate in the amateur TCP/IP network, you should be sure your TNC can operate in the KISS mode (short for *keep it simple, stupid*).

VHF/UHF Packet Operating

Today, most amateur packet-radio activity occurs at VHF, on 2 meters, but activity on 222 MHz continues to grow. The most commonly used data rate on VHF is 1200 bauds, with frequency-modulated AFSK tones of 1200 and 2200 Hz. This is referred to as the "Bell 202" telephone modem standard.

Getting on the air is usually a simple matter of turning on your radio and tuning to your favorite packet-radio frequency. On 2 meters, common packet channels are 145.01, 145.03, 145.05, 145.07 and 145.09 MHz. On 222 MHz, Novice packet activity centers on 223.4 MHz. If there is a voice repeater on that frequency in your area, ask around at a club meeting or on the repeater. Someone is bound to know where the packet activity is.

If you are conducting a direct connect (a contact without using a digipeater), move your contact to an unused simplex frequency. It is very inefficient to use a frequency where other stations, especially digipeater stations, are operating. The competition slows down your packets and, in return, you are also slowing down the other stations. You should use a frequency occupied by a digital repeater only if you are using that digital repeater. Check with other hams in your area for more information on local packet operating frequencies. Although you are operating your packet station on simplex, it is best not to use the designated FM simplex frequencies because packet and phone are not compatible modes.

What If I Still Have Questions?

The material presented here is only a very basic sketch of packet-radio operation. More material is presented in the following ARRL books: *Your Packet Companion*, *Practical Packet Radio*, *Packet: Speed, More Speed and Applications* and *The ARRL Operating Manual*. *QST* and the other ham magazines often have articles about packet radio. Finally, there are sure to be a few "packeteers" in your local club. For a list of clubs in your area, write to the Educational Activities Department at ARRL HQ, 225 Main Street, Newington, CT 06111.

[Now turn to Chapter 13 and study questions N2B04 through N2B07and N8A09. Review this section if you have any difficulty with these questions.]

[**If you are studying for the Technician exam**, turn to Chapter 14 and review questions T2B10, T2B11, T8A04 and T8B04. Review this section if you have any difficulty.]

KEY WORDS

Emergency traffic – Messages with life and death urgency or requests for medical help and supplies that leave an area shortly after an emergency.

Frequency coordinator – An individual or group that recommends repeater frequencies to reduce or eliminate interference between repeaters operating on or near the same frequency in the same geographical area.

Health and Welfare traffic – Messages about the well being of individuals in a disaster area. Such messages must wait for **Emergency** and **Priority traffic** to clear, and results is advisories to those outside the disaster area awaiting news from family and friends.

Priority traffic – Emergency-related messages, but not as important as **Emergency traffic**.

Radio Amateur Civil Emergency Service (RACES) – A part of the Amateur Service that provides radio communications for civil preparedness organizations during local, regional or national civil emergencies.

Tactical call signs – Names used to identify a location or function during local emergency communications.

2 Additional Operating Procedures for Technicians

REPEATERS

You learned the basics of repeater operation in the first part of this chapter. There are a few additional concepts that are important for Technician licensees.

You will find repeater stations on each of the VHF and UHF bands. Table 2-7 summarizes the standard frequency offset between the repeater input and output frequencies on each band. Since the offset is different for each band, it will be helpful for you to know the standard offsets. For example, on the 2-meter (144 to 148 MHz) band, most repeaters use an input/output frequency separation of 600 kHz. On the 1.25-meter (222 to 225 MHz) band, the standard offset is 1.6 MHz and on the 70-cm (420 to 450 MHz) band it is 5.0 MHz. You can see that when there is more space available on the band, a wider offset is chosen. By providing more space between the input and output frequencies there is less chance for interference or interaction between the two.

Table 2-7
Repeater Input/Output Offsets

Band	*Offset*
6 meters	1 MHz
2 meters	600 kHz
1.25 meters	1.6 MHz
70 cm	5 MHz
33 cm	12 MHz
23 cm	20 MHz

You should always consider other repeater users when selecting an operating frequency. You can easily select another frequency or move your conversation to a simplex frequency so you don't cause interference to the repeater. It is not practical to change the repeater operating frequencies, however. Keep this in mind even for testing your radios. For example, if you want to test a pair of hand-held transceivers on your work bench you should select an unoccupied simplex frequency rather than tying up the local repeater for your testing.

[Turn to Chapter 14 and study questions T2A01 through T2A03 now. Also study question T2B04. Review this section if you have difficulty with any of these questions.]

Various Modes

With a Technician license you can sample virtually every type of Amateur Radio operating activity. Even without passing a Morse code exam you can use CW on the VHF and UHF bands. You can also use single sideband (SSB) voice on those bands. There are segments of each band dedicated to CW operation as well as other "weak signal" operation such as SSB. You can sample amateur television (ATV), Earth-Moon-Earth (EME or moonbounce) communication, amateur satellites and packet radio. The only type of operating that you can't experience is on the high-frequency (HF) bands. There is certainly no reason to feel confined to FM voice on 2 meters!

Your selection of station equipment should be guided by the type of operating you want to try. If you want to create an effective weak-signal station, for example, you will want to look for a multi-mode VHF transceiver rather than just an FM rig. You'll also want to put up some type of directional antenna that you can rotate to point at other stations rather than one that sends signals equally well in all directions.

If amateur television (ATV) sounds interesting, you can sample the activity in your area with a cable-ready TV receiver connected to a good outside beam antenna for the 70-cm (420 to 450-MHz) amateur band. Tune the TV to cable channels 57 through 61. These cable channels are not the same as the UHF broadcast channels 57 to 61. Cable channel 57 is 421.25 MHz, and each channel is 6 MHz higher in frequency.

[Turn to Chapter 14 and study questions T2B09, T2B12 and T2B13 now. Review this section as needed.]

EMERGENCY OPERATIONS

When cities, towns, counties, states or the federal government call for emergency communications, amateurs answer. They press their mobile and portable radio equipment into service, using alternatives to commercial power. Generators, car batteries, windmills or solar energy can provide power for equipment during an emergency, when normal ac power is often out of service. Many amateurs have some way to operate their station without using commercial ac power. Such power lines are often knocked down and areas without power during a natural disaster such as a hurricane, tornado, earthquake or ice storm.

A dipole antenna is the best choice for a portable HF station that can be set up in emergencies — it can be installed easily, and wire is light and portable. Carry an ample supply of wire and you'll be ready to go on the air at any time and any place. If you're planning to use a battery-powered hand-held transceiver, bring along at least one spare, *charged* battery pack. One of the most important accessories you can have for your hand-held radio is several extra battery packs, as long as you keep them charged.

A hand-held transceiver is a very useful piece of equipment. It can be used in a variety of emergency situations. You can use such a radio at home, in your car and you can take it just about anywhere. You can carry it along on an emergency search and rescue mission, use it for communications from an emergency shelter or take into the field for making damage reports to government officials.

If you find yourself in a life or property-threatening situation and you want to make an emergency call for help, you should know the basic procedures for the various modes. For Morse code (CW) operation, the proper distress call is to send SOS several times and then your call sign. Pause for a reply and then repeat the procedure until you receive an answer. Notice that the letters SOS are sent as a single character, with no pause between the letters. On phone, the proper distress call is MAYDAY sent several times followed by your call sign. Again, you should pause for an answer, and repeat the procedure until you do receive an answer. You can select any frequency and any mode that you think will be most likely to bring the desired response for your emergency.

Tactical Communications

Tactical communications is first-response communications in an emergency involving a few people in a small area. This type of communications is unformatted and seldom written. It may be urgent instructions or requests such as "Send an ambulance," or "Who has the medical supplies?"

Tactical communications usually use 2-meter repeater net frequencies or the 146.52-MHz simplex calling frequency. Compatible mobile, portable and fixed-station equipment is plentiful and popular for these frequencies. Tactical communications is particularly important when working with local government and law-enforcement agencies. Use the 12-hour local-time system for times and dates when working with relief agencies. Most may not understand the 24-hour system or Coordinated Universal Time (UTC).

Another way to make tactical communications efficient is to use tactical call signs, which describe a function, location or agency. Their use promotes coordination with individuals or agencies who are monitoring. When operators

Figure 2-9 — Emergency preparedness pays off during any disaster, whether it is weather related, earthquake or man-made disaster. Here, Jan Jubon, K2HJ (l) and Joe Peters, WB4WZZ, cover the night shift at the SKYWARN Amateur Radio station in the Washington, DC, forecast office of the National Weather Service as a hurricane approaches the Florida coast. Joe holds the latest printout of a GOES satellite image of the hurricane. (*Photo courtesy of SKYWARN Net Manager Dan Gropper, KC4OCG.*)

change shifts or locations, the set of tactical call signs remains the same. Amateurs may use such **tactical call signs** as *parade headquarters*, *finish line*, *Red Cross* or *Net Control*. This procedure promotes efficiency and coordination in public-service communication activities. Tactical call signs do not fulfill the identification requirements of Section 97.119 of the FCC Rules, however. Amateurs must also identify their station operation with their FCC-assigned call sign. Identify at the end of the operation and at intervals not to exceed 10 minutes during the operation.

Health-and-Welfare Traffic

There can be a large amount of radio traffic to handle during a disaster. Phone lines still in working order are often overloaded. They should be reserved for emergency use by those people in peril. Shortly after a major disaster, **Emergency traffic** messages leave the disaster area. These have life-and-death urgency or are for medical help and critical supplies. Handle them first. Next is **Priority traffic**. These are emergency-related messages, but not as important as Emergency messages. Then, handle **Health-and-Welfare traffic**, which pertains to the well being of evacuees or the injured. This results in timely advisories to those waiting outside the disaster area.

Hams inside disaster areas cannot immediately find out about someone's relative when their own lives may be threatened — they are busy handling emergency messages. After the immediate emergency subsides, concerned friends and relatives of possible victims can send health-and-welfare inquiries into the disaster area.

RACES

The **Radio Amateur Civil Emergency Service (RACES)** is a part of the amateur service that provides radio communications only for civil defense purposes. It is active *only* during periods of local, regional or national civil emergencies.

You must be registered with the responsible civil defense organization to operate as a RACES station. RACES stations may not communicate with amateurs not operating in a RACES capacity. Restrictions do not apply when stations are operating in a non-RACES amateur capacity, such as the ARES, the Amateur Radio Emergency Service.

Only civil-preparedness communications can be transmitted during RACES operation. These are defined in section 97.407 of the FCC regulations. Rules permit tests and drills for a maximum of one hour per week. All test and drill messages must be clearly identified as such.

[Congratulations! You have studied all of the material about operating procedures for your Technician license exam. Before you go on to Chapter 3, turn to Chapter 14 and study questions T2B08, T2C01, T2C02 and T2C04 through T2C12. Review this section if you have difficulty with any of these questions.]

3 KEY WORDS

D region — The lowest region of the ionosphere. The D region contributes very little to short-wave radio propagation. It acts mainly to absorb energy from radio waves as they pass through it. This absorption has a significant effect on signals below about 7.5 MHz during daylight.

DX — Distance, foreign countries.

E region — The second lowest ionospheric region, the E region, exists only during the day. Under certain conditions, it may refract radio waves enough to return them to Earth.

F region — A combination of the two highest ionospheric regions, the F1 and F2 regions. The F region refracts radio waves and returns them to Earth. Its height varies greatly depending on the time of day, season of the year and amount of sunspot activity.

Ground-wave propagation — The method by which radio waves travel along the Earth's surface.

Ionosphere — A region of electrically charged (ionized) gases high in the atmosphere. The ionosphere bends radio waves as they travel through it, returning them to Earth. Also see sky-wave propagation.

Line-of-sight propagation — The term used to describe VHF and UHF propagation in a straight line directly from one station to another.

Maximum usable frequency (MUF) -- The highest-frequency radio signal that will reach a particular destination using **sky-wave propagation**, or *skip*. The MUF may vary for radio signals sent to different destinations.

Propagation — The study of how radio waves travel.

Reflection — Signals that travel by **line-of-sight propagation** are reflected by large objects like buildings.

Skip zone — An area of poor radio communication, too distant for ground waves and too close for sky waves.

Sky-wave propagation — The method by which radio waves travel through the ionosphere and back to Earth. Sometimes called *skip*, sky-wave propagation has a far greater range than **line-of-sight** and **ground-wave propagation**.

Sunspots — Dark spots on the surface of the sun. When there are few sunspots, long-distance radio propagation is poor on the higher-frequency bands. When there are many sunspots, long-distance HF propagation improves.

Sunspot cycle — The number of **sunspots** increases and decreases in a predictable cycle that lasts about 11 years.

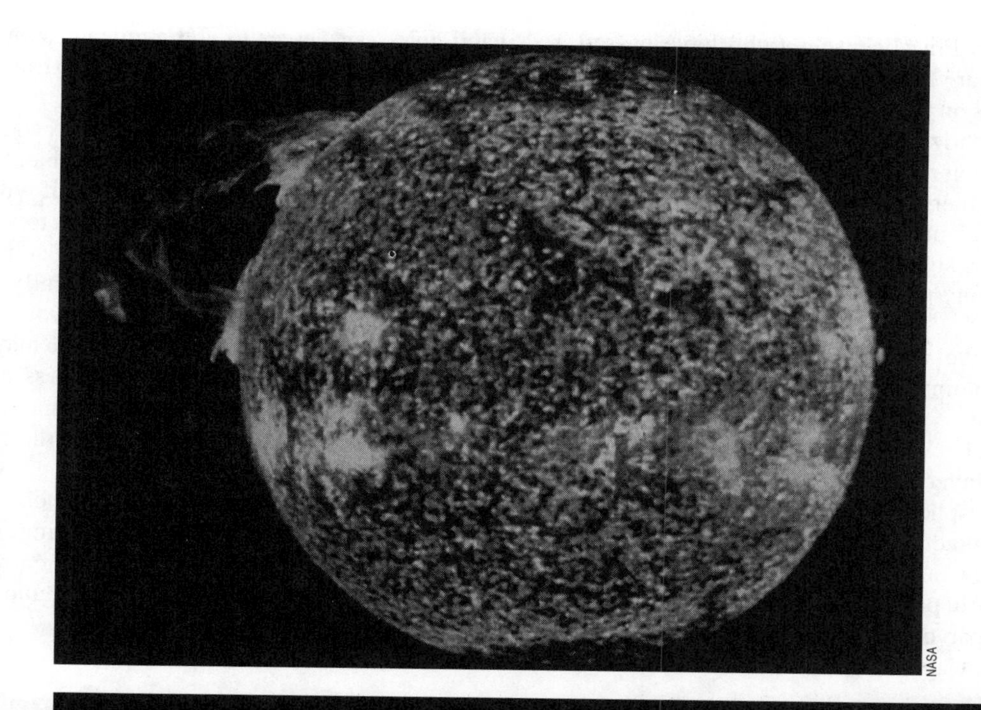

3 Radio-Wave Propagation

The sun has a large effect on long-distance radio communication on Earth. Solar flares, like the ones shown here, as well as sunspots, can make long-distance communication possible—or impossible.

Propagation: How Signals Travel

Now that we've looked at several modes of operation, it's time to discuss how radio waves travel. As we've seen, amateurs enjoy the privilege of using many different frequency bands. Some of these are scattered throughout the high-frequency (HF) spectrum. Some of these HF bands work better during the day, and some work better at night. Some frequencies are good for long-distance communications, and others provide reliable short-range communications. Amateurs who want to work a certain part of the country or world need to know which frequency to use and when to be there. Experience, combined with lots of careful listening, is a good way to gain this knowledge.

Novices have privileges on four HF bands. To keep things simple, we'll refer to them as the "Novice bands,"

Table 3-1

High-Frequency Ham Bands, with Novice and Technician Plus Privileges

HF Bands For Full Amateur Privileges (MHz)	*Novice/Tech Plus Privileges (MHz)*	*Band (Meters)*
1.8 - 2.0	None	160
3.5 - 4.0	3.675 - 3.725	80
7.0 - 7.3	7.1 - 7.15	40
10.10 - 10.15	None	30
14.00 - 14.35	None	20
18.068 - 18.168	None	17
21.00 - 21.45	21.1 - 21.2	15
28.00 - 29.70	28.1 - 28.5	10

although Technician Plus licensees (who have passed a 5-wpm code test) share these privileges. (All Technicians have other privileges on the VHF and UHF bands.)

Table 3-1 summarizes the HF privileges granted to radio amateurs. We often refer to amateur bands by approximate wavelength rather than frequency. When you hear someone refer to the 15-meter band, 15 meters is the approximate wavelength of a signal in that band. During a typical day, you can reach any part of the world using one of the Novice bands.

Radio waves travel to their destination in three ways:

- Directly from one point to another.
- Along the ground.
- Refracted, or bent, back to Earth by the **ionosphere**. (The ionosphere is a layer of charged particles, or ions, in Earth's outer atmosphere. These ionized gases make long-distance radio contacts possible on the HF bands.)

The study of how radio waves travel from one point to another is the science of **propagation**. Radio-wave propagation is a fascinating part of ham radio. Let's take a look at the different ways radio waves travel.

Line-of-Sight Propagation

Line-of-sight propagation occurs when signals travel in a straight line from the transmitting antenna to the receiving antenna. These *direct waves* are useful mostly in the very-high frequency (VHF) and ultra-high frequency (UHF) ranges. (As a Novice you will be allowed to use the 222 to 225-MHz band in the VHF range and the 1270 to 1295-MHz band in the UHF range.) TV and FM radio broadcasts (those that are broadcast to your TV or radio antenna instead of through a cable) are examples of signals that are received as direct waves. When you transmit on a local repeater frequency, direct waves generally travel in a straight line to the repeater. The repeater sends the signals, in a straight line, to other fixed, portable and mobile stations. When you use a hand-held transceiver to communicate with another ham using a hand-held radio, your signals also travel in a straight line.

Normally, you'll be contacting nearby stations—within 100 miles or so — on the VHF/UHF bands. The signal travels directly between stations, so if you're using a directional antenna, you normally point it toward the station you are trying to contact. VHF and UHF signals, however, are easily reflected by buildings (especially metal-framed ones), hills and even airplanes. Some of your signal reaches the other station by a direct path and some may be reflected. When such **reflections** occur, it is possible to contact other stations by pointing your antenna toward the reflecting object, rather than directly at the station you're trying to contact.

Reflections can cause problems for mobile operation, as the propagation path is constantly changing. The direct and reflected waves may first cancel and then reinforce each other. This causes a rapid fluttering sound called *picket fencing*.

Ground-Wave Propagation

In **ground-wave propagation**, radio waves travel along the Earth's surface, even over hills. They follow the curvature of the Earth for some distance. AM broadcasting signals travel by way of ground-wave propagation during the day. Ground wave works best at lower frequencies. You might have an 80-meter QSO with a station a few miles away, on the other side of a hill. For that contact, you are using ground-wave propagation.

Ground-wave propagation on the ham bands means relatively short-range communications. Stations at the high-frequency end of the AM broadcast band (the 1600-kHz end) generally carry less than a hundred miles during the day. Stations near the low-frequency end of the dial (540 kHz) can be heard up to 100 miles or so away. Amateur Radio frequen-

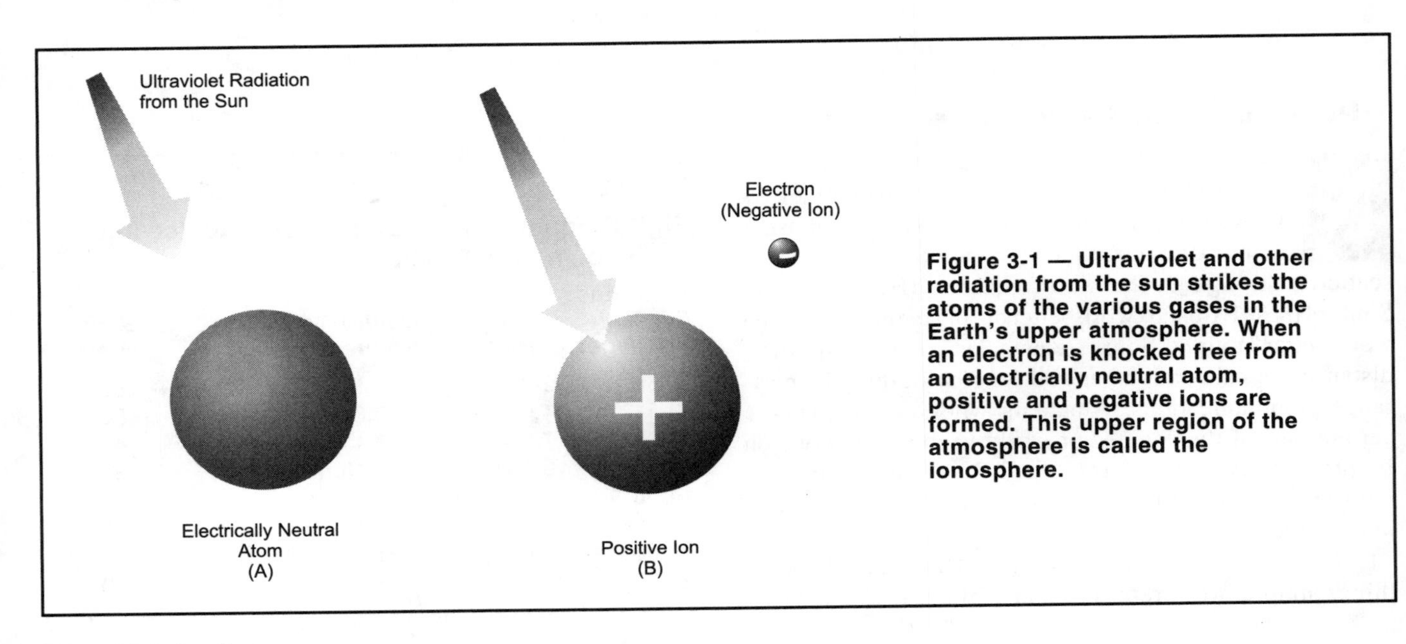

Figure 3-1 — Ultraviolet and other radiation from the sun strikes the atoms of the various gases in the Earth's upper atmosphere. When an electron is knocked free from an electrically neutral atom, positive and negative ions are formed. This upper region of the atmosphere is called the ionosphere.

cies are higher than the AM broadcast band, so the ground-wave range is even shorter.

Sky-Wave Propagation (Skip)

The Earth's upper atmosphere (25 to 200 miles above Earth) consists mainly of oxygen and nitrogen. There are traces of hydrogen, helium and several other gases. The atoms making up these gases are electrically neutral: They have no charge and exhibit no electrical force outside their own structure. The gas atoms absorb ultra-violet radiation and other radiation from the sun. This knocks electrons out of the atom. These electrons are negatively charged particles, and the remaining portions of the gas atoms form positively charged particles. The positive and negative particles are called *ions*. The process by which ions are formed is called ionization. Figure 3-1 illustrates this process.

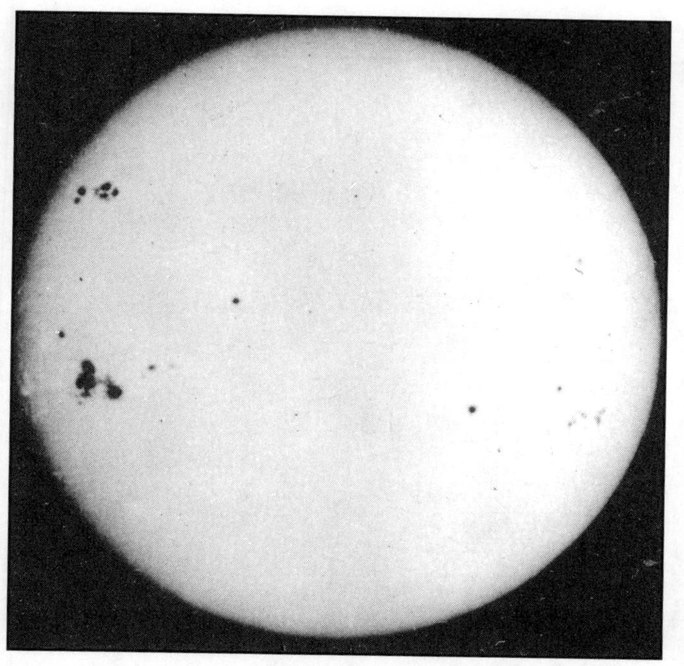

Figure 3-2 — Cool sunspots allow hot propagation on Earth!

When ionized by solar radiation, this region, called the **ionosphere**, can refract (bend) radio waves. If the wave is bent enough, it returns to Earth. If the wave is not bent enough, it travels off into space. Ham radio contacts of up to 2500 miles are possible with one *skip* off the ionosphere. Worldwide communications using several skips (or hops) can take place if conditions are right. This is the way long-distance radio signals travel.

Two factors determine **sky-wave propagation** possibilities between two points: the *frequency* in use and the level of *ionization*. The higher the frequency of the radio wave, the less it is bent by the ionosphere. The highest frequency at which the ionosphere bends radio waves back to a desired location on Earth is called the **maximum usable frequency (MUF)**. The MUF for communication between two points depends on solar radiation strength and the time of day. Radio waves that travel beyond the horizon by refraction in the ionosphere are called *sky waves*. Sky-wave propagation takes place when a signal is returned to Earth by the ionosphere.

Ionization of the ionosphere results from the sun's radiation striking the upper atmosphere. It is greatest during the day and during the summer. The amount of radiation coming from the sun varies throughout the day, season and year. This radiation is closely related to visible **sunspots** (grayish-black blotches on the sun's surface). See Figure 3-2. Hams interested in long-distance communication always know what part of the **sunspot cycle** we are in. Sunspots vary

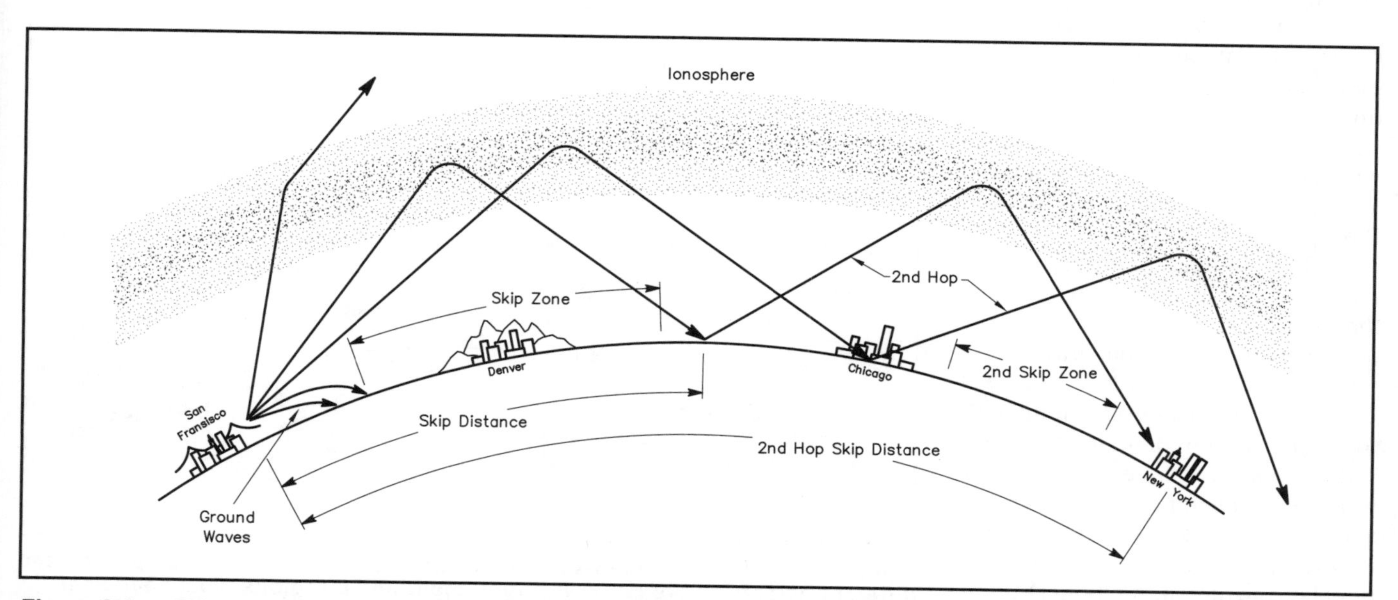

Figure 3-3 — This drawing illustrates how radio waves travel into the ionosphere and are bent back to Earth. Ground waves, skip distance and skip zone are all shown on the drawing.

in number and size over an 11-year cycle. More sunspots usually means more ionization of the ionosphere. As a result, the MUF tends also to be higher.

By contrast, when sunspots are low, radiation—and thus the MUF—is lower. That is why HF communication is enhanced during times of greater sunspot activity.

Skip propagation has both a maximum range limit and a minimum range limit. That minimum is often greater than the ground-wave range. There is an area between the maximum ground-wave distance and the minimum skip distance where radio signals on a particular frequency will not reach by normal methods. This "dead" area is called the **skip zone**. Figure 3-3 illustrates the difference between ground-wave and skip propagation. The drawing also shows the concept of skip zone and skip distance. Some radio signals may not be bent enough to bring them back to Earth. There is always some signal that reaches the skip zone, but signals are often weak.

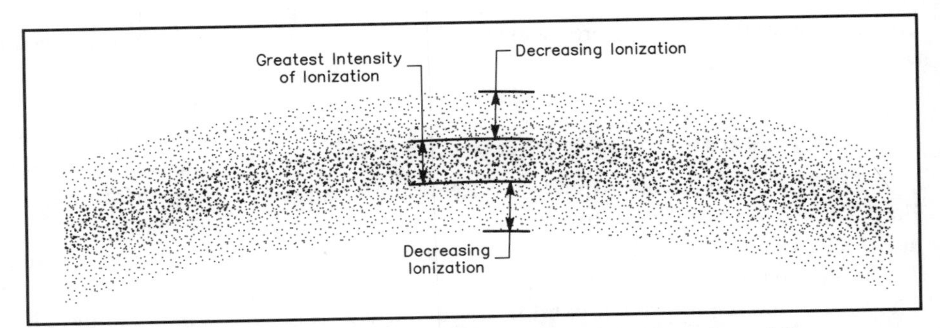

Figure 3-4 — This drawing shows a cross section of one region of the ionosphere. The intensity of the ionization is greatest in the central region and decreases above and below the central region.

The Ionosphere

Several ionized regions appear at different heights in the atmosphere. Each region has a central region where the ionization is greatest. The intensity of the ionization decreases above and below this central area in each region. See Figure 3-4.

The ionosphere consists of several regions of charged particles. These regions have been given letter designations, as shown in Figure 3-5. Scientists started with the letter D just in case there were any undiscovered lower regions. None have been found, so there is no A, B or C region.

The **D region**: The lowest region of the ionosphere affecting propagation is the D region. This region is in a relatively dense part of the atmosphere about 35 to 60 miles above the Earth. When the atoms in this region absorb sunlight and form ions, the ions don't last very long. They quickly combine with free electrons to form neutral atoms again. The amount of ionization in this region varies widely. It depends on how much sunlight hits the region. At noon, D-region ionization is maximum or very close to it. By sunset, this ionization disappears.

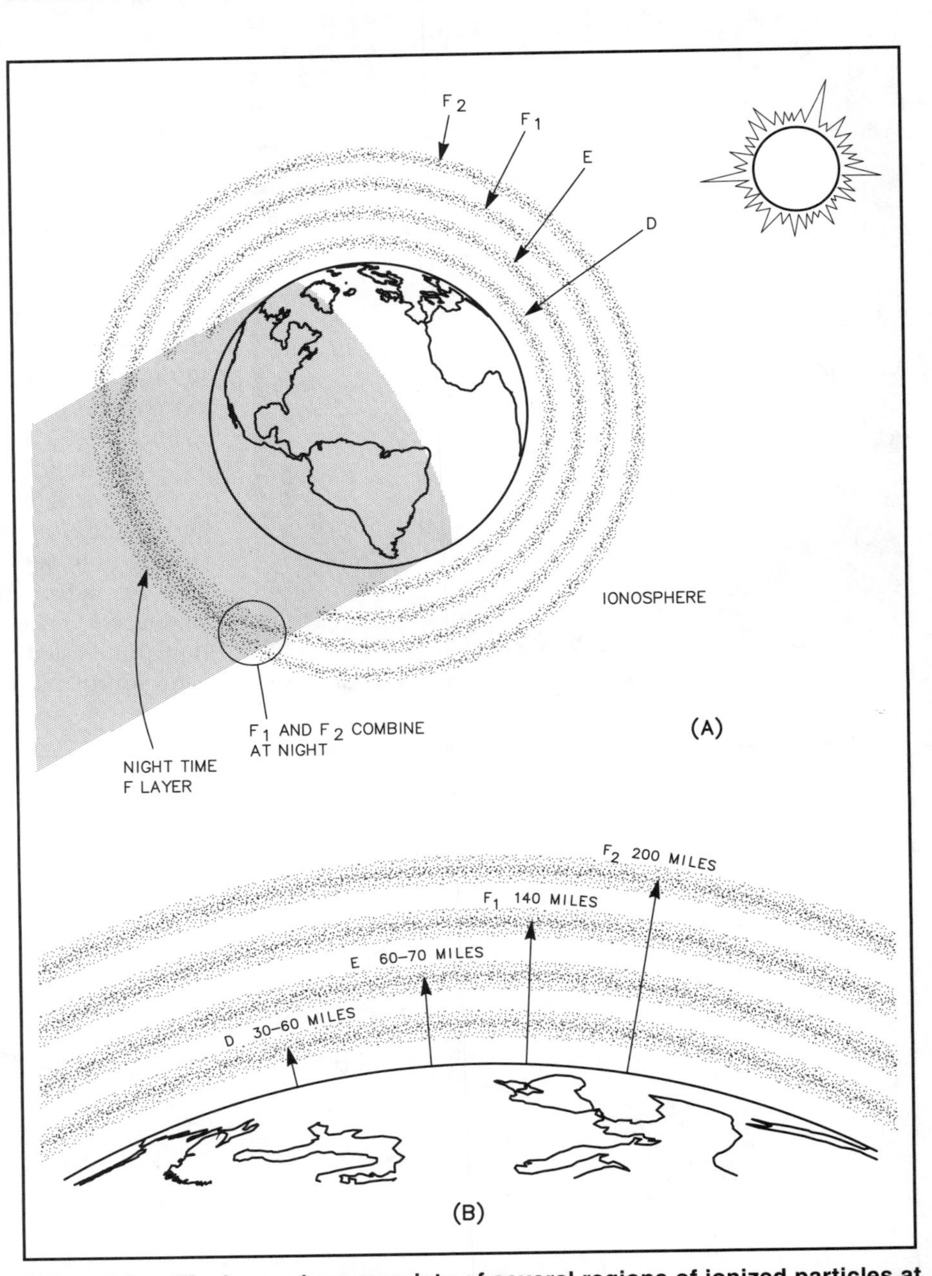

Figure 3-5 — The ionosphere consists of several regions of ionized particles at different heights above Earth. At night, the D and E regions disappear when the positive and negative ions combine to form gas atoms again. The F1 and F2 regions combine to form a single F region at night.

The D region is ineffective in refracting or bending high-frequency signals back to earth. The D region's major effect is to absorb energy from radio waves. As radio waves pass through the ionosphere, they give up energy. This sets some of the ionized particles into motion. Effects of absorption on lower frequencies (longer wavelengths) are greater than on higher frequencies. Absorption also increases when there is more ionization. The more ionization, the more energy the radio waves lose passing through the ionosphere. Absorption is most pronounced at midday. It is responsible for the short daytime communications ranges on the lower-frequency amateur bands (160, 80 and 40 meters).

The next region of the ionosphere is the **E region**, at an altitude of about 60 to 70 miles. At this height, ionization produced by sunlight does not last very long. This makes the E region useful for bending radio waves only when it is in sunlight. Like the D region, the E region reaches maximum ionization around midday. By early evening the ionization level is very low. The ionization level reaches a minimum just before sunrise, local time. Using the E region, a radio signal can travel a maximum distance of about 1250 miles in one hop. Sporadic E, or E skip, is a type of sky-wave propagation that allows long-distance communications on the VHF bands (6 meters, 2 meters and 222 MHz). Although, as its name implies, sporadic E occurs only during certain times of the year, it is the most common type of VHF sky-wave propagation.

The **F region**: The region of the ionosphere most responsible for long-distance amateur communication is the F region. This region is a very large region. It ranges from about 100 to 310 miles above the Earth. The height depends on season, latitude, time of day and solar activity. Ionization reaches a maximum shortly after noon local standard time. It tapers off very gradually toward sunset. At this altitude, the ions and electrons recombine very slowly. The F region remains ionized during the night, reaching a minimum just before sunrise. After sunrise, ionization increases rapidly for the first few hours. Then it increases slowly to its noontime maximum.

During the day, the F region splits into two parts, F1 and F2. The central part of the F1 region forms at an altitude of about 140 miles. For the F2 region, the central region forms at about 200 miles above the Earth. These altitudes vary with the season of the year and other factors. At noon in the summer the F2 region can reach an altitude of 300 miles. At night, these two regions recombine to form a single F region slightly below the higher altitude. The F1 region does not have much to do with long-distance communications. Its effects are similar to those caused by the E region. The F2 region is responsible for almost all long-distance communication on the amateur HF bands. A one-hop radio transmission travels a maximum of about 2500 miles using the F2 region.

PROPAGATION ON OUR HF BANDS

Operating on 10 Meters (28.1 to 28.5 MHz)

At times the best of bands, 10 meters also can seem the worst of bands. The highest frequency amateur HF band, 10 is more subject to the whims of the sunspot cycle. During years of high sunspot activity, 10 meters remains open all day and most of the night. Strong signals arrive from all over the world. In years of low solar activity, you may hear nothing but local signals for days at a time.

When operating on 10 meters, remember that the part of the Novice band used for CW (Morse code) is a full 200 kHz wide. (The section of a band used for a certain mode is called a *subband.*) The voice subband is also 200 kHz wide. Some signals may be weak, but perfectly readable. Tune the entire Novice subband carefully. If there aren't any signals in the Novice portion, listen in the rest of the band. A good place to listen is around 28.5 MHz. If the band is open, you'll likely hear someone.

During the sunspot minimum, you'll have to work a little harder to see if 10 is open. Several automated beacons operate between 28.1 and 28.3 MHz. They transmit continuous signals to test propagation. If you can hear one of these beacons, the band is open. You'll find these beacons listed occasionally in *QST* and other ham publications. Even if 10 seems dead, it may just be a case of everyone listening and no one calling! Try calling CQ.

Ten meters offers many Novices their first taste of **DX**: The chance to communicate with hams in foreign lands. Familiarize yourself with some of the common DX call signs. That way, you won't be surprised the first time you hear one. In a typical week, you might hear a JA (Japan), an EA (Spain) or an LA (Norway). Some countries also use call signs beginning with a number. Don't be surprised if you work a 4Z (Israel) or a 9Y (Trinidad). See Chapter 1 for more information about international call sign prefixes.

In addition to DX potential, 10 meters also provides excellent short-range communications. Many clubs and local groups meet regularly on 10 meters. Check with a local radio club to find out if there are any nets in your area.

Ten meters is also the lowest-frequency Novice and Technician Plus band that allows emissions other than CW. From 28.1 to 28.3 MHz, Novice and Technician Plus operators are allowed to use CW and frequency-shift keying (FSK). FSK modes include packet radio, Baudot radioteletype (RTTY), ASCII and AMTOR. From 28.3 to 28.5 MHz, Novice and Technician Plus operators may use CW and single-sideband (SSB) voice. The maximum power level that these licensees may use for any mode on 10 meters is 200 watts PEP output.

Operating on 15 Meters (21.1 to 21.2 MHz)

The 15-meter band shares several characteristics with the 10-meter band. It is not, however, as dependent on high sunspot activity. The 21-MHz frequency range is still high enough so that propagation conditions change dramatically with sunspot variations. Because of its lower frequency, 15 meters is open for longer periods and more often than 10. Also, the band is more stable in years of low sunspot activity.

DX is plentiful on 15 meters, even in the lean years of the solar cycle. During the day, stations in Europe, Africa and South America may be workable. You may hear signals from the Pacific and Japan in the late afternoon through early evening hours. Fifteen meters provides good communications to areas all over the US.

Many foreign stations operate within the Novice part of the band using phone transmissions. American hams are prohibited from doing this, but many other countries allow it. If you hear a foreign station on voice, give him a CW call at the end of his QSO. Many are glad to work Novices. This kind of operation won't give you much practice copying CW, but it will give you a chance to hear what phone DX stations sound like. You might even work a "new country."

Because 15 meters is so popular, you'll most likely hear stations if the band is open. Again, listen at the bottom of the CW and phone portions if you don't hear signals in the Novice band. This is where stations usually congregate.

Fifteen is one of the best bands around because of its DX possibilities. More stable than 10 meters, it is the most reliable Novice band for DX. It's usable even at the low point of the sunspot cycle. Higher in frequency than 40 and 80 meters, it provides good long-distance contacts.

Operating on 40 Meters (7.10 to 7.15 MHz)

In many respects, the 40-meter band is one of the best US ham bands. It suffers from some of the worst interference you'll ever hear, however. International treaties allow the world outside of the Americas to use the 40-meter band above 7.1 MHz for high-power shortwave broadcasting. This causes quite a racket in North America.

During the morning and early afternoon hours, 40 meters provides good, reliable communications. Skip distance ranges from about 400 to 1200 miles. Later in the afternoon, however, the MUF drops. Skip distance then increases to several thousand miles. You'll have no trouble detecting the increase in range. You'll find yourself suddenly listening to those foreign broadcast stations.

Even though it has interference from the broadcasters, 40 meters is one of the most popular Novice bands. It is open to some part of the country 24 hours a day. During the morning and afternoon hours, you'll be able to chat with other hams to your heart's content. In the late afternoon, evening and at night, you'll be able to work more distant locations. Once you learn to live with the broadcast interference, you'll find yourself using 40 meters for many of your CW contacts.

Operating on 80 Meters (3.675 to 3.725 MHz)

At 3.7 MHz, 80 meters is the lowest in frequency of all the Novice bands. It is very popular among Novices for stateside contacts. A half-wavelength dipole antenna for 80 is rather large (see Chapter 9). It need not be parallel to the ground — its shape is not critical. Most Novices manage to put up one that works fairly well.

During the daylight hours, 80 meters provides reliable communications out to about 350 miles. Although 80 is not as heavily populated as 40 during the day, you will probably find someone eager to talk with you.

Eighty comes into its own in the evening and nighttime hours. At that time, 10 and 15 meters shut down for the night and 40 opens up to foreign broadcasters. Many hams switch to 80 meters. The band provides strong, reliable local propagation. It has none of the uncertainty of 10 and 15, nor the interference of 40.

The long-range capabilities of the 80-meter band are subject to seasonal changes. During the summer months, nighttime communications may extend to 1000 miles. Atmospheric static levels, however, are often high. Contacts can be difficult to establish and maintain. The winter months are a different story. Static levels are usually very low, and cross-country and DX contacts are normal occurrences.

Like 40 meters, 80 is a 24-hour-a-day happening. It provides excellent local communications almost all the time. It offers long-distance possibilities at night during the winter months.

Eighty is often used to make and maintain schedules with friends. After operating on the band for a while, you will come to know many of your fellow 80-meter Novices. Plan to spend some of your evening operating time on 80 meters making new friends. You'll likely meet them face-to-face at local conventions or hamfests.

[You should turn to Chapter 13 now and study questions N3A01 through N3A18. Review this section if any of these questions give you difficulty.]

[**If you are studying for a Technician license,** you should also turn to Chapter 14 and study questions T3A12, T3B01 through T3B04, T3B06 through T3B11, T3B13 and T3B14. Review this section as needed.]

KEY WORDS

Critical frequency — The highest frequency at which a vertically incident radio wave will return from the ionosphere. Above the critical frequency, radio signals pass through the ionosphere instead of returning to Earth.

Doppler effect – A change in the observed frequency of a signal, as compared with the transmit frequency, caused by satellite movement toward or away from you.

Earth-Moon-Earth (EME) or Moonbounce — A method of communicating with other stations by reflecting radio signals off the Moon's surface.

Temperature inversion — A condition in the atmosphere in which a region of cool air is trapped beneath warmer air.

Troposphere — The region in Earth's atmosphere just above the Earth's surface and below the ionosphere.

Tropospheric bending — When radio waves are bent in the troposphere, they return to Earth farther away than the visible horizon.

Tropospheric ducting — A type of VHF propagation that can occur when warm air overruns cold air (a temperature inversion).

Visible horizon — The most distant point one can see by line of sight.

3 Propagation: A Closer Look for Technicians

VIRTUAL HEIGHT OF THE IONOSPHERE

The regions of the ionosphere have considerable depth, with areas of greater ionization near the center of the region and less ionization toward both edges. (See Figure 3-4.) For practical purposes, it is often convenient to think of each region as having a definite height. The height from which a simple reflection from the region would give the same effects (as observed from the ground) as the effects of the gradual bending that actually takes place is called the virtual height of the ionospheric region. See Figure 3-6.

The virtual height of an ionospheric region for various frequencies is determined with an ionosonde. An ionosonde is a variable-frequency transmitter and receiver. It directs radio energy vertically and measures the time required for the signal to make a round trip. As the frequency increases, there is a point at which no energy will return. The highest frequency at which vertically incident radio waves return to Earth is called the **critical frequency**. Signals at higher frequencies pass through the ionosphere into space.

The Scatter Modes

All electromagnetic-wave propagation is subject to scattering influences. These alter idealized patterns to a great degree. The Earth's atmosphere, ionospheric regions and any objects in the path of radio signals scatter the energy. Understanding how scattering takes place helps us use this propagation mode to our advantage.

There is an area between the outer limit of ground-wave propagation and the point where the first signals return from the ionosphere. We studied this area, called the *skip zone*, earlier in this chapter. See Figure 3-3. The skip zone is often described as if communications between stations in each other's skip zone were impossible. Actually, some of the transmitted signal is scattered in the atmosphere, so the

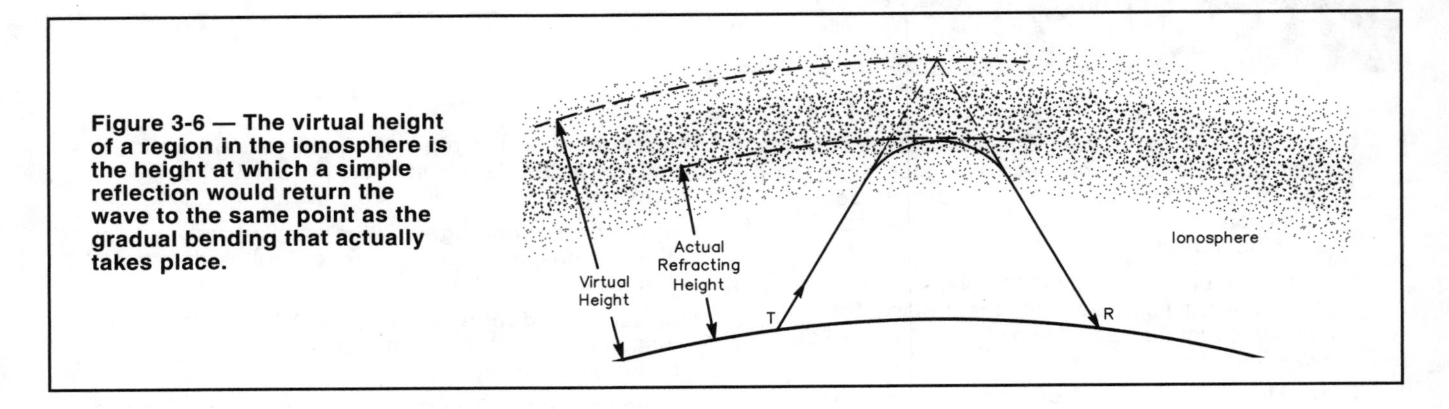

Figure 3-6 — The virtual height of a region in the ionosphere is the height at which a simple reflection would return the wave to the same point as the gradual bending that actually takes place.

signal can be heard over much of the skip zone. You usually need a very sensitive receiver and good operating techniques to hear these signals, because they are usually weak and distorted.

The troposphere is a region of the atmosphere below the ionosphere. VHF "tropo" scatter is usable out to about 500 miles from the transmitting station.

Ionospheric scatter, mostly from the height of the E region, is most apparent at frequencies up to about 60 or 70 MHz. This type of forward scatter may be usable out to about 1200 miles. Ionospheric scatter propagation is most noticeable on frequencies close to or above the MUF.

Another means of ionospheric scatter is provided by meteors entering Earth's atmosphere. As the meteor passes through the ionosphere, a trail of ionized particles forms. These particles can scatter radio energy. See Figure 3-7. This ionization is short-lived and can show up as short bursts of signals with little communication value. There may be longer periods of usable signal levels, lasting up to a minute or more. Meteor scatter is most common between midnight and dawn. It peaks between 5 and 7 AM local time. Meteor scatter is an interesting method of amateur communication at 21 MHz or higher, especially during periods of low solar activity.

You can observe a complex form of scatter when you are working very near the MUF. The transmitted wave is refracted back to Earth at some distant point. This may be an ocean area or land mass. A portion of the transmitted signal reflects back into the ionosphere. Some of this signal comes toward the transmitting station. The reflected wave helps fill in the skip zone, as shown in Figure 3-8.

Backscatter signals are generally weak and subject to distortion, because the signal may arrive at the receiver from many different directions. Backscatter is usable from just

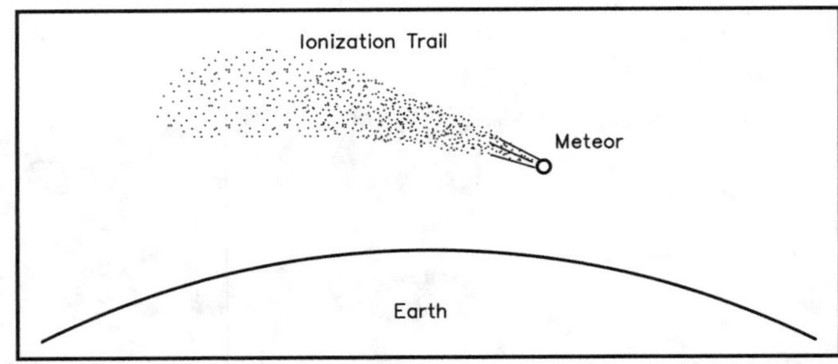

Figure 3-7 — Meteors passing through the atmosphere create trails of ionized gas. These ionized trails can be used for short-duration communications.

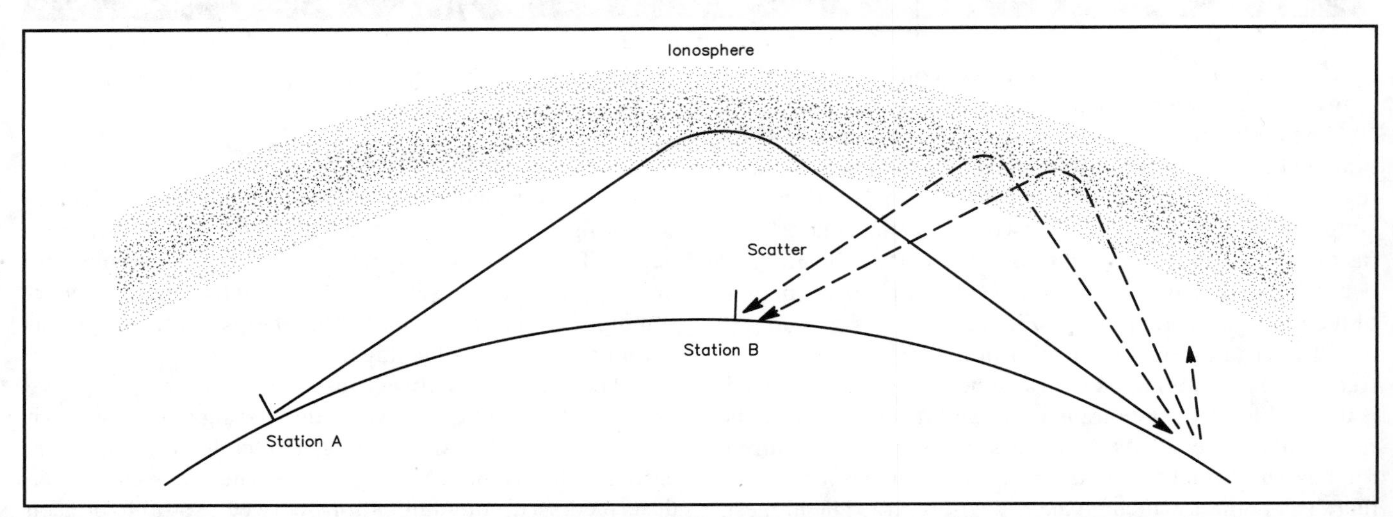

Figure 3-8 — When radio waves strike the ground after passing through the ionosphere, some of the signal may reflect back into the ionosphere. If some of the signal is reflected back toward the transmitting station, some of the energy may be scattered into the skip zone.

beyond the local range out to several hundred miles. Under ideal conditions, backscatter is possible over 3000 miles or more. The term "sidescatter" is more descriptive of what probably happens on such long paths, however.

[Now turn to Chapter 14 and study exam questions T3B05 and T3B12. Review the material in this section as needed.]

Tropospheric Bending and Ducting

The **troposphere** consists of atmospheric regions close to the Earth's surface. **Tropospheric bending** is evident over a wide range of frequencies. It is most useful in the VHF/UHF region, especially at 144 MHz and above. Instead of gradual changes in air temperature, pressure and humidity, distinct regions may form in the troposphere. Adjacent regions having significantly different densities will bend radio waves passing between regions. (In the same way, light is bent when it passes from air into water. That's why a spoon in a glass of water seems to bend when it enters the water.)

The **visible horizon** is the most distant point one can see. It is limited by the height of the observer above ground. Slight bending of radio waves occurs in the troposphere. This will cause signals to return to Earth somewhat beyond the geometric horizon, and allows you to contact stations that are somewhat farther away than would otherwise be possible. This radio-path horizon is generally about 15% farther away than the true horizon. See Figure 3-9. In general, VHF signals travel by line-of-sight propagation within the range of the visible horizon.

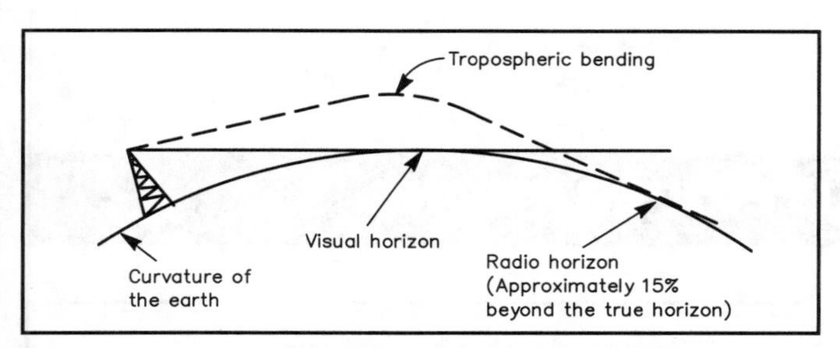

Figure 3-9 — Under normal conditions, tropospheric bending causes radio waves to return to Earth beyond the visible horizon.

Sometimes, especially during the spring, summer and fall months, it is possible to make VHF and UHF contacts over long distances — up to 1000 miles or more. This occurs during certain weather conditions that cause *tropospheric enhancement* and **tropospheric ducting**. When such "tropo" openings occur, the VHF and UHF bands are filled with excited operators eager to work DX. The troposphere is the layer of the atmosphere just below the stratosphere. It extends upward approximately 7 to 10 miles. In this region clouds form and temperature decreases rapidly with altitude.

Under normal conditions, the temperature of the air gradually decreases with increasing height above ground. When there's a stable high-pressure system, a mass of warm air may overrun cold air. When warm air covers cold air, we have a temperature inversion. Radio waves can be trapped below the warm air mass. They can travel great distances with little loss. The area between the earth and the warm air mass is known as a duct. See Figure 3-10.

The term for radio-wave propagation through a duct is **tropospheric ducting**. As with tropospheric bending, we find ducting primarily at VHF, 144 MHz and higher. Ducts usually form over water, though they can form over land as well. A widespread temperature inversion formed over an ocean may help your VHF or UHF signals travel several hundred miles.

The VHF bands experience some sky-wave propagation when the ionization in the ionosphere is high. You can experience sporadic E, or E-skip propagation on any of the VHF bands. You aren't likely to find this enhanced ionospheric propagation on the UHF bands, however. So tropospheric

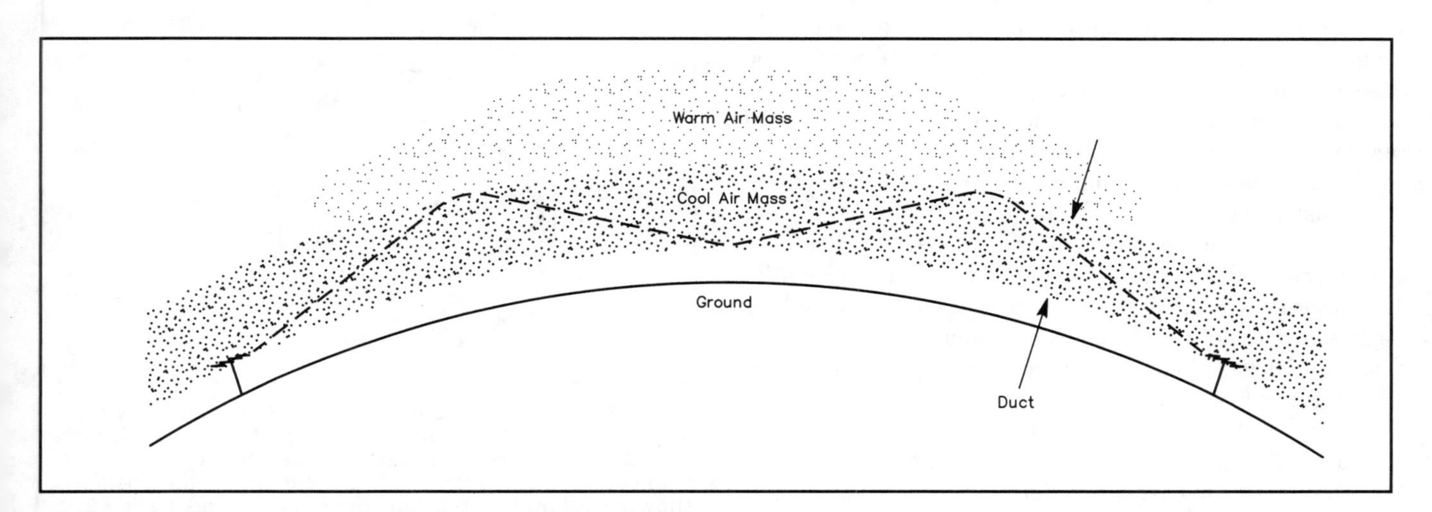

Figure 3-10 — When a cool air mass is overrun by a mass of warmer air, a "duct" may be formed, allowing VHF and UHF radio signals to travel great distances with little attenuation, or signal loss.

ducting is the most common type of enhanced propagation at UHF. Tropospheric ducting supports contacts of 950 miles or more over land and up to 2500 miles over oceans.

The lowest usable frequency for duct propagation depends on two factors: the depth of the duct and the amount the refractive index changes at the air-mass boundary. You can learn more about VHF and UHF propagation from *The ARRL Handbook*.

[At this point, turn to Chapter 14 and study questions T3A01 through T3A06, T3A08 and T3A11. Review any material necessary.]

Weak-Signal Modes and the Ionosphere

Probably the most popular type of operating on the VHF and UHF bands involves local communications through repeaters using FM. Many amateurs enjoy using the various types of enhanced propagation to make contacts with more-distant stations, however. Modes such as Morse code (CW) and single-sideband (SSB) voice are better suited to this weak-signal type of work, so they have greater potential for DX contacts on the VHF and UHF bands.

In addition to the tropospheric enhancements just discussed, you can sometimes make use of the ionosphere for your VHF DX operating. As you learned earlier, as the frequency increases you won't be able to enjoy sky-wave propagation as often. In fact you will very seldom (if ever) experience ionospheric propagation at frequencies in the UHF range!

To take advantage of the ionospheric propagation on the VHF bands you will want to learn to recognize the signs that it may occur. For example, if you are operating on the 10-meter band and notice that you are receiving very strong signals from stations that are 500 or 600 miles away, you should check the 6 and 2-meter bands for long-range skip conditions.

The E region of the ionosphere probably has the largest effect on VHF sky-wave propagation. During the summer, even when we are at a low point on the 11-year sunspot cycle, the 6-meter band (50-MHz) often experiences sporadic-E propagation. This means as a Technician licensee your best chance of experiencing sky-wave propagation will occur on the 6-meter band.

[Turn to Chapter 14 now and study questions T3A07, T3A09, T3A10 and T3A13 through T3A15. Review this section as needed.]

AMATEUR SATELLITE OPERATIONS

Amateur Radio operators have built many satellites since the first one was launched in 1961. Amateurs use these Orbiting Satellites Carrying Amateur Radio (OSCAR) to communicate with other amateurs around the world. The satellites often use the VHF and UHF bands because radio signals on those bands normally go right through the ionosphere. The satellites retransmit signals to provide greater communications range than would normally be possible on those bands. While it is possible to use HF signals for satellite operation (and some satellites do), more of the HF signal energy may be bent back to the Earth rather than going through to the satellite. (And these signals already provide long-distance communication.)

Since satellite communication uses line-of-sight propagation, two amateurs can communicate through a satellite as long as the satellite is in view of both stations. Figure 3-11 shows how a satellite can relay an amateur's signals. Notice that stations 1 and 2 can communicate through the satellite at this time, but neither station could communicate with station 3 at this time. Later, as the satellite orbits the Earth, stations 2 and 3 will be able to use the satellite, but not station 1.

It is helpful to have a directional antenna for some satellite operation so you can point the antenna in the direction of the "bird." This also requires that you be able to aim your

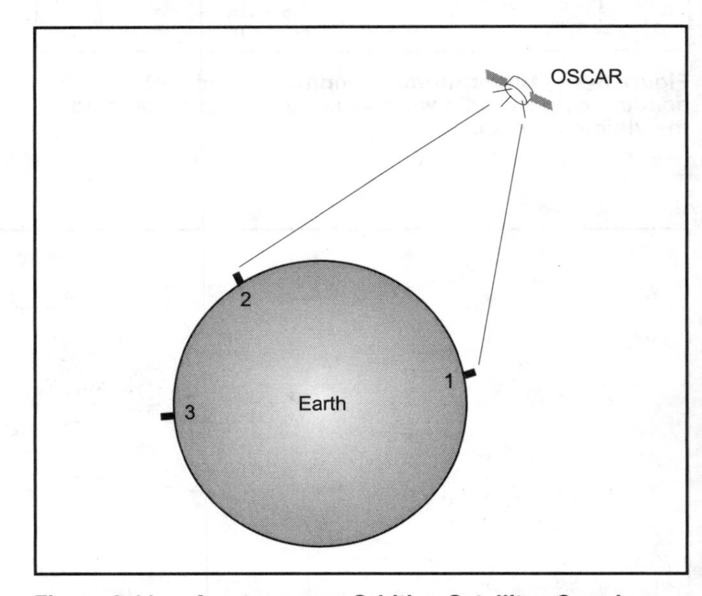

Figure 3-11 — Amateurs use Orbiting Satellites Carrying Amateur Radio (OSCARs) to relay signals. Both stations must be able to "see" the satellite, for direct communication to take place. With the satellite in the position shown, stations 1 and 2 can communicate, but station 3 cannot. As the satellite orbits the Earth, stations 2 and 3 may be able to communicate, but not station 1.

antenna in different compass directions (azimuth) as well as change the elevation angle to keep the antenna pointed at the satellite.

When the satellite is near directly overhead, and you can point your antenna at the satellite, relatively low power is required. It is even possible to operate through some satellites using simple vertical antennas and low-power hand-held radios. When the satellite is near the horizon, however, you may need more power even if you can point your gain antenna directly at the bird. This is primarily because your signal must travel a longer distance through the relatively dense air close to the Earth.

As a satellite tumbles through space, the orientation of its antennas relative to Earth changes. An Earth station using horizontally (or vertically) polarized antennas will notice deep signal fading in this case. Many satellites and satellite operators use circularly polarized antennas to reduce these fading effects. Some satellites spin to stabilize their antenna orientation, and that also reduces this fading effect.

Because the satellite is moving across the sky while you are operating through it, you may have to take one more factor into account: **Doppler effect**. Doppler effect is a change in the received signal frequency caused by the satellite movement toward or away from you. If you listen to a car horn or train whistle as the car or train passes you, you will hear the change in the pitch or tone of the horn caused by the Doppler effect. You can also hear Doppler effect in the engine sound as a race car whizzes past you on the track. You will have to adjust the receiver frequency as the satellite moves across the sky. As the satellite moves toward you, you will receive it on a frequency slightly higher than the satellite's transmit frequency, and as it moves away from you, you will receive it on a frequency slightly lower than the transmit frequency.

[Turn to Chapter 14 and study questions T3C01 through T3C04 and questions T3C10 and T3C11 now. Review this section if any of these satellite-operating questions give you difficulty.]

EARTH-MOON-EARTH (MOONBOUNCE) COMMUNICATIONS

The concept behind **Earth-Moon-Earth — EME** or **Moonbounce** — communication is really quite simple: Stations that can simultaneously see the Moon communicate by reflecting VHF or UHF signals off the Moon's surface! Those stations can be separated by more than 11,000 miles on the Earth's surface.

There is one drawback to this communications method, however. The Moon's average distance from the Earth is 239,000 miles, and EME signals must travel twice the distance to the Moon. Path losses are huge when compared to "local" VHF paths. Path loss refers to the total signal loss between the transmitting and receiving stations as compared to the total radiated signal energy. In addition to the long distance to the Moon and back, the Moon's surface is irregular and not a particularly efficient reflector of radio waves.

A typical EME station uses high-gain antennas and a high-power amplifier. For example, a high-gain array of Yagi antennas would be a good choice for a moonbounce station. You would not even want to try moonbounce with a simple ground-plane antenna, no matter how much transmitter power you had!

You might think that you would have to take Doppler effect into account for EME communications, just as you do with satellites. The Moon really doesn't move very fast relative to the Earth, however, so you don't need to be concerned with Doppler effect. Consider that a typical low-orbit satellite or space station like the Shuttle goes from horizon to horizon in 10 to 20 minutes, while it takes the Moon many hours to make a similar path. Although the Moon is actually speeding through space, it is moving slowly relative to the Earth.

[Congratulations! You have completed all of the material about radio-wave propagation for your Technician license exam. Before you go on to Chapter 4, turn to Chapter 14 and study questions T3C05 through T3C09. Review this section if you have difficulty with any of those questions.]

4 KEY WORDS

Ground connection — A connection made to the earth for electrical safety. This connection can be made inside (to a metal cold-water pipe) or outside (to a **ground rod**).

Ground rod — A copper or copper-clad steel rod that is driven into the earth. A heavy copper wire from the ham shack connects all station equipment to the ground rod.

Key-operated on-off switch — A good way to prevent unauthorized persons from using your station is to install a key-operated switch that controls station power.

Lightning protection — There are several ways to help prevent lightning damage to your equipment (and your house), among them unplugging equipment, disconnecting antenna feed lines and using a lightning arrestor.

National Electrical Code — A set of guidelines governing electrical safety, including antennas.

Safety interlock — A switch that automatically turns off ac power to a piece of equipment when the top cover is removed.

Station grounding — Connecting all station equipment to a good earth ground improves both safety and station performance.

4 Amateur Radio Practices: Assembling a Station

W1AW may be every ham's dream station, with all the latest equipment and lots of it. Don't expect your first station to look like this, but you can have plenty of fun with a simple station layout.

This chapter discusses how to set up an amateur station, and why it's important to do so with safety in mind. In addition, we'll look at some common and useful station accessories that will make operating your station more enjoyable.

First, we'll discuss setting up a home station. Later, we'll take a look at installing a mobile station.

HOME STATIONS

Location

First, give some thought to station location. Hams put their equipment in many places. Some use the basement or attic, while others choose the den, kitchen, or a spare bedroom. Some hams with limited space build their station into a small closet. A fold-out shelf and folding chair form the operating position.

Where you put your station depends on the room you have available and on your personal tastes. There are, however, several things to keep in mind while searching for the best place. The photos show several ways amateur stations can be arranged.

How you arrange your station equipment is a matter of personal taste and available space. The most important considerations are safe operation and a pleasing and convenient layout.

One often-overlooked requirement for a good station location is adequate electrical service. Eventually you will have several pieces of station equipment and accessories, many of which will require power to operate. Be sure that at least one, and preferably several, electrical outlets are located near your future operating position. Be sure the outlets provide the proper voltage and current for your rig. Most modern radios require only a few amperes. You may run into problems, however, if your shack is on the same circuit as the air conditioner or washing machine. The total current drawn at any one time must not exceed rated limits. Someday you'll probably upgrade and may want to purchase a linear amplifier. If so, you should have a 240-volt line in the shack, or the capability to add one.

Another must for your station is a good **ground connection**. A good ground not only reduces the possibility of electrical shock but also improves the performance of your station. By connecting all of your equipment to ground, you will help to avoid stray radio-frequency (RF) current in the shack. Stray RF can cause equipment to malfunction. A good ground can also help reduce the possibility of interference. The wire connecting your station to an earth ground should be as short as possible.

Basement and first-floor locations generally make it easier to provide a good ground connection for your station. You can also find a way to put your ham shack on the second or third floor or even on the top floor of a high-rise apartment, for that matter.

Your radio station will require a feed line of some sort to connect the antenna to the radio. So you will need a convenient means of getting the feed line into the shack. There are many ways of doing this, but one of the most effective requires only a window. Many hams simply replace the glass pane in a nearby window with a clear acrylic panel. You could also use a replacement panel made from wood or other composition material. The panel can be drilled to accept as many feed lines as needed. Special threaded "feed-through" connectors make the job easy. If you decide to relocate your station, the glass pane can be replaced. This will restore the window to its original condition.

A simpler, more temporary approach is to pass the feed lines through the open window and close it gently. Do this carefully, because crushed coaxial cable will give you nothing but trouble. You can add some foam or other soft material to close the gaps and keep "critters" out. Another

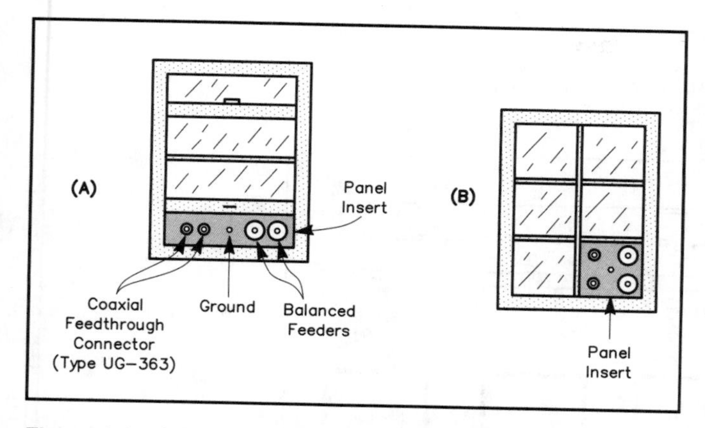

Figure 4-1—A Plexiglas, metal, wood or composition panel can be used to replace a pane or glass in a window, as shown at A. Then feed-through connectors can be installed so feed lines can be attached on either side as a way to connect your antennas. A similar panel can also be placed in the window opening, and the window then closed on the new panel. Be sure to secure stop blocks or use another method to prevent the window from simply being slid open, allowing intruders to gain access to your equipment.

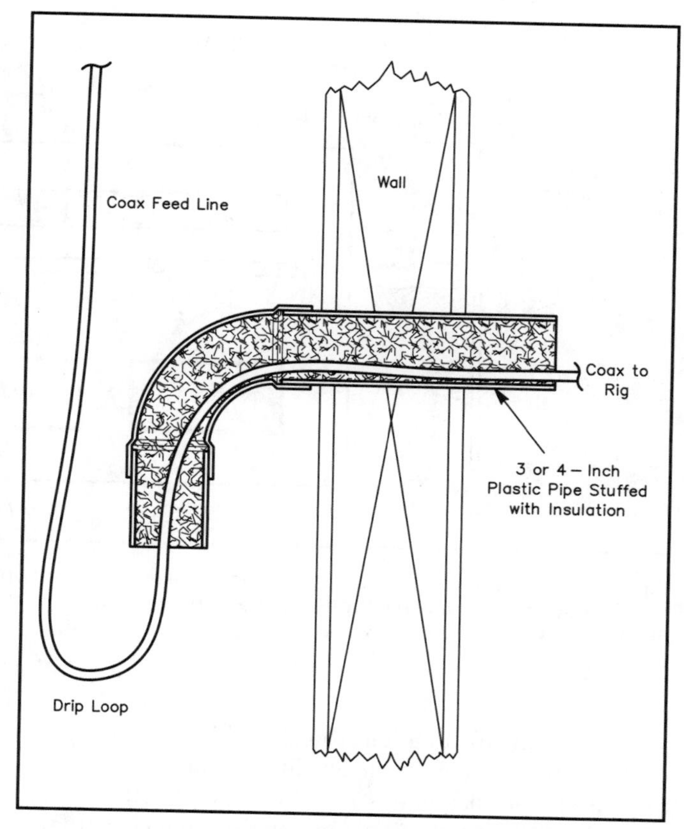

Figure 4-2—Neither insects nor rain can get into your shack when you use this method to bring cables into your shack. Install a piece of PVC or other pipe through the wall into your shack. Use an elbow and short length of pipe on the outside to prevent rain from falling into the open end of the pipe. Run the feed lines through, then stuff the tube with fiberglass insulation or similar material. If you use a 3 or 4-inch pipe there will be plenty of room to add more feed lines later. Just remove the stuffing, add the new feed lines, and then replace the stuffing.

method involves cutting a spacer from wood. acrylic or other suitable material to fit the width of the opening and drilling holes to pass the feed lines through. Place your spacer in the window opening and close the window on the spacer. Be sure to secure the window with a block of wood or some other "locking" device. Your new equipment, seen through an open window, may tempt an unwanted visitor! Figure 4-1 illustrates these techniques.

For a more permanent installation you may want to consider a feed-through pipe in the wall. A piece of PVC pipe can be fit to an opening cut in the wall, and the feed lines pass through the pipe. Stuff the pipe with fiberglass insulation or other suitable material to prevent drafts and seal out "critters." You can caulk around the tube to seal the pipe to your house siding. You can use almost any size pipe, but if you select 3 or 4-inch PVC there will be plenty of room to add more feed lines as your "antenna farm" grows! See Figure 4-2 for an example of how to make this installation.

Another important requirement for your station is comfort. The space should be large enough so you can spread out as needed. Operating from a telephone booth isn't much fun! Because you will probably be spending some time in your shack (an understatement!), be sure it will be warm in the winter and cool in the summer. It should also be as dry as possible. High humidity can cause such equipment problems as high-voltage arcing and switch-contact failure.

If your station is in an area often used by other family members, be sure they know what they shouldn't touch. You should have some means of ensuring that no "unauthorized" person can use your equipment. One way to do this is to install a **key-operated on-off switch** in the equipment power line. When the switch is turned off and you have the key in your pocket, you will be sure no one can misuse your station.

You will eventually want to operate late at night to snag the "rare ones" on 80 and 40 meters. Although a Morse code or voice contact is music to your ears, it may not endear you to a sleeping family. Putting your station in a bedroom shared with others may not be the best idea. You can keep the "music" to yourself, however, by using a good pair of headphones when operating your station.

What is a Good Ground?

Since **station grounding** is a good idea for several reasons, it's worth discussing in detail. All station equipment should be tied to a good ground. Most amateurs connect their equipment to a **ground rod** driven into the ground as close to the shack as possible. For a few dollars you can purchase a $^{1}/_{2}$ or $^{5}/_{8}$-inch-diameter, 8-foot-long ground rod at any electrical supply store. (Eight feet is the shortest

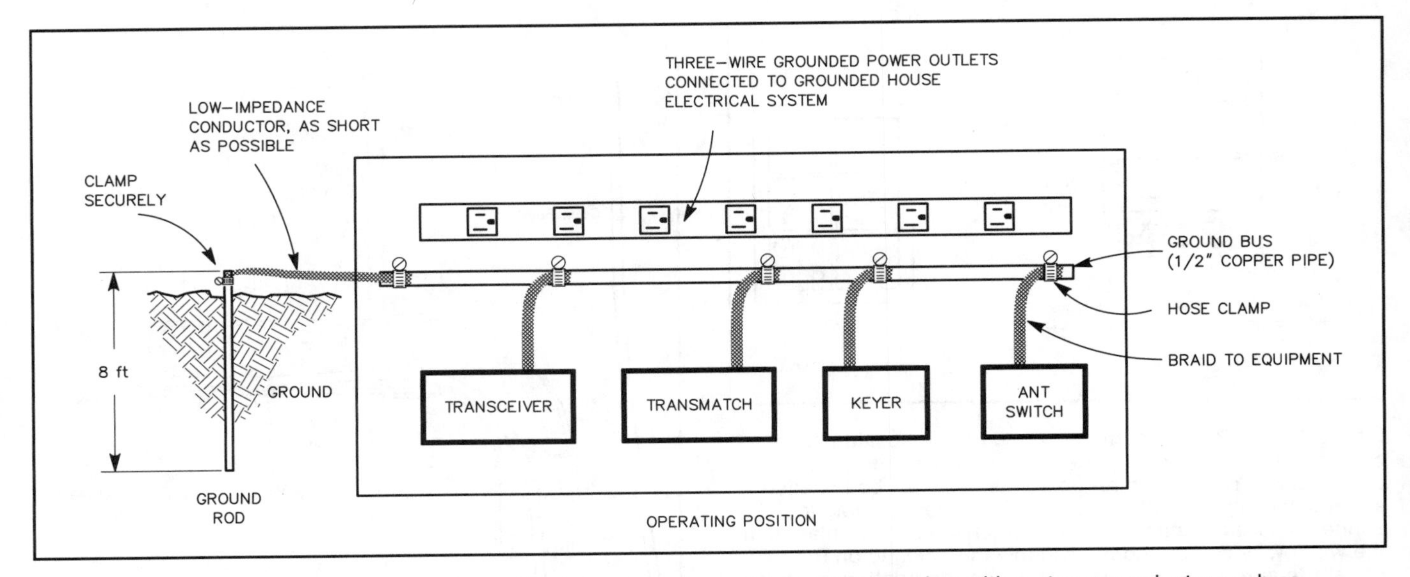

Figure 4-3—An effective station ground looks like this: all equipment is bonded together with a strong conductor such as copper flashing or coaxial copper braid. It then ties into a good earth ground—an 8-foot ground rod located as close to the station as possible.

practical length for a station ground rod.) Drive this copper-clad steel rod into the ground outside of your house, as close to your station location as possible. (Don't settle for the short, thin "ground rods" sold by some discount electronics stores. The copper cladding on the outside of most of these steel rods is very thin. They begin to rust almost immediately when they are put in the ground.)

Run a heavy copper wire (number 8 or larger) from your shack and attach it to the rod with a clamp. You can purchase the clamp when you buy the ground rod. Heavy copper strap or flashing (sold at hardware or roofing-supply stores) is even better. The braid from a piece of RG-8 coaxial cable also makes a good ground cable. Figure 4-3 shows one method of grounding each piece of equipment in your station. It's important to connect the chassis of each piece of station equipment to an effective ground connection. Keep the cable between your station and the earth ground as short as possible. Such a ground connection serves as an important safety measure to protect you from electrical shock. It also helps reduce interference problems from RF signals going places they shouldn't be.

The **National Electrical Code** requires all ground rods to be connected to form a single grounding system. These connections must use number 8 or larger copper wire. Have a qualified electrician perform this work if you are not familiar with the National Electrical Code and common wiring practices.

The National Electrical Code (NEC) describes safe grounding practices for electrical wiring, antennas and other electrical equipment. Published by The National Fire Protection Association, the NEC forms the basis for most local building codes regarding electrical wiring practices. It is a good idea to check with local building officials and the requirements of the National Electrical Code to ensure that your antenna installation meets all safety requirements.

Some hams ground their station equipment by connecting the ground wire to a cold-water pipe. Caution is in order here. If you live in an apartment or have your shack in an attic, be careful. The cold-water pipe near your transmitter may follow such a long and winding path to the earth that it may not act as a ground at all! It may, in fact, act as an antenna, radiating RF energy — exactly what you don't want it to do.

Beware, too, of nonmetallic cold-water pipes. PVC and other plastic pipes are effective insulators. There may be a piece of copper water pipe running close to your station. If there is a piece of PVC pipe connected between that spot and where the water line enters your house, however, you will not have a ground connection!

When your station is not being used, it is also important to ground all antennas, feed lines and rotator cables for effective **lightning protection**. Lightning hazards are discussed later in the chapter.

[It's time to study the following questions in Chapter 13: N4A01, N4A03, N4A06 and N4A08 through N4A11. If you are uncertain about the answers after reviewing these questions, reread the material in this section.]

[**If you are preparing for the Technician License exam,** then you should also turn to Chapter 14 and study questions T4A08 and T4A09. Review this section as needed.]

Arranging Your Equipment

Before you set everything up and hook up the cables, think about where you want each piece of equipment. While there is no one best layout for a ham station, some general rules do apply. Of course, the location you've chosen for your

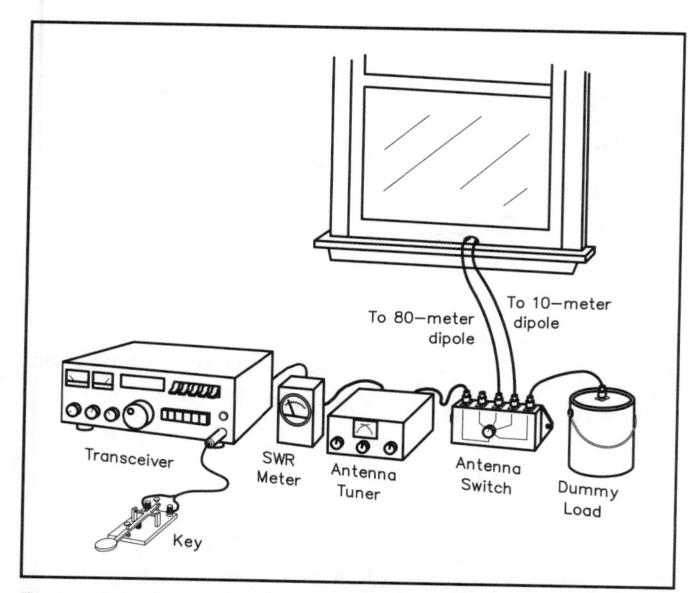

Figure 4-4—The "business side" of a typical station. Some radios have built-in antenna tuners and keyers.

station may limit your choices a bit. For example, if you're going to put the station in the basement, you may have a lot of space. If you'll be using a corner of the bedroom or den, however, you may have to keep the equipment in a small area.

Generally, the piece of gear that requires the most adjustment is the transceiver (or receiver if you have one). Make sure you can conveniently reach its controls, keeping in mind which hand you're going to use to make the adjustments. It doesn't make much sense to put the transceiver on the left side of the desk or table if you're going to adjust it with your right hand. Once you've found the best location for the transceiver, you can position the rest of your equipment around it.

If you'll be using Morse code, placement of the sending device (key, keyer or keyboard) is also very important. It must be easy to reach with the hand you send with, and placed so your arm will be supported when you're using it. Try to position it away from anything dangerous, such as sharp corners and edges, rough surfaces and electrical wires.

If you have room for a desk or table long enough to hold all your gear, you might prefer to keep everything on one surface. If you're pressed for space, however, a shelf built above the desk top is a good solution. To ensure that there is adequate air circulation above your transceiver, leave at least 3 inches of space between the top of the tallest unit on the desktop and the bottom of your shelf. Try to restrict the shelf space to lightweight items that you don't have to adjust very often, such as the station clock, SWR meter and antenna-rotator control box.

After you have a good idea of where you want everything, you can start connecting the cables. Figure 4-4 shows how an amateur station, including some common accessories, can be connected. Some of these accessories are discussed in more detail in the "Amateur Radio Practices for Technician Licensees" section of this chapter.

MOBILE STATIONS

Many VHF/UHF-oriented hams start with a hand-held transceiver. Others buy a radio designed for mobile operation. Either type can be used in a car or truck (or boat or airplane). The hand-held radio gives maximum convenience, as it can be removed easily and used anyplace. The mobile radio is ideal for those who want the advantages of a permanent installation.

Hand-held radios can be powered from your vehicle's cigarette lighter. Most hand-helds have an optional cable available for this purpose. Operating from your vehicle's electrical system saves the radio's battery for "foot mobile" operation.

Installing a mobile transceiver requires some thought. One type of mobile radio has a detachable front panel. You mount the light, thin panel on or under the instrument panel, and hide the rest of it in the trunk. The more traditional (one-piece) type of transceiver needs extra support for mounting beneath the instrument panel. Use machine screws and nuts, not sheet-metal screws.

A clean (but often challenging) way of installing a mobile radio is to build it into the dashboard or console. Keep resale value in mind as you plan your installation: The person who buys your car may not appreciate the "benefit" of your mobile installation. Figure 4-5 shows a neat in-dash installation.

It's best to connect a permanently installed radio's power cable directly to the vehicle battery. This minimizes voltage drops and electrical noise. Although one side of the

Figure 4-5—In-dash installation of a mobile transceiver is both convenient and attractive.

Figure 4-6—To protect your equipment against being zapped, install fuses on both the positive and negative leads, close to the battery terminals.

battery is connected to the car frame, you should run positive *and* negative wires, and install fuses in series with both the red and black leads. The fuses should be located *at the battery*. See Figures 4-6 and 4-7.

You'll want to prevent unauthorized persons from using your mobile radio once it's installed. You can do that

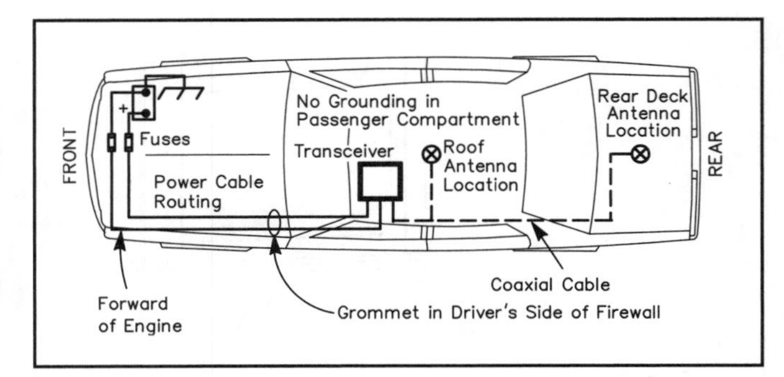

Figure 4-7—A wiring diagram for a typical mobile transceiver installation.

easily by disconnecting and removing the microphone when you leave the vehicle.

[Now turn to Chapter 13 and study question N4A02. Review this section if you have any difficulty with this question.]

[**If you're preparing for the Technician exam,** also see question T4A13 in Chapter 14.]

ELECTRICAL SAFETY

It is important to arrange your home, portable or mobile station so it is safe for you and for any visitors. One noteworthy type of electrical hazard is lightning, and there are some safety precautions later in this section.

Wiring should be neat and out of the way. Make sure there is no way you can tangle your feet in loose wires. Don't leave any voltage (regardless of how low) exposed.

A main power switch allows you to turn all station equipment on or off at once. This saves needless wear on the equipment power switches and, more importantly, is a highly recommended safety item. Make sure every member of your family knows how to turn off the power to your workbench and operating position. If you ever receive an electrical shock and cannot free yourself, the main disconnect switch will help your rescuer come to your aid quickly.

You must take special safety precautions if young children can come into your shack. Locking your station into its own room is an ideal solution. Few of us have this luxury, however. There are other ways to secure your equipment so unauthorized persons won't be able to use it. For example, you can build your station into a closet or cabinet that can be locked. If the equipment is in a nonsecure area, install a key-operated power switch and keep track of the keys. Even a simple toggle switch, if well-hidden, can help keep your station secure.

The same principles apply if you set up your station for public display. FCC rules require that only a licensed control operator may put the station on the air. To ensure that unauthorized persons will not be able to transmit, you may have to remove a microphone or control cable temporarily while the equipment is left unattended.

Whether you use commercially built equipment or homemade gear, you should never operate the equipment without proper shielding over all circuit components. Dangerous voltages may be exposed on the chassis-mounted components. Therefore, all equipment should have a protective shield on the top, the bottom and all sides. An enclosure also prevents unwanted signals from entering a receiver, or from being radiated by a transmitter.

There should be a device that turns power off automatically if you remove the shielding. Such a **safety interlock** reduces the danger of electrical shock from high voltages when you open the cabinet. The safety interlock switch should control the voltage applied to the power supply circuit. Any equipment that connects to the 120-V or 240-V ac supply should include such a safety switch, including a vacuum-tube power amplifier, a high-voltage power supply and a station-monitor oscilloscope . If you are installing such a safety switch, you should ensure that it has a voltage rating and a current capacity that is sufficient for the circuit being protected.

In arranging your station, always think of safety as

well as comfort and convenience. After you've arranged things where you think they should be, try to find fault with your layout. Don't feel satisfied with your station until you can't find anything to improve.

All the equipment in your station should be connected to earth ground to prevent electrical shocks. Earlier in this chapter, we talked about what makes a "good ground." Briefly, you should connect your equipment to the earth with as short a cable as possible. If you live in an apartment, you may have to use an indoor cold-water pipe rather than an outside ground rod.

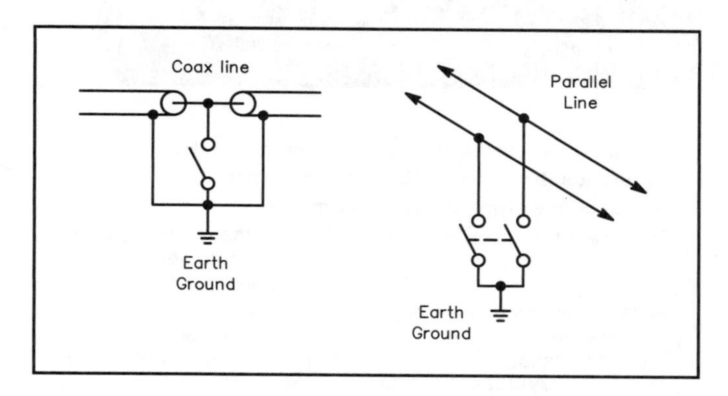

Figure 4-8—A heavy-duty knife switch can be used to connect the wires in your antenna feed line to ground.

Lightning Protection

The lightning hazard from an antenna is often exaggerated. Ordinary amateur antennas are no more likely to be hit by a direct strike than any other object of the same height in the neighborhood. Just the same, lightning does strike thousands of homes each year, so it doesn't hurt to be careful. When your station is not in use, you should ground the antenna and rotator cables and unplug your equipment. An ungrounded antenna can pick up large electrical charges from storms in the area. These charges can damage your equipment (particularly receivers) if you don't take precautions.

Most commercial beam and vertical antennas are grounded for lightning protection through the tower itself. Of course, the tower must be grounded, too. If you use a roof mount, run a heavy ground wire from the mount to a ground rod. Dipoles and end-fed wires are not grounded. Disconnect the antenna feed line from your equipment and use an alligator-clip lead to connect both sides of it to your station ground.

Storm clouds often carry dangerous electrical charges that are coupled into high objects (like amateur antennas) without visible lightning. You may be operating and suddenly hear a "SNAP!" in your antenna tuner, or your receiver may go dead. You can usually hear an increase in static crashes in your receiver well in advance of a thunderstorm. Be safe. When it sounds like a thunderstorm is headed your way, get off the air. If the weather forecast is for thunderstorms, don't operate! Snow and rain also generate static charges on antennas, but usually not enough to damage equipment.

The best protection against lightning is to disconnect all antennas and rotator control cables and connect them to ground. You should also form the habit of unplugging all power cords when you aren't on the air. It takes time to hook up everything when you want to operate again, but you will protect your station and your home if you follow this simple precaution. By the way, power companies recommend that you unplug *all* electronic appliances, including TVs, VCRs and computers, when a storm threatens.

Why unplug your equipment if the antennas are disconnected? Lightning can still find its way into your equipment through the power cord. Power lines can act as long antennas, picking up sizable charges during a storm. Simply turning off the main circuit breaker is not enough — lightning can easily jump over the circuit breaker contacts and find its way into your equipment.

You may decide to leave your antennas connected and your equipment plugged in except during peak thunderstorm months. If so, you can still protect the equipment from unexpected storms. One simple step you can take is to install a grounding switch, as shown in Figure 4-8. A small knife switch will allow you to ground your feed line when you are not on the air. It will not disturb the normal operation of your station (with the switch open, of course!) if the lead from the feed line to the switch is no more than a couple of inches long. An alligator clip can be used instead of the switch. Whatever you use, don't forget to disconnect the ground when you transmit. This precaution is useful only on the HF bands. The switch will cause high SWR if used at VHF and UHF.

Another device that can help protect your equipment in an electrical storm is a *lightning arrestor*. This device connects permanently between your feed line and the ground. When the charge on your antenna builds up to a large enough potential, the lightning arrestor will "fire." This shorts the charge to ground — not through your station. A lightning arrestor can help prevent serious damage to your equipment. Most, however, don't work fast enough to protect your station completely. Lightning arrestors are useful for commercial stations and public service (fire and police) stations that must remain on the air regardless of the weather. Unless you are actually handling emergency communications, you shouldn't rely on them alone to protect your equipment, home or life.

[Now turn to Chapter 13 and study questions N4A04, N4A05 and N4A07. Also study questions N4B01, N4B02 and N4B10 through N4B12. Review this section if you have any difficulty with these questions.]

KEY WORDS

Ammeter — A test instrument that measures current.

Crystal calibrator (see **Marker generator**).

Directional wattmeter (see **Wattmeter**)

Dummy antenna — A station accessory that allows you to test or adjust transmitting equipment without sending a signal out over the air. Also called *dummy load.*

Marker generator — A high-stability oscillator that produces reference signals at known frequency intervals. It can be used to calibrate receiver and transmitter tuning dials. Also called a **crystal calibrator**.

Multimeter — An electronic test instrument used to measure current, voltage and resistance in a circuit. Describes all meters capable of making these measurements, such as the volt-ohm-milliammeter (VOM), vacuum-tube voltmeter (VTVM) and field-effect transistor VOM (FET VOM).

National Electrical Code — A set of guidelines governing electrical safety, including antennas.

Reflectometer — A test instrument used to indicate standing wave ratio (SWR) by measuring the forward power (power from the transmitter) and reflected power (power returned from the antenna system).

S meter — A meter in a receiver that shows the relative strength of a received signal.

Signal generator — A test instrument that produces a stable low-level radio-frequency signal. The signal can be set to a specific frequency and used to troubleshoot RF equipment.

SWR meter (see **Reflectometer**)

Voltmeter — A test instrument used to measure voltage.

Wattmeter — Also called a *power meter*, a test instrument used to measure the power output (in watts) of a transmitter. A **directional wattmeter** measures both forward and reflected power.

WWV and **WWVH** — Radio stations that broadcast the precise time on precise frequencies. Hams tune in to these stations to calibrate transceiver dials (and to set their clocks).

4 Amateur Radio Practices for Technician Licensees

The remainder of this chapter covers the items related to setting up a station that you'll need to know for your Technician exam. These include electrical wiring safety, power-supply safety, test equipment, frequency accuracy, antenna measurements and station accessories.

ELECTRICAL WIRING SAFETY

Your station equipment makes use of ac line voltage. This voltage can be dangerous. The equipment also generates additional potentially lethal voltages of its own. You should be familiar with some basic precautions. Your own safety and that of others depends on it.

Power-Line Connections

In most residential systems, three wires come in from outside to the distribution board. Older systems may use only two wires. In the three-wire system, the voltage between the two "hot" wires is normally 240. The third wire is neutral and is grounded. Half of the total voltage appears between each of the hot wires and neutral, as shown in Figure 4-9. Lights, appliances and 120-V outlets are divided as evenly as possible between the two sides of this circuit. Half of the load connects between one hot wire and the neutral. The other half of the load connects between the other hot wire and neutral.

Heavy appliances, such as electric stoves, clothes dryers and most high-power amateur amplifiers, are designed for 240-V operation. Connect these across the two hot wires. Both ungrounded wires should be fused. The neutral wire should never have a fuse or switch in it.

High-power amateur amplifiers use the full 240 V. It only takes half as much current to supply the same power as with a 120-V line. The power supply circuit efficiency improves with the higher voltage and lower current. A sepa-

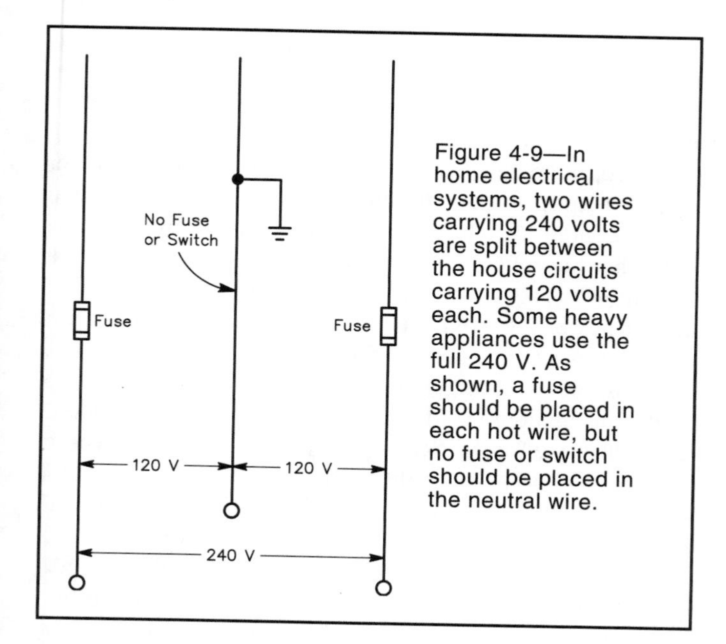

Figure 4-9—In home electrical systems, two wires carrying 240 volts are split between the house circuits carrying 120 volts each. Some heavy appliances use the full 240 V. As shown, a fuse should be placed in each hot wire, but no fuse or switch should be placed in the neutral wire.

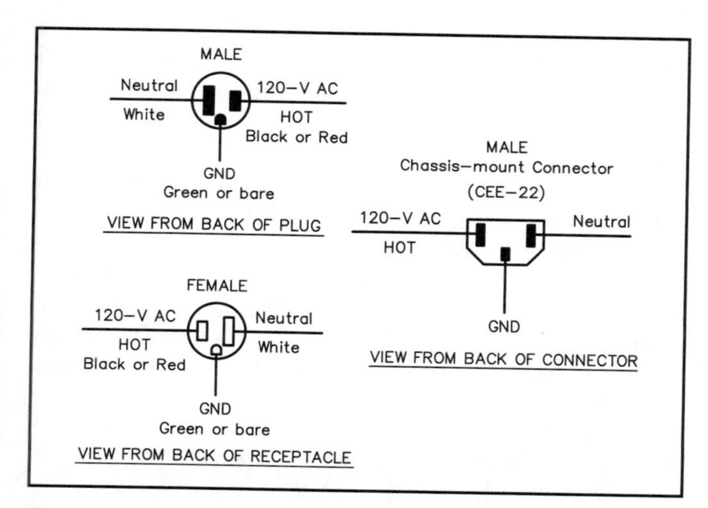

Figure 4-10—Correct wiring technique for 120-V power cords and receptacles. The white wire is neutral, the green wire is ground and the black or red wire is the hot lead. Notice that the receptacles are shown as viewed from the back, or wiring side.

rate 240-V circuit for the amplifier ensures that you stay within the capacity of the 120-V circuits in your shack. An amplifier drawing more current than the house wiring can handle could cause the house lights to dim. This is because the heavy current causes the power-line voltage to drop.

You should make sure everyone in your family knows where the main electrical box is located in your home. This box contains the fuses or circuit breakers for each electrical circuit in your home. Everyone should know how to turn off the electricity to your house in an emergency.

Three-Wire 120-V Power Cords

State and national electrical-safety codes require three-wire power cords on many 120-V tools and appliances. Two of the conductors (the "hot" and "neutral" wires) power the device. The third conductor (the safety ground wire) connects to the metal frame of the device. See Figure 4-10. The "hot" wire is usually black or red. The "neutral" wire is white. The frame/ground wire is green or sometimes bare. These colors are the most common. You may find other colors used on some equipment, however, especially if it is made for sale in Europe. There are a variety of possible color combinations, so if you don't see the red or black, white, and green or bare wires you should proceed with extreme caution.

Let's look at the power cord connections for a transmitter power supply. The power supply will have a transformer in it. When you attach the power cord, the black wire will attach to one end of the fuse. The white wire will go to the side of the transformer primary winding without a fuse. The green (or bare) wire will attach to the chassis.

When plugged into a properly wired mating receptacle, a three-contact plug connects the third conductor to an earth ground. This grounds the appliance chassis or frame and prevents the possibility of electric shock. A defective power cord that shorts to the case of the appliance will simply blow a fuse. Without the ground connection, the case could carry the full line voltage, presenting a severe shock hazard. All commercially manufactured electronic test equipment and most ac-operated amateur equipment uses these three-wire cords. Adapters are available for use where older electrical installations do not have mating receptacles. The lug from the adapter must be attached under the cover-plate screw. The outlet (and outlet box) must be grounded for this to be effective. Power wires coming into the electric box inside a flexible metal covering provide grounding through the metal covering. The common name for this type of wire is armored cable.

A "polarized" two-wire plug has one blade that is wider than the other. The mating receptacle will accept the plug only one way. This ensures that the hot and neutral wires in the appliance connect to the appropriate wires in the house electrical system. Consider what happens without this polarized plug and receptacle. The power switch in the equipment will be in the hot wire when the plug is inserted one way. It will be in the neutral wire when inserted the other way. This can present a dangerous condition. It is possible for the equipment to be "hot" even with the switch off. With the switch in the neutral line, the hot line may be connected to the equipment chassis. An unsuspecting operator could form a path to ground by touching the case, and might be electrocuted or receive a nasty shock!

Wiring an outlet or lamp socket properly is important. The black (hot) wire should be connected to the brass terminal on the lamp socket or outlet. The white (neutral) wire should be connected to the white or silver-colored terminal. This will ensure that the proper blade of the plug connects to the hot wire (that the polarity is correct). This practice is especially important when wiring lamp sockets. The brass screw of the socket connects to the center pin in the socket. With this pin as the "hot" connection, there is much less

Table 4-1
Current-Carrying Capability of Some Common Wire Sizes

Copper Wire Size (AWG)	*Allowable Ampacity (A)*	*Max Fuse or Circuit Breaker (A)*	*Permissible Load*[2] *(A)*
6	55	50	40
8	40	40	32
10	30	30	24
12	25 (20)[1]	20	16
14	20 (15)[1]	15	12

[1]The National Electrical Code limits the fuse or circuit breaker size (and as such, the maximum allowable circuit load current) to 15 A for #14 AWG and to 20 A for #12 AWG conductors.

[2]Adjustments from these values may be possible. Consult your local electrical inspector, a licensed electrician or a registered electrical engineer for assistance.

shock hazard. Anyone unscrewing a bulb from a correctly wired socket has to reach inside the socket to get a shock. In an incorrectly wired socket, the screw threads of the bulb are "hot." A dangerous shock may result when replacing a bulb.

Current Capacity

There is another factor to take into account when you are wiring an electrical circuit. It is the current-handling capability (*ampacity*) of the wire. Table 4-1 shows the current-handling capability of some common in-wall residential wire sizes. The table shows that number 14 wire could be used for a circuit carrying 15 A. You must use number 12 (or larger) wire for a circuit carrying 20 A. Wires smaller than number 14 may not be used for in-wall residential wiring.

To remain safe, don't overload the ac circuits in your home. The circuit breaker or fuse rating is the maximum load for the line at any one time. Simple addition can be used to calculate the current drawn on the circuit serving your ham radio equipment. Most equipment has the power requirements printed on the back. If not, the owner's manual should contain such information. Calculations must include any other household equipment or appliances on the same line (lighting, air-conditioning, fans, and so on). The sum of all the power requirements connected to the same circuit must not exceed the permissible loads shown in Table 4-1.

Do not put a larger fuse in an existing circuit — too much current could be drawn. (See Table 4-1 for the maximum fuse size for the existing wire size.) The wires would become hot and a fire could result. Installing too large of a fuse or circuit breaker violates the **National Electrical Code** and most local ordinances. Such violations may also invalidate your fire insurance! The National Electrical Code is used across the US as the basis for electrical safety relating to power wiring and antennas. If your transceiver blows a fuse in the main ac power line, you should find out what caused the fuse to blow and repair the problem. Replace the fuse only with another one of the same rating. For example if you replaced a 5-amp fuse with one rated for 30 amps the transceiver could draw too much current, causing the wires to overheat and even starting a fire.

Even in an automotive circuit, you must be sure all equipment has the proper current rating and appropriate fuses. If the main power switch in your radio became defective, you should check the current and voltage ratings of the original switch before replacing it. Buying an inexpensive switch rated for 1 amp is a bad idea if the transceiver draws 8 amps on transmit. Again, the new switch could overheat and become a safety hazard.

[Now turn to Chapter 14 and study questions T4A01 through T4A06. Also study questions T4B15 and T4B16. Review this section if you have any difficulty.]

POWER-SUPPLY SAFETY

Safety must always receive careful consideration during the design and construction of any power supply. Power supplies can produce potentially lethal currents and voltages. Be careful to guard against accidental exposure to these currents and voltages. Use electrical tape, insulated tubing (spaghetti) or heat-shrink tubing to cover exposed wires. This includes component leads, component solder terminals and tie-down points. Whenever possible, connectors used to mate the power supply to the outside world should be of an insulated type. They should be designed to prevent accidental contact with the voltages present. Power to the supply from the ac mains should be controlled by a clearly labeled front-panel switch. That way it can be seen and reached easily in an emergency.

All dangerous voltages in equipment should be made inaccessible. A good way to ensure this is to enclose all equipment in metal cabinets. That way no "hot" spots can be reached. Don't forget any component shafts that might protrude through the front panel. If a control shaft is hot, protect yourself from accidental contact by using an insulated shaft extension or insulated knob. Another precaution to take is to avoid wearing loose jewelry when working around potentially dangerous pieces of equipment, such as an amplifier. A metal bracelet or chain around your neck could accidentally touch a high-voltage point, giving you a shock, or possibly electrocuting you.

Table 4-2

Effects of Electric Current Through the Body of an Average Person

Current (1 Second Contact)	*Effect*
1 mA	Just perceptible.
5 mA	Maximum harmless current.
10-20 mA	Lower limit for sustained muscular contractions.
30-50 mA	Pain.
50 mA	Pain, possible fainting. "Can't let go" current.
100-300 mA	Normal heart rhythm disrupted. Electrocution if sustained current.
6A	Sustained heart contraction. Burns if current density is high.

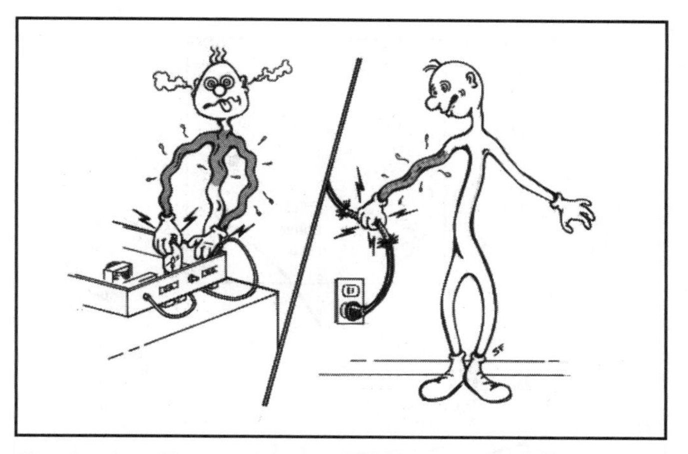
Figure 4-11—The path from the electrical source to ground affects how severe an electrical shock will be. The most dangerous path (from hand to hand directly through the heart) is shown at A. The path from one finger to the other shown at B is not quite so dangerous.

Each metal enclosure must be effectively grounded. This is generally accomplished by connections to the equipment grounding conductor (green or bare wire) contained in the power cord or premises wiring system. If a connection is made to a separate ground rod, the ground rod must be bonded to the building grounding electrode system as described earlier in this chapter. Then if a failure occurs inside a piece of equipment, the metal case will never present a shock hazard. The fuse will blow or the circuit breaker will trip instead.

You should never underestimate the potential hazard when working with electricity. Table 4-2 shows some of the effects of electric current — as little as 100 mA, or 1/10A, can be fatal! As the saying goes, "It's volts that jolts, but it's mills that kills." Low-voltage power supplies may seem safe, but even battery-powered equipment should be treated with respect. The minimum voltage considered dangerous to humans is 30 volts. These voltage and current ratings are only general guidelines. Automobile batteries are designed to provide very high current (as much as 200 A) for short periods when starting a car. This much current can kill you, even at 12 volts. You will feel pain if the shock current is in the range of 30 to 50 mA. Even a current as small as 2 mA (1/500 A) will give you a tingling sensation that might feel like pain.

A few factors affect just how little voltage and current can be considered dangerous. One factor is skin resistance. The lower the resistance of the path, the more current that will pass through it. If you perspire heavily, you may get quite a bit more severe shock than if your skin were dry. Another factor is the path through the body to ground. As Figure 4-11 shows, the most dangerous path is from one hand to the other or through one arm to the opposite leg. This path passes directly through a person's heart. Even a very minimal current can cause heart failure and death. Current passing from one finger to another on the same hand will not have quite such a serious effect. For this reason, if you must troubleshoot a live circuit, keep one hand behind your back or in your pocket. If you do slip, the shock may not be as severe as if you were using both hands.

If you should discover someone who is being burned by high voltage, immediately turn off the power, call for help and give cardio-pulmonary resuscitation (CPR). Be sure all family members know how to turn off power at the main power switch. These measures could save the life of a friend or family member.

[Now turn to Chapter 14 and study exam questions T4A07, T4A10 through T4A12 and T4A14 through T4A16. Review any topics necessary before proceeding.]

USING TEST EQUIPMENT

The Voltmeter

The **voltmeter** is an instrument used to measure voltage. It is a basic meter movement with a resistor in series, as shown in Figure 4-12. The current multiplied by the resistance will be the voltage drop across the resistance. An instrument used this way is calibrated in terms of the voltage drop across the resistor to read voltage.

A high value resistor provides a high impedance input for the meter. An ideal voltmeter would have an infinite input impedance. It would not draw any current from the circuit, and would not affect the circuit under test in any way. Real-world voltmeters, however, have a finite value

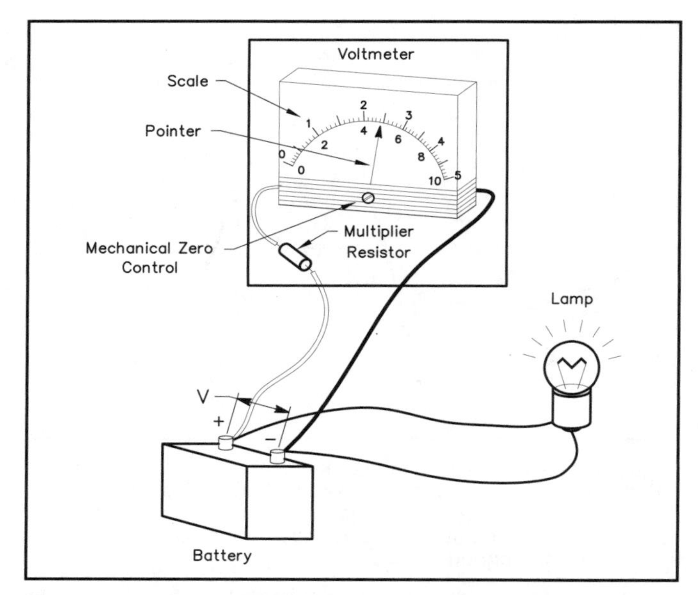

Figure 4-12—When you use a voltmeter to measure voltage, the meter must be connected in parallel with the voltage you want to measure.

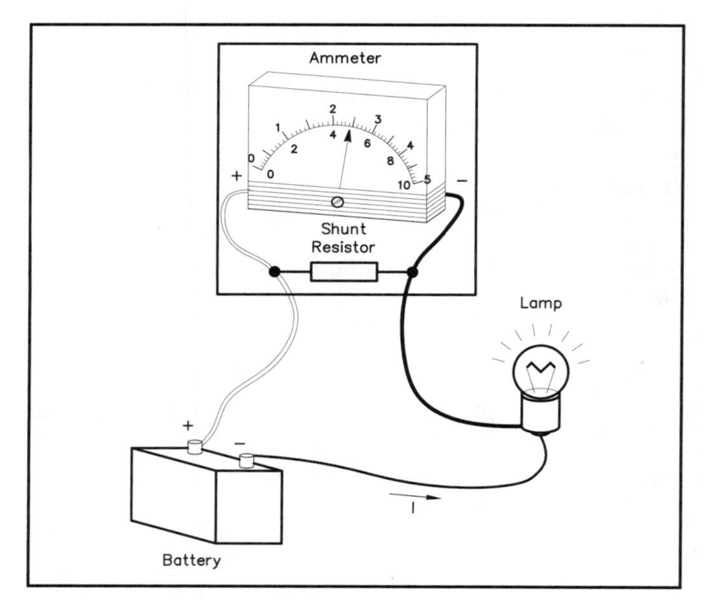

Figure 4-13—To measure current you must break the current at some point and connect the meter in series at the break. A shunt resistor expands the scale of the meter to measure higher currents than it could normally handle.

of input impedance. There is a possibility that the voltage you are measuring will change when the meter is connected to the circuit. This is because the meter adds some load to the circuit when you connect the voltmeter.

Use a voltmeter with a very high input impedance compared with the impedance of the circuit you are measuring. This prevents the voltmeter from drawing too much current from the circuit. Excessive current drawn from the circuit would significantly affect circuit operation. The input impedance of an ordinary voltmeter is about 20 kilohms per volt. Voltage measurements are made by placing the meter in parallel with the circuit voltage to be measured.

The range of a meter can be extended to measure a wide range of voltages. This is accomplished by adding resistance in series with it. Changing the multiplier resistor in Figure 4-12 changes the measuring range of the meter. You should always be sure to select a voltage range on your meter that is higher than the voltage you expect to measure. If you select a range that is much lower than the circuit voltage, the meter could draw too much current. This would burn out the meter movement, destroying the meter circuitry.

The Ammeter

The **ammeter** depends on a current flowing through it to deflect the needle. The ammeter is placed in series with the circuit. That way, all the current flowing in the circuit must pass through the meter. Many times the meter cannot handle all the current. The range of the meter can be extended by placing resistors in parallel with the meter to provide a path for part of the current. A shunt resistor is shown in Figure 4-13. Selecting different shunt resistors changes the current-measuring range. The shunt-resistor values are calculated so the total circuit current can be read on the meter. You must be sure to select a current measuring range that is greater than the current you expect to find in the circuit. For example, if you were to set your ammeter to measure microamps and then connect it to a circuit that draws 5 amps, you will probably burn out the meter, destroying the circuitry. The safest approach is to always start with the highest meter setting, and then switch to lower ranges as needed.

Most ammeters have very low resistance, but low-impedance circuits require caution. There is a chance that the slight additional resistance of the series ammeter will disturb circuit operation.

Multimeters

A **multimeter** is a piece of test equipment that most amateurs should know how to use. The simplest kind of multimeter is the volt-ohm-milliammeter (VOM). As its name implies, a VOM measures voltage, resistance and current. VOMs use one basic meter movement for all functions. The movement requires a fixed amount of current (often 1 mA) for a full-scale reading. As shown in the two previous sections, resistors are connected in series or parallel to provide the proper voltage or current meter reading. In a VOM, a switch selects various ranges for voltage, resistance and current measurements. This switch places high-value *multiplier* resistors in series with the meter movement for voltage measurement. It connects low-value *shunt* resistors in parallel with the movement for current measurement. These parallel and series resistors extend the range of the basic meter movement.

If you are going to purchase a VOM, buy one with the highest ohms-per-volt rating that you can find. Stay away from meters rated under 20,000 ohms per volt if you can.

Measuring resistance with a meter involves placing the meter leads across the component or circuit you wish to measure. Make sure to select the proper resistance scale. The full-scale-reading multipliers vary from 1 to 1000 and higher. The scale is usually compressed on the higher end of the range. See Figure 4-14. For best accuracy, keep the reading in the lower-resistance half of the scale. On most meters this is the right-hand side. Thus, if you want to measure a resistance of about 5000 ohms, select the R × 1000 scale. Then the meter will indicate 5.

A multimeter includes a battery for resistance measurements. The battery supplies a small current through the resistor or other component you are testing. You should never try to measure resistance in a circuit that has power applied to it. Ideally you will only measure resistance for components that are removed from the circuit. Trying to measure resistance of a part in a circuit will give an inexact measurement at best. If you have your meter connected in a circuit to make a voltage measurement, with power applied to the circuit, and then switch your meter to a resistance scale, you could burn out the meter movement. This would destroy the meter circuitry.

A vacuum-tube voltmeter (VTVM) operates in the same manner as an ordinary VOM. There is one important difference. The meter in a VTVM is isolated from the circuit under test by a vacuum-tube dc amplifier. As a result, the only additional circuit loading is from the tube input impedance, which is very high. The standard VTVM input impedance is 11 MΩ. This is useful for measuring voltages in high-impedance circuits, such as vacuum-tube grid circuits and FET gate circuits.

Another type of meter is a field-effect transistor volt-ohm-milliammeter (FET VOM). This instrument uses a field-effect transistor (FET) to isolate the indicating meter from the circuit to be measured. FET VOMs have an input impedance of several megohms. They are the solid-state equivalent of a VTVM.

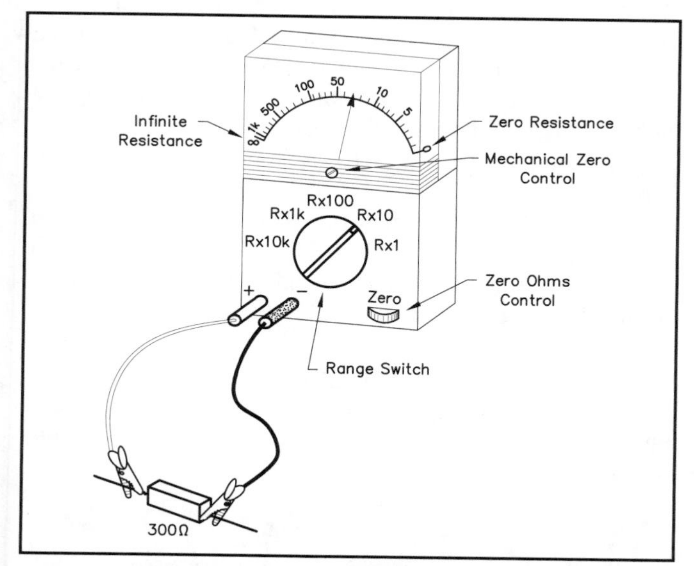

Figure 4-14—Resistance is measured across the component. For best accuracy the reading should be taken in the lower-resistance half of the scale.

Wattmeters

A **wattmeter** is a device connected in the transmission line to measure the power (in watts) coming out of a transmitter. Wattmeters are designed to operate at a certain line impedance, normally 50 ohms. Make sure the feed-line impedance is the same as the design impedance of the wattmeter. If impedances are different, any measurements will be inaccurate. Likewise, a power meter designed for use at 3-30 MHz (HF) will be inaccurate at VHF. For most accurate measurement, the wattmeter should be connected directly at the transceiver output (antenna) jack.

All wattmeters must contain some type of detection circuitry to enable you to measure power. The accuracy and upper frequency range of the wattmeter is limited by the detection device. Stray capacitance and coupling within the detection circuits can cause problems. A loss of sensitivity or accuracy occurs as the operating frequency is increased.

One type of wattmeter is a **directional wattmeter**. There are two kinds of directional wattmeters. One has a meter that reads forward power and another meter that reads reflected power. The other has a single meter that can be switched to read either forward or reflected power.

You can use a directional wattmeter to measure the output power from your transmitter. It is easy to become confused by the readings on your wattmeter, however. First measure and record the forward power, or power going from the transmitter to the antenna. Then measure the reflected power, which is any power coming back toward the transmitter from the antenna. To find the true power from your transmitter, you must subtract the reflected power from the forward reading. The power reflected from the antenna will again be reflected by the transmitter. This power adds to the forward power reading on the meter.

Suppose your transmitter power output is 80 watts and 10 watts of power is reflected from the antenna. Those 10 watts will be added into the forward reading on your meter when they are reflected from the transmitter. Your forward wattmeter reading would show 90 watts, which is incorrect. To find true forward power you must subtract the reflected measurement from the forward power measurement:

$$\text{True forward power} = \text{Forward power reading} - \text{Reflected power reading.} \quad \text{(Eq 4-1)}$$

For example, suppose you have a wattmeter connected in the line from your transmitter. It gives a forward-power reading of 96 watts and a reflected-power reading of 4 watts. What is the true power out of your transmitter?

$$\text{True forward power} = 96\text{ W} - 4\text{ W} = 92\text{ watts.}$$

The main reason for using a wattmeter in your amateur station is to ensure that you are not exceeding the maximum power allowed by your license. The FCC specifies this maximum power in terms of peak envelope power or PEP. For this reason, the most accurate way to measure your power is with a peak-reading RF wattmeter. Some wattmeters are calibrated to measure average power rather than peak power. You can use an average-reading meter, but you must be aware that if your *average power* is at or close to the maximum limit then your *peak envelope power* may be higher than the limit!

[Now turn to Chapter 14 and study exam questions T4B01 through T4B14 as well as question T4C08. Review this section as needed.]

FREQUENCY ACCURACY

It's important to ensure that your transceiver frequency dial is accurate. If it isn't, you could transmit outside your allowed frequencies and be in violation of FCC rules. There are two good ways to ensure that your dial is accurate: use of a marker generator (crystal calibrator) and checking your dial by tuning in radio stations WWV or WWVH.

Marker Generators

The **marker generator** is a high-stability oscillator that produces an RF signal with a frequency that doesn't change. It is usually built into the receiver. This oscillator generates a series of signals that mark the exact edges of the amateur bands (and subbands, in some cases). It does this by oscillating at a low frequency that has harmonics falling on the desired frequencies. Most marker generators put out harmonics at 25, 50 or 100-kHz intervals. Since marker generators normally use crystal oscillators, they are often called **crystal calibrators**.

The marker generator is very useful. You can calibrate your receiver with it. You can then determine your precise transmitter frequency. First, turn on your marker generator (calibrator). This injects the calibrator signal into the receiver circuit. Then set your transceiver or receiver dial on the proper frequency marks. Calibration instructions are usually provided in your operator's manual. When the dial is calibrated, put your transmitter in the tune or spot position and read the frequency on the receiver dial.

The marker frequencies must be accurate. Now you know that the transmitter frequency is between the markers that show the ends of the band (or subband). In addition, the transmitter frequency must not be too close to the edge of the band (or subband). If it is, the sideband frequencies, especially in a voice transmission, will extend over the edge.

WWV and WWVH

One easy way to calibrate your receiver frequency readout or dial is to tune in **WWV** (Fort Collins, Colorado) or **WWVH** (Kauai, Hawaii). These stations transmit reference signals on 2.5, 5, 10, 15 and 20 MHz. When you tune in WWV or WWVH, you'll hear a "tic" pulse every second and the precise time announced each minute. At 18 minutes past each hour these stations also transmit some measurements and information that can help you predict the propagation of your radio signals.

Signal Generators

A **signal generator** produces a stable, low-level signal that can be set to a specific frequency. There are two different types of signal generators. *Audio-frequency signal generators* create signals in the audio range. *Radio-frequency signal generators* provide radio-frequency signals. These signals can be used to align or adjust tuned circuits for best performance. One common use of a signal generator is in the alignment of receivers. The generator can be adjusted to the desired signal frequency. Then the associated tuned circuits can be adjusted for best operation. This is sometimes referred to as *aligning* the tuned circuits. Proper adjustment is indicated by the appropriate output meter (maximum signal strength, for instance).

Sometimes a band of frequencies must be covered to chart the frequency response of a filter. For this application, a swept frequency generator is used. A swept frequency generator automatically sweeps back and forth over a selected range of frequencies.

You can also use a signal generator to adjust the filter circuits in your transmitter. When you do, a dummy antenna must be connected to the transmitter output to avoid radiating a signal (dummy antennas are discussed in the "Station Accessories" section later in this chapter).

[Now turn to Chapter 14 and study questions T4C01 through T4C05 and questions T4C10 and T4C11. Review any material as necessary.]

ANTENNA SYSTEM MEASUREMENTS

A properly operating antenna system is essential for a top-quality amateur station. If you build your own antennas, you must be able to tune them for maximum operating efficiency. Even if you buy commercial antennas, there are tuning adjustments that must be made because of differences in mounting location. Height above ground and proximity to buildings and trees will have some effect on how your antenna operates. You should be able to monitor your antenna system periodically for signs of problems, and troubleshoot failures as necessary.

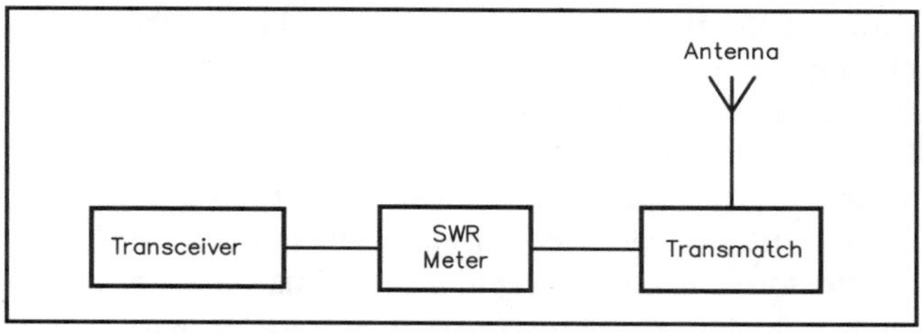

Figure 4-15—A reflectometer or SWR meter should be connected between the transmitter and Transmatch in order to adjust the Transmatch properly.

Impedance Match Indicators

A useful metering device is the **reflectometer** (**SWR meter**). A reflectometer is a device used to measure *standing-wave ratio*, or *SWR*. SWR is a measure of the relationship between the amount of power traveling to the antenna and the power reflected back to the transmitter. The reflection is caused by *impedance mismatches* in the antenna system. The energy going from the transmitter to the antenna is represented by the forward, or incident, voltage. The energy reflected by the antenna is represented by the reflected voltage.

A reflectometer is placed in the transmission line between the transmitter and antenna. In the most common amateur application, the reflectometer is connected between the transmitter and Transmatch, as shown in Figure 4-15. A Transmatch is used to cancel out the capacitive or inductive component in the antenna impedance. With a Transmatch you can match the impedance of your antenna system to the transmitter output, normally 50 ohms. The reflectometer is used to indicate minimum reflected power as the Transmatch is adjusted. This indicates when the antenna system (including feed line) is matched to the transmitter output impedance.

To obtain a valid measure of the impedance match between an *antenna* and the *feed line*, place the SWR meter at the antenna feed point. If you put the meter at the transmitter end, feed line losses make SWR readings look lower than they really are. Most SWR meters are designed for HF use (from 3 to 30 MHz). These *may* also be useful at VHF, as long as the meter properly calibrates to full scale in the set position.

[Now turn to Chapter 14 and study questions T4C06, T4C07 and T4C09. Review any material as necessary.]

STATION ACCESSORIES

Until now, we have talked about the bare necessities — what you need to set up a basic Amateur Radio station. Few hams are content with the bare necessities, however. Most add accessories that make operating more convenient and enjoyable. Let's look at several popular ones: dummy antennas, antenna switches, keyers, station clocks and logbooks.

Dummy Antenna

A **dummy antenna**, sometimes called a *dummy load*, is nothing more than a large resistor. It replaces your antenna when you want to operate your transmitter without radiating a signal. The dummy antenna safely converts the RF energy coming out of your transmitter into heat. The heat goes into the air or into the coolant, depending on the type of dummy antenna. It does all this while presenting your transmitter with a constant 50-ohm load. Dummy antennas are used to make test transmissions, or to make tuning adjustments after changing bands. Be sure the one you choose is rated for the type of operating (HF or VHF/UHF) and power level you'll be using.

Relatively inexpensive, a dummy antenna is one of the most useful accessories you can own. Every conscientious amateur should own one. You can build your own or buy one that's ready to use (see Figure 4-16). The container, often a gallon paint can, acts as a shield to keep RF energy from being radiated. When filled with transformer oil, it allows the dummy antenna to dissipate RF power. You can use an antenna switch to connect the dummy antenna to the transmitter. See Figure 4-17.

With a dummy antenna connected to the transmitter,

you can make off-the-air tests — no signal goes out over the air. For transmitter tests, remember that a dummy antenna is a resistor. It must be capable of safely dissipating the entire (continuous) power output of the transmitter. A dummy antenna will often get warm during use, since the RF energy from the transmitter is turned into heat.

A dummy antenna should provide the transmitter with a perfect load. This usually means that it has a pure resistance (with no reactance) of approximately 50 ohms. The resistors used to construct dummy antennas must be noninductive. Composition resistors are usable, but wire-wound resistors are not. A single high-power resistor is best. Several lower-power resistors can be connected in parallel to obtain a 50-ohm load capable of dissipating high power. You must be sure the dummy antenna is able to handle the full transmitter power. For example, for a 100-watt SSB transmitter, you will need a dummy antenna that is rated to handle at least 100 watts of power.

Some older amateur literature suggested using a standard light bulb as a dummy antenna. A 100-watt light bulb would be used for a 100-watt transmitter, for example. The transmitter would be adjusted for maximum brightness of the bulb. While this was probably an acceptable dummy antenna for a tube-type radio, it should not be used with a modern transistorized radio. The impedance of the light bulb changes significantly as bulb heats up.

The S Meter

S meters indicate the relative strength of a received signal. They are calibrated in S units from S1 to S9; above S9 they are calibrated in decibels (dB), as shown in Figure 4-18. In the 1940s at least one manufacturer made an attempt to establish some significant numbers for S meters. S9 was to be equal to a signal level of 50 microvolts, with each S unit equal to 6 dB. This means that for an increase of one S unit, the received signal power would have to increase by a factor of four. Such a scale is useful for a theoretical discussion. Real S-meter circuits fall far short of this ideal for a variety of reasons.

Figure 4-16—A dummy antenna like this one uses a liquid coolant, often transformer oil, to dissipate heat. Dry dummy antennas dissipate heat into the air.

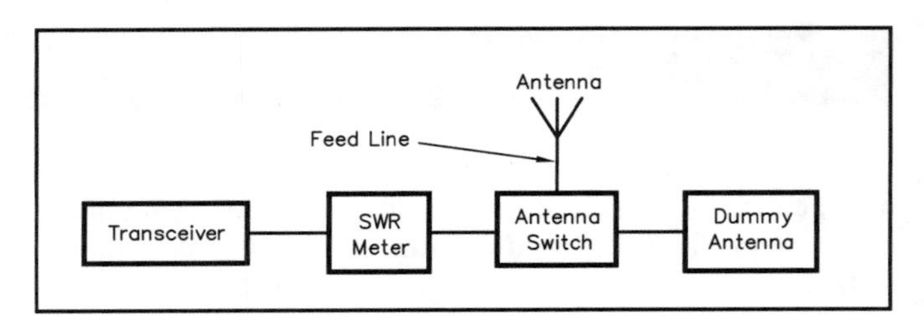

Figure 4-17—A dummy antenna is just a resistance that provides your transmitter with a proper load when you are tuning up and don't want to radiate a signal. Connect it to your antenna switch.

Suppose you were working someone who was using a 25-watt transmitter, and your S meter was reading S7. If the other operator increased power to 100 watts (an increase of four times), your S meter would read S8. If the signal was S9, the operator would have to multiply his power by a factor of 10 to make your S meter read 10 dB over S9. It would take another 10 times increase in signal strength to make the meter read 20 dB over S9.

S meters on modern transceivers may or may not respond in this manner. S-meter operation is based on the output of the automatic gain-control (AGC) circuitry; the S meter measures the AGC voltage. As a result, every S meter responds differently; no two S meters will give the same reading. The S meter is useful for giving relative signal-strength indications, however. You can see changes in signal levels on an S meter that you may not be able to detect just by listening to the audio output level.

[Now turn to Chapter 14 and study questions T4D01 through T4D11. Review the material in this section if you have any difficulty with these questions.]

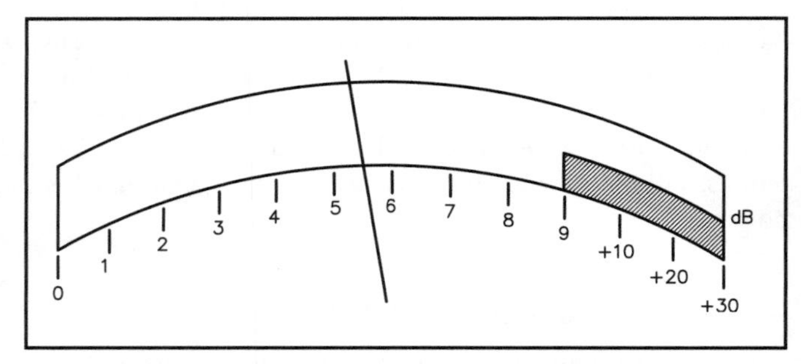

Figure 4-18—An S meter is calibrated in S units up to S9. Above S9 it is calibrated in decibels (dB) over S9.

5 KEY WORDS

Alternating current (ac) — Electrical current that flows first in one direction in a wire and then in the other. The applied voltage is also changing polarity. This direction reversal continues at a rate that depends on the frequency of the ac.

Ampere (A) — The basic unit of electrical current. Current is a measure of the electron flow through a circuit. If we could count electrons, we would find that if there are 6.24×10^{18} electrons moving past a point in one second, we have a current of one ampere.[1] We abbreviate amperes as amps.

Audio frequency (AF) — The range of frequencies that the human ear can detect. Audio frequencies are usually listed as 20 Hz to 20,000 Hz.

Battery — A device that stores electrical energy. It provides excess electrons to produce a current and the voltage or EMF to push those electrons through a circuit.

Centi — The metric prefix for 10^{-2}, or divide by 100.

Closed, or **complete circuit** — An electrical circuit with an uninterrupted path for the current to follow. Turning a switch on, for example, closes or completes the circuit, allowing current to flow.

Conductor — A material that has a loose grip on its electrons, so an electrical current can pass through it.

Current — A flow of electrons in an electrical circuit.

Deci — The metric prefix for 10^{-1}, or divide by 10.

Direct current (dc) — Electrical current that flows in one direction only.

Electromotive force (EMF) — The force or pressure that pushes a current through a circuit.

Electron — A tiny, negatively charged particle, normally found in an area surrounding the nucleus of an atom. Moving electrons make up an electrical current.

Energy — The ability to do work; the ability to exert a force to move some object.

Frequency — The number of complete cycles of an alternating current that occur per second.

Giga — The metric prefix for 10^9, or times 1,000,000,000.

Hertz (Hz) — An alternating-current frequency of one cycle per second. The basic unit of frequency.

Insulator — A material that maintains a tight grip on its electrons, so that an electric current cannot pass through it (within voltage limits).

Kilo — The metric prefix for 10^3, or times 1000.

Mega — The metric prefix for 10^6, or times 1,000,000.

Metric prefixes — A series of terms used in the metric system of measurement. We use metric prefixes to describe a quantity as compared to a basic unit. The metric prefixes indicate multiples of 10.

Metric system — A system of measurement developed by scientists and used in most countries of the world. This system uses a set of prefixes that are multiples of 10 to indicate quantities larger or smaller than the basic unit.

Micro — The metric prefix for 10^{-6}, or divide by 1,000,000.

Milli — The metric prefix for 10^{-3}, or divide by 1000.

Ohm — The basic unit of electrical resistance, used to describe the amount of opposition to current.

Ohm's Law — A basic law of electronics. Ohm's Law gives a relationship between voltage (E), current (I) and resistance (R). The voltage applied to a circuit is equal to the current through the circuit times the resistance of the circuit ($E = IR$).

Open circuit — An electrical circuit that does not have a complete path, so current can't flow through the circuit.

Pico — The metric prefix for 10^{-12}, or divide by 1,000,000,000,000.

Power — The rate of energy consumption. We calculate power in an electrical circuit by multiplying the voltage applied to the circuit times the current through the circuit ($P = IE$).

Power supply — An electrical circuit that provides excess electrons to flow into another circuit. The power supply also supplies the voltage or EMF to push the electrons along. Power supplies convert a power source (such as the ac mains) to a form useful for various circuits.

Radio frequency (RF) — The range of frequencies that can be radiated through space in the form of electromagnetic radiation. We usually consider RF to be those frequencies higher than the audio frequencies, or above 20 kilohertz.

Resistance — The ability to oppose an electric current.

Resistor — Any material that opposes a current in an electrical circuit. An electronic component especially designed to oppose current.

Short circuit — An electrical circuit in which the current does not take the desired path, but finds a shortcut instead. Often the current goes directly from the negative power-supply terminal to the positive one, bypassing the rest of the circuit.

Volt (V) — The basic unit of electrical pressure or EMF.

Voltage — The EMF or pressure that causes electrons to move through an electrical circuit.

Voltage source — Any source of excess electrons. A voltage source produces a current and the force to push the electrons through an electrical circuit.

Watt (W) — The unit of power in the metric system. The watt describes how fast a circuit uses electrical energy.

Wavelength — The distance an ac signal will travel during the time it takes the signal to go through one complete cycle.

[1]Numbers written as a multiple of some power are expressed in exponential notation. This notation is explained in detail on page 5-2.

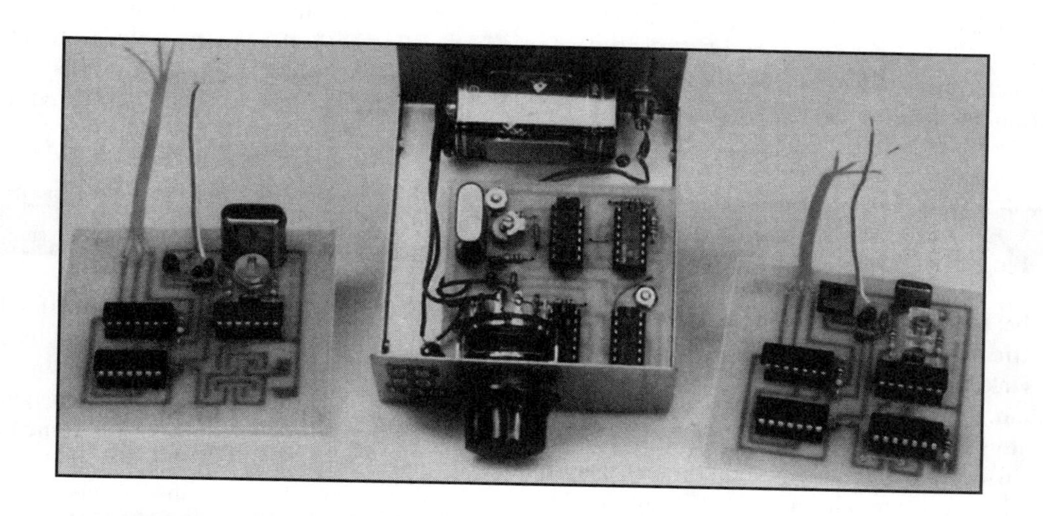

5 Understanding Basic Electrical Principles

We can't see electricity as it flows through a circuit, yet we need some understanding of how it works to design even a simple circuit.

This chapter introduces you to basic electronics and radio theory. First we present the theory that you'll need to pass your Novice exam. This basic theory will be a foundation for you to build on. Later in the chapter is a section devoted to the theory needed to pass the Technician exam. Even if you don't plan to try for your Technician license right away, you'll probably want to study this section, too. The theory you learn in this chapter will also help you to assemble and operate your amateur station.

There is much for you to learn, and most of the material presented here will be new to you. We will start with a description of the metric system of measure. Many of the measurements we will make in electronics are based on metric units, so an understanding of this measuring system is important. Then we will cover some basic electrical principles. Here you'll learn the basics like voltage, current, conductors and insulators. Then we'll move on to electronics fundamentals, where you learn about resistance, Ohm's Law and power. You'll also learn about direct and alternating currents. The Technician sections of this chapter explain more about alternating current, as it applies to inductance and capacitance.

To get the most from this chapter, you should take it one section at a time. Study the material in each section and really know it before you go on to the next section. The sections build on each other, so you may find yourself referring to sections you've already studied from time to time.

We use many technical terms in electronics. We have provided definitions that are as simple and to-the-point as possible. You will want to refer often to the **Key Words** at the beginning of this chapter. Don't be afraid to turn back to any section you have already studied. This review is helpful if you come across a term that you are not sure about. You will probably not remember every bit of this theory just by reading the chapter once.

This chapter includes many drawings and illustrations to help you learn the material. Pay attention to these graphics, and you'll find it easier to understand the text. We'll direct you to the Question Pools (Chapters 13 and 14) at appropriate points in the text. Use these directions to help you study the Novice and Technician Question Pools. When you think you've learned the section, you are ready to move on.

If you have trouble understanding parts of this chapter, ask your instructor or another experienced ham for help. Many other books can help, too. ARRL's *Understanding Basic Electronics* is written for students with no previous electronics background. The booklet *First Steps in Radio* is a collection of articles from a *QST* series by Doug DeMaw. *First Steps* covers a wide range of technical topics written especially for a beginner. To study more advanced theory, you may want to purchase a copy of *The ARRL Handbook*. These publications are available from your local ham radio dealer or from ARRL Headquarters.

Remember! Take it slowly, section by section. Before you know it, you'll have learned what you need to know to pass your test and get on the air. Good luck!

THE METRIC SYSTEM

We'll be talking about the units used to describe several electrical quantities later in this chapter. Before we do that, let's take a few minutes to become familiar with the **metric system**. This simple system is a standard system of measurement used all over the world. All the units used to describe electrical quantities are part of the metric system.

In the US, we use a measuring system known as the US Customary system for many physical quantities, such as distance, weight and volume. In this system there is no logical progression between the various units. For example, we have 12 inches in 1 foot, 3 feet in 1 yard and 1760 yards in 1 mile. For measuring the volume of liquids we have 2 cups in 1 pint, 2 pints in 1 quart and 4 quarts in 1 gallon. To make things even more difficult, we use some of these same names for different volumes when we measure dry materials! As you can see, this system of measurements can be very confusing. Even those who are very familiar with the system do not know all the units used for different types of measurements. Not many people know what a *slug* is, for example.

It is exactly this confusion that led scientists to develop the orderly system we know today as the metric system. This system uses a basic unit for each different type of measurement. For example, the basic unit of length is the meter. (This unit is spelled metre nearly everywhere in the world except the US!) The basic unit of volume is the liter (or litre). The unit for mass (or quantity of matter) is the gram. The newton is the metric unit of force, or weight, but we often use the gram to indicate how "heavy" something is. We can express larger or smaller quantities by multiplying or dividing the basic unit by factors of 10 (10, 100, 1000, 10,000 and so on). These multiples result in a standard set of prefixes, which can be used with all the basic units. Table 5-1 summarizes the most-used **metric prefixes**. These same prefixes can be applied to any basic unit in the metric system. Even if you come across some terms you are unfamiliar with, you will be able to recognize the prefixes.

We can write these prefixes as powers of 10, as shown in Table 5-1. The power of 10 (called the *exponent*) shows how many times you must multiply (or divide) the basic unit by 10. For example, we can see from the table that **kilo** means 10^3. Let's use the meter as an example. If you multiply a meter by 10 three times, you will have a *kilo*meter. (1 meter $\times 10^3 = 1$ m $\times 10 \times 10 \times 10 = 1000$ meters, or 1 kilometer.) If you multiply 1 meter by 10 six times, you have a **mega**meter. (1 meter $\times 10^6 = 1$ m $\times 10 \times 10 \times 10 \times 10 \times 10 \times 10 = 1{,}000{,}000$ meters or 1 megameter.)

Notice that the exponent for some of the prefixes is a negative number. This indicates that you must *divide* the basic unit by 10 that number of times. If you divide a meter by 10, you will have a **deci**meter. (1 meter $\times 10^{-1} = 1$ m $\div 10 = 0.1$ meter, or 1 decimeter.) When we write 10^{-6}, it means you must divide by 10 six times. (1 meter $\times 10^{-6} = 1$ m $\div 10 \div 10 \div 10 \div 10 \div 10 \div 10 = 0.000001$ meter, or 1 **micro**meter.)

We can easily write very large or very small numbers with this system. We can use the metric prefixes with the basic units, or we can use powers of 10. Many of the quantities used in basic electronics are either very large or very small numbers, so we use these prefixes quite a bit. You should be sure you are familiar at least with the following prefixes and their associated powers of 10: **giga** (10^9), **mega** (10^6), **kilo** (10^3), **centi** (10^{-2}), **milli** (10^{-3}), **micro** (10^{-6}) and **pico** (10^{-12}).

Let's try an example. We have a receiver dial calibrated in megahertz (MHz), and it shows a signal at a frequency of 3.525 MHz. Where would a dial calibrated in kilohertz show the signal? From Table 5-1 we see that kilo means times 1000, and mega means times 1,000,000. That means that our signal is at 3.525 MHz $\times$ 1,000,000 =

Table 5-1
International System of Units (SI) — Metric Units

Prefix	*Symbol*	*Multiplication Factor*
tera	T	10^{12} = 1,000,000,000,000
giga	G	10^9 = 1,000,000,000
mega	M	10^6 = 1,000,000
kilo	k	10^3 = 1,000
hecto	h	10^2 = 100
deca	da	10^1 = 10
(unit)		10^0 = 1
deci	d	10^{-1} = 0.1
centi	c	10^{-2} = 0.01
milli	m	10^{-3} = 0.001
micro	m	10^{-6} = 0.000001
nano	n	10^{-9} = 0.000000001
pico	p	10^{-12} = 0.000000000001

10^{18} 10^{15} 10^{12} 10^{9} 10^{6} 10^{3} 10^{2} 10^{1} 10^{0} 10^{-1} 10^{-2} 10^{-3} 10^{-6} 10^{-9} 10^{-12} 10^{-15} 10^{-18}

E • • P • • T • • G • • M • • k h da U d c m • • µ • • n • • p • • f • • a

Figure 5-1—This chart shows the symbols for all metric prefixes, with the power of 10 that each represents. Write the abbreviations in decreasing order from left to right. The dots between certain prefixes indicate there are two decimal places between those prefixes. You must be sure to count a decimal place for each of these dots when converting from one prefix to another. When you change from a larger to a smaller prefix, you are moving to the right on the chart. The decimal point in the number you are changing also moves to the right. Likewise, when you change from a smaller to a larger prefix, you are moving to the left, and the decimal point also moves to the left.

3,525,000 hertz. There are 1000 hertz in a kilohertz, so 3,525,000 divided by 1000 gives us 3525 kHz.

How about another one? If we have a current of 3000 milliamperes, how many amperes is this? From Table 5-1 we see that milli means multiply by 0.001 or divide by 1000. Dividing 3000 milliamperes by 1000 gives us 3 amperes. The metric prefixes make it easy to use numbers that are a convenient size simply by changing the units. It is certainly easier to work with a measurement given as 3 amperes than as 3000 milliamperes!

Notice that it doesn't matter what the units are or what they represent. Meters, hertz, amperes, volts, farads or watts make no difference in how we use the prefixes. Each prefix represents a certain multiplication factor, and that value never changes.

By now you should begin to understand how to change prefixes in the metric system. First write the number and find the proper power of 10 (from memory or Table 5-1), and then move the decimal point to change to the basic unit. Then divide by the multiplication factor for the new prefix you want to use. With a little more practice you'll be changing prefixes with ease.

There is another method you can use to convert between metric prefixes, but it involves a little trick. Learn to write the chart shown in Figure 5-1 on a piece of paper when you are going to make a conversion. Always start with the large prefixes on the left and go toward the right with the smaller ones. Sometimes you can make an abbreviated list, using only the units from kilo to milli. If you need the units larger than kilo or smaller than milli, be sure to include the dots as shown in Figure 5-1. (They mark the extra decimal places between the larger and smaller prefixes, which go in steps of 1000 instead of every 10.) Once you learn to write the chart correctly, it will be very easy to change prefixes.

Let's work through an example to show how to use this chart. For this example, we'll use a term that you will run into quite often in your study of electronics: **hertz**. Hertz (abbreviated Hz) is a unit that refers to the frequency of a radio or television wave.

Change 3725 kilohertz to hertz. Since we are starting with kilohertz (kilo), begin at the k on the chart. Now count each symbol to the right, until you come to the basic unit (U). Did you count three places? Well that's how many places you must move the decimal point to change from kilohertz (kHz) to hertz (Hz). Which way do you move the decimal point? Notice that you counted to the right on the chart. Move the decimal point in the same direction. Now you can write the answer: 3725 kHz = 3,725,000 Hz!

Suppose a meter indicates a voltage of 3500 millivolts (abbreviated mV) across a circuit. How many volts (abbreviated V) is that? First, write the list of metric prefixes. Since you won't need those smaller than milli or larger than kilo, you can write an abbreviated list. You don't have to write the powers of 10, if you remember what the prefixes represent. To change from milli to the unit, we count 3 decimal places toward the left. This tells us to move the decimal point in our number three places to the left.

3500 mV = 3.5 V

Let's try one more example, for some extra practice changing metric prefixes by moving the decimal point in a number. What if someone told you to tune your radio receiver to 3,725,000 Hz? You probably won't find any radio receiver with a dial marking like this! To make the number more practical, we'll write the frequency with a prefix that's more likely to appear on a receiver dial.

Our first step is to select a new prefix to express the number. We can write the number with only one or two digits to the left of the decimal point. It looks like we'll need the entire prefix chart for this one, so write it down as described earlier. (You can look at the chart in Figure 5-1, but you should practice writing it for those times when you don't have the book — like your exam!)

The next job is to count how many places you can move the decimal point. The number you end up with should have one or two digits to the left of the decimal point. Remember that metric prefixes larger than kilo represent multiples of 1000, or 10^3. Did you count six places to move the decimal point in our example, 3,725,000 Hz? That would leave us with 3.725×10^6 Hz.

Now go back to the chart and count six places to the left. (This is the same number of places and the same direction as we moved the decimal point.) The new spot on the chart indicates our new metric prefix, mega, abbreviated M. Replacing the power of 10 with this prefix, we can write our frequency as 3.725 MHz.

[Before you go on to the next section, turn to Chapter 13. Be sure you can answer all of the questions in the Novice Question Pool with numbers between N5A01 and N5A13. Review this section if you have any difficulty.]

BASIC ELECTRICAL PRINCIPLES

In this section, you will learn what electricity is and how it works. We'll introduce you to the atom and the electron, the basic elements of electricity. There are no questions about atoms and electrons on the Novice or Technician exams, but understanding them will help you understand the rest of this chapter.

Electricity

The word is a spine-tingling mystery. It's the force behind our space-age civilization. It's one of nature's greatest powers. We love it; we fear it. We use it in our work and play. But what is it?

It's a mystery only in our minds. Actually, electricity is the marvelous stuff which, when untamed, we call lightning. One lightning bolt produces enough electricity to supply your needs for a lifetime. In another form, electricity is the power in a battery that cranks the engine to start your car. Electricity also ignites the gasoline in the engine. Yet, with all its power, electricity is the careful messenger carrying information from your brain to your muscles, enabling you to move your arms and legs. You can buy a small container of electricity no bigger than a dime (a battery). Electric utilities generate and transmit huge amounts of electricity every day. From lightning bolts to brain waves, it's all the same stuff: **electrons**.

Inside Atoms

Everything you can see and touch is made up of *atoms*. Atoms are the building blocks of nature. Atoms are too small to see, but the *subatomic particles* inside atoms are even smaller.

A phenomenon that has fascinated mankind for ages, lightning is simply a natural source of electricity.

Each atom has a *nucleus* in its center. Other particles orbit around this central core. Think of the familiar maps of our solar system: planets orbit the sun. In an atom, charged particles orbit the central core (the nucleus). Other charged particles make up the nucleus. Figure 5-2 is a simplified illustration of an atom's structure.

Some particles have *negative charges* while others have *positive charges*. The core of an atom contains positively charged particles. Negative particles, called **electrons**, orbit around the nucleus. Scientists have identified more than 100 different kinds of atoms. The number of positively and negatively charged particles in an atom determines what type of element that atom is. Different kinds of atoms combine to form various materials. For example, a hydrogen atom has one positively charged particle in its nucleus and one electron around the outside. An oxygen atom has eight positive particles in the nucleus and eight electrons around the outside. When two hydrogen atoms combine with one oxygen atom, we have water.

Have you ever tried to push the north poles of two magnets together? Remember that soft, but firm, pressure holding them apart? Similar poles in magnets repel each other; opposite poles attract each other. You can feel this if you experiment with a pair of small magnets.

Charged particles behave in a way similar to the two magnets. A positively charged particle and a negatively charged particle attract each other. Two positive or two negative particles repel each other. Like repels like; opposites attract.

Electrons stay near the central core, or nucleus, of the atom. The positive charge on the nucleus attracts the negative electrons. Meanwhile, since the electrons are all nega-

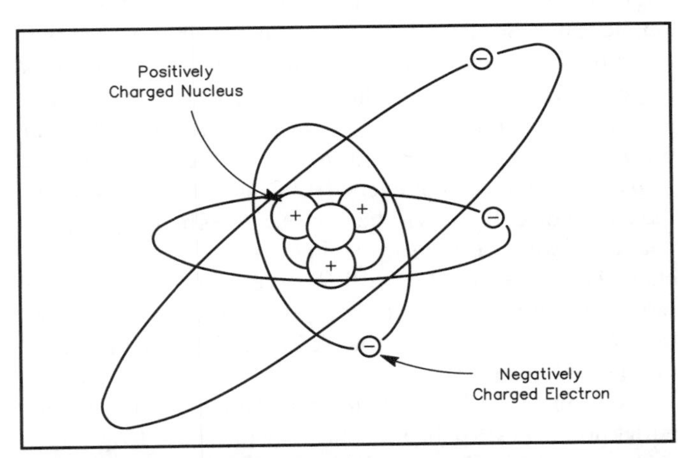

Figure 5-2—Each atom is a microscopic particle composed of a central, dense, positively charged nucleus, surrounded by tiny negatively charged electrons. There are the same number of positive particles in the nucleus and negative electrons outside the nucleus.

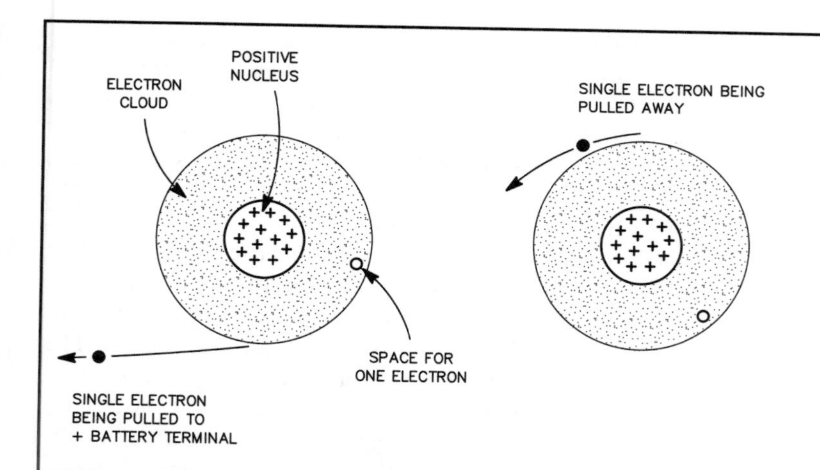

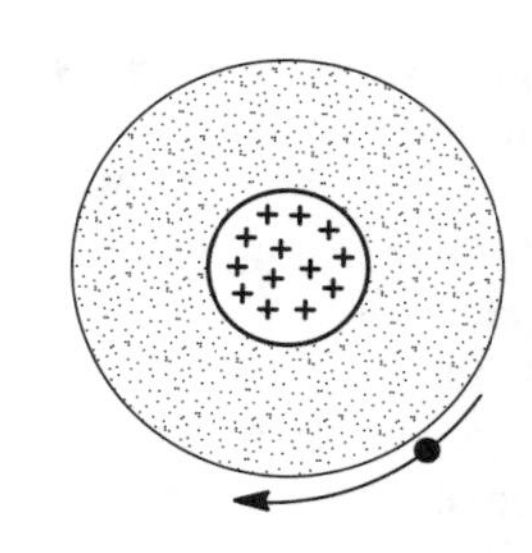

Figure 5-3 — An ion is an elec-trically charged particle. When an electron (a negative ion) moves from one atom to another, the atom losing the electron becomes a positive ion. This electron movement represents an electric current. The electrons are moving from right to left in this drawing.

tively charged, they repel each other. This makes the electrons move apart and fill the space around the nucleus. (Scientists sometimes refer to this area as an electron "cloud.") An atom has an equal balance of negative and positive charges and shows no electrical effect to the outside world. We say the atom is *neutral*.

Electron Flow

In many materials, especially metals, it's easy to dislodge an electron from an atom. When the atom loses an electron, it upsets the stability and electrical balance of the atom. With one particle of negative charge gone, the atom has an excess of positive charges. The free electron has a negative charge. We call the atom that lost the electron a positive *ion* because of its net positive charge. (An ion is a charged particle.) If there are billions of similar ions in one place, the quantity of charge becomes large enough to cause a noticeable effect.

Positively charged ions can pull negative particles (electrons) from neutral atoms. These electrons can move across the space between the atom and the ion and orbit the positively charged ion. Now the positively charged ion has become a neutral atom again, and another atom has become a positively charged ion! If this process seems confusing, take a look at Figure 5-3. Here we show a series of atoms and positive ions, with electrons moving from one atom to the next. The electrons are moving from right to left in this diagram. We call this flow of electrons *electricity*. Electricity is nothing more than the flow of electrons.

How difficult is it for a positive ion to rip an electron away from an atom? That depends on the individual atoms making up a particular material. Some atoms hold firmly to their electrons and won't let them flow away easily. Other atoms keep only a loose grip on electrons and let them slip away easily. This means that some materials carry electricity better than others. **Conductors** are materials that keep only a loose grip on their electrons. We call those materials that hold tightly to their electrons **insulators**.

DEMONSTRATING ELECTRON FLOW

Here's an easy way to see the power of electrical charges. On a dry day hang some metal foil from dry thread. Rub one end of a plastic comb on wool (or run it through your hair). Then bring the end of the comb near the metal foil.

Rubbing the comb on wool detaches electrons from the wool fibers, depositing them on the surface of the comb. When you bring the charged end of the comb near the foil, free electrons in the foil will be repelled by the negative charge on the comb. The free electrons will move as far from the comb as they can—on the edge farthest from the comb. This leaves the near edge of the foil with a shortage of electrons (in other words, a positive charge). The near edge of the foil is then attracted to the negatively charged comb.

As soon as the foil touches the comb, some of the excess electrons on the comb flow onto the foil. (Electricity!) The foil will then have a net negative charge, as the comb does, and the foil will be repelled by the comb. This simple experiment shows the attraction, repulsion and flow of electrons—the heart of electricity.

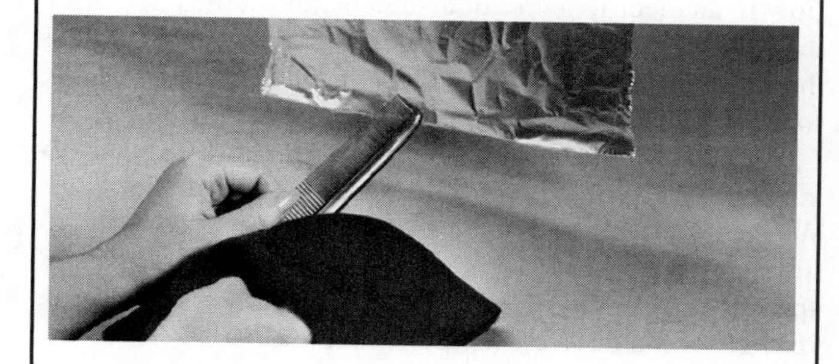

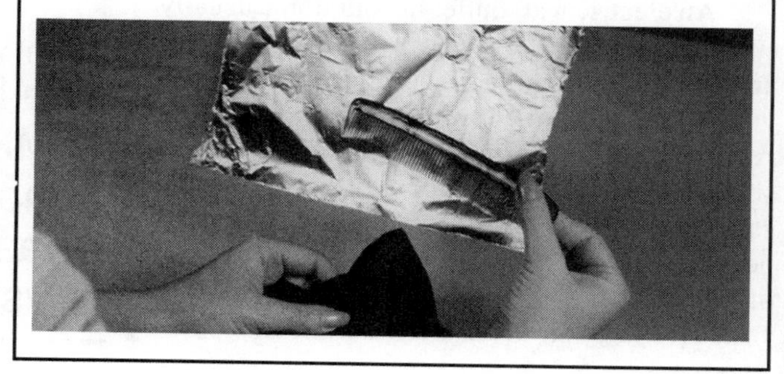

Why Do Electrons Flow?

There are many similarities between electricity flowing in a wire and water flowing through a pipe. Most people are familiar with what happens when you open a faucet and water comes out. We can use this to make a useful comparison between water flow and electron flow (electricity). Throughout this chapter we use examples of water flowing in a pipe to help explain electronics.

Do you know how your town's water system works? Chances are, there is a large supply of water stored somewhere. Some towns use a lake or river. Other towns get their water from wells and store it in a tank or reservoir. The system then uses gravity to pull the water down from the tank or reservoir. The water travels through a system of pipes to your house. Because the force of gravity is pulling down on the water in the tank, it exerts a pressure on the water in the pipes. This makes the water flow out of the faucets in your house with some force. If you have a well, you probably have a storage tank. The storage tank uses air pressure to push the water up to the top floor of your house.

In these water systems, a pump takes water from the large supply and puts it into a storage tank. Then the system uses air pressure or the force of gravity to push the water through pipes to the faucets in your home.

We can compare electrons flowing through wire to water flowing through a pipe. We need some force to make water flow through a pipe. What force exerts pressure to make electrons flow through a wire?

Voltage

The amount of pressure that it takes to push water to your house depends on the path the water has to take. If the water has to travel over hills along the way, more pressure will be required than if the water simply has to flow down off a mountain. The pressure required to make electrons flow in an electrical circuit also depends on the opposition that the electrons must overcome. The pressure that forces the electrons through the circuit is known as **electromotive force** or, simply, **EMF**.

EMF is similar to water pressure. More pressure moves more water. Similarly, more EMF moves more electrons. We measure EMF in a unit called the **volt (V)**, so we sometimes refer to the EMF as a **voltage**. If more voltage is applied to a circuit, more electrons will flow. We measure voltage with a device called a *voltmeter*.

An electric wall outlet in your home usually supplies about 120 volts. If you have an outlet for an electric stove or an electric clothes dryer in your home, that outlet probably provides 240 V. A car **battery** is normally rated at 12 volts. A single D-cell battery supplies 1.5 V. **Voltage sources** come in a wide variety of ratings, depending on their intended use. With our system of metric prefixes, we can express a thousand volts as 1 kilovolt or "1 kV."

Another way to think about this electrical voltage pushing electrons through a circuit is to remember that like-charged objects repel. If we have a large group of electrons, the negative charge of these electrons will act to repel, or push, other electrons through the circuit. In a similar way, a large group of positively charged ions attract, or pull, electrons through the circuit.

Because there are two types of electric charge (positive and negative), there are also two polarities associated with a voltage. A **voltage source** always has two terminals, or poles: the positive terminal and the negative terminal. The negative terminal repels electrons (negatively charged particles) and the positive terminal attracts electrons. If we connect a piece of wire between the two terminals of a voltage source, electrons will flow through the wire. We call this flow an electrical **current**.

Batteries

In our water system, a pump supplies pressure to pull water from the source and force it into the pipes. Similarly, electrical circuits require an electron source and a "pump" to move the electrons along.

A **battery** is one example of a **power supply**. We use a battery as both the source of electrons and the pump that moves them along. The battery is like the storage tank in our water system. A battery provides pressure to keep the electrons moving.

There is an excess of electrons at the negative terminal of a power supply. At the positive terminal there is an excess of positive ions. With a conducting wire connected between the two terminals, the electrical pressure (voltage) generated by the power supply will cause electrons to move through the conductor.

Batteries come in all shapes and sizes. Some batteries are tiny, like those used in hearing aids and cameras. Other batteries are larger than the one in your car. A battery is one kind of voltage source.

Current

You have probably heard the term "current" used to describe the flow of water in a stream or river. Similarly, we call the flow of electrons an electric **current**. Each electron is extremely small. It takes quintillions and quintillions of electrons to make your toaster heat bread or your TV draw

The basic unit of electromotive force (EMF) is the volt. The volt was named in honor of Alessandro Giuseppe Antonio Anastasio Volta (1745-1827). This Italian physicist invented the electric battery.

pictures. (A quintillion is a one with 18 zeros after it—1,000,000,000,000,000,000. Using powers of 10, as described earlier, we could also write this as 1×10^{18}.)

When water flows from your home faucet, you don't try to count every drop. The numbers would be very large and unmanageable, and the drops are coming out much too fast to count! To measure water flow, you count larger quantities such as gallons and describe the flow in terms of gallons per minute. Similarly, we can't deal easily with large numbers of individual electrons, nor can we count them conveniently. We need a shorthand way to measure the number of electrons. So, as with gallons per minute of water, we have amperes of electric current. We measure current with a device called an *ammeter*.

Suppose you are looking through a "window" into a wire, and can count electrons as they move past you. (See Figure 5-4.) If you count 6,240,000,000,000,000,000 (6.24×10^{18}) electrons moving past your "window" each second, the circuit has a current of one **ampere**. (Don't worry! You won't have to remember this number.) So when you express a circuit's current in amperes, remember that it is a measure of the number of electrons flowing through the circuit. A circuit with a current of 2 amperes has twice as many electrons flowing out of the supply as a circuit with a current of 1 ampere.

Write "2 A" for two amperes or "100 mA" (milliamps) for 0.1 ampere (sometimes also abbreviated amp or amps). You can use all the metric prefixes with the ampere. Most of the time you will see currents expressed in amps, milliamps and microamps. See Table 5-1 to review the list of metric prefixes.

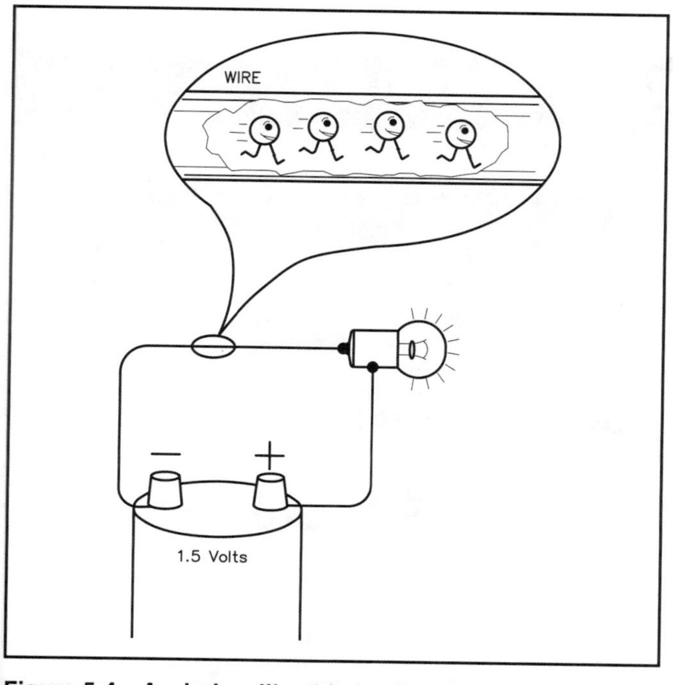

The action of electric current on a magnet was first applied to telegraphy by André Marie Ampère (1775-1836) in 1820. An ampere is the basic unit of electrical current.

[Now study questions N5B01 through N5B08 in Chapter 13. Review this section if you have trouble answering these questions.]

WIRE

1.5 Volts

Figure 5-4—A window like this into a wire might allow you to count electrons as they flow past, to measure the current. Even so, you would have to count 6.24×10^{18} electrons per second for a current of only 1 ampere!

Conductors

As we pointed out earlier, some atoms have a firm grasp on their electrons and other atoms don't. More current can flow in materials made of atoms that have only a weak hold on their electrons. Some materials, then, conduct electricity better than others.

Silver is an excellent **conductor**. The loosely attached electrons in silver atoms require very little voltage (pressure) to produce an electric current. Copper is much less expensive than silver and conducts almost as well. We can use copper to make wire needed in houses, and in radios and other electronic devices. Steel also conducts, but not as well as copper. In fact, most metals are fairly good conductors, so aluminum, mercury, zinc, tin and gold are all conductors.

Insulators

Other materials keep a very firm grip on their electrons. These materials do not conduct electricity very well, and are called **insulators**. Materials such as glass, rubber, plastic, ceramic, mica, wood and even air are poor conductors. Pure distilled water is a fairly good **insulator**. Most tap water is a good conductor, however, because it has minerals and other impurities dissolved in it. Figure 5-5 lists some common insulators and conductors.

The electric power company supplies 120 V on the wires into your home. That voltage is available for your use at electrical outlets or sockets. Why don't the electrons spill out of the sockets? The insulation between the two sides of the outlet prevents the electrons from flowing from one side to the other. The air around the socket acts as an insulator to stop them from flowing into the room.

Because insulators are *poor* conductors rather than *non*conductors, every insulator has a *breakdown voltage*. A voltage higher than the breakdown voltage will force electrons to move through the insulator. The insulator will start to conduct electricity. Depending on the material, it may be damaged if you exceed the breakdown voltage. Better insulators have higher breakdown voltages. Breakdown voltage also depends on the thickness of the insulating material. A thin layer of one insulating material (like Teflon or mica) may be just as good as a much thicker layer of another material (like paper or air).

A good example of an insulator that will conduct at very high voltage is air. Air is a fine insulator at the voltages normally found in homes and industry. When a force of millions of volts builds up, however, there's enough pressure to send a bolt of electrons through the air — *lightning*. You can produce a voltage large enough to make a spark jump through a thin air layer by shuffling your feet across the carpet on a dry day. When you reach for a metal object like a doorknob, you can often feel the spark jump from your finger. If the room is darkened, you can also see the spark.

When you are insulating wires or components, always be sure to use the right insulating material. Make sure you use enough insulation for the voltages you're likely to encounter. Heat-shrinkable tubing or other insulating tubing is often convenient for covering a bare wire or a solder connection. You can also wrap the wire with electrical tape. Several layers of tape, wrapped so it overlaps itself, will provide enough insulation for up to a few hundred volts.

[Before you go on to the next section, turn to Chapter 13 and study questions N5B12 through N5B14. If you have any difficulty, review this section.]

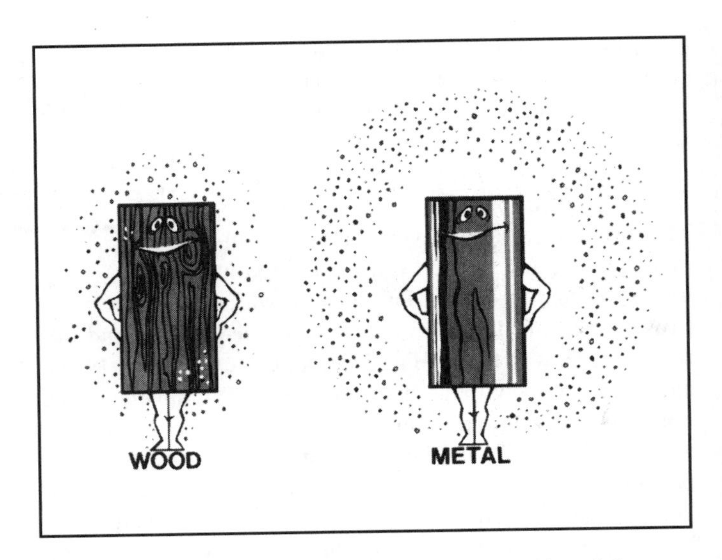

Figure 5-5—Here's one way to show how an insulator differs from a conductor. Wood, on the left, holds onto its electrons pretty tightly, keeping them flowing between its atoms. Metals, on the other hand, are more generous with their electrons. Electrons are more easily pulled away from the metal atoms, and the metal atoms are then left with a positive charge. If the metal atoms attract extra electrons from neighboring atoms, they become negatively charged.

ELECTRONICS FUNDAMENTALS

In this section, you'll learn about **resistance**. You'll see how resistance fits into one of the most fundamental laws of electronics, *Ohm's Law*. We will also briefly discuss *open* and *short circuits* as well as *closed* or *complete circuits*. Learning about these circuits will make it much easier for you to understand some of the other concepts you will be studying. Be sure you understand these circuits.

This section shows you how to do some basic circuit calculations. We have tried to keep the arithmetic simple, and to explain all of the steps in the solution. Associating numbers with a concept often makes it easier to understand. Read these examples and then try the calculations yourself. Be sure you can work the problems in the question pool when you study those questions.

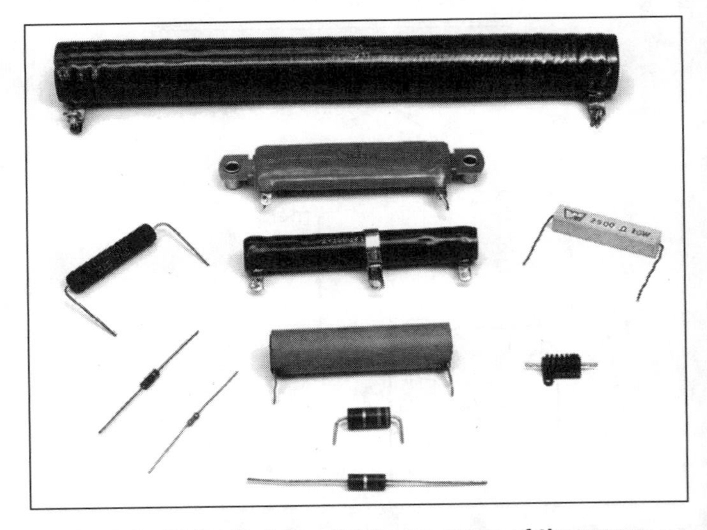

Figure 5-6—This photograph shows some of the many types of resistors. Large power resistors are at the top of the photo. The small resistors are used in low-power transistor circuits.

Resistance

What if you partially blocked a water pipe with a sponge? Eventually, the water would get through the sponge, but it would have less pressure than before. The sponge opposes, or resists the water trying to flow through the pipe, and it takes pressure to overcome that resistance.

The basic unit of resistance is the ohm, named in honor of Georg Simon Ohm (1787-1854).

Similarly, materials that conduct electrical current also present some opposition, or **resistance**, to the movement of electrons. **Resistors** are devices that are especially designed to make use of this opposition. Resistors limit the amount of current that flows through a circuit because they oppose the flow of electrons. Figure 5-6 shows some common resistors.

In a water pipe, increasing the pressure forces more water through the sponge (the resistance). In an electrical circuit, increasing the voltage forces more current through the resistor. The relationship between voltage, current and resistance is predictable. We call this relationship **Ohm's Law** and it is a basic electronics principle.

The **ohm** is the basic unit used to measure resistance. The abbreviation for ohms is Ω, the Greek capital letter omega. This unit is named for Georg Simon Ohm, a German physics teacher and mathematician. Of course we also use the metric prefixes with ohms, when appropriate. So you will often see resistors specified as having 47 kilohms or 1.2 megohms of resistance. Written with abbreviations, these resistors would be 47 kΩ and 1.2 MΩ. As you could probably guess, when you want to measure an amount of resistance, you will use an *ohmmeter*.

[Turn to Chapter 13 and study questions N5B09 through N5B11 now. Review this section if you can't answer them.

If you are preparing for the Technician exam, also study questions T5A01, T5A03 and T5A04 in Chapter 14.]

Ohm's Law

The amount of water flowing through a pipe increases as we increase the pressure and decreases as we increase the resistance. If we replace "pressure" with "voltage," this same statement describes current through an electric circuit. We can write a mathematical relationship for an electric circuit:

$$\text{Current} = \frac{\text{Voltage}}{\text{Resistance}} \qquad \text{(Eq 5-1)}$$

This equation tells us the current through a circuit equals the voltage applied to the circuit divided by the circuit resistance.

If the voltage stays constant but more current flows in the circuit, we know there must be less resistance. The relationship between current and voltage is a measure of the resistance:

$$\text{Resistance} = \frac{\text{Voltage}}{\text{Current}} \qquad \text{(Eq 5-2)}$$

We can state this equation in words as: The circuit resistance is equal to the voltage applied to the circuit divided by the current through the circuit.

There Really Was an Ohm (1787-1854)

Although we take Ohm's Law for granted, it wasn't always so widely accepted. In 1827, Georg Simon Ohm finished his renowned work, *The Galvanic Circuit Mathematically Treated*. Scientists at first resisted Ohm's techniques. The majority of his colleagues were still holding to a non-mathematical approach. Finally, the younger physicists in Germany accepted Ohm's information in the early 1830s. The turning point for this basic electrical law came in 1841 when the Royal Society of London awarded Ohm the Copley Medal.

As the oldest son of a master locksmith, Georg received a solid education in philosophy and the sciences from his father. At the age of 16 he entered the University of Erlangen (Bavaria) and studied for three semesters. His father then forced him to withdraw because of alleged overindulgences in dancing, billiards and ice skating. After 4½ years, his brilliance undaunted, Ohm returned to Erlangen to earn his PhD in mathematics.

Enthusiasm ran high at that time for scientific solutions to all problems. The first book Ohm wrote reflected his highly intellectualized views about the role of mathematics in education. These opinions changed, however, during a series of teaching positions. Oersted's discovery of electromagnetism in 1820 spurred Ohm to avid experiments, since he taught Physics and had a well-equipped lab. Ohm based his work on direct scientific observation and analyses, rather than on abstract theories.

One of Ohm's goals was to be appointed to a major university. In 1825 he started doing research with the thought of publishing his results. The following year he took a leave of absence from teaching and went to Berlin.

In Berlin, Ohm made his now-famous experiments. He ran wires between a zinc-copper battery and mercury-filled cups. While a Coulomb torsion balance (voltmeter) was across one leg of the series circuit, a "variable conductor" completed the loop. By measuring the loss in electromagnetic force for various lengths and sizes of wire, he had the basis for his formulas.

Because *The Galvanic Circuit* produced near-hostility, Ohm withdrew from the academic world for nearly six years before accepting a post in Nuremberg. English and French physicists do not seem to have been aware of the profound implications of Ohm's work until the late 1830s and early 1840s.

Following his belated recognition, Ohm became a corresponding member of the Berlin Academy. Late in 1849 he went to the University of Munich and in 1852, only two years before his death, he achieved his lifelong dream of a full professorship at a major university.

Finally, we can determine the voltage if we know how much current is flowing and the resistance in the circuit:

Voltage = Current × Resistance (Eq 5-3)

The voltage applied to a circuit is equal to the current through the circuit times the circuit resistance.

Scientists are always looking for shorthand ways of writing these relationships. They use symbols to replace the words: E represents voltage (remember EMF?), current is I (from the French word *intensité*) and resistance is R. We can now express **Ohm's Law** in a couple of letters:

$E = IR$ (volts = amperes × ohms) (Eq 5-4)

This is the most common way to express Ohm's Law, but we can also write it as:

$I = \frac{E}{R}$ (amperes = volts divided by ohms) (Eq 5-5)

and

$R = \frac{E}{I}$ (ohms = volts divided by amperes) (Eq 5-6)

E is EMF in volts, I is the current in amperes and R is the resistance in ohms. If you know two of the numbers, you can calculate the third. If a circuit has an EMF of 1 volt applied to it, and the current through the circuit is 1 ampere, then the resistance of that circuit will be 1 ohm.

Figure 5-7 shows a diagram to help you solve Ohm's Law problems. Simply cover the symbol of the quantity that you do not know. If the remaining two are side-by-side, you must multiply them. If one symbol is above the other, then you must divide the quantity on top by the one on the bottom.

If you know current and resistance in a circuit, Ohm's Law will give you the voltage (Eq 5-4). For example, what is the voltage applied to the circuit if 2 amperes of current flows through 50 ohms of resistance? From Eq 5-4 or Figure 5-7, we see that we must multiply 2 amperes times 50 ohms to get the answer, 100 volts. The EMF in this circuit is 100 volts.

$E = I R$ (Eq 5-4)

$E = 2$ amperes × 50 ohms

$E = 100$ volts

Suppose you know voltage and resistance — 200 volts in the circuit to push electrons against 100 ohms of resistance? Eq 5-5 gives the correct equation, or you can use

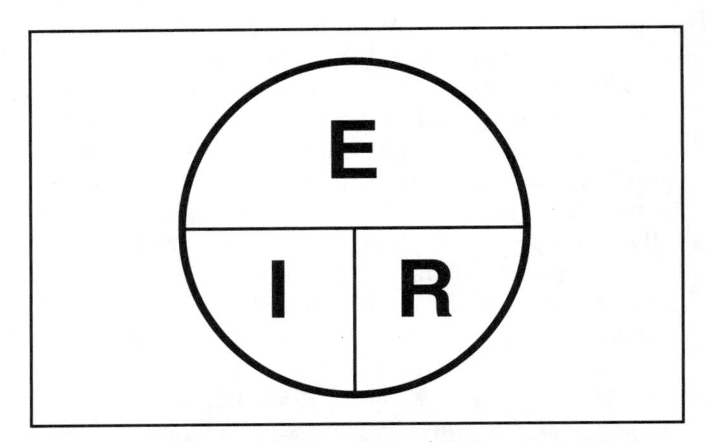

Figure 5-7—This simple diagram will help you remember the Ohm's Law relationships. To find any quantity if you know the other two, simply cover the unknown quantity with your hand or a piece of paper. The positions of the remaining two symbols show if you have to multiply (when they are side by side) or divide (when they appear one over the other as a fraction).

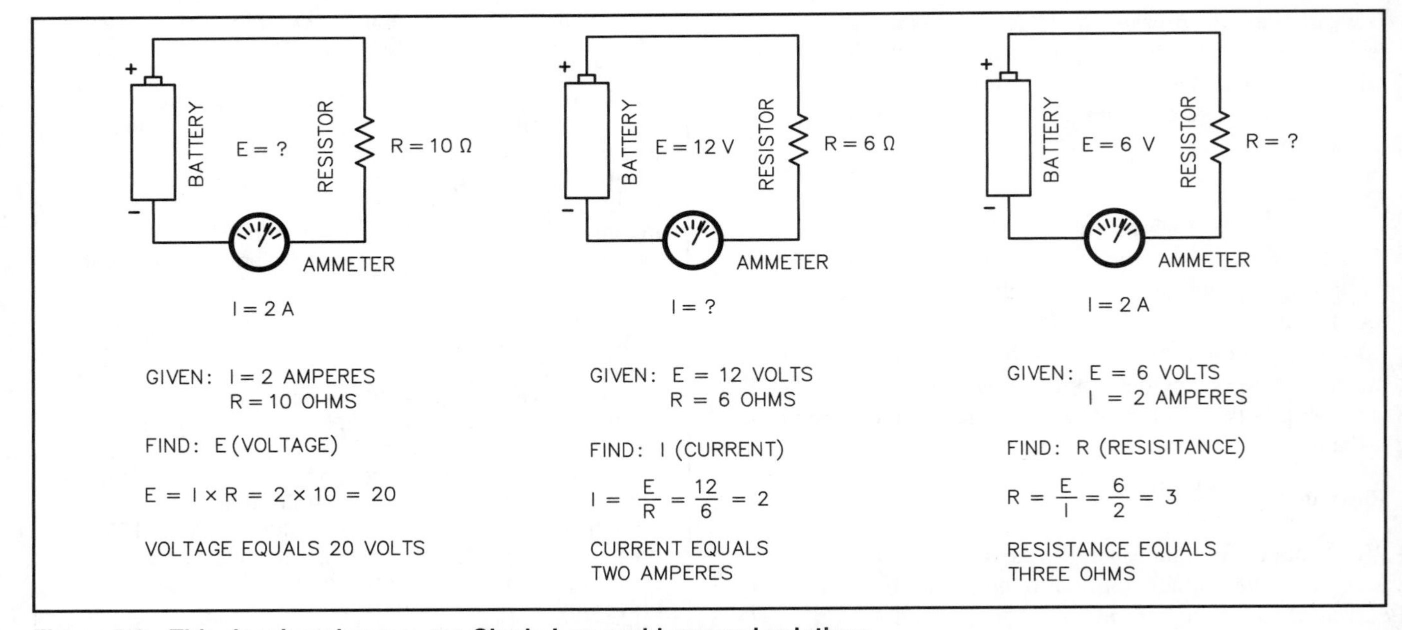

Figure 5-8—This drawing shows some Ohm's Law problems and solutions.

the diagram in Figure 5-7. You must divide 200 volts by 100 ohms to find that 2 amperes of current is flowing.

$$I = \frac{E}{R} \quad \text{(Eq 5-5)}$$

$$I = \frac{200 \text{ volts}}{100 \text{ ohms}}$$

I = 2 amperes

If you know voltage and current in a circuit, you can calculate resistance. For example, suppose a current of 3 amperes flows through a resistor connected to 90 volts. This time Eq 5-6 is the one to use, and you can also find this from Figure 5-7. 90 divided by 3 equals 30, so the resistance is 30 ohms.

$$R = \frac{E}{I} \quad \text{(Eq 5-6)}$$

$$R = \frac{90 \text{ volts}}{3 \text{ amperes}}$$

R = 30 ohms

If you know E and I, you can find R. If you know I and R, you can calculate E. If E and R are known, you can find I.

Put another way, if you know volts and amperes, you can calculate ohms. If amperes and ohms are known, volts can be found. Or if volts and ohms are known, amperes can be calculated. Figure 5-8 illustrates some simple circuits and how Ohm's Law can be used to find an unknown quantity in the circuit. Make up a few problems of your own and test how well you understand this basic law of electricity. You'll soon find that this predictable relationship, symbolized by the repeatable equation of Ohm's Law, makes calculating values of components in electrical circuits easy. Ohm's Law is one electrical principle that you will use when working with almost any electronic circuit!

[Before you read further, turn to Chapter 13 and study questions N5C01 through N5C05. Review this section if you don't understand any of these questions.]

[**If you are preparing for the Technician exam**, also study questions T5A02 and T5B01 through T5B09 in Chapter 14. Review the material in this section if you have difficulty with any of these questions.]

Open and Short Circuits

You've probably heard the term **short circuit** before. A short circuit happens when the current flowing through the components doesn't follow the path we expect it to. Instead, the current finds another path, a shorter one, between the terminals of the power source. This is why we call this path a short circuit. Because there is less opposition to the flow of electrons, there is a larger current. Often the current through the new (short) path is so large that the wires or components can't handle it. When this happens, the wires and components can be damaged. Figure 5-9 B shows a bare wire causing a short circuit.

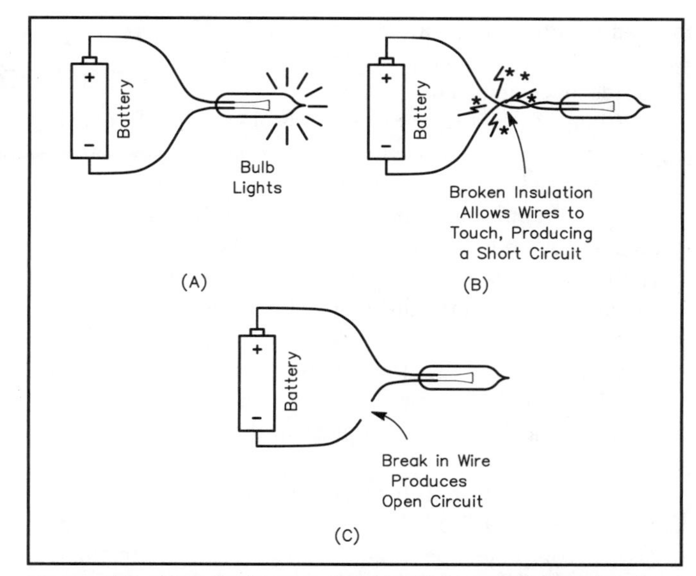

Figure 5-9—Part A shows a light bulb in a working circuit. In B, the insulation covering the wires has broken, and the two wires are touching, so we have a short circuit. In C, one wire has broken, preventing the current from flowing through the bulb. This is an example of an open circuit.

Some people think that a short circuit occurs when there is no resistance in the connection between the positive and negative terminals of the power supply. Actually there will always be *some* resistance but it may be such a small amount that it can be ignored.

In the extreme case, if a short circuit develops in our house wiring, the wire may overheat and can even start a fire. This is why it is important to have a properly rated fuse connected in line with a circuit. We will describe fuses in more detail in Chapter 6.

Current flows *through* the circuit, but the same current that flows out of a battery or other power supply will flow back into the supply at the other terminal. You can imagine electrons leaving the negative battery terminal and flowing through the circuit. All the electrons that leave the battery eventually return to the positive terminal. A short circuit usually allows more current than the circuit was designed to handle.

Too much current is a relative term. A short circuit draws more current than is intended. If you connect two or three electric space heaters and a few lights to one 15-amp household circuit, they may draw too much current. This will blow a fuse or trip a circuit breaker, but it does not mean you have a short circuit. Each heater is drawing the current it is designed to draw.

The opposite of a short circuit is an **open circuit**. In an open circuit the current is interrupted, just as it is when you turn a light switch off. There is no current through an open circuit. The switch *breaks* (opens) the circuit, putting a layer of insulating air in the way so no current can flow. This break in the current path presents an extremely high resistance. An open circuit can be good, as when you throw the on/off switch to off. An open circuit can be bad if it's an unwanted condition caused by a broken wire or a bad component. Figure 5-9 C illustrates an open circuit.

When a fuse blows or a circuit breaker trips it creates an open circuit. We use fuses or circuit breakers in our house wiring and in electrical equipment to protect against the large current drawn by a short circuit or an overloaded circuit. It is much better to blow the fuse or trip the circuit breaker and create an open circuit than to cause a fire because the wires overheated!

The basic unit of power is the watt. This unit is named after James Watt (1736-1819), the inventor of the steam engine.

Sometimes the resistance in the circuit path is so large that it is impractical to measure it. When this is true, we say that there is an infinite resistance in the path, creating an open circuit.

A **closed**, or **complete circuit** has an uninterrupted path for the current to follow. Turning a switch on, for example, closes or completes the circuit, allowing current to flow.

[At this time you should turn to Chapter 13 and study the questions N5C09 through N5C11. If you have any difficulty with these questions, review this section.]

Energy and Power

We define **energy** as the ability to do work. An object can have energy because of its position (like a rock ready to fall off the edge of a cliff). An object in motion also has energy (like the same rock as it falls to the bottom of the cliff). In electronics, a power supply or battery is the source of electrical energy. We can make use of that energy by connecting the supply to a light bulb, a radio or other circuit.

A voltage source pushes electrons through the resistance in an electric circuit. Suppose we have a circuit with two resistors connected so the current goes from the battery, through one resistor, then through the other and finally back to the battery. If we know how much current flows through the circuit, we can use Ohm's Law to calculate the voltage across each resistor. The example shown in Figure 5-10 has two equal-value resistors, and each resistor has half the battery voltage because the total battery voltage is applied across both resistors. The voltage that appears across each individual resistor is called a *voltage drop*. In our example, after the current has gone through the first resistor, the voltage has dropped from 10 V to 5 V, and after the second resistor it has dropped to 0 V because it is back to the battery terminal at that time.

These voltage drops occur because electrical **energy** is "used up" or "consumed." Actually, we can't lose the electrical energy; it is just changed to some other form. The resistor heats up because of the current through it; the resistance converts electrical energy to heat energy. More current produces still more heat, and the resistor becomes warmer. If too much current flows, the resistor might even catch fire!

As electrons flow through a light bulb, the resistance of the bulb converts some electrical energy to heat. The filament in the bulb gets so hot that it converts some of the electrical energy to light energy. Again, more current produces more light and heat.

You should get the idea that we can "use up" a certain amount of energy by having a small current go through a resistance for a long time or by having a larger current go through it for a shorter time. When you buy electricity from a power company, you pay for the electrical energy that you use each month. You might use all of the energy in one day, but you probably use a small amount every day. Your bill would be the same in either case, because the electric meter on your house just measures how much energy you use. The power company sends someone around to read the meter each month to determine how much electrical energy you used.

Sometimes it is important to know how fast a circuit can use energy. You might want to compare how bright two different light bulbs will be. If you're buying a new freezer, you might want to know how much electricity it will use in a month. You will have to know how fast the freezer or the light bulbs use electrical energy. We use the term **power** to define the rate of energy consumption. The basic unit for measuring power in the metric system is the **watt**. You have probably seen this term used to rate electrical appliances. You know that a light bulb rated at 75 watts will be brighter than one rated at 40 watts. (Sometimes we abbreviate watts with a capital W.) These same numbers tell us which light bulb uses more electrical energy each minute (or hour) that they are turned on. For example, if you turn on light bulbs of 60 watts, 75 watts and 100 watts, the 100-watt light bulb will use the most electrical energy in an hour.

[Before you go on to the last Novice section in this chapter, turn to Chapter 13 and study questions N5C06, N5C07 and N5C08. Review this section if you have any problems.]

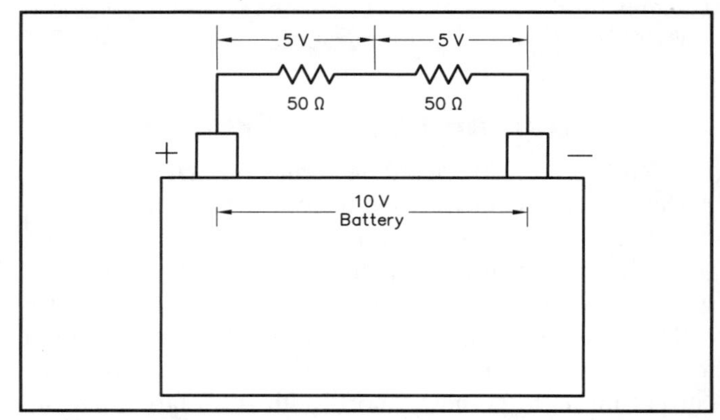

Figure 5-10—When two resistors are connected in series with a battery, part of the battery voltage appears across each of the resistors. Here each resistor has half the voltage because the two resistors have equal values.

DIRECT AND ALTERNATING CURRENT

In this section, you will learn what we mean by **direct current** and **alternating current**. You will also learn the meaning of some important terms that go along with alternating current, such as **frequency** and **wavelength**.

Two Types of Current

Until now, we have been talking about **direct current** electricity, known as **dc** for short. In direct current, the electrons flow in one direction only — from negative to positive. Batteries are the most common source of direct current. Lead-acid car batteries, nickel-cadmium (NiCd) rechargeable batteries and alkaline batteries are all examples. We can also get dc from a solar panel.

In our water-flow analogy, we normally think of water flowing one direction through the pipes. But water could actually flow in either direction through a pipe. We know that water can flow in more than one direction. The tides in the ocean are a good example of water flowing in one direction, then reversing and flowing in the opposite direction.

There is a second kind of electricity called **alternating current**, or **ac**. In ac, the terminals of the power supply change from positive to negative to positive and so on. Because the poles change and electrons always flow from negative to positive, ac flows first in one direction, then the other. The current *alternates* in direction.

We call one complete round trip a *cycle*. The **frequency** of the ac is the number of complete cycles, or alternations, that occur in one second. We measure frequency in hertz (abbreviated Hz). Frequency is a measure of the number of times in one second the alternating current flows back and forth. One cycle per second is 1 Hz. 150 cycles per second is 150 Hz. One thousand cycles per second is one kilohertz (1 kHz). One million cycles per second is one megahertz (1 MHz). If a radio wave makes 3,725,000 cycles in one second, this means it has a frequency of 3,725,000 hertz (Hz), 3725 kilohertz (kHz) or 3.725 megahertz (MHz).

More AC Terminology

Batteries provide direct current. To make an alternating current from this direct-current source, you would have to switch the polarity of the voltage source rapidly. Imagine trying to turn the battery around so the plus and minus terminals changed position very rapidly. This would not be a very practical way to produce ac! You must have a power supply in which the polarity is constantly changing. The terminals must be positive and negative one moment, and then negative and positive, constantly switching back and forth.

The power company has a more practical way to create ac: they use a large machine called an *alternator* to produce power at their generating stations. The ac supplied to your home goes through 60 complete cycles each second. Thus, the electricity from the power company has a frequency of 60 Hz.

This 60-hertz ac electricity builds slowly to a peak current or voltage in one direction, then decreases to zero and reverses to build to a peak in the opposite direction. If you plot these changes on a graph, you get a gentle up-and-down curve. We call this curve a *sine wave*. Figure 5-11 shows two cycles of a sine-wave ac signal.

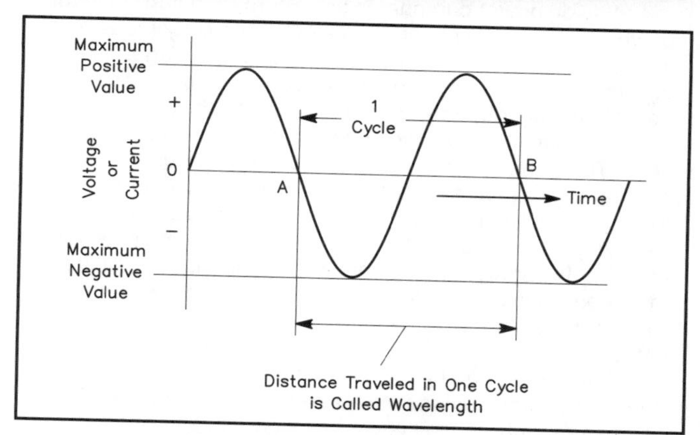

Figure 5-11—The sine wave is one way to show alternating current. Let's follow one cycle starting on line "0" at point A, indicated near the center of the graph. The wave goes in a negative direction to its most negative point, then heads back up to zero. After the wave goes through zero, it becomes more and more positive, reaches the positive peak, then goes back to zero again. This is one full cycle of alternating current.

Alternating current can do things direct current can't. For instance, a 120-V ac source can be increased to a 1000-V source with a *transformer*. Transformers can change the value of an ac voltage, but not a dc voltage. The power company supplies 120-V ac and 240-V ac to your house.

[Now study the following questions in Chapter 13: N5D01 through N5D06, N5D11 and N5D12. Review this section if you have difficulty answering any of these questions.]

The basic unit of frequency is the hertz. This unit is named in honor of Heinrich Rudolf Hertz (1857-1894). This German physicist was the first person to demonstrate the generation and reception of radio waves.

Why Use Alternating Current in Our Homes?

Why do power companies use alternating current in the power lines that run to your home? One important reason is so they can use transformers to change the voltage. This allows the company to use an appropriate voltage for each part of their distribution system. In this way the power company can minimize the power losses in the transmission lines. The gebnerator at the power station produces ac by moving a wire (actually many turns of wire) through a magnetic field in an alternator. The resulting output has a relatively low voltage. Why don't the power companies send this directly through the power lines to your house? At first, this seems like a good idea. It would eliminate the many transformers and power stations that often clutter our landscape.

The answer can be found in Ohm's Law. Even a very good conductor, such as the copper used in the power company's high-voltage lines, has a certain amount of resistance. This factor becomes very important when we consider the very long distances the generated electricity must travel.

Remember that the voltage drop across a resistance is given by the formula E=IR, where I is the value of current and R is the value of resistance. If we can reduce either the resistance of the wire or the value of the current through the wire, we can reduce the voltage drop. The resistance of the wire is relatively constant, although we can reduce it somewhat by using a very large diameter wire. If we increase the voltage,a smaller current will be required for the same power transfer from the generating station to your home.

Using a very high voltage also provides more "overhead." If the power company starts with 750,000 volts, and the voltage has dropped to 740,000 volts by the time it reaches the first substation, they just use a transformer rated for 740,000-V input to give the desired output. If they send 50,000 volts on to the next substation, there is still plenty of overhead. By the time it gets to the power lines outside your house, the voltage has dropped to around 3000 volts. A pole transformer then steps it down to 240 volts to supply power to your house. This voltage is normally split in half to provide two 120-V circuits to your house.

Frequency and Wavelength

We discussed the frequency of an ac signal earlier. From this discussion, you must realize that alternating currents and voltages can change direction at almost any rate imaginable. Some signals have low frequencies, like the 60-Hz-ac electricity the power company supplies to your house. Other signals have higher frequencies; for example, radio signals can alternate at more than several million hertz.

When we talk about such a wide range of frequencies, it is common to describe several smaller ranges. For example, we often describe an alternating current as an audio-frequency signal or a radio-frequency signal. If you connected an ac signal having a frequency anywhere between 20 Hz and 20,000 Hz (20 kHz) to a loudspeaker, you would hear a sound. Because these signals can produce sounds, they are called **audio-frequency (AF)** signals. The higher the frequency of the signal, the higher the pitch of the sound you would hear. (**Caution — DO NOT** connect the 60-Hz power from a household receptacle to a speaker even though 60 Hz is in the audio-frequency range! You may be seriously injured or **KILLED** by the voltage of this signal!)

Not all people can hear this full range of signals from 20 Hz to 20 kHz. Some people hear the low frequencies better than the high frequencies, and others hear the high frequencies better. This is the general range of frequencies that humans can expect to hear, however. (Dogs can hear signals at a much higher frequency, which is why dog whistles, used for training, may not produce a sound you can hear.)

Signals that have a frequency higher than 20,000 Hz (20 kHz) are called **radio-frequency (RF)** signals. Signals in the RF range can also be broken into smaller groups, such as very-low frequency (VLF), high frequency (HF), very-high frequency (VHF), ultra-high frequency (UHF) and so on. Don't worry about the names of these ranges for your exam. You will probably hear the terms as you listen in on other hams' discussions, though. The Novice bands are in the HF, VHF and UHF ranges.

If we know the frequency of an ac signal, we can use that frequency to describe the signal. We can talk about 60-Hz power or a 3725-kHz radio signal. **Wavelength** is another quality that can be associated with every ac signal. As its name implies, wavelength refers to the distance that the wave will travel through space in a single cycle. All such signals (sometimes called electromagnetic waves) travel though space at the speed of light, 300,000,000 meters per second (3.00×10^8 m/s). We use the lower-case Greek letter lambda (λ) to represent wavelength.

The faster a signal alternates, the less distance the signal will be able to travel during one cycle. There is an equation that relates the frequency and the wavelength of a signal to the speed of the wave:

$$c = f \lambda \qquad \text{(Eq 5-7)}$$

where:

c is the speed of light, 3.00×10^8 meters per second
f is the frequency of the wave in hertz
λ is the wavelength of the wave in meters

We can solve this equation for either frequency or wavelength, depending on which quantity we want to find.

$$f = \frac{c}{\lambda} \qquad \text{(Eq 5-8)}$$

and

$$\lambda = \frac{c}{f} \qquad \text{(Eq 5-9)}$$

From these equations you may realize that as the frequency increases the wavelength gets shorter. As the frequency decreases the wavelength gets longer. Suppose you are transmitting a radio signal on 7.125 MHz. What is the wavelength of this signal? We can use Eq 5-9 to find the answer. First we must change the frequency to hertz:

7.125 MHz = 7,125,000 Hz.

Then we use this value in Eq 5-9.

$$\lambda = \frac{c}{f} = \frac{3.00 \times 10^8 \frac{m}{s}}{7.125 \times 10^6 \text{ Hz}}$$

$$\lambda = \frac{300,000,000 \frac{m}{s}}{7,125,000 \text{ Hz}} = 42 \text{ meters}$$

Of course, you already knew that this frequency was in the 40-meter Novice band, so this answer should not surprise you.

As another example, what is the wavelength of a signal that has a frequency of 3.725 MHz? (3.725 MHz = 3,725,000 Hz.)

$$\lambda = \frac{c}{f} = \frac{3.00 \times 10^8 \frac{m}{s}}{3.725 \times 10^6 \text{ Hz}}$$

$$\lambda = \frac{300,000,000 \frac{m}{s}}{3,725,000 \text{ Hz}} = 80.5 \text{ meters}$$

Table 5-2

Novice-Band Frequencies and Wavelengths

Frequency Range (megahertz)	*Approximate Wavelength (meters)*
3.675 — 3.725	80
7.1 — 7.15	40
21.1 — 21.2	15
28.1 — 28.5	10
222 — 225	1.25
1270 — 1295	0.23 (23 centimeters)

Even if you have trouble with this arithmetic, you should be able to learn the frequency and wavelength relationships for the six Novice bands, as shown in Table 5-2.

Notice that higher-frequency signals have shorter wavelengths. Lower-frequency signals have longer wavelengths. As you increase a signal frequency the wavelength gets shorter. As a signal's wavelength increases, the frequency goes down.

[Congratulations! You are now well on your way to knowing all the electronics you will need to pass your Novice exam. You have completed all of the material on basic electronics principles for your Novice exam. Before going on to the next chapter, turn to Chapter 13 and study questions N5D07 through N5D10 and N5D13 through N5D15. Don't hesitate to come back to this chapter to review any sections that you are still a little uncertain about.]

[**If you are preparing for the Technician exam**, you should continue with this chapter.]

KEY WORDS

Capacitor — An electrical component usually formed by separating two conductive plates with an insulating material. A capacitor stores energy in an *electric field*.

Capacitance — A measure of the ability of a capacitor to store energy in an *electric field*.

Farad — The basic unit of capacitance.

Henry — The basic unit of inductance.

Induced EMF — A voltage produced by a change in magnetic lines of force around a conductor. When a magnetic field is formed by current in the conductor, the induced voltage always opposes change in that current.

Inductance — A measure of the ability of a coil to store energy in a *magnetic field*.

Inductor — An electrical component usually composed of a coil of wire wound on a central core. An inductor stores energy in a *magnetic field*.

Parallel circuit — An electrical circuit in which the electrons follow more than one path in going from the negative supply terminal to the positive terminal.

Series circuit — An electrical circuit in which all the electrons must flow through every part of the circuit. There is only one path for the electrons to follow.

5 Electronics Theory for Technicians

SERIES AND PARALLEL CIRCUITS

There are two basic ways that you can connect the parts in an electric circuit. If we hook several resistors together in a string, we call it a **series circuit**. If we connect several resistors side-by-side to the same voltage source, we call it a **parallel circuit**.

In our water-pipe example, what would happen to the current through the pipe if we placed another sponge in it? You're right: The second sponge would further reduce the flow. We could do the same thing with a single, larger sponge. The total resistance in a series circuit is the sum of all the resistances in the circuit.

In a series circuit, the same current, I, flows through each resistor, since it has no other path to follow. When a voltage source (like a battery) is hooked to our string of resistors, part of the battery voltage appears across each resistor. We can calculate the voltage across each resistor. How? Using Ohm's Law, of course.

Remember E = IR? The voltage across any resistor in

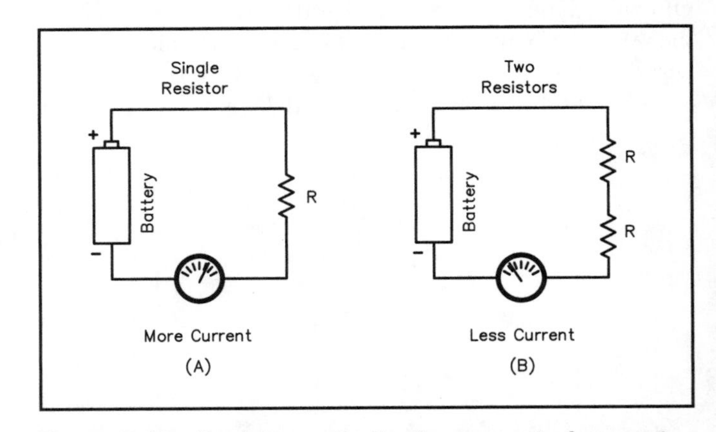

Figure 5-12—Resistance limits the amount of current that can flow in a circuit. Adding a second resistance reduces the current because the total resistance is larger. The total resistance of a string of series-connected resistors is the sum of all the individual resistances.

the circuit will be its value in ohms multiplied by the current in amperes. We call the voltage across the resistor the *voltage drop*. If you add the voltage drop across each resistor in a simple series circuit like the one shown in Figure 5-12, they will equal the total of the battery voltage. This figure also shows that with more resistors in a series circuit, there will be less current.

Now let's look at the case of a parallel circuit. This is similar to having two water pipes running side by side. More water can flow through two parallel pipes of the same size than through a single one. With two pipes, the current is greater for a given pressure.

If these pipes had sponges in them, the flow would be reduced in each pipe. There are still two paths for the water to take. More water will flow than if there was a single pipe with a similar sponge in it. Now let's go back to electrical resistors and voltage. Adding a resistor in parallel with another one provides two paths for the electrical current to follow. This reduces the total resistance. You can connect more than two resistors in parallel, providing even more paths for the electrons. This will reduce the resistance still more. Figure 5-13 shows that as we add resistors in parallel, we provide more current paths. The result is less total resistance in the circuit, and more current.

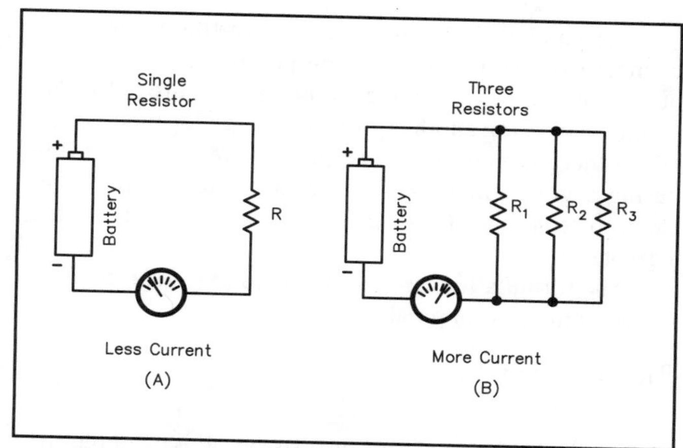

Figure 5-13—When three resistors are connected in parallel, the electrical current splits into three separate paths through the resistors. The full battery voltage is applied to each resistor. Current through each individual resistor is independent of the other resistors. For example, if R_2 and R_3 are changed to a value different than the original value, the current in R_1 would remain unchanged. Even if R_2 and R_3 are removed, the current in R_1 remains unchanged.

CALCULATING SERIES AND PARALLEL RESISTANCE

Sometimes it's necessary to calculate the total resistance of resistors connected in a **series circuit**. Resistors in series are connected end to end like a string of sausages. At other times you must calculate total resistance in a parallel circuit. In a **parallel circuit**, resistors are side by side like a picket fence. There will be times when you need a certain amount of resistance somewhere in a circuit. There may be no standard resistor value that will give the necessary resistance. Sometimes you may not have a certain value on hand. By combining resistors in parallel or series you can obtain the desired value.

When you connect resistors in series, as in Figure 5-14, the total resistance is simply the sum of all the resistances. Let's use the analogy of a sponge in a water pipe again. Connecting resistors in series is like putting several sponges in the same pipe. The total resistance to the water would be the sum of all the individual resistances. Resistors in series add.

$$R_{TOTAL} = R_1 + R_2 + R_3 + ... + R_n \qquad \text{(Eq 5-10)}$$

where n is the total number of resistors.

The total resistance of a string of resistors in series will always be greater than any individual resistance in the string. All the circuit current flows through each resistor in a series circuit. A series circuit with resistor values of 2 ohms, 3 ohms and 5 ohms would have a total resistance of 10 ohms. If a circuit has two equal-value resistors connected in series, the total resistance will be twice the value of either resistor alone.

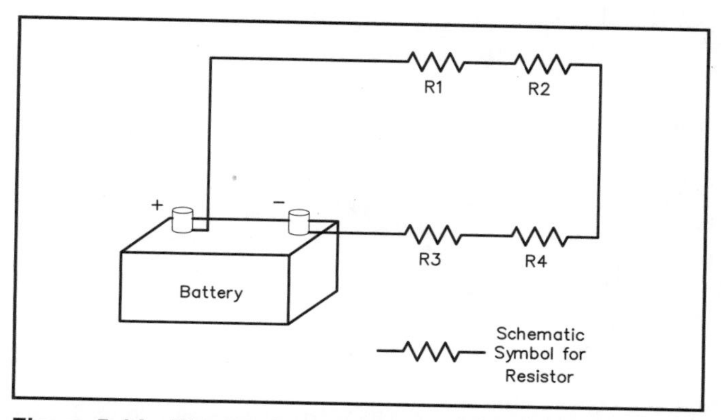

Figure 5-14—The total resistance of a string of series-connected resistors is the sum of all the individual resistances.

In a parallel circuit, things are a bit different. When we connect two or more resistors in parallel, more than one path for current exists in the circuit. See Figure 5-15. This is like connecting another pipe into our water-pipe circuit. When there is more than one path, more water can flow during a given time. With more than one resistor, more electrons can flow. This means there is less resistance and a greater current.

The formula for calculating the total resistance of resistors connected in parallel is:

$$R_{TOTAL} = \frac{1}{\frac{1}{R_1} + \frac{1}{R_2} + \frac{1}{R_3} + \cdots + \frac{1}{R_n}} \quad \text{(Eq 5-11)}$$

where n is the total number of resistors. For example, if we connect three 100-ohm resistors in parallel, their total resistance is:

$$R_{TOTAL} = \frac{1}{\frac{1}{100\ \Omega} + \frac{1}{100\ \Omega} + \frac{1}{100\ \Omega}}$$

$$R_{TOTAL} = \frac{1}{\frac{3}{100\ \Omega}} = \frac{1}{0.03}\ \Omega = 33.3\ \Omega$$

If we connect a 50-ohm resistor in parallel with another 50-ohm resistor, our equation becomes:

$$R_{TOTAL} = \frac{1}{\frac{1}{50\ \Omega} + \frac{1}{50\ \Omega}}$$

$$R_{TOTAL} = \frac{1}{\frac{2}{50\ \Omega}} = \frac{1}{\frac{1}{25\ \Omega}} = \frac{1}{0.04}\ \Omega = 25\ \Omega$$

To calculate the total resistance of two resistors in parallel, Equation 5-11 reduces to the "product over sum" formula. Divide the product of the two resistances by their sum:

$$R_{TOTAL} = \frac{R_1 \times R_2}{R_1 + R_2} \quad \text{(Eq 5-12)}$$

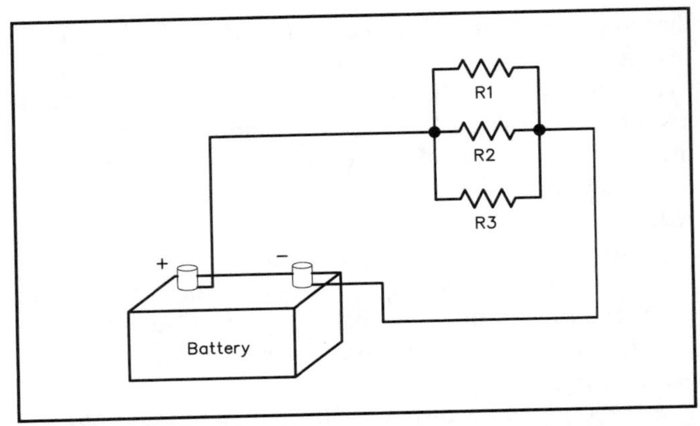

Figure 5-15—In this circuit, the electrical current splits into three separate paths through the resistors. Each resistor is exposed to the full battery voltage. Current through each individual resistor is in–dependent of the other resistors. For example, if R_2 and R_3 are changed to a value different than the original value, the current in R_1 would remain unchanged. Even if R_2 and R_3 are removed, the current in R_1 remains unchanged.

If we connect two 100-ohm resistors in parallel, the total resistance would be:

$$R_{TOTAL} = \frac{100\ \Omega \times 100\ \Omega}{100\ \Omega + 100\ \Omega} = \frac{1000\ \Omega^2}{200\ \Omega} = 50\ \Omega$$

From these two examples you can see that the total resistance of two equal-value resistors connected in parallel is always half the value of one of the resistors.

Now let's consider any parallel combination of resistors. The total resistance is always less than the smallest value of the parallel combination. You can use this fact to make a quick check of your calculations. The result you calculate should be smaller than the smallest value in the parallel combination. If it isn't, you've made a mistake somewhere!

[You should turn to Chapter 14 now and study questions T5B10 and T5B11. Review this section if you don't understand either of those questions.]

INDUCTANCE

The motion of electrons produces magnetism. Every electric current creates a *magnetic field* around the wire in which it flows. (A magnetic field represents the invisible magnetic force, such as the attraction and repulsion between magnets.) Like an invisible tube, the magnetic field is positioned in concentric circles around the conductor. See Figure 5-16A. The field is established when the current flows, and collapses back into the conductor when the current stops. The field increases in strength when the current increases and decreases in strength as the current decreases. The force produced around a straight piece of wire by this magnetic field is usually very small. When the same wire is formed into a coil, the force is much greater. In coils, the magnetic field around each turn also affects the other turns. Together, the combined forces produce one large magnetic field, as shown in Figure 5-16B. Much of the energy in the magnetic field concentrates in the material in the center of the coil (the *core*). Most practical **inductors** consist of a length of wire wound on an iron core or a core made from a mixture of iron and other materials.

An inductor stores energy in a magnetic field. This property of a coil to store energy in a magnetic field is called **inductance**. Magnetic fields can also set electrons in motion. When a magnetic field increases in strength, the voltage on a conductor within that field increases. When the field strength decreases, so does the voltage.

Let's apply a dc voltage to an inductor. As there is no current to start with, the current begins to increase when the voltage is applied. This will establish an electrical current in the inductor, and this current produces a magnetic field. The magnetic field in turn *induces*, or creates, a voltage in the wire. The voltage induced by the magnetic field of the inductor opposes the applied voltage. Therefore, the inductor will oppose the increase in current. This is a basic property of inductors. Any changes in current through the inductor, whether increasing or decreasing, are opposed. A voltage is induced in the coil that opposes the applied voltage, and tries to prevent the current from changing. This is called **induced EMF** (voltage) or back EMF. Gradually a current will be produced by the applied voltage. (The term "gradually" is relative. In radio circuits, the time needed to produce the current in the circuit is often measured in microseconds.) The important point is that the current doesn't increase to its final value instantly when the voltage is first applied.

The "final" current that flows through the inductor is limited only by any resistance that might be in the circuit. There is very little resistance in the wire of most coils. The current will be quite large if there is no other resistance.

In the process of getting this current to flow, energy is stored. This energy is in the form of a magnetic field around the coil. When the applied current is shut off, the magnetic field collapses. The collapsing field returns energy to the circuit as a momentary current that continues to flow in the same direction as the original current. This current can be quite large for large values of inductance. The induced voltage can rise to many times the applied voltage. This can cause a spark to jump across switch or relay contacts when the circuit is broken to turn off the current. This effect is called "inductive kickback."

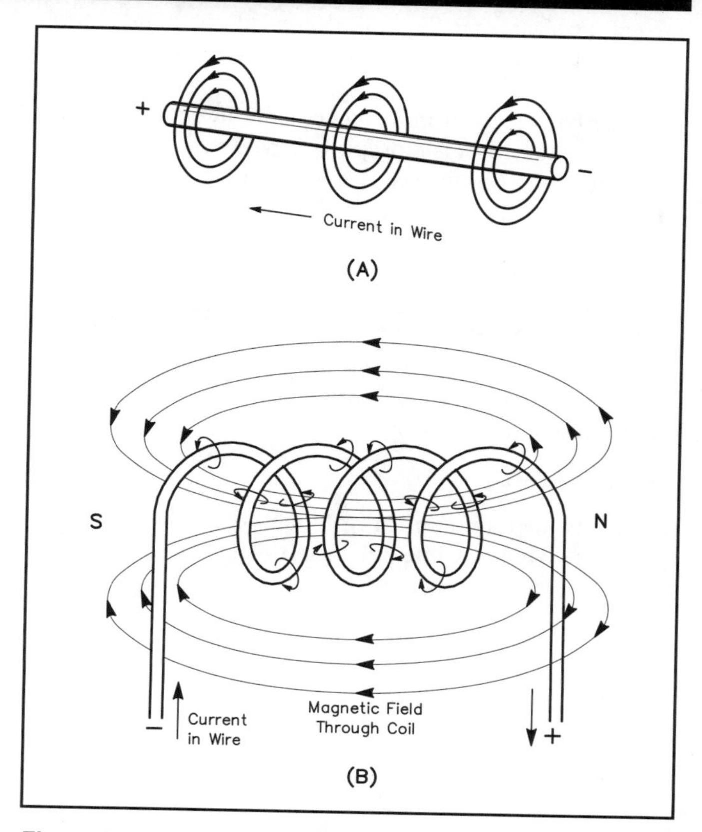

Figure 5-16—A magnetic field surrounds a wire with a current in it. If the wire is formed into a coil, the magnetic field becomes much stronger as the lines of force reinforce each other.

When an ac voltage is applied to an inductor, the current through the inductor will reverse direction every half cycle. This means the current will be constantly changing. The inductor will oppose this change. Energy is stored in the magnetic field while the current is increasing during the first half cycle. This energy will be returned to the circuit as the current starts to decrease. A new magnetic field will be produced during the second half cycle. The north and south poles of the field will be the reverse of the first half cycle. The energy stored in that field will be returned to the circuit as the current again starts to decrease. Then a new magnetic field will be produced on the next half cycle. This process keeps repeating, as long as the ac voltage is applied to the inductor.

Inductors play several very important roles in electronic circuits. When an ac signal is applied to the inductor, the inductance will act to oppose or reduce the flow of ac.

As we learned earlier, however, when a dc signal is applied to an inductor, the inductance will have little effect on the current, at least after that initial opposition. Inductors reduce the flow of ac signals but allow dc signals to flow freely.

The basic unit of inductance is the **henry**, named for the American physicist Joseph Henry. The henry is often too large for practical use in measurements. We use the millihenry (10^{-3}) abbreviated mH, or microhenry (10^{-6}) abbreviated μH.

Inductors in Series and Parallel

In circuits, inductors combine like resistors. The total inductance of several inductors connected in series is the sum of all the inductances:

$$L_{TOTAL} = L_1 + L_2 + L_3 + ... + L_n \qquad \text{(Eq 5-13)}$$

where n is the total number of inductors.

For two equal-value inductors connected in series, the total inductance will be twice the value of one of the inductors.

For parallel-connected inductors:

$$L_{TOTAL} = \frac{1}{\frac{1}{L_1} + \frac{1}{L_2} + \frac{1}{L_3} + \cdots \frac{1}{L_n}} \qquad \text{(Eq 5-14)}$$

You should recognize this equation from our study of combining resistors in parallel, and realize that for two parallel-connected inductors, Equation 5-14 reduces to:

$$L_{TOTAL} = \frac{L_1 \times L_2}{L_1 + L_2} \qquad \text{(Eq 5-15)}$$

For two equal-value inductors connected in parallel, the total inductance will be half the value of one of the components.

[Turn to Chapter 14 now and study questions T5A05 through T5A08 and questions T5B12 and T5B13. Review this section as needed.]

CAPACITANCE

A simple **capacitor** is formed by separating two conductive plates with an insulating material. Connect one plate to the positive terminal of a voltage source. Connect the other plate to the negative terminal. We can build up a surplus of electrons on one plate, as shown in Figure 5-17. At some point, the voltage across the capacitor will equal the applied voltage, and the capacitor is said to be charged. This stored electric charge produces an *electric field*, which is an invisible electric force of attraction or repulsion acting between charged objects. (In this case the electric field is between the capacitor plates.) The capacitor stores energy as an electric field between the capacitor plates. Once the capacitor has charged to the full voltage of the applied signal, no more charge will flow onto the capacitor plates, so the current stops.

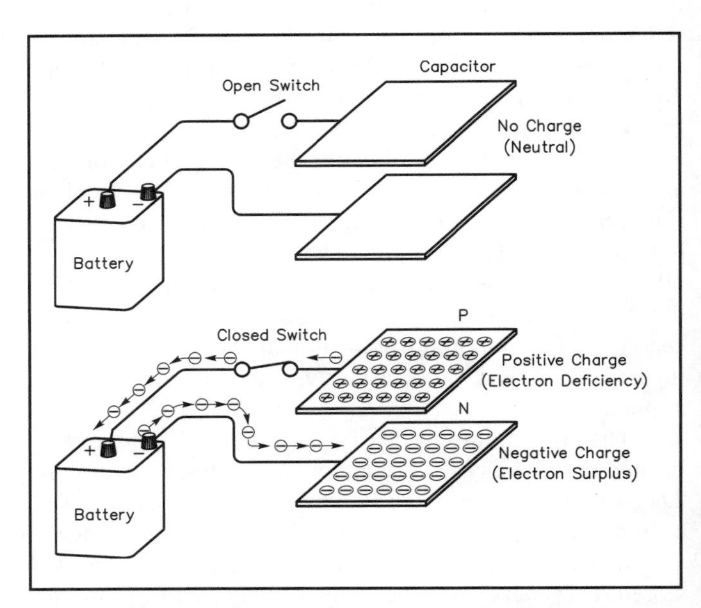

Figure 5-17—When a voltage is applied to a capacitor, an electron surplus (negative charge) builds up on one plate, while an electron deficiency forms on the other plate to produce a positive charge.

If we connect a load to a charged capacitor, it will discharge through the load, releasing stored energy. The basic property of a capacitor (called **capacitance**) is this ability to store a charge in an electric field.

If we connect an ac signal to a capacitor the plates will charge during one part of the ac cycle. After the signal reaches the peak voltage, however, the charge will start to flow back into the circuit in the opposite direction until the capacitor is charged with the opposite polarity during the second half of the ac cycle. This process of charging first one direction and then the other will continue as long as the ac signal is applied to the capacitor plates.

Capacitors play a role that is opposite to that of inductors. A capacitor will block direct current because as soon as the capacitor charges to the applied voltage level, no more current can flow. A capacitor will pass alternating current with little or no opposition, however.

The basic unit of capacitance is the **farad**, named for Michael Faraday. Like the henry, the farad is usually too large a unit for practical measurements. For convenience, we use microfarads (10^{-6}), abbreviated µF, or picofarads (10^{-12}), abbreviated pF.

Capacitors in Series and Parallel

We can increase the value of capacitance by increasing the total plate area. We can effectively increase the total plate area by connecting two capacitors in parallel. The two parallel-connected capacitors act like one larger capacitor, as shown in Figure 5-18A and B. The total capacitance of several parallel-connected capacitors is the sum of all the values.

$$C_{TOTAL} = C_1 + C_2 + C_3 + ... + C_n \quad \text{(Eq 5-16)}$$

where n is the total number of capacitors.

If you connect two equal-value capacitors in parallel, the total capacitance of this combination will be twice the value of either capacitor alone.

Connecting capacitors in series has the effect of increasing the distance between the plates, thereby reducing the total capacitance, as shown in Figure 5-18C and D. For capacitors in series, we use the familiar reciprocal formula:

$$C_{TOTAL} = \frac{1}{\frac{1}{C_1} + \frac{1}{C_2} + \frac{1}{C_3} + \cdots + \frac{1}{C_n}} \quad \text{(Eq 5-17)}$$

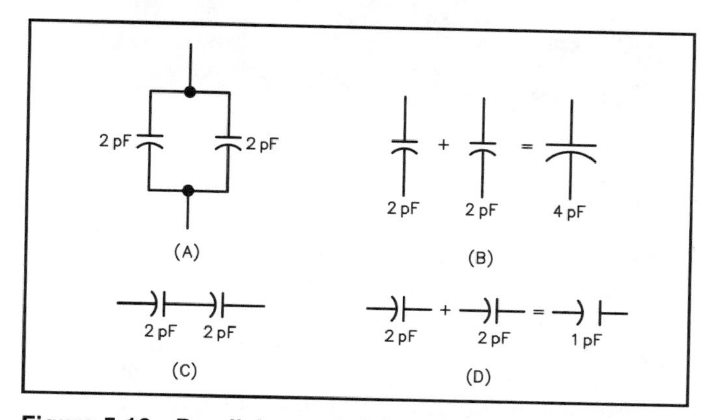

Figure 5-18—Parallel-connected capacitors are shown at A. This connection has the effect of increasing the total plate area, as shown at B. This increases the capacitance. Series connection, shown at C, has the effect of increasing the spacing between the plates, as shown at D. This decreases the capacitance.

When there are only two capacitors in series, Equation 5-17 reduces to the product over sum formula:

$$C_{TOTAL} = \frac{C_1 \times C_2}{C_1 + C_2} \quad \text{(Eq 5-18)}$$

Using this formula for two equal-value resistors or two equal-value inductors in *parallel*, we discovered the total value is half the value of either one. Two equal-value capacitors connected in *series* equals one-half the value of either single capacitor. The total capacitance will always be less than any of the individual capacitance values when you connect capacitors in series.

[This completes your study of Chapter 5. Now turn to Chapter 14 and study exam questions T5A09 through T5A12 and questions T5B14 and T5B15. Before proceeding to the next chapter, review the material in this section if you have difficulty with any of these questions.]

6 KEY WORDS

Battery — A device that converts chemical energy into electrical energy.

Chassis ground — The common connection for all parts of a circuit that connect to the negative side of the power supply.

Double-pole, double-throw (DPDT) switch — A switch that has six contacts. The DPDT switch has two center contacts. The two center contacts can each be connected to one of two other contacts.

Double-pole, single-throw (DPST) switch — A switch that connects two contacts to another set of contacts. A DPST switch turns two circuits on or off at the same time.

Earth ground — A circuit connection to a ground rod driven into the Earth or to a cold-water pipe made of copper that goes into the ground.

Fuse — A thin metal strip mounted in a holder. When too much current passes through the fuse, the metal strip melts and opens the circuit.

Integrated circuit (IC) — A modern electronics component that consists of many transistor elements on a single wafer of silicon.

NPN Transistor — A transistor that has a layer of P-type semiconductor material sandwiched between layers of N-type semiconductor material.

Pentode — A vacuum tube with five active elements: cathode, plate, control grid, screen grid and suppressor grid.

PNP Transistor — A transistor that has a layer of N-type semiconductor material sandwiched between layers of P-type semiconductor material.

Potentiometer — Another name for a **variable resistor**. The value of a potentiometer can be changed without removing it from a circuit.

Resistor — A circuit component that controls current through a circuit.

Schematic symbol — A drawing used to represent a circuit component on a wiring diagram.

Single-pole, double-throw (SPDT) switch — A switch that connects one center contact to one of two other contacts.

Single-pole, single-throw (SPST) switch — A switch that only connects one center contact to another contact.

Switch — A device used to connect or disconnect electrical contacts.

Transistor — A solid-state device made of three layers of semiconductor material. See **NPN transistor** and **PNP transistor**.

Triode — A vacuum tube with three active elements: cathode, plate and control grid.

Variable resistor — A resistor whose value you can change without removing it from a circuit.

6 Circuit Components

Batteries, switches, fuses, resistors, capacitors, inductors, tubes and transistors are all an important part of the electronic devices we use every day. How do these components work and what do they do?

Before we look at the operation of electronic circuits, let's discuss some basic information about the parts that make up those circuits. This chapter presents the information about circuit components that you need to know for your Novice and Technician exams. You will find descriptions of several types of fuses, switches, resistors and semiconductor devices. We combine these components with other devices to build practical electronic circuits.

Every circuit component has a **schematic symbol**. A schematic symbol is nothing more than a drawing used to represent a component. We use these symbols when we are making a circuit diagram, or wiring diagram, to show how the components connect for a specific purpose. You will learn the schematic symbols for the circuit components discussed in this chapter. As you discover more about electronics, you will learn how these symbols can be used to illustrate practical circuit connections.

Resistors

Resistors are important components in electronic circuits. We talked about the concept of resistance in Chapter 5. A resistor opposes the flow of electrons. We can control the electron flow (the current) by varying the resistance in a circuit.

Most resistors have standard fixed values, so they are called **fixed resistors**. **Variable resistors**, also called **potentiometers**, allow us to change the value of the resistance without removing and changing the component.

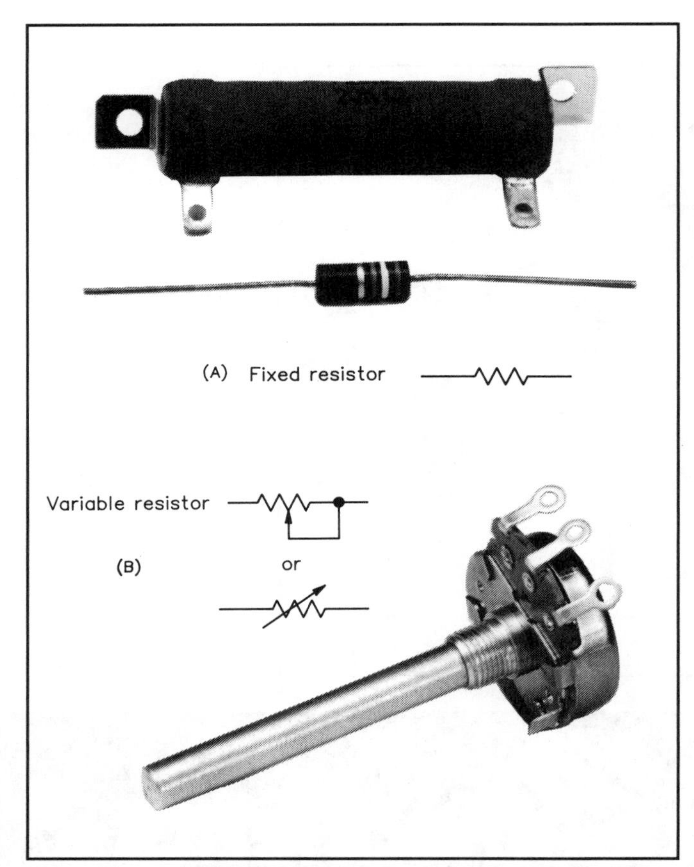

Figure 6-1 — Fixed resistors come in many standard values. Most of them look something like the ones shown at A. Variable resistors (also called potentiometers) are used wherever the value of resistance must be adjusted after the circuit is complete. Part B shows a potentiometer.

Potentiometers are used as the volume and tone controls in most stereo amplifiers. Figure 6-1 shows two types of fixed resistors, a potentiometer and their schematic symbols. When you look at these resistor symbols and see the zigzag lines you can just imagine how difficult it will be for the electrons to fight their way through these peaks and valleys!

[Now turn to Chapter 13 and study questions N6A05, N6A06 and N6A07. Review this section if you have any problems.]

Switches

How do you control the lights in your house? What turns on your car radio? A **switch**, of course.

The simplest kind of switch just connects or disconnects a single electrical contact. Two wires connect to the switch; when you turn the switch on, the two wires are connected. When you turn the switch off, the wires are disconnected. This is called a **single-pole, single-throw switch**. It connects a single pair of wires (single pole) and has only two positions, on or off (single throw). Sometimes we abbreviate single pole, single throw as **SPST**. Figure 6-2A shows a simple SPST switch and the symbol we use to represent it on schematic diagrams.

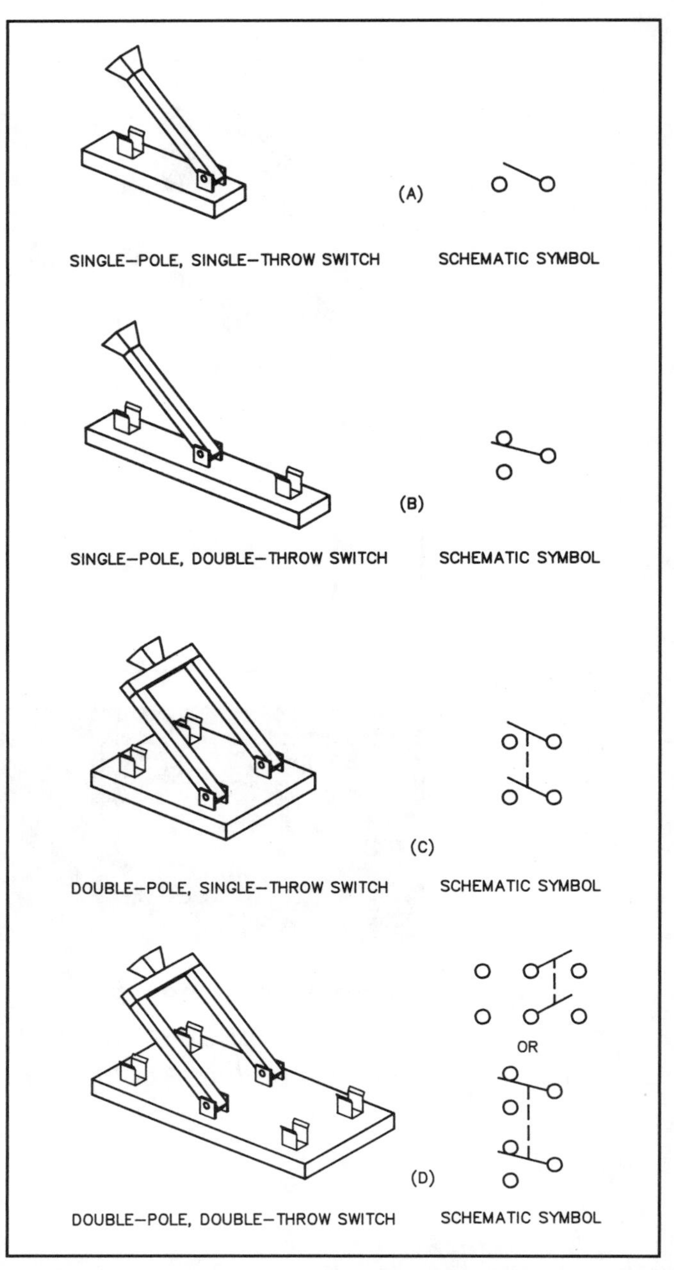

Figure 6-2 — A single-pole, single-throw (SPST) switch can connect or disconnect one circuit. A single-pole, double-throw (SPDT) switch can connect one center contact to one of two other contacts. A double-pole, single-throw switch connects or disconnects two circuits at the same time. A double-pole, double-throw (DPDT) switch is like two SPDT switches in one package. Each half of the DPDT switch can connect one contact to two other contacts.

The switches shown in Figure 6-2 are called *knife switches* because a metal blade pivots at one end to make or break the contact. These switches show the basic operation, and physically look like the schematic symbols. They can be dangerous if used with high-voltage circuits, however. Do not use a knife switch in a circuit that carries 120-volt house current unless it is enclosed and operated with an insulated handle designed for this purpose. Knife switches *are* available, however, and can be used in low-voltage circuits.

Most of the time you'll want to use switches with contacts sealed inside a protective case. When mounted in the wall or on a control panel, the switch lever is safely insulated from the circuit.

If we want to control more devices with a single switch, we need more contacts. If we add a second contact to a single-pole, single-throw switch we can select between two devices. This kind of switch is called a **single-pole, double-throw switch**. Sometimes we abbreviate single pole, double throw as **SPDT**. An SPDT switch connects a single wire (single pole) to one of two other contacts (double throw). The switch connects a center wire to one contact when the switch is in one position. When you flip the switch to the other position, the switch connects the center wire to the other contact. An SPDT switch connects one input to either of two outputs. Electric circuits that allow you to turn lights on or off from either of two locations use SPDT switches. Figure 6-2B shows an SPDT knife switch and the schematic symbol for any SPDT switch.

Suppose you want to switch two circuits on and off at the same time. Then you use a **double-pole, single-throw (DPST) switch**. This type of switch connects two inpu lines to their respective output lines at the same time. See Figure 6-2C for an example of this switch and its schematic-diagram symbol. When you look at the knife-switch drawing or the schematic symbol you can easily see that the DPST switch connects two inputs at the same time, one input to one output, and the other input to the other output.

We can add even more contacts to the switch. A **double-pole, double-throw switch** has two sets of three contacts. We use the abbreviation **DPDT** for double pole, double throw. You can think of a DPDT switch as two SPDT switches in the same box with their handles connected. A DPDT switch has two center contacts. The switch connects each of these

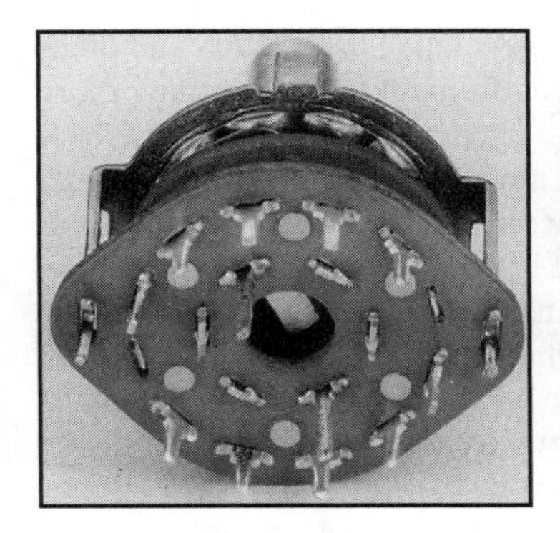

Figure 6-3 — A rotary switch can connect one wire to several contacts. Most antenna switches use a rotary switch to connect a transceiver to several antennas. This photograph shows a two-pole, five-position switch; it can connect each center contact to five outside contacts.

Figure 6-4 — The schematic symbol for a single-pole, six-position rotary switch.

two center contacts (double pole) to one of two other contacts (double throw). Figure 6-2D shows a DPDT switch and its schematic symbol.

All these switches are very useful, but they only connect a center contact to one or two other contacts. What if we want to connect a single contact to *several* other contacts? We might want to use one switch to connect our transmitter to several different antennas. We can do this with a *rotary switch*. As its name implies, a rotary switch turns around a central shaft to connect one center contact to several outer contacts.

Switches like this can have many contacts. They can also have more than one center contact, or pole. We specify the particular kind of switch by the number of contacts (positions) it has around the outside, and by the number of center contacts and switch arms (poles) it has. Figure 6-3 shows a photograph of a two-pole, five-position rotary switch; it has five separate contacts and two rotary arms. The schematic symbol in Figure 6-4 is a single-pole, six-position rotary switch.

[You should turn to Chapter 13 now. Study questions N6A01, N6A02, N6A10, N6A11, N6A12 and N6A13. Review this section if you have any problems with these questions.]

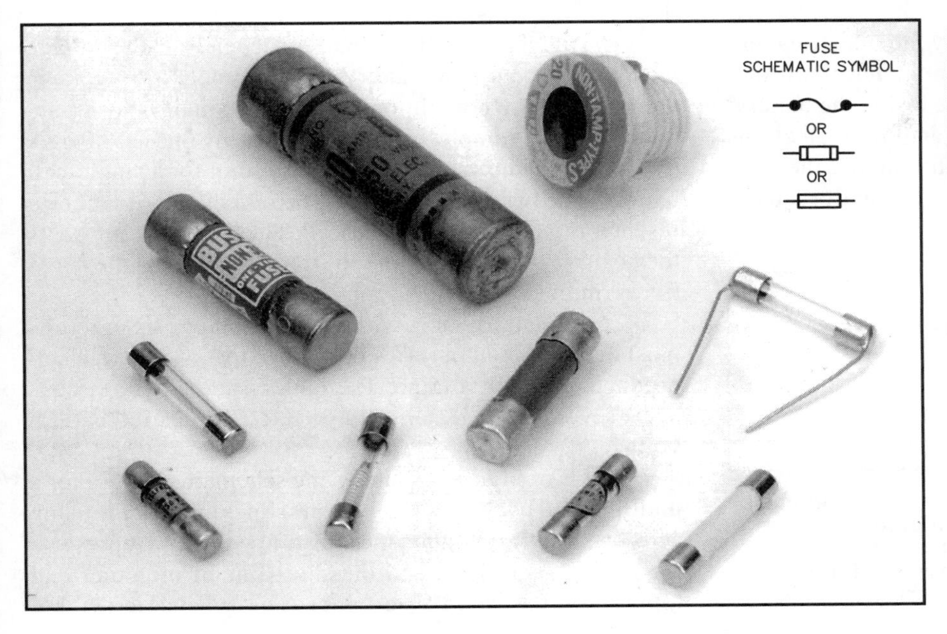

Figure 6-5 — Some common fuses. Fuses protect circuits from excessive current. The inset drawing shows common fuse schematic symbols.

Fuses

What would happen if part of your receiver suddenly developed a short circuit? The current, flowing without opposition through portions of the circuit, could easily damage components not built to withstand such high current.

To protect against unexpected short circuits and other problems, most electronic equipment includes one or more fuses. A **fuse** is simply a device made of metal that will heat up and melt when a certain amount of current flows through it. The amount of current that causes each fuse to melt (or "blow") is determined by the manufacturer. When the fuse (usually placed in the main power line to the equipment) blows, it creates an open circuit, stopping the current.

Fuses come in many shapes and sizes. Figure 6-5 shows some of the more common fuse types. This figure also shows the three common symbols used to represent a fuse on schematic diagrams. Notice that two of these symbols seem to have a small wire through the center and one is open through the center. That should help remind you that if too much current flows through the fuse, the wire will melt and produce an open circuit.

A fuse in a transistor radio using little power may be designed to blow at 250 mA or less. The fuses for your home's 120-V circuits may be designed to blow at 15 or 20 A. The principle is the same: When excessive current flows through the fuse it melts, creating an open circuit to protect your equipment. Remember that fuses are designed to protect against too much current, not too much voltage.

Most people use circuit breakers instead of fuses to protect their house circuits from overload. Circuit breakers use a spring-loaded mechanism that unlatches and pops open if too much current flows through the breaker. To reset the breaker you simply move the lever to the off position and then back to the on position. Circuit breakers are also available for use on circuit boards and other applications where you need to protect against having too much current in the circuit.

[To check your understanding of this section, study exam questions N6A03 and N6A08. Review this section if you have any problems.]

Batteries

We talked about batteries briefly in Chapter 5. Simply put, a **battery** changes chemical energy into electrical energy. When we connect a wire between the terminals of a battery, a chemical reaction takes place inside the battery. This reaction produces free electrons, and these electrons flow through the wire from the negative terminal to the positive terminal. Batteries may be small or large, round or square.

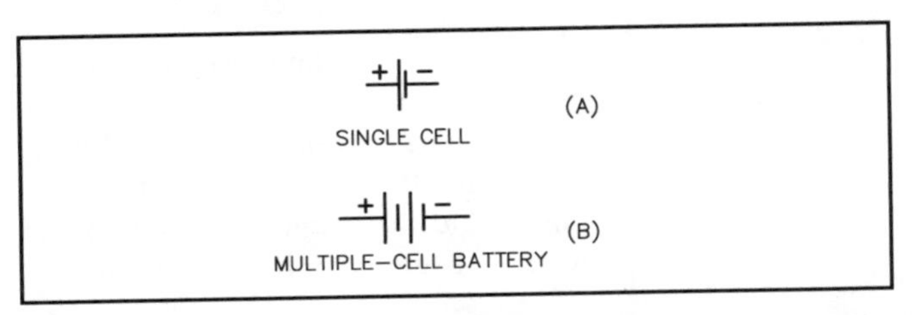

Figure 6-7 — Some small batteries contain only one cell. We use the symbol at A for a single-cell battery. Manufacturers add several cells in series to produce more voltage. Part B is the schematic symbol for a multiple-cell battery.

Figure 6-6 — Batteries come in all shapes and sizes. A battery changes chemical energy into electrical energy.

Hearing aid and calculator batteries are tiny. The battery that starts your car is large by comparison, but batteries can be even larger than that. Figure 6-6 shows several different batteries.

Batteries are made up of *cells*. Each cell has a positive electrode and a negative electrode. The cells produce a small voltage. The voltage a cell produces depends on the chemical process taking place inside the cell. Rechargeable nickel-cadmium cells produce about 1.2 volts per cell. Common zinc-acid and alkaline flashlight cells produce about 1.5 volts per cell. The lead-acid cells in a car battery each produce about 2 volts.

The number of cells in a battery depends on the voltage we want to get out of the battery. If we only need a low voltage the battery may contain only one cell. Small hearing-aid batteries, for example, usually contain only one cell. Part A of Figure 6-7 shows the schematic symbol for a single-cell battery. The two vertical lines in the schematic symbol represent the two electrodes in the cell. The long line represents the positive terminal and the short line represents the negative terminal.

To produce a battery with a higher voltage, several cells must be connected in series so their outputs add. Each cell produces a small voltage. The battery manufacturer connects several cells in series to produce the desired battery voltage.

Part B of Figure 6-7 shows the schematic symbol for a multiple-cell battery. We use several lines to show the many cells in the battery. Again, the long line at one end represents the positive terminal, and the short line at the other end represents the negative terminal. The schematic symbol does not indicate the number of cells in the battery. Two sets of long and short lines represent any multiple-cell battery.

[Make a quick trip to Chapter 13 now and look at questions N6A04 and N6A09. Review this section if those questions confuse you.]

Antennas and Grounds

When you use your receiver or transmitter, you must connect it to an antenna. You should also connect all the equipment in your station to a good Earth ground. There is more information about these connections in Chapter 4. There are really two kinds of ground connections: **chassis ground** and **Earth ground**. The metal box that your radio is built on is called a chassis. Most manufacturers use the chassis as a common connection for all the places in the circuit that connect to the negative side of the power supply. This common connection is called the chassis ground. Figure 6-8A shows the schematic symbol for a chassis ground. When you look at this symbol you can visualize the box shape that represents the chassis.

To keep your station safe, you should also connect the chassis ground to an Earth ground. You can provide this Earth ground by driving an 8-foot or longer copper-coated steel rod into the Earth. (Such *ground rods* are available at electrical-supply stores.) Then connect a heavy copper strap or wire braid between the ground rod and the equipment chassis.

Another way to provide an Earth ground is to connect the copper strap or wire braid to a copper cold-water pipe that goes into the Earth. You have to be certain that the copper pipe continues into the Earth, however. If the pipe changes to plastic pipe it will not provide a suitable Earth ground connection.

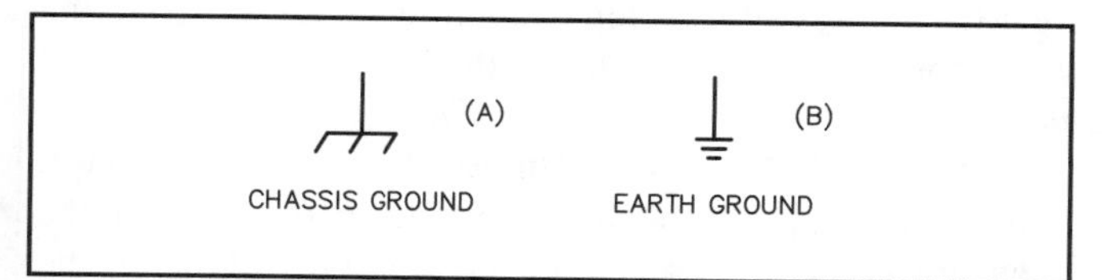

Figure 6-8 — The metal chassis of a radio is sometimes used to make a common ground connection for all the circuit points that connect to the negative side of the power supply. We use the symbol at A to show those connections on a schematic diagram. Your radio equipment should be connected to a ground rod, or a copper cold-water pipe that goes into the ground, for safety. We use the symbol at B to show an earth-ground connection.

Figure 6-9 — Most receivers and transmitters are useless without an antenna. This schematic symbol represents the antenna in a circuit.

This connection is called an Earth ground because it goes into the Earth. An Earth ground has a different schematic symbol than a chassis ground. Figure 6-8B shows the schematic symbol for an Earth ground. When you look at the lines at the bottom of this symbol it might remind you of the point on a garden spade shovel. If you think about pushing that spade into the ground, you will remember the Earth ground symbol.

Your receiver won't hear signals and no one will hear your transmitter without an antenna. The antenna is a very important part of any radio installation. We use the symbol shown in Figure 6-9 to represent the antenna on a schematic diagram.

An antenna radiates the radio-frequency energy from a transmitter. This means the radio energy leaves the antenna and travels off into the air around it. We call this energy a *radio wave*. When the radio wave travels past another antenna it produces a small current in the wire, creating a signal for a receiver. Antennas radiate and receive radio energy. When you look at the symbol for an antenna you can imagine the radio signal leaving the wires or metal conductors of the antenna and radiating into space.

There are many types of antennas. You will learn more about the various kinds of antennas in Chapter 9. You will also learn how to choose antennas for various kinds of radio equipment.

[Now turn to Chapter 13 and study questions numbered N6B03 through N6B06. If you have any trouble answering the questions, review this section.]

Transistors

Many of the great technological advances of recent times — men and women in space, computers in homes, ham radio stations tiny enough to be carried in a shirt pocket — all have been made possible by *semiconductor* electronics. Not simply a partial conductor as the name implies, a semi-

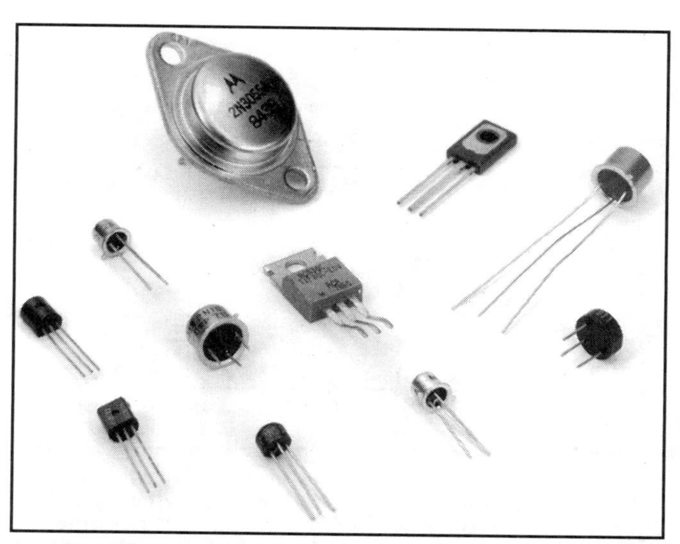

Figure 6-10 — Transistors are packaged in many different cases.

conductor has some of the properties of a conductor and some properties of an insulator.

Diodes and **transistors** are two types of semiconductors, the *solid-state devices* that have replaced vacuum tubes in most uses and created many new applications. Most semiconductor devices are much smaller than comparable tubes, and they produce less heat. Semiconductors are also usually less expensive than tubes.

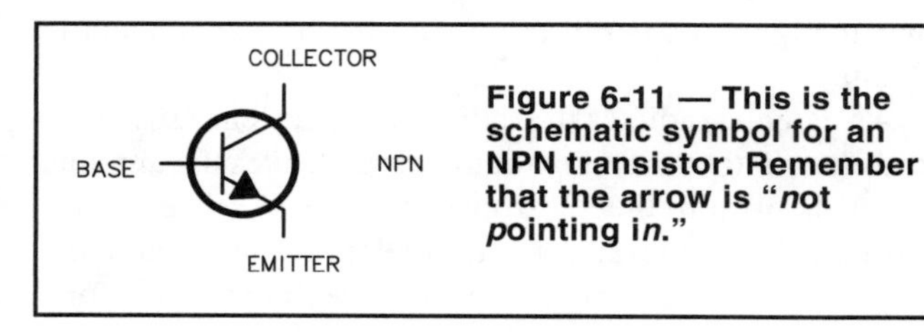

Figure 6-11 — This is the schematic symbol for an NPN transistor. Remember that the arrow is "*n*ot *p*ointing i*n*."

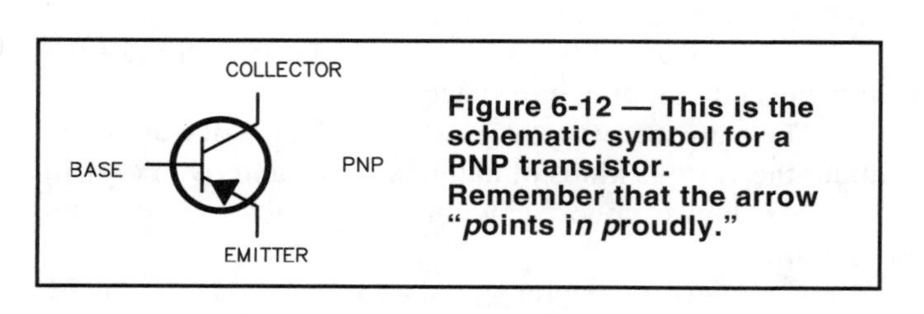

Figure 6-12 — This is the schematic symbol for a PNP transistor. Remember that the arrow "*p*oints i*n p*roudly."

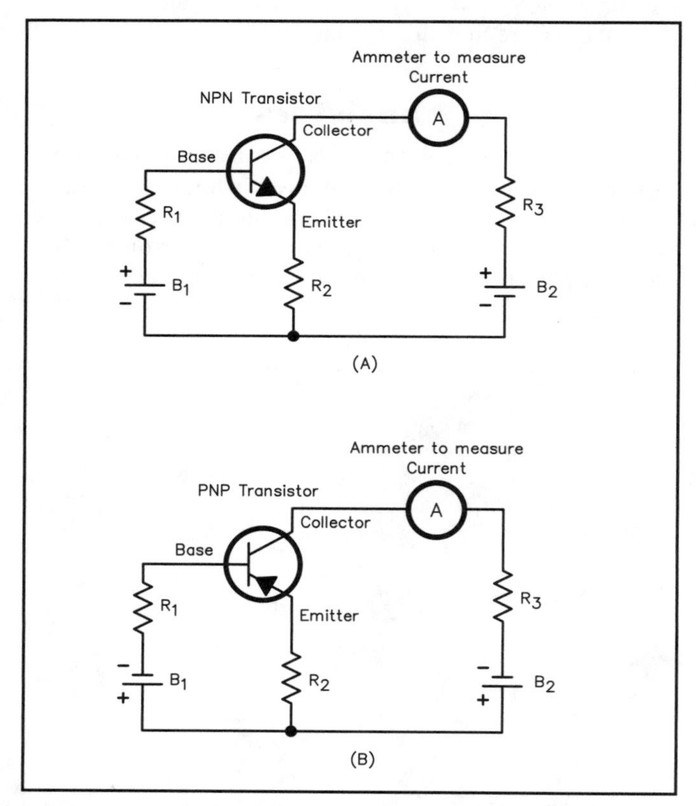

Figure 6-13 — These diagrams show simple transistor circuits for NPN and PNP transistors. Batteries B_1 and B_2 provide the proper operating voltages and polarities to produce a current through the collector/emitter part of each circuit.

Portable broadcast-band radios, which weighed several pounds and were an armful shortly after World War II, weigh ounces and can be carried in your pocket today. A complete ham radio station, which would have filled a room 50 years ago, now can be built into a container the size of a shoebox, or smaller. Solid-state technology has made all this possible.

Transistors come in many shapes and sizes. Figure 6-10 shows some of the more common case styles for transistors. The most common type of transistor is the *bipolar* **transistor**. Bipolar transistors are made of two different kinds of material (*bi* means two, as in *bi*cycle). The two kinds of semiconductor material are known as *N-type material* and *P-type material*. The N refers to the way electrons, or negative electric charges, move through the material. The P refers to the way positive charges move through this material. (Don't worry about the details of how this works now. You can learn about how transistors actually work after you pass your license exam.)

There also are two kinds of bipolar transistors. Each kind of bipolar transistor has a separate schematic symbol. Figure 6-11 shows the schematic symbol for an **NPN transistor**. You can remember this symbol by remembering that the arrow is "***n***ot ***p***ointing i***n***." An NPN transistor has a layer of P-type semiconductor material sandwiched between two layers of N-type material.

Figure 6-12 shows the symbol for the other kind of bipolar transistor, the **PNP transistor**. Remember this symbol by saying that the arrow "***p***oints i***n p***roudly." A PNP transistor has a layer of N-type semiconductor material sandwiched between two layers of P-type material.

The schematic symbols show that transistors have three leads, or electrodes. Each of the electrodes connects to a different part of the transistor. Transistors can amplify small signals; this is what makes them so useful. Usually, we use a low-voltage signal applied to the *base* of the transistor to control the current through the *collector* and *emitter*.

Figure 6-13A shows a simple NPN transistor circuit. Battery B_1 applies a positive voltage to the base of the transistor and battery B_2 applies a positive voltage to the collector. The emitter has a negative voltage applied. A small current flows through the emitter and base portion of the circuit. This small current controls a (usually larger) current from the negative emitter to the positive collector. If the base current decreases the collector current also decreases. If the base current increases, so will the collector current. If you remove B_1 or otherwise open the base circuit, there will be no collector current.

Figure 6-13B is a similar circuit for a PNP transistor. Notice this time B_1 applies a negative voltage to the transistor base lead and B_2 applies a negative voltage to the collector. Once again a small current through the base and emitter portion of the circuit controls the current through the collector and emitter part of the circuit.

[Time for another look at the question pool. Study the questions in Chapter 13 with numbers N6B01, N6B02, N6B07 and N6B08. Review this section if you have problems.]

Vacuum Tubes

The development of the vacuum tube was an important milestone in the history of radio. The vacuum tube was the first *active* electronic device — that is, the vacuum tube can *amplify*, or produce an enlarged version of the input signal.

From the outside, a tube looks like a glass bulb with pins sticking out of the bottom. Sometimes there are also leads coming out of the top or sides. Some tubes have a metal collar or band around the base, and some tubes have ceramic or metal envelopes (outer shell). Tubes are quite fragile, and will break easily if mishandled. They usually plug into a socket wired into a circuit. Figure 6-14 shows some common tubes.

Figure 6-14 — Here are some common vacuum tubes. This figure shows tiny receiving tubes and large transmitting tubes.

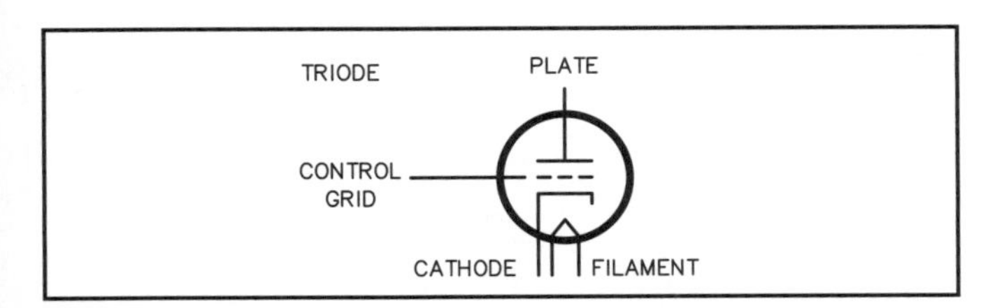

Figure 6-15 — A triode vacuum tube has three active elements. This is the schematic symbol for a triode. We don't count the filament or heater because it only serves to heat the cathode and produce the free electrons that will flow inside the tube. Triodes are used in some amateur power amplifiers.

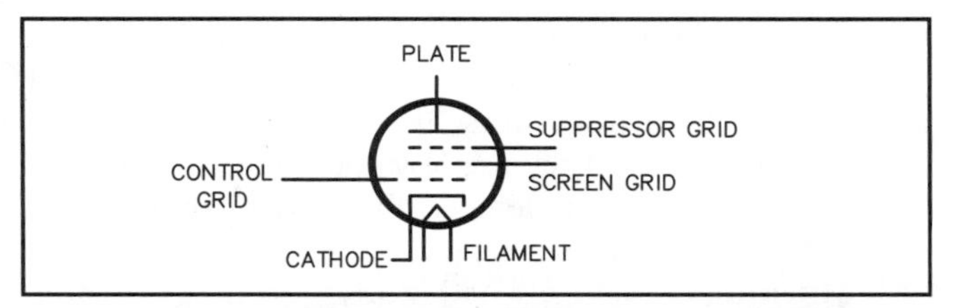

Figure 6-16 — A pentode vacuum tube has five active elements. This is the schematic symbol for a pentode. Pentodes are used in some amateur power amplifiers.

Tubes are named for the number of elements they have inside them. All tubes have at least two elements, the *plate* and the *cathode*. A tube with only two elements has a limited usefulness, however. A two-element tube is called a *diode* (*di* means two). You may have heard the name diode, but it was probably used with regard to a semiconductor diode rather than a tube.

Tubes became much more useful when inventors added a third element, the *control grid*. This grid controls the flow of electrons through the tube. By controlling the flow of electrons through the tube, you can use it as an amplifier. Tubes generally use voltages that are much higher than the voltages used with transistors.

A tube with three elements is called a **triode** (*tri* means three, as in *tri*cycle). Figure 6-15 shows the schematic symbol for a triode vacuum tube. The *filament* gets hot when a current flows through it, and the heat knocks electrons loose from the cathode. (The filament is also sometimes called the *heater*.) These electrons are then pulled toward the plate, and collected there. As you can guess, the cathode connects to the negative side of the voltage supply and the plate connects to the positive side. Notice that we don't include the filament when we are counting elements in a tube.

While transistors have replaced tubes in many modern applications, tubes are still used in many types of electronic circuits. Radio and television sets using tubes are still in operation. The *picture tube* or *cathode-ray tube* (*CRT*) is a vacuum tube. Tubes can handle high power and high voltage. This makes them ideal for use in the output stage of an amateur transmitter or as high-power amplifiers.

Vacuum tubes often have additional grids added to improve their performance for specific applications. Triodes are often used in amateur high-power amplifiers, but many amplifiers also use **pentode** vacuum tubes. (*Penta* means five: the Pentagon is a five-sided building used as the US military headquarters.) Figure 6-16 shows the schematic symbol for a pentode vacuum tube.

Integrated Circuits

No discussion of modern electronics would be complete without at least a brief mention of **integrated circuits (ICs)**. An IC usually performs several circuit functions in one package, and sometimes an entire application requires only the addition of a few external components. Like transistors, ICs are made with semiconductor materials. They use low operating voltages and are generally low-power devices. Figure 6-17 shows some examples of typical IC packages.

[Turn to Chapter 13 and study questions N6B09 through N6B13. If you're only preparing for the Novice exam now, that's it for this chapter! By now, you should have a basic understanding of how all of the circuit components included on the Novice exam work. If you had trouble with any of the related questions in the question pool, review those sections before you proceed to the next chapter.]

[**If you're preparing for the Technician exam**, continue with this chapter after you review what you've learned so far.]

Figure 6-17 — Integrated circuits (ICs) come in a variety of package styles.

KEY WORDS

Capacitor — An electronic component composed of two or more conductive plates separated by an insulating material.

Color code — A system in which numerical values are assigned to various colors. Colored stripes are painted on the body of resistors and sometimes other components to show their value.

Core — The material used in the center of an inductor coil, where the magnetic field is concentrated.

Electrolytic capacitor — A polarized capacitor formed by using thin foil electrodes and chemical-soaked paper.

Electric field — An invisible force of nature. An electric field exists in a region of space if an electrically charged object placed in the region is subjected to an electrical force.

Film resistor — A resistor made by depositing a thin layer of resistive material on a ceramic form.

Inductor — An electrical component usually composed of a coil of wire wound on a central core.

Potentiometer — A resistor whose resistance can be varied continuously over a range of values.

Reactance — The property of an inductor or capacitor (measured in ohms) that impedes current in an ac circuit without converting power to heat.

Resistor — Any material that opposes a current in an electrical circuit. An electronic component specifically designed to oppose current.

Toroidal core — A donut-shaped form used to hold the turns of an inductor. Toroidal cores are usually made from ferrite or powdered-iron materials.

Variable capacitor — A capacitor that can have its value changed within a certain range.

Variable resistor — A resistor whose value can be adjusted over a certain range.

Wire-wound resistor — A resistor made by winding a length of wire on an insulating form.

6 Circuit Components For Technicians

This section presents the additional information about circuit components that you will need to know to pass your Technician written exam. You will find descriptions of resistors, capacitors and inductors. You can combine these components with other devices to build practical electronic circuits. We describe some of these circuits in Chapter 7.

Turn to the Element 3A questions in Chapter 14 when directed to do so. How well you understand the questions will show you where you need to do some extra studying. You should thoroughly understand how these components work. When you do, you will have no problem learning to connect them to make a circuit perform a specific task.

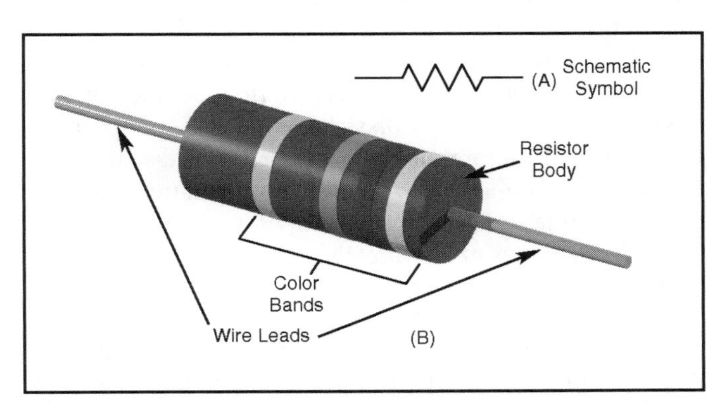

Figure 6-18 — Part A shows the schematic symbol for a resistor. Part B shows a typical resistor, including the colored stripes that identify the resistor value. Resistors made in this way range in power rating from 1/8 watts to 2 watts.

Resistors

When electrons flow from one point in a circuit to an-

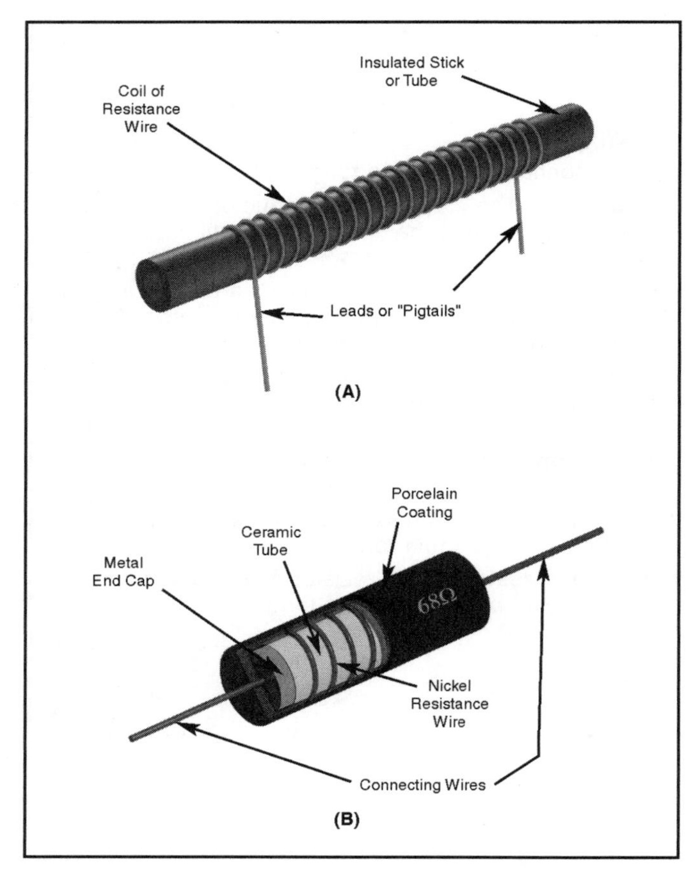

Figure 6-19 — A simple wire-wound resistor is shown at A. At B is a commercially produced wire-wound resistor. These resistors are produced by winding resistive wire around a nonconductive form. Most wire-wound resistors are then covered with a protective coating.

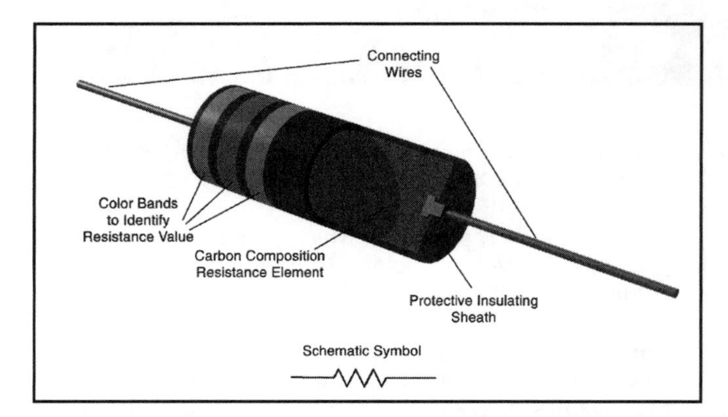

Figure 6-20 — Composition resistors are composed of a mixture of carbon and clay. The resistance is controlled by varying the amount of carbon in the material when the resistor is manufactured.

other point, there is *current* in that circuit. A perfect insulator would allow no electrons to flow through it (zero current), while a perfect conductor would allow infinite current (unlimited electrons flow). In practice, however, there is no such thing as a perfect conductor nor a perfect insulator. Partial opposition to electron flow occurs when the electrons collide with other electrons or atoms in the conductor. The result is a reduction in current, and the conductor produces heat.

Resistors allow us to control the current in a circuit by controlling the opposition to electron flow. As they oppose the flow of electrons they dissipate electrical energy in the form of heat. The more energy a resistor dissipates, the hotter it will become. Figure 6-18A shows the schematic symbol for all fixed-value resistors. Part B shows a common resistor. You will learn how to interpret the colored stripes on a resistor later in this chapter.

All conductors exhibit some resistance. Thick (heavy) conductors have only a little resistance. Thinner conductors have more resistance. This is just what we would expect. Think of electron flow in a conductor as being like water flow in a pipe. If we reduce the size of the pipe, less water can flow through it. Different conductor materials also have different amounts of resistance. Copper is a good conductor, and so has a small resistance. A nickel wire has more resistance than a copper wire, if both have the same diameter.

We use numbers in the *American Wire Gauge (AWG)* scale to measure wire sizes. Smaller numbers in this scale indicate larger diameter wires. A number 12 wire (a common size for house wiring) has a larger diameter than a number 30 wire (which compares to the thickness of a hair).

One simple way of producing a resistor is to use a length of wire. For example, the resistance of 1000 feet of number 28 nickel wire is 337 ohms. Therefore, if we need a resistor of 3.4 ohms, we can use 10 feet of this size nickel wire. You might think 10 feet of wire makes a pretty inconvenient resistor, but we can wind it over a form of some sort. See Figure 6-19A. This makes a unit of a much more convenient size. This is precisely how **wire-wound resistors** are constructed. See Figure 6-19B.

There is a problem with this type of construction. At radio frequencies, the wire-wound construction causes the resistor to act as an **inductor**. (You will learn more about inductors later in this chapter.) It does not behave as a pure resistance. For RF circuits, resistors must be noninductive. Wire-wound resistors are common in dc circuits that carry large currents, because they can get rid of large amounts of heat.

There is another way to form a resistor. This is to connect leads to both ends of a block or cylinder of material that has a high resistance. Figure 6-20 shows a resistor made in this way. It uses a mixture of carbon and clay as the resistive element. This type of resistor is a *carbon-composition resistor*. The proportions of carbon and clay determine the value of the resistor.

An advantage of carbon-composition resistors is the wide range of available values. Other advantages are low inductance and capacitance plus good surge-handling capability. The ability to withstand small power overloads without being completely destroyed is another advantage. The main disadvantage is that the resistance of the composition resistor will vary widely. Variations are caused by operating temperature changes and resistance changes as the resistor gets older.

There are two types of **film resistors**. These are made by depositing a thin layer or film of resistive material on a ceramic form. The ceramic form is usually shaped as a cylinder. Metal end caps provide a connection between the resistive film and the resistor leads used to connect the part in a circuit. The entire assembly then gets a protective plastic or epoxy coating.

You can easily recognize the "dog bone" shape of film resistors. They are thinner in the middle than at the ends. This is because of the metal end caps used to connect the leads to the resistive element.

Carbon-film resistors are manufactured by using high temperatures to break down certain gaseous hydrocarbons. The resulting carbon is then deposited in a thin layer or film on the ceramic form. The thickness of the deposited film provides a way to control the final resistance.

The major advantages of carbon-film resistors are low cost and improved stability with age and temperature changes. These resistors cannot withstand electrical overloads or surges. They can be used as fuses for some applications.

The *metal-film resistor* has replaced the carbon-composition type in many low-power applications today. Metal-film resistors are formed by depositing a thin layer of resistive alloy on a cylindrical ceramic form. Nichrome (an alloy made of nickel and chromium) or other materials are used. See Figure 6-21. This film is then trimmed away in a spiral fashion to form the resistance path. The trimming can be done on a mechanical lathe or by using a laser. The resistor is then covered with an insulating material to protect it.

Because of the spiral shape of the resistance path, metal-film resistors will exhibit some inductance. The effects of this inductance increase at higher frequencies. The higher the resistance of the unit, the greater the inductance will be. This is caused by the increased path length. Metal-film resistors provide much better temperature stability than other types.

Most resistors have standard fixed values. They are called *fixed resistors*. **Variable resistors** are sometimes called **potentiometers**. They can be used to adjust the voltage, or potential, in a circuit. Potentiometers are also used a great deal in electronics. The construction of a wire-wound variable resistor and the schematic symbol for all variable resistors is shown in Figure 6-22. Variable resistors are also made with a ring of carbon compound in place of the wire windings. A connection is made to each end of the resistance ring. A third contact is attached to a movable arm, or wiper. The wiper can be moved across the ring. As the wiper moves from one end of the ring to the other, the resistance varies from minimum to maximum.

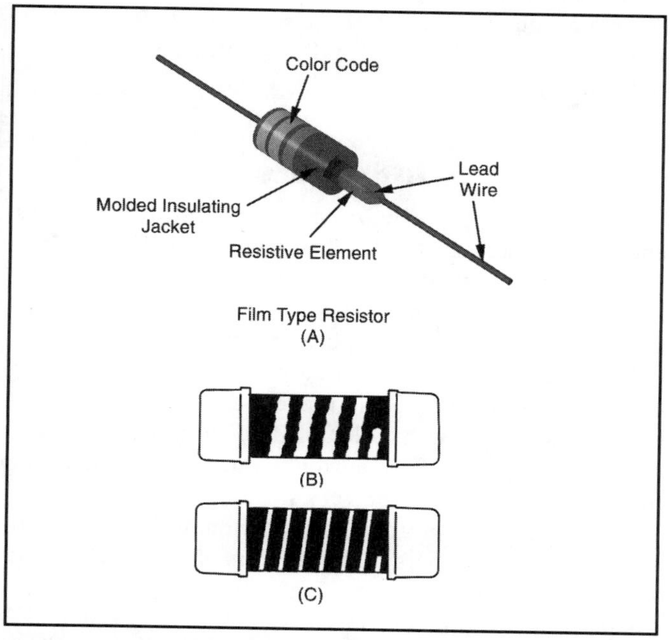

Figure 6-21 — The construction of a typical metal-film resistor is shown at A. Film resistors are formed by depositing a thin film of material on a ceramic form. Material is then trimmed off in a spiral to produce the specified resistance for that particular resistor. The excess material may either be trimmed with a mechanical lathe or a laser. A lathe produces a rather rough spiral, like that shown in part B. The laser produces a finer cut, shown at C.

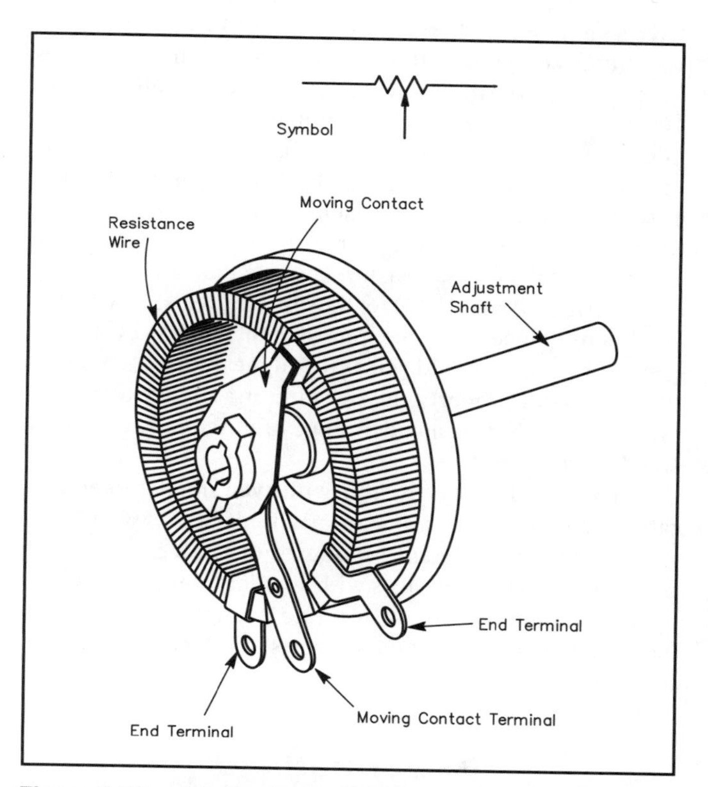

Figure 6-22 — The typical construction of a wire-wound variable resistor. As the shaft is rotated, the moving contact is electrically connected to different parts of the wire winding. This effectively changes the length of wire between the end terminal and moving-contact terminal. Increasing or decreasing wire length between terminals thus increases or decreases resistance.

Color Codes

It is not always practical to print the resistance values on the side of a small resistor. A **color code** shows the value of the resistor. As shown in Figure 6-23, the color of the first three bands shows the resistor value. The color of the fourth band indicates the resistor tolerance.

Standard fixed resistors are usually found in values ranging from 2.7 Ω to 22 MΩ (22 megohms, or 22,000,000 ohms). Resistance tolerances on these standard values can be ±20%, ±10%, ±5%, and ±1%. A tolerance of 10% on a 200-ohm resistor means that the actual resistance of a particular unit may be anywhere from 180 Ω to 220 Ω. (Ten percent of 200 is 20, so the resistance can be 200 + 20 or 200 – 20 ohms.)

Here is another example to help you understand this calculation. Find the range of possible values for a 100 Ω resistor that has a 10% tolerance rating.

First calculate the range of possible variation:

$$100\ \Omega \times 10\% = 100\ \Omega \times 0.10 = 10\ \Omega$$

Now use this value to calculate the possible range of resistances:

$$100\ \Omega - 10\ \Omega = 90\ \Omega$$

$$100\ \Omega + 10\ \Omega = 110\ \Omega$$

A 100-ohm resistor that has a 10% tolerance rating will have a value somewhere between 90 and 110 Ω.

Suppose you are building a sensitive circuit that requires a resistor that is no more than 105 Ω. Would you select a 10%, 100-Ω resistor for this circuit? Probably not, because when you go to your local electronics parts store to buy the resistor, you may get a unit that has a resistance as high as 110-Ω. What tolerance component should you choose? A 5% tolerance resistor will suit your purpose in this circuit. You might also select a 1% resistor, because the value will only vary between 99 and 101 Ω for those parts. Resistors with the smaller tolerance rating are more expensive, however. As a general rule you should select the highest tolerance rating that will work properly in your circuit. (Higher tolerance ratings mean there is a wider range of possible values for a standard stock part.)

You can also buy resistors with even smaller tolerance ratings. *Precision resistors* are specified with 0.5%, 0.25%, 0.1% and 0.05% tolerance ratings. For a price, you could probably even buy resistors with smaller tolerances. If you have an ohmmeter that is accurate enough, you also can measure the values of standard parts to select those within the limits you require.

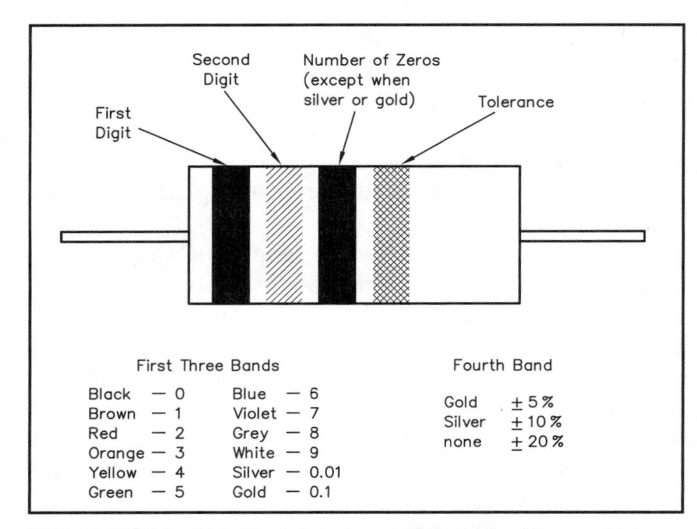

Figure 6-23 — Small resistors are labeled with a color code to show their value. For example, proceeding from left to right, a resistor with color bands of yellow/violet/brown/gold is a 470-Ω resistor with a 5% tolerance.

Power Ratings

Resistors are also rated according to the amount of power they can safely handle. Power ratings for wire-wound resistors typically start at 1 W and range to 10 W or larger. Resistors with higher power ratings are also physically large. Greater surface area is required to dissipate the increased heat generated by large currents. Carbon-composition and film resistors are low-power components. You will find these resistors in 1/8-W, 1/4-W, 1/2-W, 1-W and 2-W power ratings. A resistor that is being pushed to the limit of its power rating will feel very hot to the touch. Sometimes a resistor gets very hot during normal operation. You should probably replace it with a unit having a higher power-dissipation rating.

[Now turn to Chapter 14 and study questions all of the questions from T6A01 through T6A14. Review this section as needed.]

Inductors

In Chapter 5 we learned that an **inductor** stores energy in a magnetic field. When a voltage is first applied to an inductor, with no current flowing through the circuit, the inductor will oppose the current. Remember from Chapter 5 that inductors oppose *any* change in current. A voltage that opposes the applied voltage is induced in the coil, and this voltage tries to prevent a current. Gradually, though, the current will build up. Only the resistance in the circuit will limit the final current value. The resistance in the wire of the coil will be very small.

In the process of getting this current to flow, the coil stores energy. The energy is in the form of a magnetic field around the wire. The increasing field strength produces a voltage that opposes the current. When you shut off the applied current, the magnetic field collapses. The collapsing field returns the stored energy to the circuit. The energy will have the form of a momentary current in the same direction as the original current. This current is produced by the EMF generated by the collapsing magnetic field. Again, this voltage opposes any change in the current already flowing in the coil.

We can summarize the actions of an inductor by considering two factors. Inductors store energy in magnetic fields, and they oppose any change in current through the inductor. Because an *electric current* produces a *magnetic field* we say the action of an inductor storing electrical energy is an *electromagnetic* process.

The amount of opposition of an inductor to changes in current is called *inductive* **reactance**. The reactance of the inductor, the amount of energy stored in the magnetic field and the back EMF induced in the coil all depend on the amount of inductance.

We use a capital L to represent the amount of inductance that a coil exhibits. Inductance depends on four things:

1) the type of material used in the **core**, and its size and location in the coil,
2) the coil diameter,
3) the length of the coil and
4) the number of turns of wire used to wind the coil.

Changing any of these factors changes the inductance.

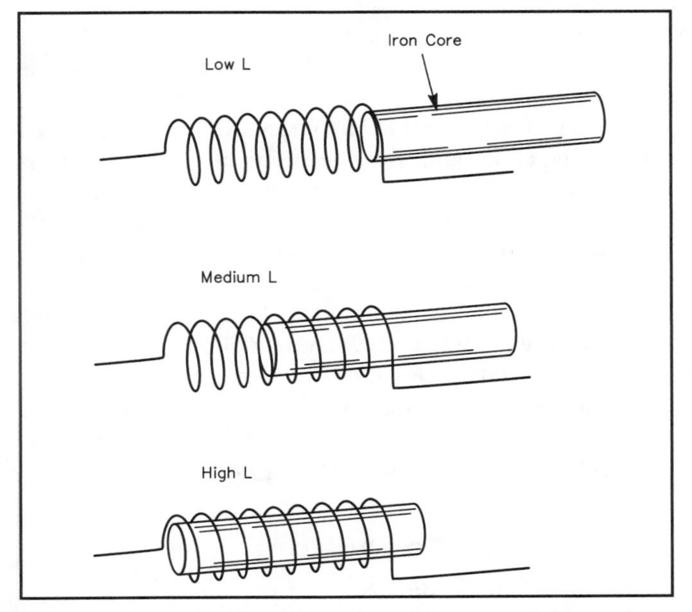

Figure 6-24 — The inductance of a coil increases as an iron core is inserted. Many practical inductors of this type use screw threads to allow moving the core in precise amounts when making small changes in inductance.

An inductor's **core** refers to the area inside the inductor, where the magnetic field is concentrated. If we add an *iron* or *ferrite core* to the coil, the inductance increases. We can also use brass as a core material in radio-frequency variable inductors. It has the opposite effect of iron or ferrite: The inductance decreases as the brass core enters the coil. A movable core can be placed in a coil, as shown in Figure 6-24. The amount of inductance can be varied as the core moves in and out of the coil. The common schematic symbols for various inductors are shown in Figure 6-25.

The adjustable coil shown in Figure 6-25 has a threaded core so you can adjust it into and out of the coil. Such inductors usually serve to adjust the operating frequency of a circuit. These adjustments often require you to turn the core material into or out of the coil while the circuit is operating. You must use a plastic tool to make these adjustments rather than a metal tool like a screwdriver. Bringing a metal tool close to the coil would change the inductance. This would result in an incorrect circuit adjustment.

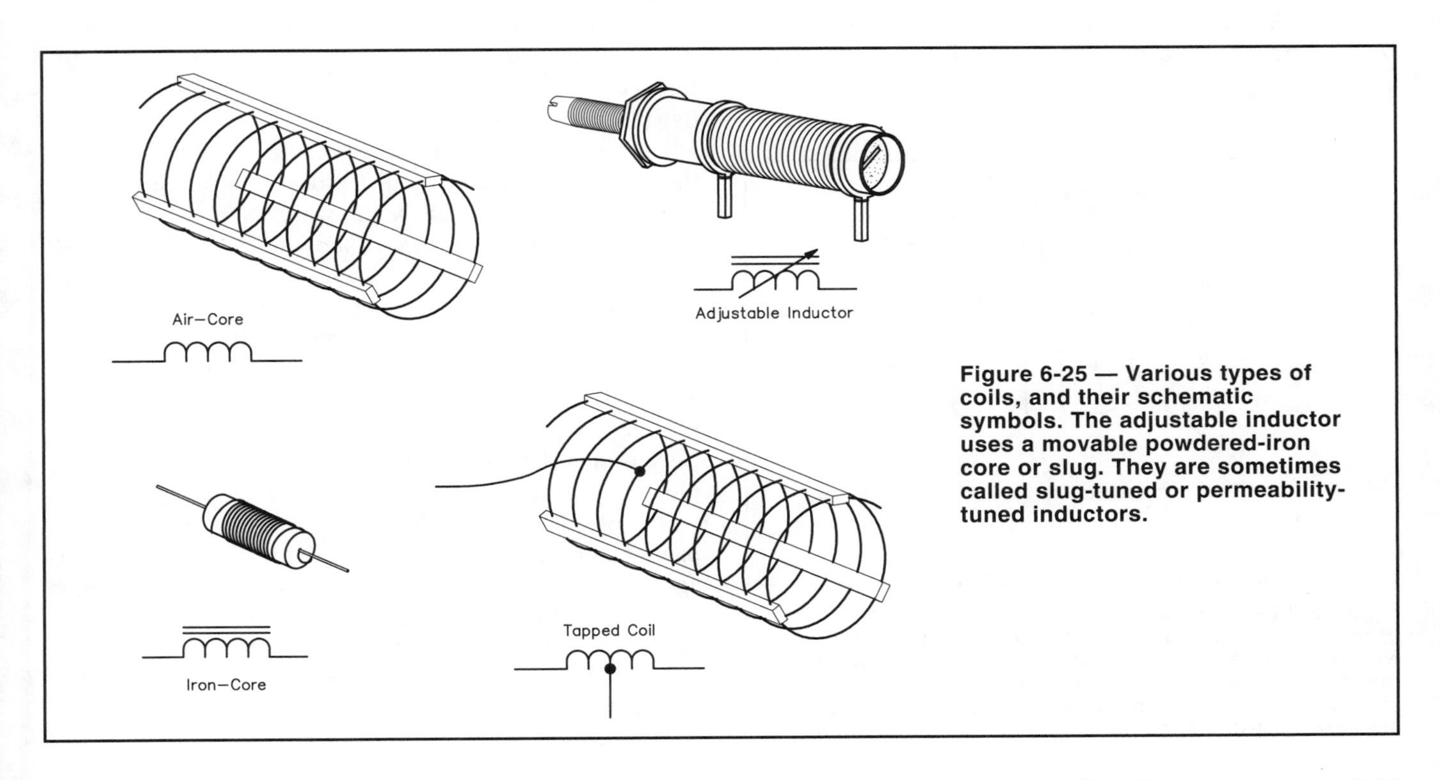

Figure 6-25 — Various types of coils, and their schematic symbols. The adjustable inductor uses a movable powdered-iron core or slug. They are sometimes called slug-tuned or permeability-tuned inductors.

Ferrite comes from the Latin word for iron. So what is the difference between an iron and a ferrite core? Well, iron cores can be made from iron sheet metal. The layers of material that make up an ac power transformer core are made that way. Cores can also be made from powdered iron mixed with a bonding material. This holds the powdered iron together so it can be molded into the desired shape. Manufacturers mix other metal alloys with the iron "ferrous" material when they make some cores. We call these ferrite cores. They can select the proper alloy to provide the desired characteristics, producing cores designed to operate best over a specific frequency range. We use iron cores most often at low frequencies. The audio-frequency range and ac-power circuits are good applications for these cores. Ferrite cores are manufactured in a wide range of compositions. This optimizes their operation for specific radio-frequency ranges.

It takes some amount of energy to magnetize the core material for any inductor. This energy represents a loss in the coil. Winding the coil on an iron material increases the inductance of the coil. The energy needed to magnetize the core also increases. Some coils are wound on thin plastic forms, or made self supporting. Then the only material inside the coil is air. These are air-core coils. They are among the lowest loss types of inductors. Air-core coils are normally used in high-power RF circuits where energy loss must be kept to a minimum.

Toroid Cores

A *toroidal inductor* is made by winding a coil of wire on a donut-shaped core called a toroid. **Toroidal core** materials are usually powdered-iron or ferrite compounds. The toroidal inductor is highly efficient. This is because there is no break in the circular core. All the magnetic lines of force remain inside the core. This means there is very little *mutual coupling* between two toroids mounted close to each other in a circuit.

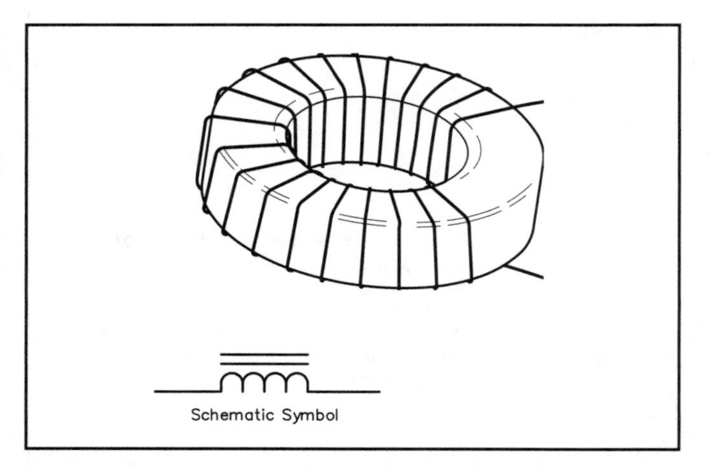

Figure 6-26 — A toroidal coil has self-shielding properties. Some inductors in some circuits will interact with the magnetic fields of other inductors. To prevent this a metal shield is used to enclose them. Toroidal inductors confine magnetic fields so well that shields are usually not necessary.

Figure 6-26 shows a toroidal inductor and the schematic symbol used to represent them. Notice that the schematic symbol is the same one used for any inductor that has an iron or ferrite core material.

Mutual coupling occurs when the magnetic flux of one coil passes through the windings of the other coil. When coils have mutual coupling, a current in one coil will induce a voltage (and a current) in the other. Mutual coupling can be a problem, especially in an RF circuit. There, signals from one stage could be coupled into another stage without following the proper signal path. Toroid cores can often help eliminate the mutual coupling that would occur between other types of coils.

In theory, a toroidal inductor requires no shield to prevent its magnetic field from spreading out and interfering with outside circuits. This self-shielding property also helps keep outside forces from interfering with the toroidal coil. Toroids can be mounted so close to each other that they can almost touch, but because of the way they are made, there will be almost no inductive coupling between them. Toroidal coils wound with only a small amount of wire on ferrite or iron cores can have a very high inductance value.

[Turn to Chapter 14 and study exam questions T6B02 through T6B04 and T6B07 through T6B11. Review any material you have difficulty with before going on.]

Capacitors

A **capacitor** is composed of two or more conductive plates with an insulating material between them. In Chapter 5 we learned that the basic property of a capacitor is the ability to store an electric charge. In an uncharged capacitor, the potential difference between the two plates is zero. As we charge the capacitor, this potential difference increases until it reaches the full applied voltage. The difference in potential creates an **electric field** between the two plates. See Figure 6-27.

A *field* is an invisible force of nature. We put energy into the capacitor by charging it. Until we discharge it, or the charge leaks away somehow, the energy is stored in the electric field. When the field is not moving, we sometimes call it an *electrostatic field.*

When we apply a dc voltage to a capacitor, current will flow in the circuit until the capacitor is fully charged to the applied voltage. After the capacitor has charged to the full applied voltage, no more current will flow in the circuit. For this reason, capacitors can be used to block dc in a circuit. Because ac voltages are constantly reversing polarity, we can block dc with a capacitor while permitting ac to pass.

When the voltage across a capacitor increases, the capacitor stores more charge in reaction to the increased voltage. When the voltage decreases, the capacitor returns some of its stored charge. These actions tend to oppose any change in the voltage applied to a capacitor. This opposition to changes in *voltage* (and current) is called *capacitive* **reactance**.

We can summarize the operation of a capacitor by con-

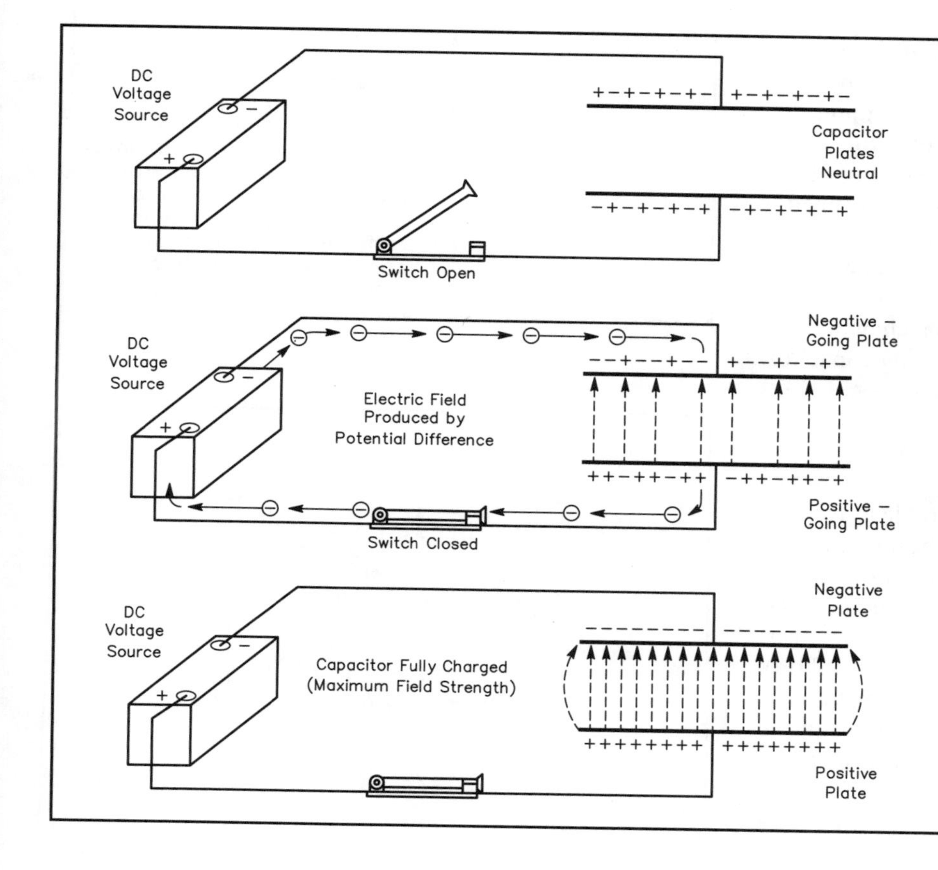

Figure 6-27 — Capacitors store energy in an electric field. The final amount of energy stored in a fully charged capacitor depends on the capacitor value and the applied voltage. If this circuit is used to fully charge the capacitor from a 6-V battery, the amount of energy stored in the capacitor could be increased by using a 12-volt battery. Another way to increase the amount of stored energy is to use a larger value capacitor.

sidering two factors. A capacitor stores electric energy in the form of an electrostatic field and it opposes any change in the applied voltage.

Factors That Determine Capacitance

The value of a capacitor is determined by three factors:

1) the type of insulating material or **dielectric** used between the plates,
2) the area of the plate surfaces and
3) the distance between the plates.

When we described a simple capacitor, we said there are two conducting plates, one connected to each polarity of the voltage supply. By adding additional sets of plates, each separated by insulating material, you can increase the effective plate area. Figure 6-28 shows the construction of a capacitor with two plates for each side. Additional layers of foil and insulating material can be added to increase the capacitance. Alternate foil layers connect to opposite sides of the capacitor in this construction.

As the surface area increases, the capacitance increases. For a given plate area, as the spacing between the plates decreases, the capacitance increases.

Practical Capacitors

Practical capacitors are described by the material used for their dielectric. Mica, ceramic, plastic-film, polystyrene, paper and electrolytic capacitors are in common use today.

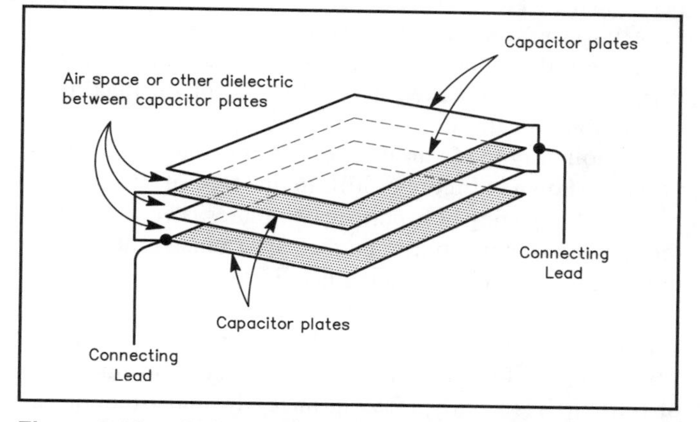

Figure 6-28 — This drawing shows how the total plate surface area can be increased by stacking several plates, with insulating material between them. This increases the capacitance of the unit, compared to the capacitance if only two plates of the same size are used. The leads connect to alternate plates.

They each have properties that make them more or less suitable for a particular application. Figure 6-29A shows the symbol used to represent capacitors on schematic diagrams.

Electrolytic capacitors have a dielectric that is formed after the capacitor is manufactured. Two common types of electrolytic capacitors are aluminum-electrolytic capacitors and tantalum-dielectric capacitors.

Electrolytic capacitors can be made with a very high

capacitance value in a small package. Electrolytic capacitors are polarized—dc voltages must be connected to the positive and negative capacitor terminals with the correct polarity. The positive and negative electrodes in an electrolytic capacitor are clearly marked. Connecting an electrolytic capacitor incorrectly causes gas to form inside the capacitor and the capacitor may actually explode. This can be very dangerous. At the very least the capacitor will be destroyed by connecting it incorrectly.

When we draw electrolytic capacitors on schematic diagrams, we must indicate the proper polarity for the capacitor. We do this by adding a + sign to the capacitor symbol, as Figure 6-29B shows.

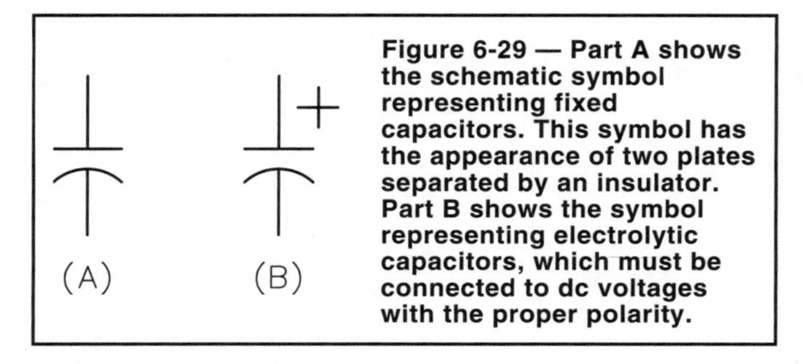

Figure 6-29 — Part A shows the schematic symbol representing fixed capacitors. This symbol has the appearance of two plates separated by an insulator. Part B shows the symbol representing electrolytic capacitors, which must be connected to dc voltages with the proper polarity.

Variable Capacitors

In some circuits it is necessary or desirable to vary the capacitance at some point. An example is in the tuning circuit of a receiver or transmitter VFO. We could use a rotary switch to select one of several different fixed-value capacitors. It is much more convenient to use a **variable capacitor**, however. A basic air-dielectric variable capacitor is shown in Figure 6-30A. This figure also shows the schematic symbol for a variable capacitor. One set of plates is fixed in one position to the capacitor frame. These plates form the *stator*. The other set of plates rotates by turning the control shaft. The movable plates form the *rotor*. The rotation controls the amount of plate area shared by the two sets of plates. When the capacitor is fully meshed, all the rotor plates are down in between the stator plates. In this position, the capacitor will have its largest value of capacitance. When the rotor is unmeshed, the value of capacitance is at a minimum. The capacitance can be varied smoothly between the maximum and minimum values.

Another type of variable capacitor is the compression variable shown in Figure 6-30B. In this type of variable capacitor, the spacing between the two plates is varied. When the spacing is at its minimum, the capacitance is at a maximum value. When the spacing is adjusted to its maximum, the value of capacitance is at a minimum. This type of variable capacitor is usually used as a *trimmer capacitor*. A trimmer capacitor is a control used to peak or fine tune a part of a circuit. It is usually left alone once adjusted to the correct value. You should use a plastic screwdriver-like adjusting tool when tuning a trimmer capacitor. The metal of a normal screwdriver can affect the capacitance and cause false readings as you try to adjust it.

[This completes your study of Chapter 6. Now turn to Chapter 14 and study questions T6B01, T6B05, T6B06 and T6B12 through T6B16. Review any material you have difficulty with before going on.]

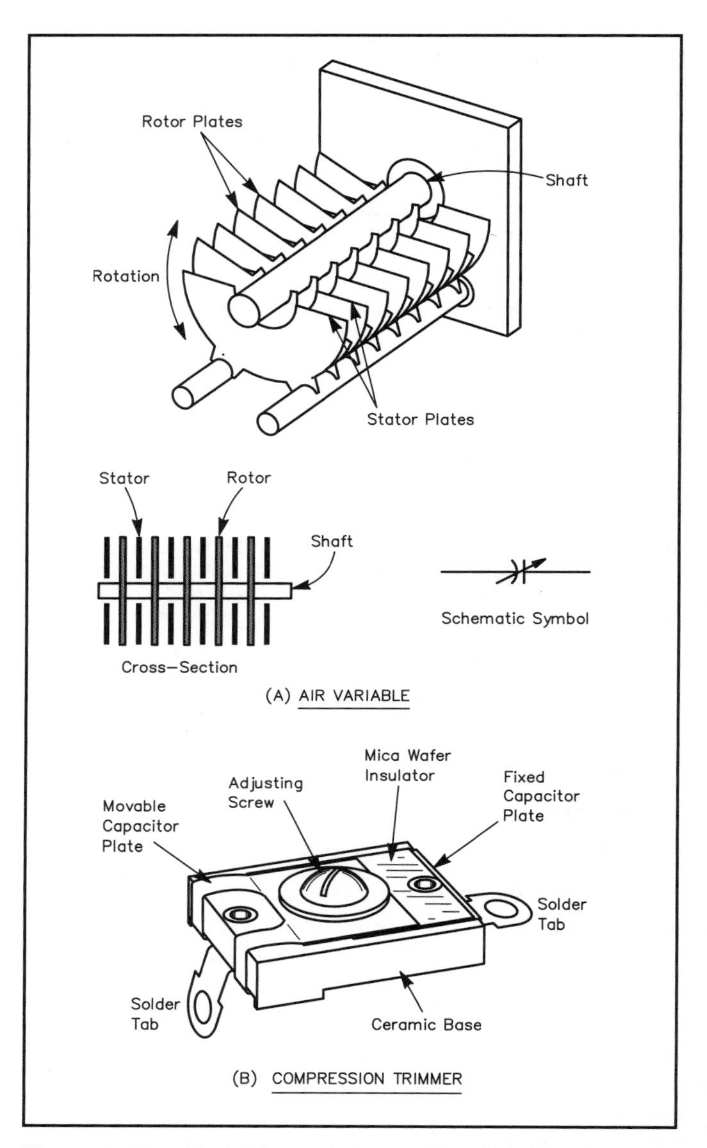

Figure 6-30 — Variable capacitors can be made with air as the dielectric, or with mica or ceramic dielectrics.

7 KEY WORDS

Antenna-matching network — A device that matches the antenna system input impedance to the transmitter, receiver or transceiver output impedance. Also called an **antenna tuner**, **impedance-matching network** or *Transmatch*.

Antenna tuner — See **antenna-matching network**.

Antenna switch — A switch used to connect one transmitter, receiver or transceiver to several different antennas.

Block diagram — A drawing using boxes to represent sections of a complicated device or process. The block diagram shows the connections between sections.

Electronic keyer — A device that generates Morse code dots and dashes electronically.

Feed line — The wires or cable used to connect a transmitter, receiver or transceiver to an antenna.

Hand key — A simple switch used to send Morse code. Also called a **telegraph key**.

Impedance-matching network — A device that matches the impedance of an antenna system to the impedance of a transmitter or receiver. Also called an **antenna-matching network**, **antenna tuner** or *Transmatch*.

Modem — Short for *mo*dulator/*dem*odulator. A modem modulates a radio signal to transmit data and demodulates a received signal to recover transmitted data.

Microphone — A device that converts sound waves into electrical energy.

Packet radio — A communications system in which information is broken into short bursts. The bursts (packets) also contain addressing and error-detection information.

Power supply — A circuit that provides a direct-current output at some desired voltage from an ac input voltage.

Radioteletype (RTTY) — Radio signals sent from one teleprinter machine to another machine. Anything that one operator types on his teleprinter will be printed on the other machine.

Receiver — A device that converts radio signals into audio signals.

SWR meter — A measuring instrument that can indicate when an antenna system is working well.

Telegraph key — See **hand key**.

Teleprinter — A machine that can convert keystrokes (typing) into electrical impulses. The teleprinter can also convert the proper electrical impulses back into text. Computers have largely replaced teleprinters for amateur radioteletype work.

Terminal node controller (TNC) — A TNC accepts information from a computer and converts the information into packets. The TNC also receives packets and extracts information to be displayed by a computer.

Transceiver — A radio transmitter and receiver combined in one unit.

Transmitter — A device that produces radio-frequency signals.

7 Practical Circuits

These students are learning how a receiver, a transmitter and an antenna combine to form an Amateur Radio station. They are also learning about our microwave bands.

In the last chapter, you learned about some of the components that make up electronic circuits. In this chapter, we'll introduce some of the basic equipment that goes into an Amateur Radio station. We will briefly cover receivers, transmitters, transceivers, antenna switches and other equipment. We'll show you how to connect the equipment to make a fully functional ham radio station.

Some of the terms may be confusing at first. Don't worry if you don't understand everything. Most of the ideas presented in this chapter are reinforced later in the book.

Throughout this chapter we use **block diagrams**. In a block diagram, each part of a station is shown as a box. The diagram shows how all the boxes connect to each other. Study the block diagrams carefully and remember to turn to Chapters 13 and 14 when instructed to study exam questions.

BASIC STATION LAYOUT

Figure 7-1 shows a block diagram of a very simple Amateur Radio station. Let's discuss the blocks one by one.

Transmitters

A **transmitter** is a device that produces a radio-frequency (RF) signal. Television and radio broadcast stations use powerful transmitters to put their signals into the air. Radio amateurs use lower-powered transmitters to send signals to each other. A transmitter produces an electrical signal that can be sent to a distant receiver.

We call the signal from a transmitter the *radio-frequency carrier* or *RF carrier*. To transmit Morse code, you could use a telegraph key as a switch to turn the carrier

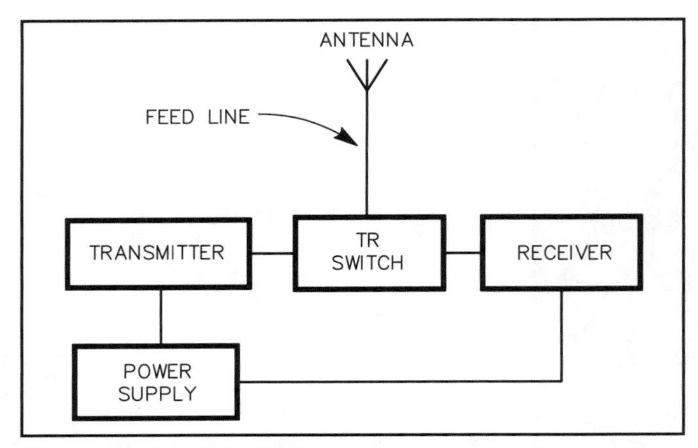

Figure 7-1 — A TR switch connects between the transmitter, receiver and antenna.

on and off in the proper code pattern. If you want to transmit voice signals, you need extra circuitry in the transmitter to add voice content to the carrier. We call this extra circuitry a *modulator*.

Receivers

The transmitter is a sending device. It sends a radio-frequency (RF) signal to a transmitting antenna, and the antenna radiates the signal into the air. Some distance away, the signal produces a voltage in a receiving antenna. That ac voltage goes from the receiving antenna into a **receiver**. The receiver converts the RF energy into an audio-frequency (AF) signal. You hear this AF signal in headphones or from a loudspeaker.

Just about everyone is familiar with receivers. Receivers take electronic signals out of the air and convert them into signals that we can see or hear. Your clock radio is a receiver and so is your television set. If you look around the room you're in right now, you'll probably see at least one receiver. The receiver is a very important part of an Amateur Radio station.

Transceivers

In many modern Amateur Radio stations, the transmitter and receiver are combined into one box. We call this combination a **transceiver**. It's really more than just a transmitter and receiver in one box, though. Some of the circuits in a transceiver are used for both transmitting and receiving. Transceivers generally take up less space than a separate transmitter and receiver.

Many modern radios require 12 V dc to operate. This makes them ideal for use in a car as part of a *mobile* radio station. You will need a separate **power supply** to operate such radios in your house. The power supply (usually) converts the 120 V ac from your wall sockets into 12 V dc to power the radios.

A 100-watt transceiver may draw 20 amps or more of current when it is transmitting. A *heavy-duty* power supply is usually required to provide the current needed to operate the radio in transmit. Some radios have built-in power supplies while others are designed only for 12-V operation, and require an extra power supply if you want to use them on 120 V ac. If you find that your radio works fine in your car, but does not work when you move it into your home, you should suspect a problem with the power supply.

Switches

Now you know the basic parts of an amateur station. The folks at the radio club gave you an old receiver, transmitter and power supply. You bring them home and set them on your desk. Now what?

Well, you know that you need to connect an antenna to the receiver if you want to hear signals. You also have to connect an antenna to the transmitter when you send out a signal. But you only have one antenna. What should you do?

You could disconnect the antenna from the receiver and connect it to the transmitter when you want to transmit. You'd have to do this every time you switch over from receive to transmit and again to switch from transmit to receive. Most radios have their antenna connectors on the rear panel, however. Besides, these connectors aren't designed to make it easy to remove and reconnect cables rapidly.

What you need is a *transmit-receive switch* (also known as a *TR switch*). We use a TR switch to connect one antenna to a receiver and a transmitter. Remember our discussion of switches from Chapter 6? The simplest TR switch is a *single-pole, double-throw (SPDT) switch*, like the one in Figure 7-2. The antenna connects to the center arm, and the receiver and transmitter connect to the outside contacts. If you throw the switch one way, the transmitter connects to the antenna. The other switch position connects the receiver to the antenna.

Many TR switches use relays to switch the antenna. A relay uses a magnetic coil to make or break contacts. You can think of a relay as a remotely operated switch. When you throw a switch at your operating position, current passes through the relay coil and the relay arm switches from one set of contacts to another. A transmit-receive relay may have several contacts. Extra contacts can be used to quiet the receiver in the transmit mode or to switch accessory devices.

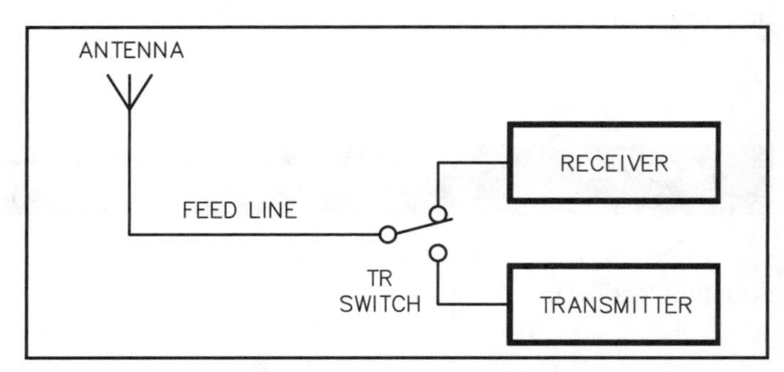

Figure 7-2 — The simplest TR switch is an SPDT switch with the center arm connected to the antenna. One position connects the antenna to the receiver; the other position connects the antenna to the transmitter.

Transceivers have a TR switch built into the radio, so only one antenna connection is needed. The transceiver switches the antenna between the receiver and transmitter sections. Some amateurs may still want to use a separate receiver with their transceiver. In this case they often use an external TR switch. A receiver should ***never*** connect to the output from a transmitter or transceiver. The transmitted signal would most likely destroy the receiver.

The Simplest Station

The block diagram in Figure 7-1 shows how we connect all this equipment. The power supply connects to the transmitter. The receiver and transmitter both connect to the TR switch. The TR switch connects the transmitter and receiver to the antenna, one at a time. This is just about all you need for the most basic Amateur Radio station.

What kind of wire should you use to make all these interconnections? The power-supply wiring can be just about any insulated wire with a diameter large enough to carry the maximum current the equipment needs. Of course it needs to be insulated to prevent short circuits and shock hazards.

The wires between the transmitter and receiver or transceiver and the antenna are called the **feed line**. Feed lines, or *transmission lines,* are specially constructed cables. There are several common feed-line types. Coaxial cable, or coax (pronounced ko'-aks) may be the most common feed line type. You will learn more about coaxial cable and other feed lines in Chapter 9.

[Turn to Chapter 13 now and study questions N7A03, N7A06, N7A07, N7A18 and N7A19. Review this section if you don't understand any of those questions.]

Connecting Many Antennas

What if you have more than one antenna? Again, you could disconnect the antenna from your transmitter or receiver and reconnect another feed line. This can be very inconvenient. A simpler technique is to use an **antenna switch**. We mentioned antenna switches in Chapter 4. An antenna switch connects one transmitter, receiver or transceiver to several antennas. You can switch from one antenna to another with a simple flick of the switch.

The antenna switch connects at the point where the feed lines from all the antennas come into the station. An antenna switch connects one receiver, transmitter or transceiver to one of several antennas. Even if you only have one antenna you may find an antenna switch useful. Many amateurs use an antenna switch to select between the station antenna and a *dummy antenna*, used for transmitter tuning and testing. See Figure 7-3.

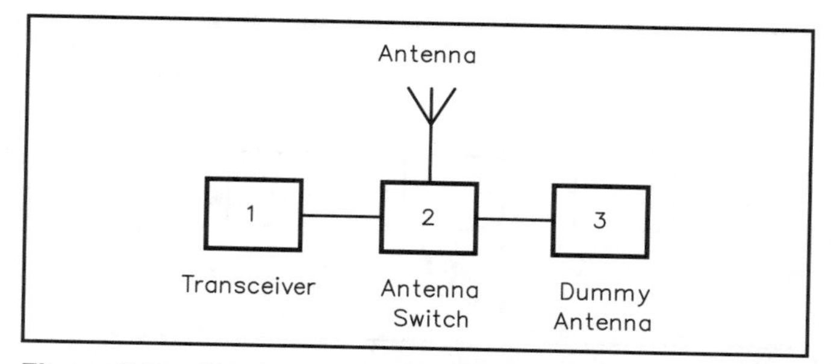

Figure 7-3 — You can use an antenna switch to connect either your station antenna or a dummy antenna (sometimes called a dummy load) to your transmitter or transceiver. A dummy antenna is useful for transmitter testing or adjustment without putting a radio signal on the air.

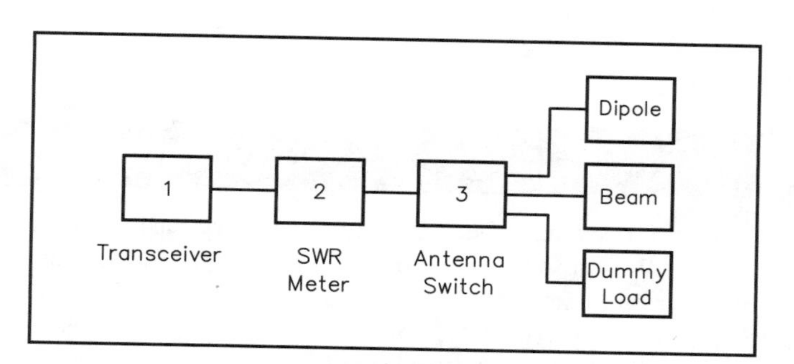

Figure 7-4 — An antenna switch can connect a transmitter to one of several antennas. Many operators place an SWR meter between the transmitter or transceiver and the antenna switch.

Monitoring the System

You may want to add an **SWR meter** to your station. This device is also called an SWR bridge. The SWR meter measures something called the *standing-wave ratio*. You don't need to know too much about SWR right now. We will go over it in more detail later in the book.

Standing-wave ratio is a good indicator of how well your antenna system is working. If you install an SWR meter in your station, you can keep an eye out for problems with your antenna. If you spot the problems early you can head them off before they damage your equipment.

The SWR meter can be connected at several points in your station. One good place to connect the meter is between the antenna switch and transceiver, as Figure 7-4 shows. If you use a separate receiver and transmitter, you can connect the SWR meter between the TR switch and the rest of the antenna system. Most amateurs connect their SWR meter as close to the transmitter output as possible.

Impedance-Matching Networks

Another useful accessory that you will see in many ham shacks is an **impedance-matching network**. (Impedance is similar to resistance.) This device may let you use one antenna on several bands. The matching network may also allow you to use your antenna on a band it is not designed for. Sometimes we call the impedance-matching network an **antenna-matching network**, **antenna tuner** or *Transmatch*. These names indicate the main function of the impedance-matching network. The network matches (tunes)

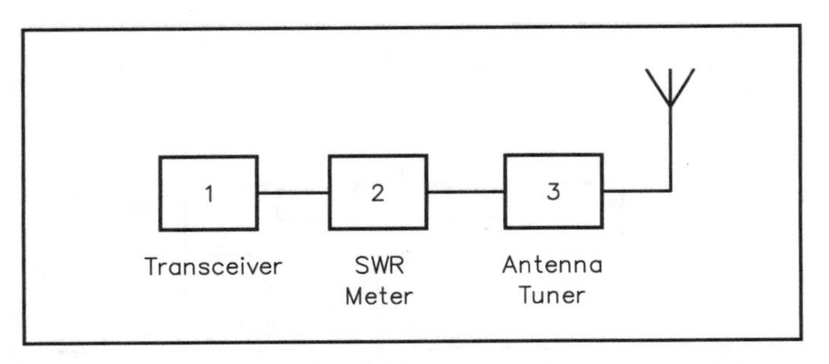

Figure 7-5 — An antenna tuner (or impedance-matching network) connects directly to the antenna feed line. Placing an SWR meter between the tuner and the transmitter or transceiver lets you see when the tuner is adjusted properly.

the impedance of the load (the antenna and feed line) to the impedance of your transmitter. We usually connect the impedance-matching network right where the antenna comes into the station.

If you are using an impedance-matching network to tune your antenna system, you will need an SWR meter. Connect the SWR meter between the impedance-matching network and the transmitter or transceiver. See Figure 7-5. The SWR meter then indicates when the matching network is adjusted properly.

[Now turn to Chapter 13. Study questions N7A01, N7A02, N7A04, N7A05 and N7A08 through N7A17. Review this section if you have any problems.]

STATION ACCESSORIES

So far, we have been talking about very basic station layout. We showed you how to connect a transmitter, receiver and antenna switch to make a simple station. To communicate effectively, you also will need a few accessories. Let's look at what you need.

Morse Code Operation

Morse code is transmitted by switching the output of a transmitter on and off. Inventive radio operators have developed many devices over the years to make this switching easier.

The simplest kind of code-sending device is one you're probably already familiar with: the **hand key**, *straight key* or **telegraph key**. These are all different names for the same piece of equipment. See Figure 7-6. A hand key is a simple switch. When you press down on the key, the contacts meet and the transmitter produces a signal.

The code you make with a hand key is only as good as your "fist," or your ability to send well-timed code. An **electronic keyer**, like the one in Figure 7-7, makes it easier to send well-timed code. You must connect a *paddle* to the keyer. The paddle has two switches, one on each side. When you press one side of the paddle, one of the switches closes and the keyer sends a continuous string of dots. When you press the other side of the paddle, the keyer sends dashes. With a little practice and some rhythm, you can send perfectly timed code with a keyer. You may want to start out with a hand key, however. Using a hand key can help you develop the rhythm you need to send good code. When you can send good code with a hand key, you're ready to try a keyer.

Both the hand key and the electronic keyer connect directly to the transmitter. The key in the block diagram in Figure 7-8 connects to the transmit section of the transceiver.

Microphones

If you want to transmit voice, you'll need a **microphone**. A microphone converts sound waves into electrical signals

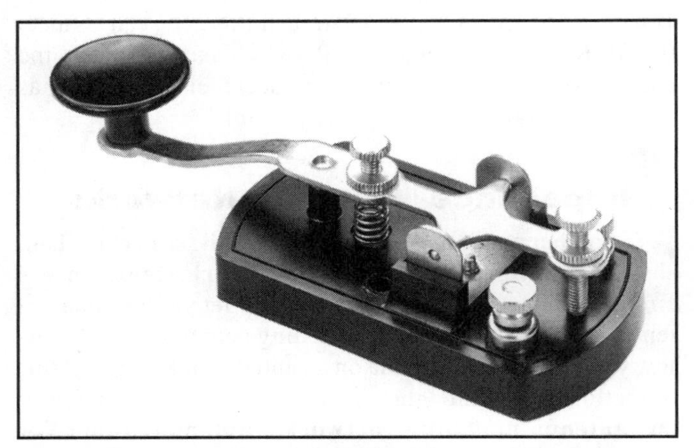

Figure 7-6 — A hand key (straight key or telegraph key) is the simplest type of Morse code sending device.

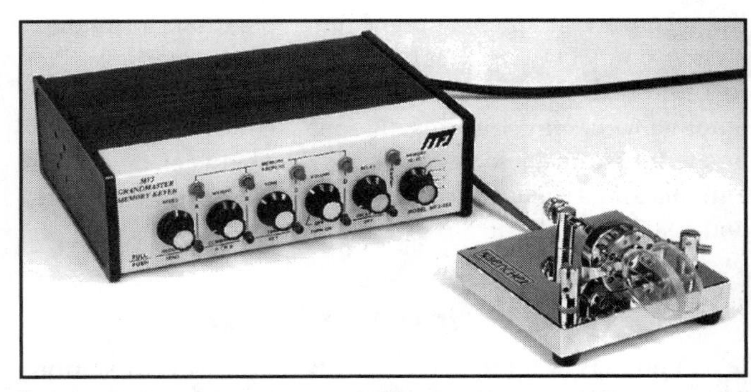

Figure 7-7 — You can produce perfect Morse code characters with an electronic keyer and a little practice.

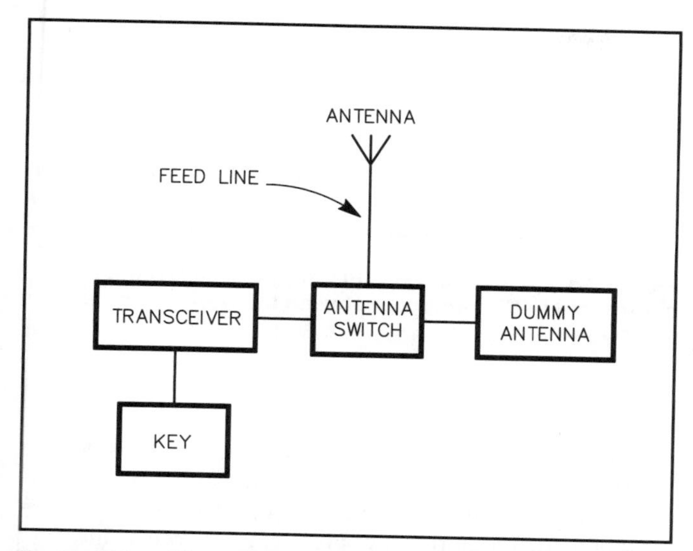

Figure 7-8 — The keying device connects directly to the transmitting section of the transceiver.

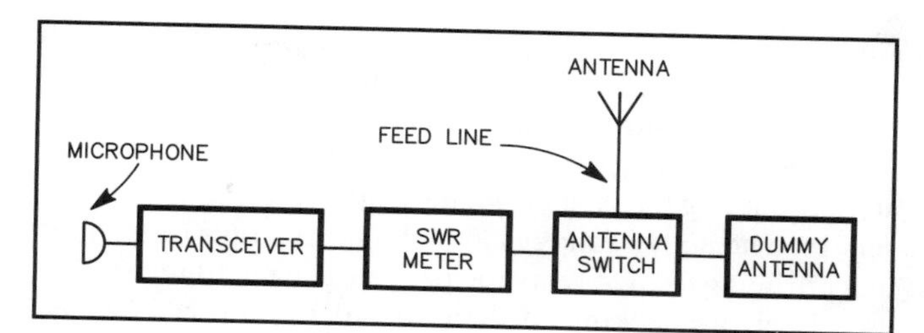

Figure 7-9 — A microphone connects directly to the transmitter or transceiver. You will need a microphone to transmit voice.

that can be used by a transmitter. All voice transmitters require a microphone of some kind. Like a code key, the microphone connects directly to the transmitter. The microphone in Figure 7-9 connects to the transmit section of the transceiver.

[Time for a trip to Chapter 13. Study questions N7B01 through N7B05. Review this section if you have problems.]

RADIOTELETYPE AND DATA COMMUNICATIONS

Radioteletype (RTTY) and data communications are Amateur Radio transmissions that are designed to be received and printed automatically. Sometimes called *digital communications*, they often involve direct transfer of information (data) between computers. When you type information into your computer, the computer (with the help of some accessory equipment) processes the information. The computer then sends the information to your transmitter and the transmitter sends it out over the air. The station on the other end receives the signal, processes it and prints it out on a computer screen or printer. RTTY and data communications have become a popular form of ham-radio communication. Here we will talk about setting up a station for **radioteletype (RTTY)** and **packet radio** communications.

Radioteletype

Radioteletype (RTTY) communications have been around for a long time. You may have seen big noisy **teleprinter** machines in old movies. A teleprinter is something like an electric typewriter. When you type on the teleprinter keys, however, the teleprinter sends out electrical codes that represent the letters you are typing. If we send these codes to another teleprinter machine, the second machine reproduces everything you type. Hams have been converting this equipment and using it on the air for years. You can also send and receive radioteletype with a computer. These days computers are so cheap and readily available that they have just about replaced the old noisy teleprinters.

We use a **modem** for Amateur Radio digital communications. Modem is short for *mo*dulator-*dem*odulator. The modem accepts information from your computer and uses the information to modulate a transmitter. The modulated transmitter produces a signal that we send out over the air. When another station receives the signal, the other station uses a similar modem to demodulate the signal. The modem then passes the demodulated signal to a computer. The computer processes and displays the signal.

Sometimes hams use an older teleprinter instead of a computer. The teleprinter converts and displays information from the modem. A complete radioteletype station must have a computer or teleprinter, a modem and a transceiver. The modem connects between the computer and the transmitter, as shown in Figure 7-10.

Packet Radio

Packet radio uses a **terminal node controller (TNC)** as an interface between your computer and transceiver. We might call a TNC an "intelligent" modem. The TNC accepts information from your computer and breaks the data into

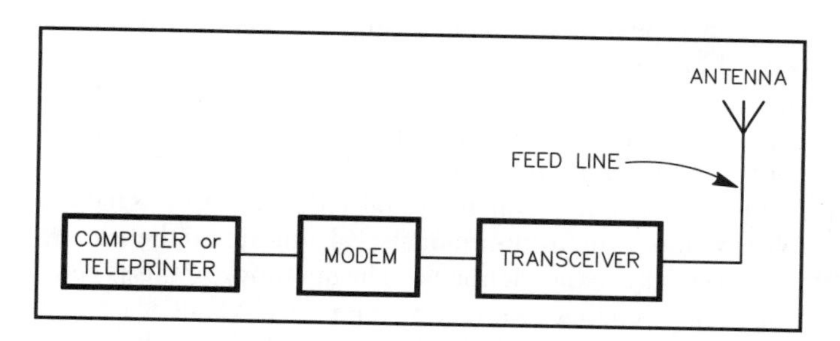

Figure 7-10 — In a typical radioteletype station the modem connects between the transceiver and the computer or teleprinter.

small pieces called *packets*. Along with the information from your computer, each packet contains addressing, error-checking and control information.

The addressing information includes the call signs of the station sending the packet and the station the packet is being sent to. The address may also include call signs of stations that are being used to relay the packet. The receiving station uses the error-checking information to determine whether the received packets contain any errors. If the received packet contains errors, the receiving station asks for a retransmission. The retransmission and error checking continue until the receiving station gets the packet with no errors.

Breaking up the data into small parts allows several users to share a channel. Packets from one user are transmitted in the spaces between packets from other users. The addressing information allows each user's TNC to separate packets for that station from packets intended for other stations. The addresses also allow packets to be relayed through several stations before they reach their final destination. The error-checking information in each packet assures perfect copy.

Packet radio and some other digital modes used on HF place an interesting requirement on your radio. The sending station must receive an automatic acknowledgment from your station to confirm that the previous data was received correctly. After sending a packet of information the transmitting station pauses to wait for that acknowledgment. If it isn't received within the specified time, the sending station repeats the data. (The actual time varies from a few hundred milliseconds to a few seconds depending on the exact mode.) In addition to the time it takes the radio signals to travel between the two stations you must also consider the time it takes your radio to switch from transmit to receive and from receive to transmit. For example, when you stop transmitting there is a short delay before you can receive any signals. During this time the radio changes from being a transmitter to being a receiver. When you decide it is time to transmit again there is another delay while the radio switches back to the transmit circuitry. So to use a radio on the digital modes, especially on the HF bands, the radio must have a fast T/R switching time.

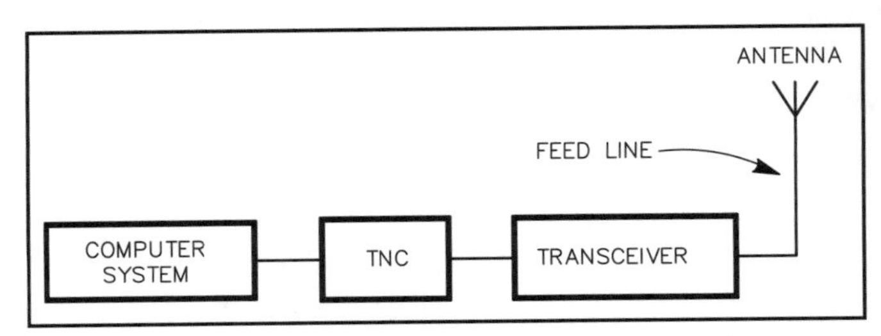

Figure 7-11 — In a packet radio station, the terminal node controller (TNC) connects between the transceiver and a computer.

A TNC connects to your station the same way a modem does. The TNC goes between the radio and the computer, as shown in Figure 7-11.

[Turn to Chapter 13 and study the questions numbered N7B06 through N7B12. If you have any problems, review this last section.]

CROWDED BAND CONDITIONS

There are several accessories that can help you enjoy Amateur Radio when the bands are crowded with other stations. You will probably want some type of *filter* to reduce the number of signals getting to your ears at one time. A filter is just a device that allows certain signals to go through while blocking others. A good receiver will have some type of filter for the radio-frequency (RF) and intermediate frequency (IF) stages of the radio. You might also want to add an external *audio filter* that will work on the audio signal coming from the receiver, to further remove interfering signals before the audio goes to your headphones or speaker.

One type of filter that is getting a lot of attention lately is the digital signal processing (DSP) filter. The signals are converted to a digital form (a series of ones and zeros that a computer can understand) and fed through a circuit controlled by a microprocessor. The microprocessor performs some operation on the signal and then converts it back to an analog form (a continuously varying signal that can be fed to a speaker or headphones). The operation of this type of filter can be changed simply by changing the software that controls the microprocessor. You can add a DSP filter as an external accessory for your station, you can buy or write software that uses the DSP features of a sound card in your computer, and some radios even come with built-in DSP filters.

Such filters can be especially helpful for receiving various types of signals under crowded band conditions. For example, a DSP filter can be made to allow only a very narrow range of frequencies through, and that can be quite effective for receiving Morse code (CW). The filter can also be adjusted to allow a wider range of frequencies to get through, so the same DSP circuit can also be used for RTTY and even voice reception.

[Turn to Chapter 13 now and study exam question N7B13. If you don't understand the answer to that question, review this last section.]

If you're preparing for the Novice exam, that's all for this chapter. Don't worry if you're a bit confused by some of the terms. We covered a lot of information fairly quickly. We'll go over a lot of the operating information later in the book. For now, become familiar with the terms, and study the block diagrams carefully. If you are preparing only for the Novice exam, you don't have to continue with this chapter.

[If you are preparing for the Technician exam, continue with this chapter *after* you review the questions in Chapter 13 listed above.]

KEY WORDS

Amplifier — A device usually employing electron tubes or transistors to increase the voltage, current or power of a signal. The amplifying device may use a small signal to control voltage and/or current from an external supply. A larger replica of the small input signal appears at the device output.

Band-pass filter — A circuit that allows signals to go through it only if they are within a certain range of frequencies. It attenuates signals above and below this range.

Beat-frequency oscillator (BFO) — An oscillator that provides a signal to the product detector. In the product detector, the BFO signal and the IF signal are mixed to produce an audio signal.

Detector — The stage in a receiver in which the modulation (voice or other information) is recovered from the RF signal.

Filter — A circuit that will allow some signals to pass through it but will greatly reduce the strength of others.

Frequency modulation (FM) — The process of varying the frequency of an RF carrier in response to the instantaneous changes in the modulating signal.

Harmonic — A signal from an oscillator or transmitter that occurs on whole-number multiples (2x, 3x, 4x, etc) of the desired operating frequency.

High-pass filter — A filter that allows signals above the cutoff frequency to pass through. It attenuates signals below the cutoff frequency.

Intermediate frequency (IF) — The output frequency of a mixing stage in a superheterodyne receiver. The subsequent stages in the receiver are tuned for maximum efficiency at the IF.

Low-pass filter — A filter that allows signals below the cutoff frequency to pass through and attenuates signals above the cutoff frequency.

Modulate — To vary the amplitude, frequency, or phase of a radio-frequency signal.

Modulation — The process of varying a radio wave in some way to send information.

Phase modulation (PM) — Varying the phase of an RF carrier in response to the instantaneous changes in the modulating signal.

Radio-frequency (RF) carrier — A constant-amplitude, unmodulated, radio-frequency signal.

Reactance modulator — A device capable of modulating an ac signal by varying the reactance of a circuit in response to the modulating signal. (The modulating signal may be voice, data, video, or some other kind depending on what type of information is being transmitted.)

Variable-frequency oscillator (VFO) — An oscillator used in receivers and transmitters. The frequency is set by a tuned circuit using capacitors and inductors. The frequency can be changed by adjusting the components in the tuned circuit.

7 Practical Circuits for Technician Licensees

In this section we will discuss low-pass, high-pass and band-pass filter circuits. We'll show you block diagrams of complete transmitters and receivers, and investigate how the stages connect to make them work.

Keep in mind that entire books have been written on each topic covered in this section. You may not understand some of the circuits from our brief discussion. It would be a good idea to consult some other reference books. ARRL's *Understanding Basic Electronics* is a good starting point. *The ARRL Handbook for Radio Amateurs* includes plenty of detail about these topics. Even that book won't tell you everything about a topic, though. The discussion in this chapter will help you understand the circuits well enough to pass your Technician license exam, however.

FILTERS

Harmonic interference to entertainment equipment is a problem some hams face. Harmonics are whole number multiples of a given frequency. For example the second harmonic of a 3.5 MHz signal is 7.0 MHz. (We just multiplied by 2 to get this answer.) Your transmitter may radiate undesired harmonics along with your signal. In the high-frequency amateur bands (3.5 to 29.7 MHz), the frequency you're transmitting on is much lower than the TV or FM channels. Some of your harmonics may fall within the home entertainment bands, though. The entertainment receiver cannot distinguish between the TV or FM signals that are supposed to be there and your harmonics, which are not. If your harmonics are strong enough, they can interfere with the broadcast signal. Modern Amateur Radio transceivers have filters built in to *attenuate* (reduce) harmonic radiation so they seldom create harmonic-related interference.

Harmonic interference must be cured at your transmitter. It is your responsibility as a licensed amateur. You must see that harmonics from your transmitter are not strong enough to interfere with other services. As mentioned before, all harmonics generated by your transmitter must be attenuated well below the strength of the fundamental frequency. If harmonics from your transmitting equipment exceed these limits, you are at fault.

You can usually tell harmonic interference when you see it on a TV set. This type of interference shows up as crosshatch or a herringbone pattern on the TV screen. See Figure 7-12. Interference from radiated harmonics seldom affects all channels. Rather, it may bother the one channel that is frequency-related to the band you're on. Generally, the low TV channels (channels 2-6) are most affected by harmonics from amateur transmitters operating below 30 MHz. Channels 2 and 6 are especially affected by 10-meter transmitters, and channels 3 and 6 experience trouble from 15-meter transmitters.

There are several possible cures for harmonic interference. We will discuss a few of them here. Try each step in the order we introduce them. Chances are good that your problem will be quickly solved. You should be familiar with three **filters**: the **low-pass filter**, the **high-pass filter** and the **band-pass filter**.

Filters allow certain frequencies through the circuit and block others. All modern radio communication devices use filter circuits. These filter circuits allow various kinds of equipment to operate on different frequencies without interfering with each other. Under certain conditions, some equipment may need a little extra help. A transmitter may need extra filtering of the output signal to reduce interference to television receivers. In other cases, the transmitter may already be well filtered (clean), but a TV receiver may need extra filtering of its input signals. A basic understanding of filters and their applications will often allow you to solve interference problems.

Low-Pass Filters

The first step you should take is installation of a **low-pass filter** in the feed line between your transmitter and antenna. A low-pass filter is one that passes all frequencies below a certain frequency, called the *cutoff frequency*. We measure the filter cutoff frequency by putting a variable-frequency signal into the filter. The input signal power is kept

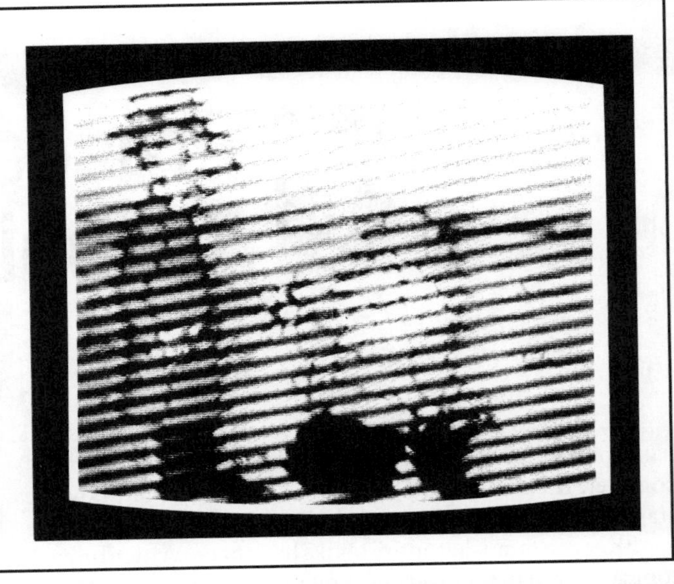

Figure 7-12 — Harmonic interference may be visible as cross-hatching of the TV screen. More severe interference can completely destroy the picture. The width of the cross-hatch lines will vary, depending on the transmitter frequency.

constant while the frequency is increased. At the same time we measure the filter output power. At some frequency the output begins to decrease. When the output power has decreased to $^{1}/_{2}$ the input power, we have found the cutoff frequency. Frequencies above the cutoff frequency are *attenuated,* or significantly reduced in amplitude. See Figure 7-13.

The cutoff frequency depends on the design of the low-pass filter. The capacitors are chosen to provide a path to ground for the unwanted high frequencies. The inductor has a value that passes the low frequencies you want, but blocks the higher frequencies.

One other important trait that must be considered in the design of a low-pass filter is the *characteristic impedance.* This is the circuit impedance for which the filter is designed. Most external low-pass filters intended for amateur use will have a characteristic impedance of 50 ohms.

The low-pass filter is always connected between the transmitter and the antenna, as close to the transmitter as possible. The filter should have the same impedance as the feed line connecting the transmitter to the antenna. Filters should be used only in a feed line with a low standing-wave ratio. The cutoff frequency has to be higher than the highest frequency used for transmitting. A filter with a 45-MHz cutoff frequency would be fine for the high-frequency (HF) bands. This filter would significantly attenuate 6-meter (50-MHz) signals, so it should not be used for 50 MHz or higher-frequency operation. Most modern transmitters have a low-pass filter built into their output circuitry to prevent excess harmonic radiation.

High-Pass Filters

Sometimes entertainment devices experience interference from amateur transmissions even though the transmitting device is operating properly. The harmonics can be well below levels necessary to cause interference, yet every time you transmit the device receives interference. This happens because the design of the entertainment device is inadequate when it is operating in the presence of strong signals. In this situation, it may be necessary to reduce the level of the amateur signal reaching the entertainment device (TV or FM stereo receiver, or VCR). The desired higher-frequency signals must be allowed to pass unaffected. A **high-pass filter** can do this.

A high-pass filter allows all frequencies above the *cutoff frequency* to pass through, and *attenuates* those below it. See Figure 7-14. A high-pass filter should be connected to a television set, stereo receiver or other home-entertainment device that is being interfered with. It will attenuate the signal from an amateur station. The inductors have a value that allows them to conduct the lower frequencies to ground. The capacitor tends to block these low-frequency signals. At the same time it allows the higher-frequency television or FM-broadcast signals to pass through to the receiver. This is useful in reducing the amount of lower-frequency signal (at the ham's operating frequency). The lower-frequency signal might overload a television set, causing interference.

For best effect, a high-pass filter should be connected as close to the television-set tuner as possible. Put it inside the TV if it is your own set and you don't mind opening the cabinet. You can also attach it directly to the antenna terminals on the back of the television. If the set belongs to your neighbors, have them contact a qualified service technician to install the filter. That way you are not held responsible if something goes wrong with the set later on.

You should always try to maintain a friendly relationship with your neighbors. If they mention an interference problem investigate it and let them know you are interested in helping solve the problem. Begin by trying to prove whether your station is causing the interference. You may be able to demonstrate that you aren't even on the air when they experience the interference. It is always helpful if you can show them that your own entertainment equipment does not experience interference while you are transmitting.

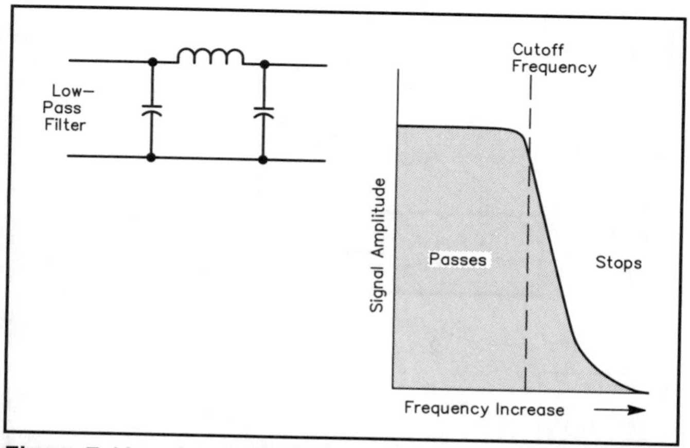

Figure 7-13 — A low-pass filter schematic diagram and its output-versus-frequency curve. The low-pass filter passes signals below the cutoff frequency and attenuates signals above the cutoff frequency.

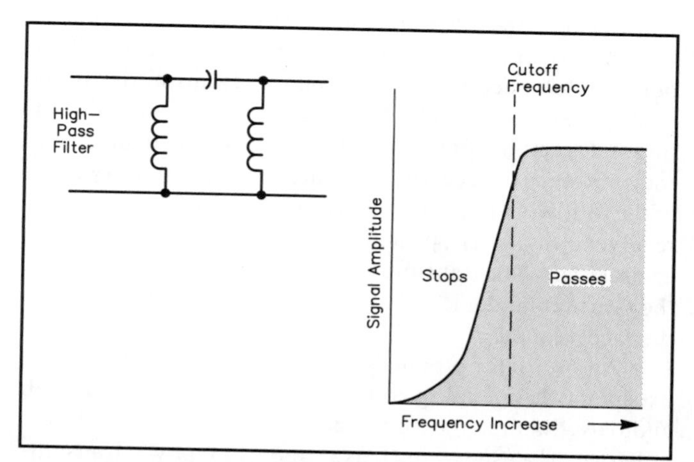

Figure 7-14 — A high-pass filter schematic diagram and its output-versus-frequency curve are shown. The high-pass filter attenuates only those signals below the cutoff frequency.

Band-Pass Filters

A **band-pass filter** is a combination of a high-pass and low-pass filter. It passes a desired range of frequencies while rejecting signals above and below the *passband*. This is shown in Figure 7-15.

Band-pass filters are commonly used in the **intermediate-frequency (IF)** stage of a receiver to provide different degrees of rejection. Very narrow filters are used for CW reception. Wider filters are switched in for SSB and AM double-sideband reception. The band-pass filter allows the receiver IF stages to select signals within a certain frequency range and block signals outside that range.

[Now turn to Chapter 14 and study exam questions T7A16, T7A17 and T7A18. Review this section if you have trouble with any of these questions.]

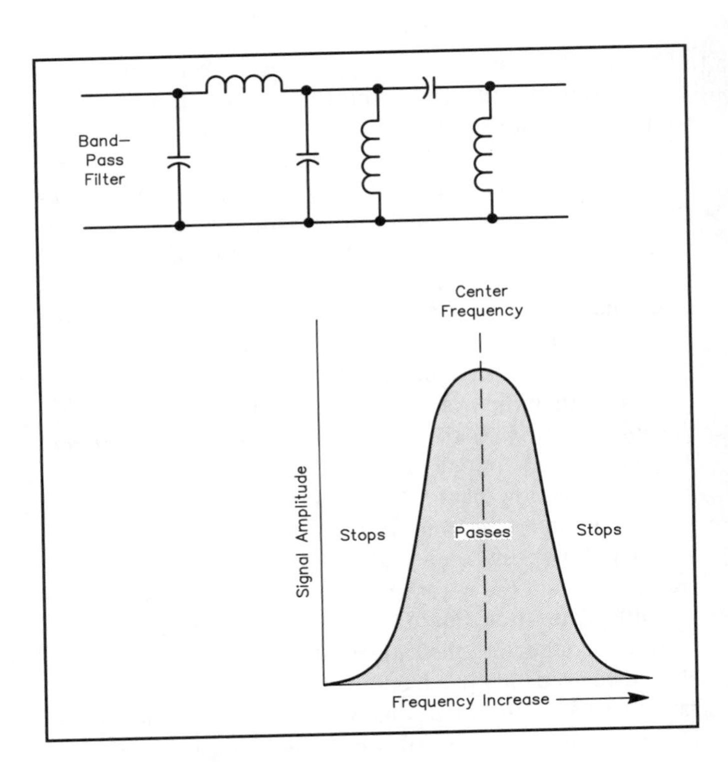

Figure 7-15 — A band-pass filter schematic diagram and its output-versus-frequency curve are shown. The band-pass filter passes only those signals within the passband.

TRANSMITTERS

Amateur transmitters range from the simple to the elaborate. They generally have two basic stages, an *oscillator* and a *power amplifier (PA)*. This section introduces the basic operation of a CW transmitter and an FM transmitter.

Separate receivers and transmitters have been replaced by transceivers in most amateur stations. A transceiver combines circuits necessary for receiving and transmitting in one package. Some circuit sections may perform both transmitting and receiving functions. Other sections are dedicated to transmit-only or receive-only functions. To simplify some of the following information, we will speak of transmitter or receiver circuits. The principles of operation are the same in transceivers. The transmitter topics apply to transceivers in the transmit mode. Receiver topics apply to transceivers in the receive mode.

An oscillator produces an ac waveform with no input except the dc operating voltages. You may be familiar with audio oscillators, like a code-practice oscillator. An RF oscillator can be used by itself as a simple low-power transmitter, but the power output is very low. See Figure 7-16. In a practical transmitter the signal from the oscillator is usually fed to one or more **amplifier** stages.

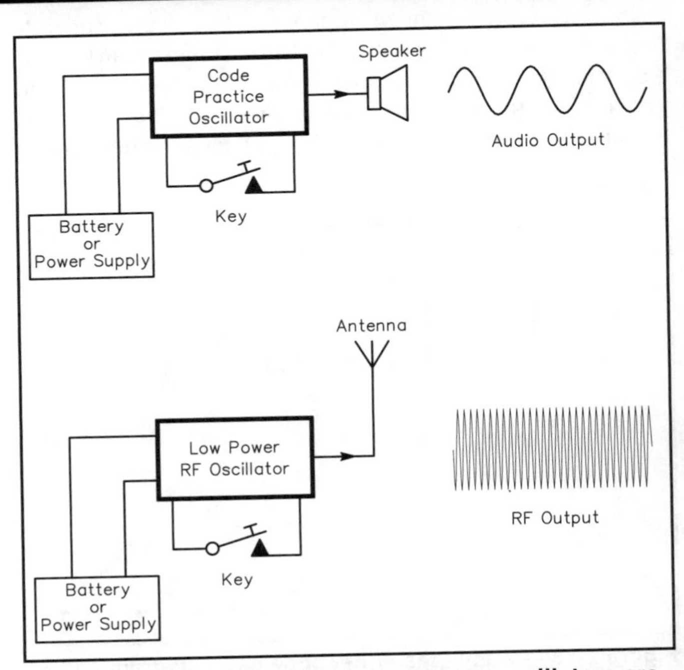

Figure 7-16 — Audio and radio-frequency oscillators are shown. Even a low-power RF oscillator connected to a good antenna can send signals hundreds of miles.

CW Transmitters

Figure 7-17A shows a block diagram of a simple amateur transmitter. It produces CW (continuous-wave) radio signals when the key is closed. The simple transmitter consists of a crystal oscillator followed by a driver stage and a power amplifier. A crystal oscillator uses a quartz crystal to keep the frequency of the radio signal constant.

To transmit information, we must **modulate** the radio signal. This means we are making some change to the radio signal, and adding information to it. Pressing the key of a CW transmitter and holding it down produces a **radio-frequency (RF) carrier**. This is also called an *unmodulated* carrier. An RF carrier is a constant-amplitude, unmodulated, radio-frequency signal. When you use the key to turn the signal on and off to produce Morse code dots and dashes, however, you modulate the signal. **Modulation** is the process of varying an RF carrier in some way to send information.

Crystal oscillators may not be practical in all cases. A different crystal is needed for each operating frequency. After a while, this becomes quite expensive, as well as impractical. If we use a **variable-frequency oscillator (VFO)** in place of the crystal oscillator, as shown in Figure 7-17B, we can change the transmitter frequency whenever we want. So we can use either a crystal oscillator or a variable frequency oscillator to control the transmitter operating frequency.

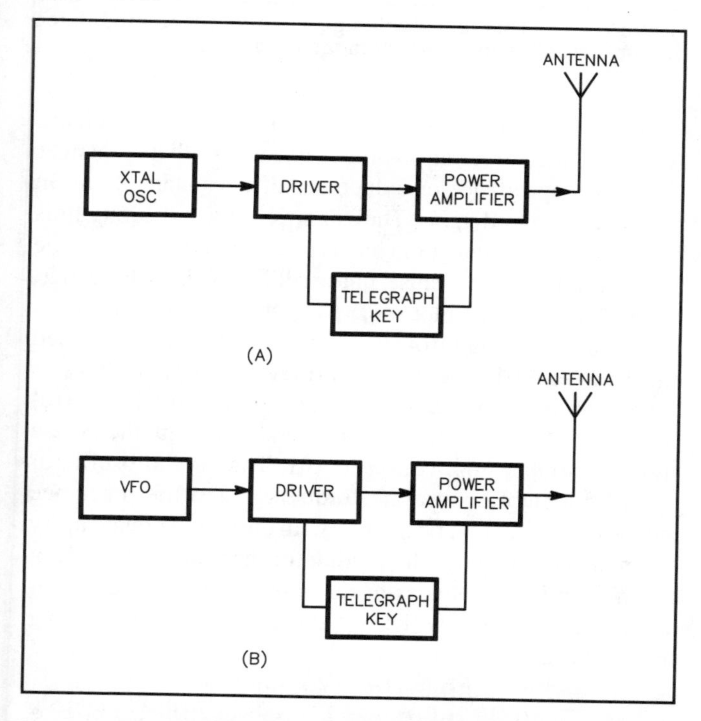

Figure 7-17 — Part A shows a block diagram of a basic crystal-controlled CW transmitter. Part B shows a simple VFO-controlled CW transmitter.

The telegraph key may connect to one or more transmitter sections. Figure 7-17 shows the key connected to both the driver and power amplifier stages. You would not normally key the oscillator stage because this may cause the oscillator to change frequency slightly.

All transmitters also require some source of operating voltage. We have left the power supply block off the Figure 7-17 drawings for simplicity.

Great care in design and construction is required if the stability of a VFO is to compare with that of a crystal oscillator. Stability means the ability of a transmitter to remain on one frequency without drifting, or changing frequency in an undesired way. The frequency-determining components of a VFO are very susceptible to changes in temperature, supply voltage, vibration and changes in the amplifiers following the VFO. These factors have far less effect on the frequency of a vibrating quartz crystal.

FM Transmitters

When a radio signal or carrier is modulated, some characteristic of the radio signal is changed in order to convey information. We transmit information by modulating any property of a carrier. For example, we can modulate the frequency or phase of a carrier. Frequency modulation and phase modulation are closely related. The phase of a signal cannot be varied without also varying the frequency, and vice versa. **Phase modulation (PM)** and **frequency modulation (FM)** are especially suited for channelized local UHF and VHF communication. They feature good audio fidelity and high signal-to-noise ratio.

A simple FM transmitter is shown in Figure 7-18. Remember that a circuit containing capacitance and inductance is resonant at some frequency. A resonant circuit in the feedback path of an oscillator can control the oscillator frequency. By changing the resonant frequency of the tuned circuit, we can change the oscillator frequency.

A capacitor microphone is just a capacitor with one movable plate (the diaphragm). When you speak into the microphone, the diaphragm vibrates and the spacing between the two plates of the capacitor changes. When the spacing changes, the capacitance changes. We can connect the microphone to the resonant circuit in our oscillator. Speaking into the microphone varies the oscillator frequency. Notice that if the microphone on your FM transmitter stops working, you will transmit an *unmodulated* **RF carrier**.

We can add frequency multipliers to bring the oscillator frequency up to our desired operating frequency. Adding an amplifier to increase the power provides a simple FM transmitter.

In practical FM systems, the carrier frequency is varied or modulated by changes in voltage that represent information to be transmitted. This information may originate from a microphone, a computer modem or even a video camera. The carrier frequency changes in proportion to the rise and fall of the modulating voltage.

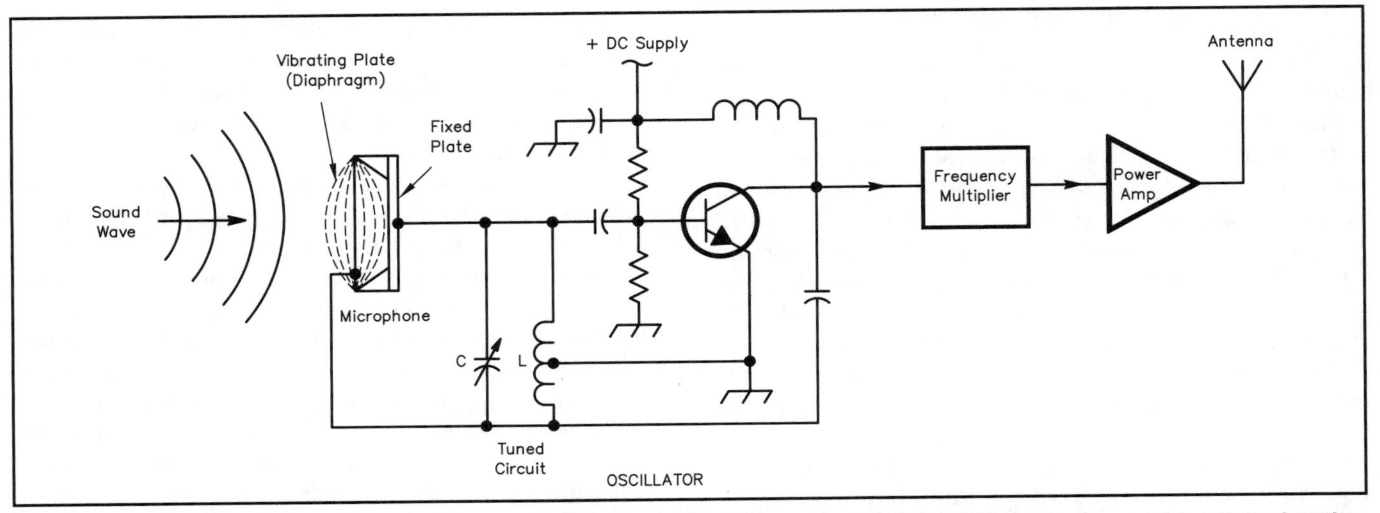

Figure 7-18 — This diagram shows a simple FM transmitter. The frequency of the oscillator is changed by changing the capacitance in the resonant circuit. The vibrating plate in the microphone is part of the total capacitance in the resonant circuit.

In other words, a modulating voltage that is becoming more positive increases the carrier frequency. As the voltage becomes more negative, the carrier frequency decreases. For some purposes, reversing this relationship will work just as well (FM voice transmitters, for instance). The main point is, the carrier frequency is increased and decreased by the modulating signal voltage. The amount of increase or decrease depends on how much the voltage of the modulating signal changes. Speaking loudly into the microphone causes larger variations in carrier frequency than speaking in a normal voice. Speaking *too* loudly, though, may cause distortion.

One way to shift the frequency of the oscillator is to use a **reactance modulator**. A reactance modulator uses a vacuum tube or transistor. It is connected so that it changes either the capacitance or inductance of the oscillator resonant circuit. The changes occur in response to an input signal.

Phase Modulation

One problem with direct-modulated FM transmitters is frequency stability. Designs that allow the oscillator frequency to be easily modulated may have an unfortunate side effect. It becomes more difficult to minimize unwanted frequency shifts arising from changes in temperature, supply voltage, vibration and so on. The frequency multiplying stages also multiply any drift or other instability problems in the oscillator. With phase modulation, the modulation takes place after the oscillator stage. **Phase modulation** produces what is called *indirect FM*. You won't be able to tell the difference between a phase-modulated signal and a frequency-modulated signal by listening to both types on your receiver. So the only real difference is in the electronics of how the signal is produced.

The most common method of generating a phase-modulated telephony signal is to use a reactance modulator. A vacuum tube or a transistor is connected so that it changes either the capacitance or inductance of a resonant circuit in response to an input signal. The RF carrier is passed through this resonant circuit. Changes in the resonant circuit caused by the reactance modulator cause phase shifts in the RF carrier. Figure 7-19 shows the blocks of a phase-modulated transmitter.

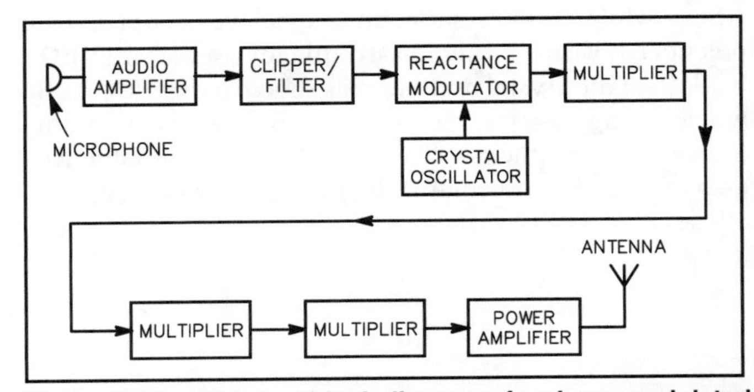

Figure 7-19 — This is a block diagram of a phase-modulated transmitter, showing the reactance modulator.

The weak signal from the microphone is first amplified and then passed through a clipper filter. The clipper filter sets the maximum amplitude of the audio signal going to the reactance modulator. If the signal exceeds the limit, the excess signal is simply *clipped*, or cut off. The crystal oscillator provides a stable operating frequency, and the reactance modulator varies that frequency with the signal picked up by the microphone. After the modulator there are usually several multiplication stages to reach the final desired operating frequency. Finally the signal is amplified before being sent to the antenna and radiated for others to receive.

[Now turn to Chapter 14 and study questions T7A01, T7A04, T7A05, T7A06 and T7A13 through T7A15. Also study questions T8A09, T8A10 and T8A12 as well as T8B03. Review this section as needed.]

RECEIVERS

The main purpose of any radio receiver is to change radio-frequency signals (which we can't hear or see) to signals that we can hear or see. A good receiver can *detect* weak radio signals. It separates them from other signals and interference. Also, it stays tuned to one frequency without drifting. The ability of a receiver to detect weak signals is called *sensitivity*. *Selectivity* is the ability to separate (select) a desired signal from undesired signals. *Stability* is a measure of the ability of a receiver to stay tuned to a particular frequency. In general, then, a good receiver is very sensitive, selective and stable.

Amateur receivers can be simple or complex. You can build a simple solid-state receiver that will work surprisingly well. *The ARRL Handbook for Radio Amateurs* has receiver plans, including sources for parts and circuit boards.

Detection

The **detector** is the heart of a receiver. It is where we collect the information we want from the signal. "Crystal sets" using galena crystals and "cat's whiskers" were an early form of *amplitude modulation (AM)* detector. AM is generated by varying the amplitude of an RF signal in response to a microphone or other signal source. The cat's whisker is just a thin, stiff piece of wire. With the galena crystal it forms a point-contact diode. See Figure 7-20.

Today, however, more sensitive and selective receivers are required. In addition, crystal sets cannot receive single sideband (SSB) and CW signals properly. The crystal set shows how simple a receiver can be, though. It is an excellent example of how detection works. Every receiver has some type of detector. Figure 7-21A shows a very simple receiver.

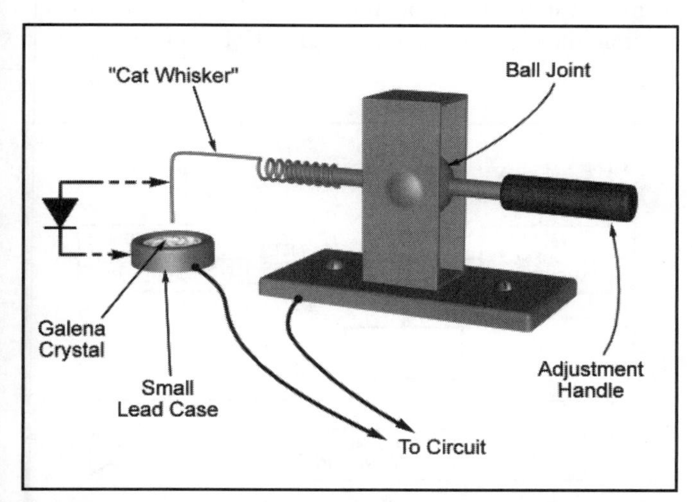

Figure 7-20 — Here is an example of a galena crystal and a cat's whisker, the detector in early crystal sets.

Diode D1 does the same job as the cat's-whisker detector. If there is a strong AM broadcast station in your area you should hear the signals from that station in your earphones with this receiver.

A more practical receiver would have some way to tune different frequencies. The circuit shown in Figure 7-21B will be a better receiver because there is a tuned circuit at the input to select signals on different frequencies. This is a tuned-radio-frequency (TRF) receiver. An incoming signal causes current to flow from the antenna, through the resonant tuned circuit, to ground. The current induces a voltage with the same waveform in L2. The L1-C1 circuit resonates at the frequency of the incoming signal and tends to reject signals at other frequencies. The diode rectifies the RF signal, allowing only half the waveform to pass through. Capacitor C2 fills in and smoothes out the gaps between the cycles of the RF signal. Only the audio signal passes on to the headphones. Amplification can improve the sensitivity of this receiver, but other receiver types have much better selectivity.

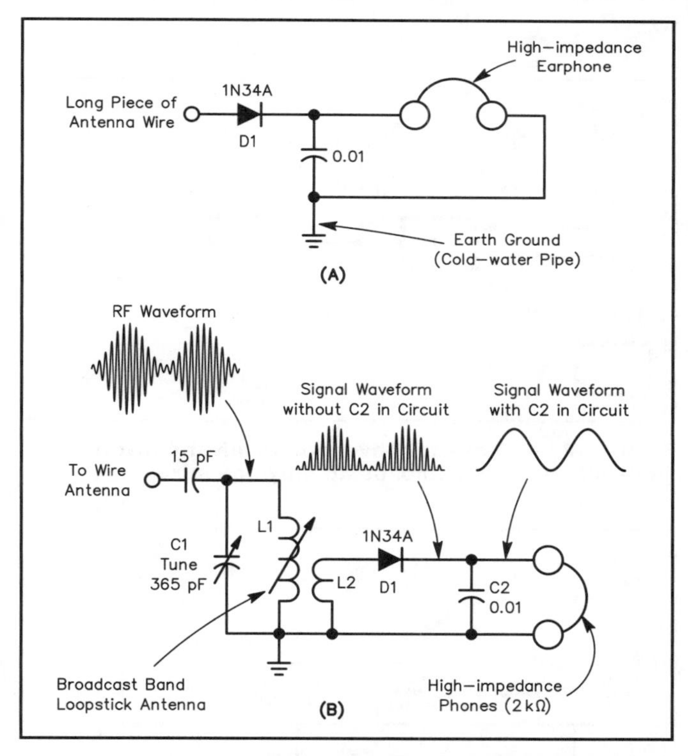

Figure 7-21 — Illustration A shows a simple AM receiver using a crystal detector. It consists only of a wire antenna, detector diode, capacitor, earphone and Earth ground. The circuit at B has a tuned circuit that helps separate the broadcast-band signals, but otherwise operates the same as the circuit at A. If you have never listened to a simple crystal radio, this can be a fun project!

Direct-Conversion Receivers

The next step up in receiver complexity is shown in Figure 7-22. Direct-conversion means the RF signal is converted directly to audio, in one step. The incoming signal is combined with a signal from a **variable-frequency oscillator (VFO)** in the *mixer* stage. In this example, we mix an incoming signal at 7040 kHz with a signal from the VFO at 7041 kHz. The output of the mixer contains signals at 7040 kHz, 7041 kHz, 14,081 kHz and 1 kHz (the original signals and their sum and difference frequencies). One of these signals (1 kHz) is within the range of human hearing. We use an audio amplifier and hear it in the headphones.

By changing the VFO frequency, other signals in the range of the receiver can be converted to audio signals. Direct-conversion receivers are capable of providing good usable reception with relatively simple, inexpensive circuits.

Superheterodyne Receivers

A block diagram of a simple superheterodyne receiver for CW and SSB is shown in Figure 7-23. The mixer produces a signal at the **intermediate frequency** or **IF**, often 455 kHz in a single-conversion receiver. The amplifier after the mixer is designed for peak efficiency at the IF. The superhet receiver solves the selectivity/bandwidth problem by converting all signals to the same IF before filtering and amplification. To receive SSB and CW signals, a second mixer, called a product detector, is used. The product detector mixes the IF signal with a signal from the **beat-frequency oscillator (BFO)**. The BFO converts the received-signal information from the intermediate frequency to the audio range. The product-detector output contains audio that can be amplified and sent to a speaker or headphones.

Most modern commercially manufactured receivers (and the receiver stage in transceivers) have two or three intermediate frequency stages. Signals are often converted first to a high frequency, such as around 73 MHz, then — after some filtering and amplifications — converted to a second IF, often around 9 MHz. After further filtering and amplification the signal may be converted to a third IF — usually around 455 kHz — and then to audio. Some receivers may convert the signal to audio after the second IF. These extra conversions help solve some problems and limitations of single-conversion methods.

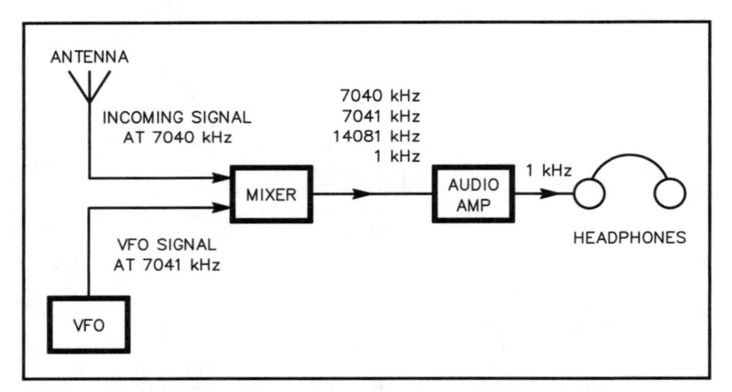

Figure 7-22 — A direct-conversion receiver converts RF signals directly to audio, using only one mixer.

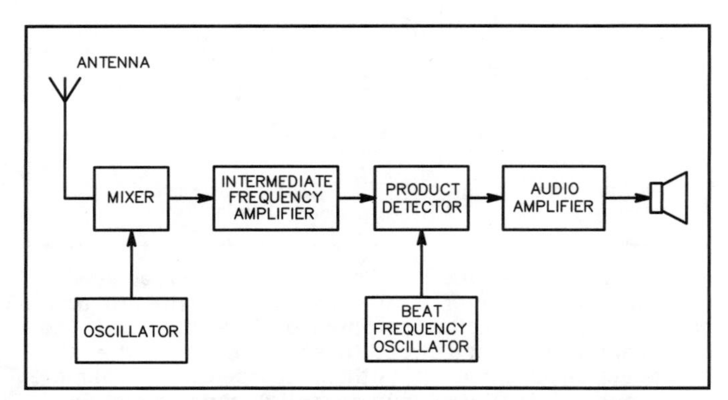

Figure 7-23 — This block diagram shows a superheterodyne SSB/CW receiver. A mixer converts the incoming signal to the intermediate frequency (IF). A second mixer, called a product detector, recovers the audio or Morse code and converts it to an audio signal.

FM Receivers

An FM superheterodyne receiver is similar, but with a few different stages. It has a wider bandwidth filter and a different type of **detector**. One common FM detector is the *frequency discriminator*. The discriminator output varies in amplitude as the frequency of the incoming signal changes. You can see that it performs the opposite job of the frequency or phase modulator we studied earlier in this chapter as part of an FM transmitter.

Most FM receivers also have a *limiter* stage between the IF amplifier and the detector. The limiter makes the receiver less sensitive to amplitude variations and pulse noise than AM or SSB/CW receivers. As its name suggests, the limiter output remains almost constant when the signal level fluctuates. Noise pulses are also amplitude-modulated signals. The limiter does not pass them on to the detector. This feature makes FM popular for mobile and portable communications. Figure 7-24 is a block diagram showing the main stages of an FM receiver.

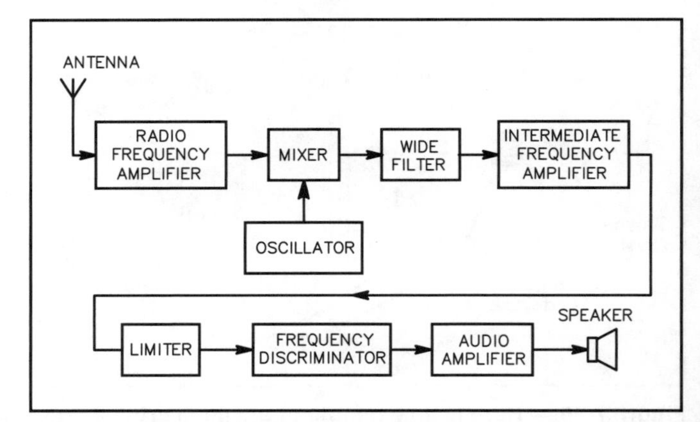

Figure 7-24 — This block diagram shows the main stages of a simple FM receiver. All receivers have some type of detector, and in the FM receiver the frequency discriminator serves as the detector, converting the received signal to audio.

AN ADDITIONAL STATION ACCESSORY

Multiband VHF/UHF radios have become quite popular in recent years. You can buy one radio and operate on the popular 2-meter (144 to 148 MHz) band as well as either the 1.25-meter (222 to 225 MHz) band or the 70-cm (420 to 450 MHz) band. Some of these radios have separate antenna connectors for each band and others have one connector used for both. With this type of radio you also have the option of using separate antennas for each band or a single antenna that is designed to operate on both bands. Your choices can lead to some interesting complications. How do you connect two antennas to a radio with only one antenna connector? Or how do you connect your dual-band antenna if your radio has two antenna connectors?

The solution to this problem may be a *duplexer*. You can think of a duplexer as a frequency-sensitive signal-steering device. Suppose you have a VHF/UHF radio with two antenna connectors, and you want to connect them both to a single dual-band antenna. A duplexer will ensure that the signal coming out the 2-meter antenna connector goes to the antenna, but not back into the radio through the 70-cm antenna connector. When you are transmitting on 70 cm, it also protects the 2-meter section of the radio in the same way. Received signals coming in on the antenna are directed to the proper antenna connector on your radio. See Figure 7-25.

[This completes your study of practical circuits for your Technician license exam. Now turn to Chapter 14 and study questions T7A02, T7A03, T7A07 through T7A12 and questions T7A19 and T7A20. Review any material you're unsure of before you go on to the next chapter.]

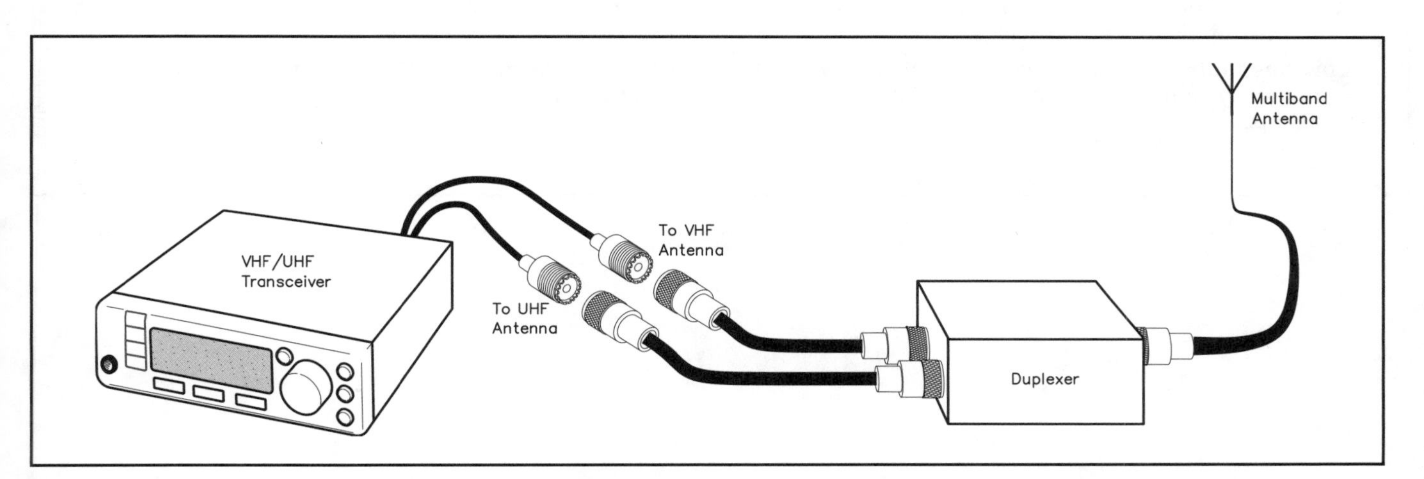

Figure 7-25 — You can use a duplexer to connect a single antenna to a VHF/UHF radio that has two antenna connectors.

8 KEY WORDS

Chirp — A slight shift in transmitter frequency each time you key the transmitter.

Front-end overload — Interference to a receiver caused by a strong signal that overpowers the receiver RF amplifier ("front end"). See also **receiver overload**.

Harmonics — Signals from a transmitter or oscillator occurring on whole-number multiples of the desired operating frequency.

High-pass filter — A filter designed to pass high-frequency signals, while blocking lower-frequency signals.

Key click — A click or thump at the beginning or end of a CW signal.

Key-click filter — A circuit in a transmitter that reduces or eliminates key clicks.

Low-pass filter — A filter designed to pass low-frequency signals, while blocking higher-frequency signals.

Radio-frequency interference (RFI) — Disturbance to electronic equipment caused by radio-frequency signals.

Receiver overload — Interference to a receiver caused by a strong RF signal that forces its way into the equipment. A signal that overloads the receiver RF amplifier (front end) causes **front-end overload**. Receiver overload is sometimes called **RF overload**.

RF overload — Another term for **receiver overload**.

Splatter — A type of interference to stations on nearby frequencies. Splatter occurs when a transmitter is overmodulated.

Spurious emissions — Signals from a transmitter on frequencies other than the operating frequency.

Superimposed hum — A low-pitched buzz or hum on a radio signal.

Television interference (TVI) — Interruption of television reception caused by another signal.

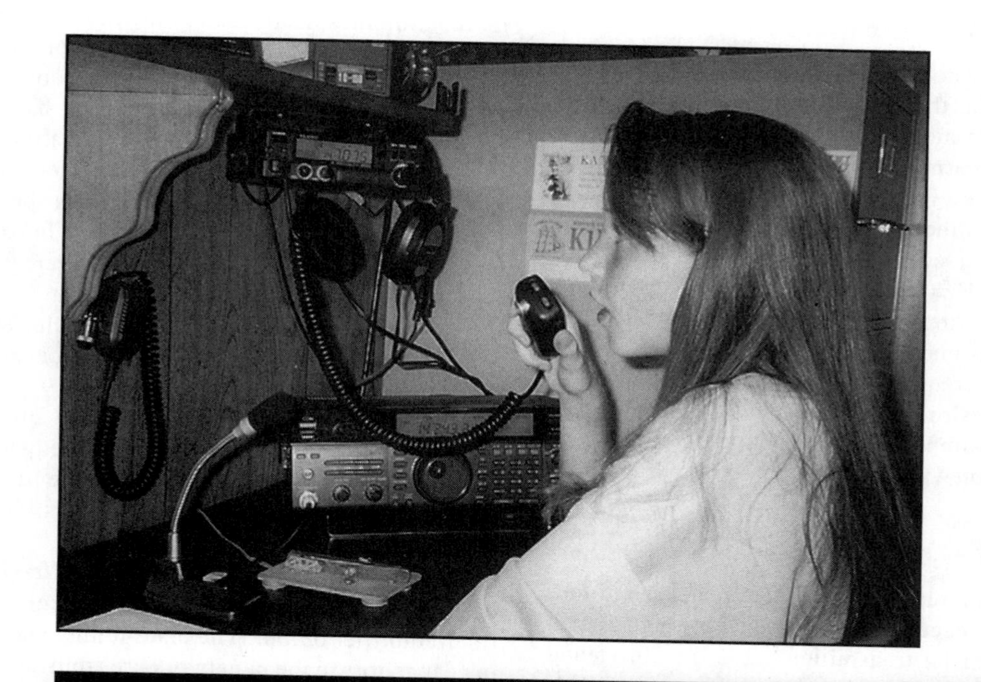

8 Amateur Radio Signals and Emissions

Amateurs may use a variety of radios to transmit on many frequency bands and use a number of different operating modes.

In Chapter 2 you learned about the common Amateur Radio modes and how to operate using those modes. The signals you transmit from your station to communicate using the various operating modes are called *emissions*. In this chapter you will learn about some of the ways the emissions from your station might interfere with other amateurs or consumer electronics equipment. Although you may never have any trouble, you should be aware of possible problems and know how to cure them.

SPURIOUS SIGNALS

An ideal transmitter emits a signal only on the operating frequency and nowhere else. Real-world transmitters radiate undesired signals, or **spurious emissions**, as well. Any transmitter can produce spurious emissions: it doesn't matter if you are using a 100-watt HF transceiver or a 2-watt hand-held VHF or UHF radio. Any signal produced by the radio that falls outside the band on which you are operating is a spurious emission. Using good design and construction practices, manufacturers and home builders can reduce spurious emissions so they cause no problems. Spurious emissions fall into two general categories: harmonics and parasitic oscillations. On the HF bands, you are more likely to have harmonic problems. Parasitic oscillations can occur in equipment designed for HF or VHF/UHF operation.

Harmonics

Harmonics are whole-number multiples of a given frequency. For example, the second harmonic of 100 Hz is 200 Hz. The fifth harmonic of 100 Hz is 500 Hz. Every oscillator generates harmonics in addition to a signal at its *fundamental frequency*. For example, consider an oscillator tuned to 7125 kHz in the 40-meter band. It also generates signals at 14,250 kHz (second harmonic), 21,375 kHz (third harmonic), 28,500 kHz (fourth harmonic), and so on. Figure 8-1 shows a spectrum-analyzer display of the output of an oscillator that has many harmonics.

Calculating the frequency of the various harmonics of a fundamental, or desired, frequency is easy. Simply multiply by the whole number of the particular harmonic. For example, suppose you want to know the fourth harmonic of a 7160-kHz signal.

7160 kHz × 4 = 28,640 kHz

Harmonics can interfere with other amateurs or other users of the radio spectrum. The second through fourth harmonics of a 40-meter transmitter fall in the 20, 15 and 10-meter amateur bands. Imagine the interference (QRM) that would result if everyone transmitted two, four, six or more harmonics in addition to the desired signal! Suppose you receive a report from another amateur that your signals were heard on 28,640 kHz when you were operating your station on 7160 kHz. You should suspect that your transmitter is radiating excessive harmonic radiation.

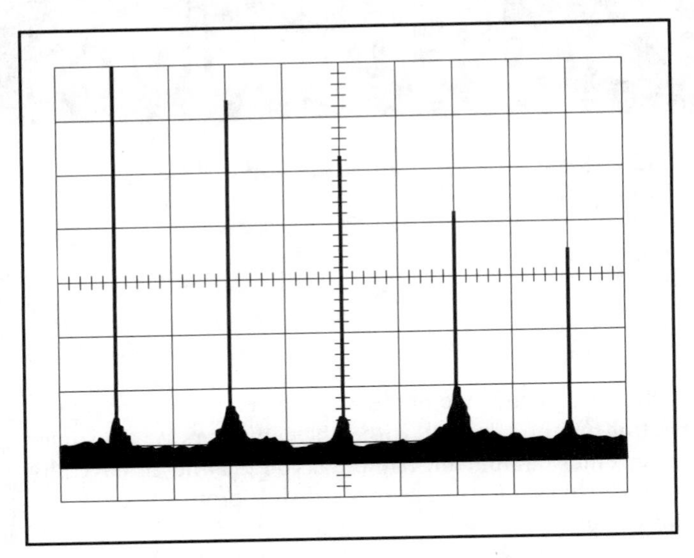

Figure 8-1—Harmonics are signals that appear at whole number multiples of the resonant, or fundamental, frequency. This drawing represents a spectrum analyzer display screen, and shows a 2-MHz signal and some of its harmonics. A spectrum analyzer is an instrument that allows you to look at energy radiated over a wide range of frequencies. Here, the analyzer is adjusted to display any RF energy between 1 MHz and 11 MHz. Each vertical line in the background grid denotes an increment of 1 MHz. The first thick black vertical "pip" represents energy from the fundamental signal of a 2-MHz oscillator. The next pip, two vertical divisions later, is the second harmonic at 4 MHz (twice the fundamental frequency). The pip at the center of the photo is the third harmonic at 6 MHz (three times the fundamental frequency). This figure shows the second, third, fourth and fifth harmonics.

To prevent the chaos that would occur if everyone transmitted harmonics, FCC regulations specify limits for harmonic and other spurious radiation. For example, if you are operating a 100-W-output transmitter, your harmonic signals can total no more than 10 milliwatts. As you can see, a transmitter that complies with the rules still generates some harmonic energy. Fortunately, that energy is so small that it is not likely to cause problems.

Good engineering calls for tuned circuits between stages in transmitters. These circuits reduce or eliminate spurious signals such as harmonics. The tuned circuits allow signals at the desired frequency to pass, but they *attenuate* (reduce) harmonics.

Transmitters with vacuum tubes in the final amplifier often use a *pi-network* impedance-matching circuit at the final-amplifier output. The plate-tuning and antenna-loading capacitors in the pi network are adjustable. When properly tuned, the pi network allows maximum power to pass at the frequency of operation but reduces harmonics.

Transmitters with solid-state final amplifiers usually operate over a wide range of frequencies with no external tuning controls. In these transmitters, the band switch activates a separate tuned *band-pass filter* for each band. As the name implies, this filter allows energy to pass at the desired frequency and attenuates spurious signals above and below.

How can you be sure that your transmitter does not generate excessive harmonics? The FCC requires all transmitter manufacturers to prove that their equipment complies with its regulations. This means that commercially manufactured equipment usually produces clean signals. If you build a transmitter from a magazine or book article, check for information on harmonic radiation. ARRL requires that all transmitter projects published in *QST* and our technical books meet the FCC specifications for commercially manufactured equipment.

Parasitic Oscillations

Parasitic oscillations in the final amplifier occur on frequencies unrelated to those for which the amplifier is designed. These spurious signals, called *parasitics*, are not harmonics. They result from an unintentional tuned feedback path in the final-amplifier circuit. Parasitics happen because an amplifier tube or semiconductor often works well at frequencies much higher than those it is used on. Parasitic oscillations usually take place in the VHF/UHF range. Most transmitters have special tuned circuits built into the final-amplifier circuit to prevent parasitic oscillations. They are called *parasitic suppressors*. Figure 8-2 shows examples of parasitic suppressors connected in amplifier circuits.

A defective component in your VHF or UHF trans-

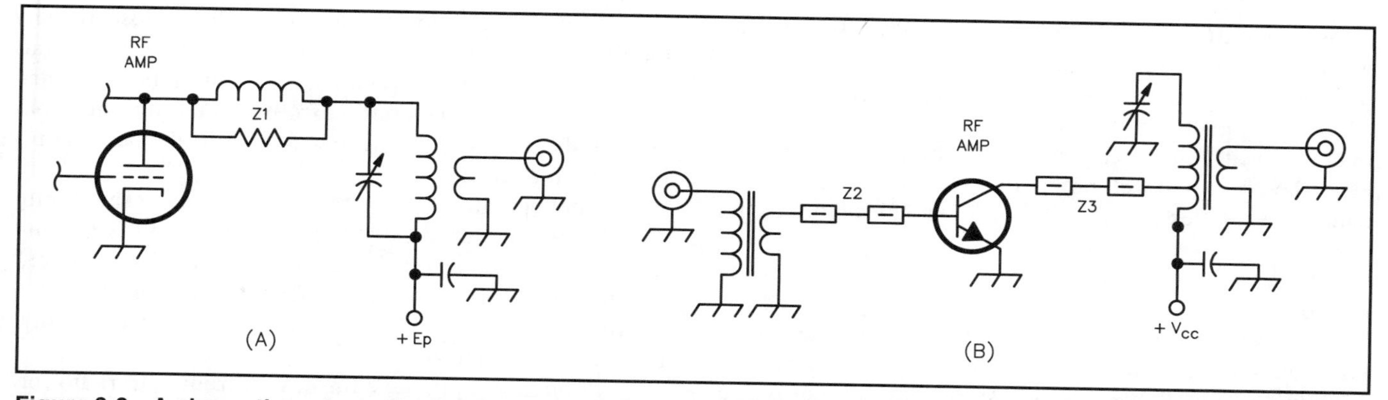

Figure 8-2—A shows the schematic diagram of a vacuum tube amplifier. The coil and resistor combination Z1 is a parasitic suppressor. B shows the schematic diagram of a transistor power amplifier. Z2 and Z3 are ferrite beads that also function as parasitic suppressors.

ceiver may cause it to radiate spurious signals. These signals may interfere with other hams on the same bands, or even users of other bands (including public safety and commercial bands).

Neutralization

Spurious signals may result from improper *neutralization* of the transmitter. An oscillator is an amplifier that has some of the output signal fed back to the input. This *positive feedback* is fine for oscillators, but we don't want positive feedback in an amplifier. Positive feedback in an audio amplifier for example, causes a howling noise. You may have heard this noise from a public-address system with the gain set too high.

We don't want positive feedback in radio amplifiers, either. In an improperly neutralized amplifier, some of the output feeds back to the input and causes the amplifier to oscillate. The signal generated may be transmitted. If the signal is strong enough, it may generate more power than the transmitter was designed to handle, and may damage the transmitter. Neutralization eliminates or neutralizes this positive feedback by applying negative feedback. If you suspect that you need to neutralize your transmitter, you may have to adjust an internal control. Figure 8-3 shows a vacuum tube amplifier circuit with a neutralizing capacitor. The neutralizing capacitor should be identified on the schematic diagram.

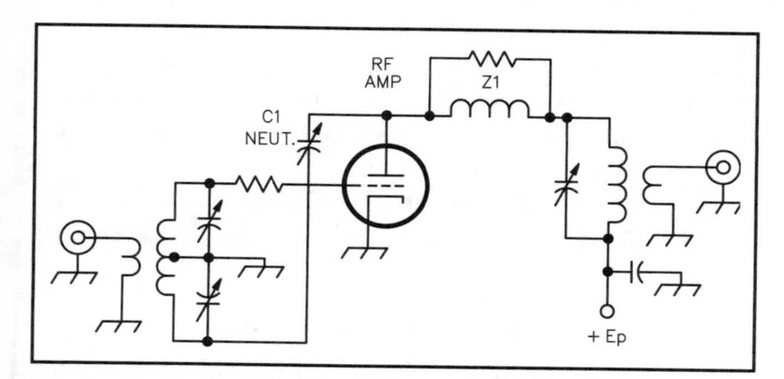

Figure 8-3—Schematic diagram of a vacuum tube amplifier. C1 is a neutralizing capacitor. It applies negative feedback to the amplifier input and prevents oscillation.

An amplifier that requires neutralization oscillates on the frequency band it is designed to amplify. Compare this with parasitics, which happen at much higher frequencies. You may need to operate your transmitter to check for proper neutralization, and for other transmitter adjustments and tests. You should do this without transmitting a signal on the air. The best way to do so is to use a *dummy load.*

Modern solid-state transmitters don't normally require neutralization. If your transmitter or transceiver has a vacuum-tube final amplifier, you may need to adjust the neutralization if you change the tubes. Often, even replacing tubes doesn't upset the adjustment. The procedure is not the same for all transmitters. Follow the instructions in your transceiver or transmitter manual, if you have it. Otherwise, ask your instructor or a local ham to help you.

Neutralization requires you to make the adjustment with the transmitter operating. If you have to remove part of the cabinet, remember that you have exposed the high-voltage components of your rig. Be extra careful to avoid the possibility of electric shock.

[Now turn to Chapter 13 and study question N4D02 and questions N8B01 through N8B05. Review this section if you have any difficulty with those questions.]

Other Causes for Spurious Emissions

We have been discussing spurious emissions caused by circuit problems in your transmitter. Your equipment can also cause spurious emissions if you operate it with some controls adjusted improperly. For example, if you operate an SSB transmitter with the microphone gain set too high you can cause splatter, or interference to frequencies near the one on which you are operating. Talking too loud into the microphone or having the microphone gain set too high causes the transmitter to overmodulate the signal. This means your transmitter may be putting out spurious emissions or **splatter** that could interfere with other stations when you operate it this way.

Many SSB transmitters include a speech processor to add extra "punch" to your voice, which will help another operator hear you under poor band conditions or interference. Too much speech processing can distort your audio and cause splatter interference on frequencies close to the one on which you are operating.

Even a hand-held FM transceiver can cause interference on nearby frequencies if the microphone gain or deviation control is set too high. On an FM transmitter the microphone gain or deviation control is usually inside the radio. You don't normally have to adjust this control, but if you consistently get reports that your audio is distorted or that you are causing splatter interference to nearby frequencies you may have to make an adjustment. You may also hear other operators that you are *over-deviating*. This all means one thing: you need to correct the problem.

It may be that your voice characteristics and the microphone you are using require a small adjustment to the deviation control. This would be especially true if you change microphones from a mobile microphone to a base-station type microphone. Any time you change microphones you should make an on-the-air check with another station to ensure the quality of your signal.

Generally you should hold a microphone close to your mouth and speak in a normal voice. If you get reports that your FM transceiver is over deviating, you should try holding the microphone a bit farther from your mouth when you are talking. You may not need to adjust the deviation control if you try this.

Many operators tend to talk louder or even shout into the microphone, especially if the other station is having difficulty hearing you. This won't normally help, however, and may even make it more difficult. Shouting into the microphone may cause the radio to over deviate, distorting the transmitted audio.

If you have to remove the covers from your radio for any reason, be sure to reinstall them and tighten all the screws before operating the radio again. In addition to protecting the electronic components from physical damage, the covers also provide shielding to the circuit. Such shielding stops any spurious emissions or unwanted RF signals from being radiated.

[Turn to Chapter 13 and study questions N4D05, N8B06 through N8B11 and N8B15. Review this section if you have difficulty with any of these questions.]

[If you are studying for the Technician license exam, you should also turn to Chapter 14 and study questions T8B11 and T8B12. Review this section as needed.]

INTERFERENCE WITH OTHER SERVICES

Radio frequency interference (RFI) has given radio amateurs headaches for years. It can occur whenever an electronic device is surrounded by RF energy. Your rig emits RF energy each time you transmit. This RF energy may interfere with your own or your neighbor's television set (causing **television interference — TVI**). You may also have problems with a stereo system, electronic organ, videocassette recorder, telephone or any other piece of consumer electronic equipment.

If you have a very visible antenna in your yard, your neighbors may blame you for any interference they experience, even when you're not on the air! If you have any problems with interference to your own equipment, it's a good bet that your neighbors do too. On the other hand, if you can show your neighbors that you don't interfere with *your* television, they may be more open to your suggestions for curing problems.

Figure 8-4—The boom in home electronic devices continues. Many amateur antennas are close to a growing number of home-entertainment devices. The result: an RFI problem that shows little sign of disappearing.

So what should you do if someone complains of interference? First, make sure that your equipment is operating properly. If the complaint is TVI, check for interference to your own TV. If you see it, stop operating and cure the problem before you go back on the air.

Even if you don't interfere with your own TV, don't stop there. Simply telling your neighbors you're not at fault can cause even more problems. Try to work with your neighbors to determine if your rig is actually causing the interference. If so, try to help solve the problem. A more-experienced ham can be a great help. If you don't know any other hams in your area, write to ARRL HQ. We'll try to help you find a knowledgeable local ham.

[Now go back to Chapter 13 and study question N4D13. Review this section if you have any problems.]

Receiver Overload

Receiver overload is a common type of TV and FM-broadcast interference. It happens most often to consumer electronic equipment near an amateur station or other transmitter. When the RF signal (at the fundamental frequency) enters the receiver, it overloads one or more circuits. The receiver front end (first RF amplifier stage after the antenna) is most commonly affected. For this reason, we sometimes call this interference **front-end overload** or **RF overload**.

A strong enough RF field may produce spurious signals in the receiver, which cause the interference. Receiver overload interference may occur in your neighbor's house or just your own. Receiver overload can result from transmitters operating on any frequency. It is the most-common interference problem caused by VHF and UHF transmitters. If you receive an RFI complaint, and determine that the problem exists no matter what frequency you operate on, you should suspect receiver overload.

Receiver overload usually has a dramatic effect on the television picture. Whenever you key your transmitter, the picture may be completely wiped out. The screen may go black, or it might just become light with traces of color. The sound (audio) will probably be affected also. In an FM receiver, the audio may be blocked each time you transmit. More often than not, overload affects only TV channels 2 through 13. In cases of severe interference, however, it may also affect the UHF channels.

The objective in curing receiver overload is to prevent the amateur signal from entering the front end of the entertainment receiver. It is important to realize that there is nothing you can do to your transmitter to cure receiver overload. It is a fundamental problem with the receiving system, and the primary responsibility for curing the problem is with the equipment manufacturer and the owner.

The first step in trying to cure such a problem with a cable-TV receiver is to have the owner or a service technician tighten all connectors and inspect the cable system transmission line. Any loose connector or break in the transmission line of a cable TV system can allow amateur signals to leak into the line, causing interference to TV receivers. Such a leak in the system can also allow Cable TV signals to leak out of the system and cause interference to amateur receivers using that frequency. Cable TV systems use some amateur frequencies in the VHF/UHF range to carry the signals along the cable. This causes no problems as long as there are no leaks in the system.

If the TV is not connected to a cable system, or if it is a stereo or other consumer electronic device receiving the interference, there are other steps you should take. Have the equipment owner or a qualified service technician install a **high-pass filter**. See Figure 8-5. The filter will block the amateur signal from coming in through the antenna feed line and reaching the receiver front-end components. This would also be the next step if tightening the cable TV connectors didn't help.

Install the filter at the TV or FM receiver input. The best location is where the antenna feed line connects to the TV or FM tuner. It is *not* usually a good idea for an Amateur Radio operator to install a filter on a neighbor's entertainment equipment. Only the owner or a qualified technician should install the filter. If you install the filter, you might later be blamed for other problems with the TV set. A high-pass filter is a tuned circuit that passes high frequencies (TV channels start at 54 MHz). The filter blocks low frequencies (the HF amateur bands are in the range of 1.8-30 MHz).

[Now study questions N4D01, N4D07, N4D08, questions N4D10 through N4D12 and question N4D15 in Chapter 13. Review this section as needed.]

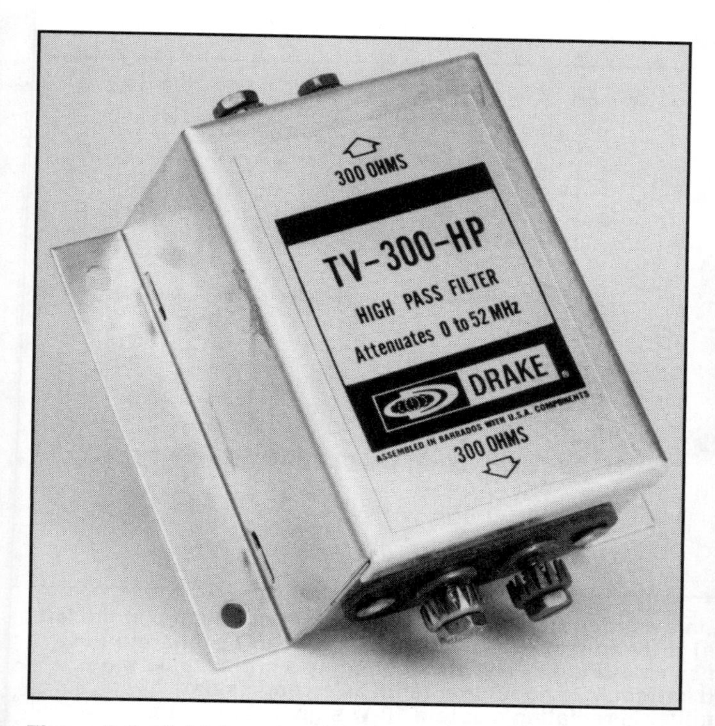

Figure 8-5—A high-pass filter can prevent fundamental energy from an amateur signal from entering a television set. This type of high-pass filter goes in the 300-ohm feed line that connects the television with the antenna.

Harmonic Interference

Another problem for hams is harmonic interference to entertainment equipment. As we learned before, harmonics

Figure 8-6—Harmonics radiated from an amateur transmitter may cause "crosshatching."

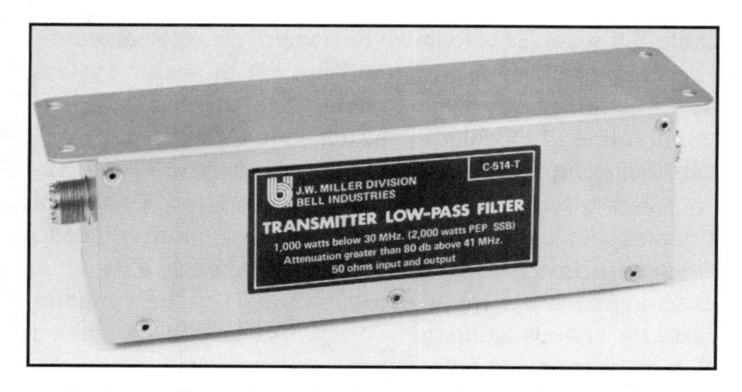

Figure 8-7—A low-pass filter. When connected in the coaxial cable feed line between an amateur transmitter and the antenna, a low-pass filter can reduce the strength of transmitted harmonics.

are multiples of a given frequency. Your HF transmitter radiates undesired harmonics along with your signal. Your transmitting frequency is much lower than the TV or FM channels. Some harmonics will fall within the home entertainment bands, however.

The entertainment receiver cannot distinguish between the TV or FM signals (desired signals) and your harmonics (undesirable intruders) on the same frequency. If your harmonics are strong enough, they can seriously interfere with the received signal. Harmonic interference shows up as a crosshatch or a herringbone pattern on the TV screen. See Figure 8-6.

Unlike receiver overload, harmonic interference seldom affects all channels. Rather, it may bother the one channel that has a harmonic relationship to the band you're on. Generally, harmonics from amateur transmitters operating below 30 MHz affect the lower TV channels (2 through 6). Ten-meter transmitters usually bother channels 2 and 6, and channels 3 and 6 experience trouble from 15-meter transmitters.

Harmonic interference must be cured at your transmitter. As a licensed amateur, you must take steps to see that harmonics from your transmitter do not interfere with other services. All harmonics generated by your transmitter must be attenuated well below the strength of the fundamental frequency. If harmonics from your transmitting equipment exceed these limits, you are at fault.

In this section we will discuss some of the several possible cures for harmonic interference. Try each step in order and the chances are good that your problem will be solved quickly.

The first step you should take is to install a **low-pass filter** like the one shown in Figure 8-7. The filter goes in the transmission line between your transmitter and antenna or antenna tuner. As the name implies, a low-pass filter is the opposite of a high-pass filter. A low-pass filter allows RF energy in the amateur bands to pass freely. It blocks very high frequency harmonics that can fall in the TV and FM bands. Low-pass filters usually have a specified cutoff frequency, often 40 MHz, above which they severely attenuate the passage of RF energy.

Even if your transmitter is working well within FCC specifications, you may need additional attenuation to re-

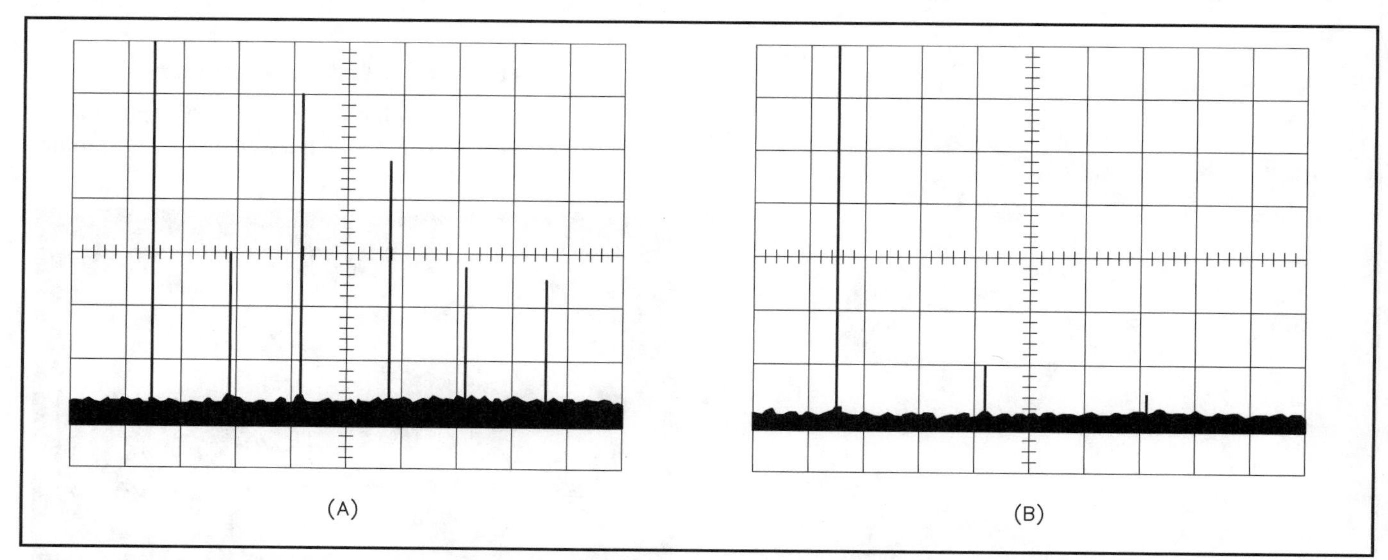

Figure 8-8—A shows a spectrum-analyzer display of the signals emitted from an amateur transmitter. The pip at the left of the display (the one that extends to the top horizontal line) is the fundamental. All other pips represent harmonics. This particular transmitter generates several harmonics. On an analyzer display, stronger signals create taller pips. Here, the harmonic signals are quite strong. In fact, the third harmonic is about one-tenth as strong as the fundamental. The fundamental signal is 100 W, so the transmitter is radiating a potent 10-W signal at the third harmonic. Most of these harmonics will cause interference to other services. B shows the output from the same transmitter, operating at the same power level at the same frequency, after installation of a low-pass filter between the transmitter and the analyzer. The harmonics have all but disappeared from the display. The stronger of the two remaining harmonics is only about 100 microwatts—weak enough that it is unlikely to cause interference.

duce harmonics. Remember, your goal is to eliminate interference. Good-quality low-pass filters often attenuate signals falling in the entertainment bands by 70 or 80 dB. This is significantly better than the 40 to 50 dB typical of amateur transmitters. A decibel (dB) is a number (the logarithm of a ratio) used to describe how effective the filter is. Larger numbers indicate better filtering. Figure 8-8 shows the output of a transmitter before and after filtering.

Another source of interference is RF energy from your transmitter that enters the ac power lines. The *ac power-line filter* is another kind of low-pass filter. It prevents RF energy from entering the ac line and radiating from power lines inside and near your house.

Multiband Antennas

You can also run into trouble if you use a multiband antenna. If your antenna works on two or three different bands, it will radiate any harmonics present on those frequencies. After all, we *want* the antenna to radiate energy at a given frequency. It cannot tell the difference between desired signal energy and unwanted harmonic energy. This problem does not usually affect home entertainment equipment. It may cause interference to other amateurs or to other radio services operating near the amateur bands, however.

For example, a multiband dipole antenna that covers 80 and 40 meters may radiate 40-meter energy while you operate on 80 meters. The second harmonic of a 3.7-MHz signal falls above the 40-meter amateur band, at 7.4 MHz. If your transmitter is free of excessive harmonic output, you will probably not have a problem. (Note: Older, tube-type equipment can sometimes radiate excessive harmonics if it isn't "tuned up" properly. If you use a rig with a "plate tuning" control, be sure to adjust it according to the manufacturer's instructions.)

Proper shielding and grounding are essential to reduce harmonic and other spurious radiation. The only place you want RF to leave your transmitter is through the antenna connection. Your transmitter must be fully enclosed in a metal cabinet. The various shields that make up the metal cabinet should be securely screwed or welded together at the seams. You must also connect the transmitter to a good earth ground connection.

Remember: A low-pass filter will only block harmonics from reaching your antenna. It will do nothing for a poorly shielded transmitter that leaks stray RF from places other than the antenna connector. Figure 8-9 summarizes the steps you can take to reduce harmonic radiation from your station.

More and more amateurs are using computers in their ham shacks. Most of the same steps you would take with your transceiver will also help reduce or eliminate interference from your computer. Yes, that's right! Your computer can cause interference to your receiver. A computer has a clock oscillator circuit that operates at a "radio" frequency. Do you have a '386/33, a '486/66 or a Pentium 150? The 33, 66 or 150 represents the clock frequency in megahertz!

Your computer should be in a metal cabinet with all screws securely attached, and you should use shielded cables with the shield connected to the equipment chassis. Be sure all your equipment is properly grounded. These steps will reduce the possibility of interference from your computer.

[Now turn to Chapter 13 and study questions N4D03, N4D04, N4D06, N4D09 and N4D14. You should also study question N8B16. Review this section as necessary.]

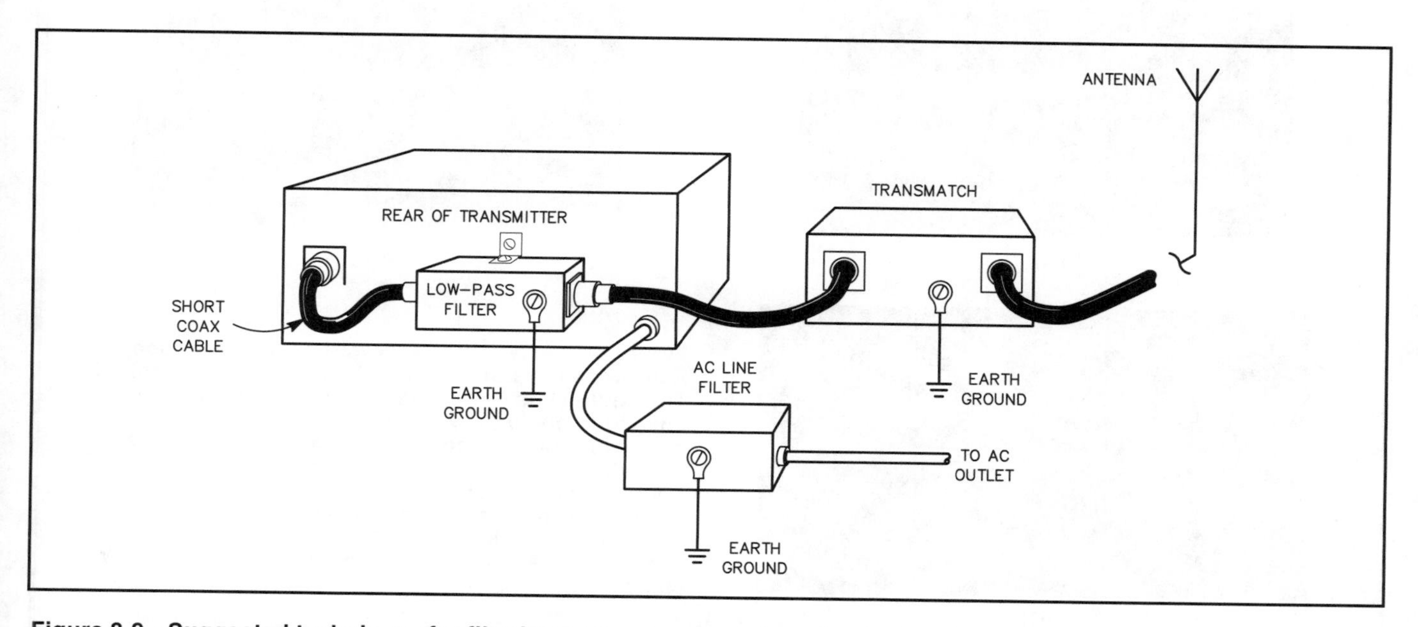

Figure 8-9—Suggested techniques for filtering harmonic energy from the leads of an amateur transmitter.

SIGNAL PURITY

As a licensed Amateur Radio operator, you are responsible for the quality of the signal transmitted from your station. The rules require your transmitted signal to be stable in frequency and pure in tone or modulation. If your signal is not "clean," it is unpleasant to listen to. It may also cause interference to others using the band or other services.

Unfortunately, some problems can creep up and give you a dirty signal. Three of the most common problems with CW signals are **key clicks**, **chirp** and **superimposed hum**. Superimposed hum is sometimes a problem with voice signals, also.

Key Clicks

As you listen on the air, you will notice that some signals have a click or thump on *make* (the instant the key contacts close) or *break* (the instant they open) or both. This **key click** is more than just annoying; it can interfere with other stations.

If you use an oscilloscope (a type of test equipment resembling a TV set) to monitor a CW signal with key clicks, you will see excessively square CW keyed waveforms. Imagine a good, stable transmitter sending a CW signal on 3.720 MHz. What happens when we turn the transmitter on and off with a telegraph key? You'd think that the only frequency the output energy could have would be 3.720 MHz. If you turn the transmitter output on and off rapidly, however, something else happens. Unwanted energy — in the form of key clicks — appears for as much as several kilohertz on either side of the operating frequency. The transmitter creates these clicks during the instant that it is turned on and off.

When we say "keyed rapidly," we don't mean that we're sending fast (like 35 wpm CW). We mean that the transmitter goes from zero power to full power and back again very quickly and abruptly. We call the time it takes to go from no power to full power the *rise time*. We call the return trip from full power to zero power the *fall time*.

An oscilloscope display of a CW signal with short rise and fall times in shown in Figure 8-10A. This signal has key clicks. A click-free signal is shown in part B. Notice how the beginning and ending of this CW pulse are soft and round. Compare this with the square shoulders of the signal with clicks. Part C shows a string of dots and dashes that will generate interference from key clicks.

We use a **key-click filter** or *shaping filter* to eliminate this unwanted and bothersome energy. The filter makes the transmitter output increase to maximum more slowly on make, and fall from full output more slowly on break. The keying waveform is softer, limiting the output energy to a few hundred hertz on either side of the operating frequency.

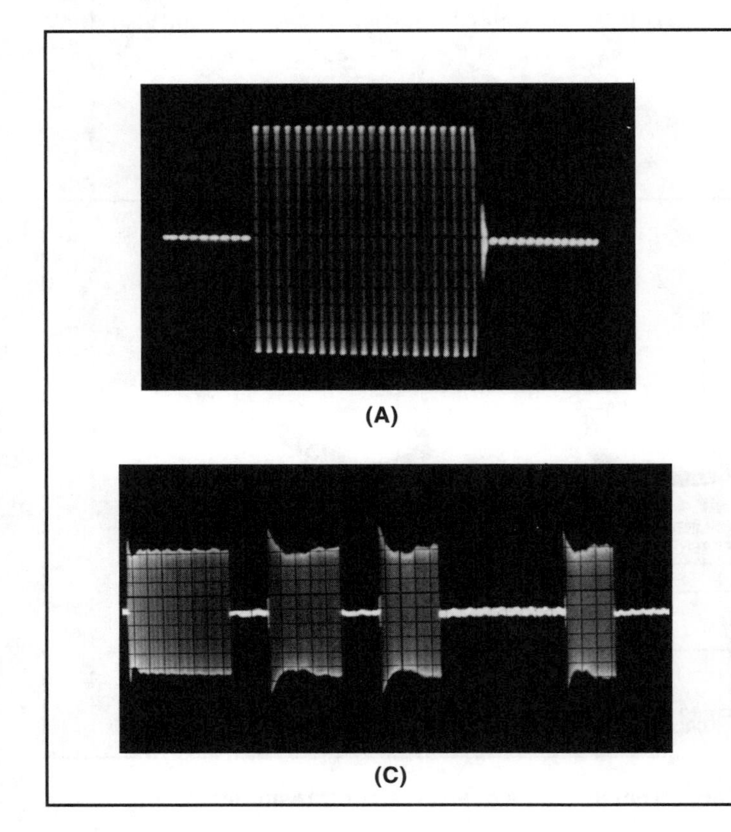

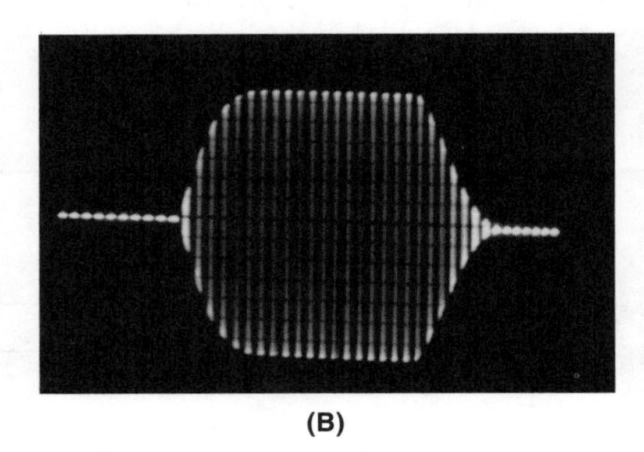

(A) (B) (C)

Figure 8-10—CW signals viewed on an oscilloscope. A shows a CW dot with no intentional wave shaping. This signal has key clicks. B shows a similar dot, shaped to eliminate key clicks. Notice how the signal at B builds and decays gradually. Compare this with the abrupt on-and-off characteristics at A. C shows a string of unshaped CW characters. Notice the sharp *spikes* at the beginning of each pulse. These spikes are audible in a receiver as key clicks. Such spikes can appear at either end of a CW character.

Chirp

Chirp is another common transmitter problem. Chirp occurs when the oscillator in your transmitter shifts frequency slightly whenever you close your telegraph key. The result is that other stations receive your transmitted signal as a chirping sound rather than as a pure tone. Your "dahdidahdit" will sound like "whoopwhiwhoopwhip." It isn't very much fun to copy a chirpy CW signal!

Chirp usually happens when the oscillator power supply voltage changes as you transmit. Your transmitter may also chirp if the load on the oscillator changes when you transmit. If the supply voltage changes, you must improve the *voltage regulation*. With better regulation, the voltage won't shift when you key your transmitter.

If it's not a voltage problem, then what? Amplifier stages after the oscillator may be loading it down and pulling its frequency. You may need a better buffer (isolation) or driver stage between the oscillator and the next stage in your transmitter. Some oscillators are sensitive to temperature changes. If there is too much current through the frequency-determining components, their temperature may increase and the resonant frequency will change.

Superimposed Hum

The last kind of signal problem we discuss is **superimposed hum**. Power supplies contain filters that remove the ac present in the rectifier output. The result is a pure, filtered dc output. If a filter capacitor fails, then the filtering action will be incomplete.

If the filter doesn't work properly, ac will be present in the power supply output. This ac will also work its way into the transmitter output. Instead of hearing a pure tone or your voice, stations receiving your signal will hear a low-pitched hum as well. If enough ac is present, the signal will have a raspy tone. Your CW signal may even buzz as if you were keying an electric razor! To cure hum, check the power-supply filter circuit. A bad filter can cause superimposed hum.

How to Get Help

Transmitting a signal with chirp, key clicks, hum or spurious emissions is a violation of both the letter and spirit of the regulations governing Amateur Radio. The best way to find out if you have a bad-sounding signal is from the stations you work. If another operator tells you about a problem with your signal, don't be offended. He cares about the image of the Amateur Service and only wants to help you. You might never be aware of a problem otherwise.

Don't let this chapter scare you. Spurious signals, unwanted harmonics, poor neutralization, key clicks, parasitic oscillations, television interference (TVI) or other equipment difficulties can all be solved. Two ARRL publications will be helpful: *The ARRL Handbook for Radio Amateurs* and *Radio Frequency Interference*. Check with your Amateur Radio instructor or any experienced ham for help. There's no substitute for experience. Another ham may know just how to solve your problem and return you to the air as soon as possible. If you'd like to set up a station in an area where RFI may be a real problem, you'll want to read *Low Profile Amateur Radio*. This book tells you how to enjoy Amateur Radio with few problems from an apartment, condo, dorm room, car or in the field.

[This completes your study of the material about amateur signals and emissions. Before you go on to the next chapter, you should study questions N8B12 through N8B14 in Chapter 13. Review this section as necessary.]

[If you are preparing for a Technician license exam, you should go on to the next section before turning to Chapter 9.]

KEY WORDS

Bandwidth — A range of associated frequencies (measured in hertz). Bandwidth describes the range of frequencies that a radio transmission occupies.

Emission type — The different kinds of radio signals, such as CW, RTTY, SSB and FM.

8 Signals and Emissions for Technician Licensees

BANDWIDTH

The amount of space in the radio-frequency spectrum that a signal occupies is called its **bandwidth**. The bandwidth of a transmission is determined by the information rate. Thus, a pure, continuous, unmodulated carrier has a very small bandwidth with no sidebands. A television transmission, which contains a great deal of information, is several megahertz wide.

Receiver Bandwidth

Receiver bandwidth determines how well you can receive one signal in the presence of another signal that is very close in frequency. The enjoyment you'll experience will depend greatly on how well you can isolate the signal you are receiving from all the others nearby.

Bandwidth is a measure of *selectivity*: how wide a range of frequencies is received with the receiver tuned to one frequency. For example, if you can hear signals as much as 3 kHz above and 3 kHz below the frequency to which you are tuned, your receiver has a bandwidth of at least 6 kHz. If you cannot hear signals more than 200 Hz above or below the frequency to which you are tuned, the bandwidth is only 400 Hz. The narrower the bandwidth, the greater the selectivity and the easier it is to copy one signal when there is another one close by in frequency.

Selectivity is determined by special intermediate-frequency (IF) filters built into the receiver. Some receivers have several filters so you can choose different bandwidths. They are necessary because different **emission types** occupy a wider frequency range than others. A 250-Hz-bandwidth filter is excellent for separating CW signals on a crowded band, but it's useless for listening to SSB, AM or FM transmissions. A wider filter is needed to allow all the transmitted information to reach the detector.

Most receivers designed for single-sideband voice operation come standard with a filter selectivity of around 2.8 kHz. This is ideal for SSB, which usually has a bandwidth between 2 and 3 kHz. Your voice contains frequencies higher than 3 kHz, but all of the sounds necessary to understand speech are between about 300 Hz and 3000 Hz. Most amateur voice transmitters limit the bandwidth of a transmitted audio signal to between 200 and 3000 Hz. The difference between these limits is the bandwidth, 2700 Hz. By using a filter selectivity of 2.8 kHz (2800 Hz), you can see that your receiver will reproduce the full range of transmitted audio.

Although a bandwidth of 2.8 kHz is also usable on CW, a narrower bandwidth is needed to prevent adjacent CW signals from getting through at the same time. Many amateurs prefer a filter bandwidth of 500 Hz or even 250 Hz for CW operation. A radioteletype signal has a bandwidth that is a little wider than a CW signal, but a 500 Hz or 250-Hz filter serves nicely for that mode, too.

CW signals have the narrowest bandwidth of any amateur emissions. Radioteletype emissions are wider than CW, and SSB signals are even wider than that. Figure 8-11 illustrates the relative bandwidths of CW, RTTY and SSB signals and the bandwidth of IF filters that might be used to

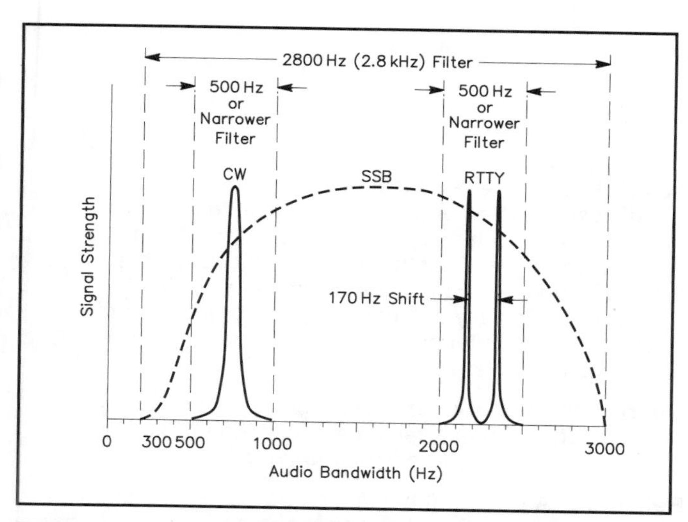

Figure 8-11—This drawing illustrates the relative bandwidths of CW, RTTY and SSB signals. The bandwidths of filters that might be used in a receiver's intermediate frequency (IF) section to receive these signals are also shown.

receive these signals. FM and PM, which we will consider next, can occupy even more bandwidth.

Bandwidth in FM and PM

When you transmit FM or PM phone (voice) emissions, the frequency of the transmitted RF signal varies an amount that depends on the strength of your voice. When you speak louder, the frequency changes a greater amount than when you speak softly. The frequency or pitch of your voice (or other signal used to modulate the transmitter) controls how fast the frequency changes. Higher-frequency tones make the frequency vary at a faster rate than low-frequency tones. *Frequency deviation* is the instantaneous change in frequency for a given signal. The frequency swings just as far in both directions, so the total frequency swing is equal to twice the deviation. In addition, there are sidebands that increase the bandwidth still further. A good estimate of the bandwidth is twice the maximum frequency deviation plus the maximum modulating audio frequency:

$$Bw = 2 \times (D + M) \qquad \text{(Eq 8-1)}$$

where:

Bw = bandwidth

D = maximum frequency deviation

M = maximum modulating audio frequency

An FM transmitter using 5-kHz deviation and a maximum audio frequency of 3 kHz uses a total bandwidth of about 16 kHz. The actual bandwidth of a typical FM signal may be somewhat greater than this. A good approximation is that the bandwidth of an FM voice signal is between 10 and 20 kHz.

Amateur Television

Fast-scan television is a popular activity on some of our UHF bands. Fast-scan TV uses the same standards as broadcast TV, so you can transmit full motion pictures on this mode. Such signals must carry all the picture information to draw a complete image on the screen 30 times each second. In addition, TV signals include a voice signal. All this adds up to a signal with a bandwidth of about 6 MHz.

[Congratulations! You have completed your study of the amateur signals and emissions for your Technician license exam. Now turn to Chapter 14 and look at questions T8B06 through T8B10. Review this section if you have difficulty.]

9 KEY WORDS

Antenna—A device that picks up or sends out radio waves.

Balun—Contraction for ***bal***anced to ***un***balanced. A device to couple a balanced load to an unbalanced source, or vice versa.

Beam antenna—A directional antenna. A beam antenna must be rotated to provide coverage in different directions.

Coaxial cable—Coax (pronounced kō-aks). A type of feed line with one conductor inside the other.

Dipole antenna—See **Half-wave dipole**. A dipole need not be ½ wavelength long.

Director—An element in front of the driven element in a Yagi and some other directional antennas.

Driven element—The part of an antenna that connects directly to the feed line.

Feed line—See **Transmission line**.

Gain—A measure of the directivity of an antenna, as compared with another antenna such as a dipole.

Half-wave dipole—A basic antenna used by radio amateurs. It consists of a length of wire or tubing, opened and fed at the center. The entire antenna is ½ wavelength long at the desired operating frequency.

Impedance—The opposition to electric current that an antenna feed line presents. Impedance includes factors other than resistance, and applies to alternating currents. Ideally, the characteristic impedance of a feed line is the same as the transmitter output impedance and the antenna input impedance.

Impedance-matching device—A device that matches one impedance level to another. For example, it may match the impedance of an antenna system to the impedance of a transmitter or receiver. Amateurs also call such devices a Transmatch, impedance-matching network or antenna tuner.

Ladder line—Another name for **open-wire feed line**.

Open-wire feed line—Parallel-conductor feed line that has air as its primary insulation material.

Parallel-conductor feed line—Feed line with two conductors held a constant distance apart.

Polarization—The electrical-field characteristic of a radio wave. An antenna that is parallel to the surface of the earth, such as a dipole, produces horizontally polarized waves. One that is perpendicular to the earth's surface, such as a quarter-wave vertical, produces vertically polarized waves. An antenna that has both horizontal and vertical polarization is said to be circularly polarized.

Quarter-wavelength vertical antenna—An antenna constructed of a quarter-wavelength long radiating element placed perpendicular to the earth.

Reflector—An element behind the driven element in a Yagi and some other directional antennas.

Resonant frequency—The desired operating frequency of a tuned circuit. In an antenna, the resonant frequency is one where the feed-point impedance contains only resistance.

Standing-wave ratio (SWR)—Sometimes called voltage standing-wave ratio (VSWR). A measure of the impedance match between the feed line and the antenna. Also, with a Transmatch in use, a measure of the match between the feed line from the transmitter and the antenna system. The system includes the Transmatch and the line to the antenna. VSWR is the ratio of maximum voltage to minimum voltage along the feed line. Also the ratio of antenna impedance to feed-line impedance when the antenna is a purely resistive load.

SWR meter—A device used to measure SWR.

Transmission line—The wires or cable used to connect a transmitter or receiver to an antenna.

Vertical antenna—A common amateur antenna, usually made of metal tubing. The radiating element is vertical. There are usually four or more radial elements parallel to or on the ground.

Wavelength—Often abbreviated λ. The distance a radio wave travels in one RF cycle. The wavelength relates to frequency. Higher frequencies have shorter wavelengths.

Yagi antenna—The most popular type of amateur directional (beam) antenna. It has one driven element and one or more additional elements.

9 Choosing an Antenna

As a new ham, you should quickly learn two antenna truths: 1) ***Any*** *antenna is better than* ***no*** *antenna! 2) Time, effort and money invested in your antenna system generally will provide more improvement to your station than an equal investment to any other part of the station.*

We know that a transmitter generates radio-frequency energy. We convert this electrical energy into radio waves with an **antenna**. An antenna may be just a piece of wire or other conductor designed to *radiate* the energy. An antenna converts current into an electromagnetic field (radio waves). The radio waves spread out or *propagate* from the antenna. You might relate their travel to the ever-expanding waves you get when you drop a pebble in water. Waves from an antenna radiate in all directions, though, not just in a flat plane.

It also works the other way. When a radio wave crosses an antenna, it generates a voltage in the antenna. That voltage isn't very strong, but it's enough to create a small current. That current travels through the **transmission line** to the receiver. The receiver detects the radio signal. In short, the antenna converts electrical energy to radio waves and radio waves to electrical energy. This process makes two-way radio communication possible with just one antenna.

Your success in making contacts depends heavily on your antenna. A good antenna can make a fair receiver seem like a champ. It can also make a few watts sound like a whole lot more. Remember, you'll normally use the same antenna to transmit *and* receive. Any improvements to your antenna makes your transmitted signal stronger, and increases the strength of the signals you receive.

Assembling an antenna system gives you a chance to be creative. You may discover, for example, that property size or landlord restrictions rule out a traditional antenna. If so, you can innovate. *Low Profile Amateur Radio* is an ARRL publication full of ideas for just these conditions. *The ARRL Antenna Book*, *The ARRL Antenna Compendiums*, and similar publications also offer lots of antenna suggestions. This chapter contains a few suggestions, too.

Some antennas work better than others. Antenna design and construction have kept radio amateurs busy since the days of Marconi. You'll probably experiment with different antenna types over the years. Putting up a better antenna is an inexpensive but rewarding way to improve your station.

WAVELENGTHS

We sometimes talk about antenna lengths in wavelengths. A **wavelength** relates to the operating frequency. When you construct an antenna for one particular amateur band, you cut it to the proper wavelength. Often you'll see the Greek letter lambda (λ) used as an abbreviation for wavelength. For example, $^1/_2$ λ means "one-half wavelength."

Most popular ham antennas are less than 1 λ long. (A very popular antenna is a $^1/_2$-λ **dipole antenna**. You'll learn how to build one later in this chapter.) There is a simple relationship between operating frequency and wavelength. The wavelength is shorter at the higher frequencies. Wavelength is longer at the lower frequencies.

If numbers interest you, use this equation to find the wavelength for a specific frequency.

$$\lambda \text{ (in feet)} = \frac{984}{\text{f(in MHz)}} \qquad \text{(Eq 9-1)}$$

This equation gives the wavelength in feet. The frequency is given in megahertz (MHz). Let's say we wanted to know the wavelength for 7.15 MHz. We divide 984 by 7.15, and the answer is about 137.6 feet.

$$\lambda \text{ (in feet)} = \frac{984}{\text{f(in MHz)}} = \frac{984}{7.15} = 137.6 \text{ feet}$$

Whenever we talk about an antenna, we specify its design frequency or the amateur band it covers. We could talk about a "40-meter dipole," for example, as an antenna intended for operation in the 40-meter band. Antennas are tuned circuits. A simple antenna such as a dipole or a $^1/_4$-λ vertical has a **resonant frequency**. Such antennas do best at their resonant frequency, as do most other tuned circuits.

To change the resonant frequency of a tuned circuit, you vary the capacitor value or the inductor value. You can change the resonant frequency of an antenna by changing its length, which affects its capacitance *and* inductance.

FEED LINES

To get RF energy from your transmitter to an antenna you use **transmission line**. A transmission line is a special cable or arrangement of wires. Such lines commonly go by the name **feed line**. They feed power to the antenna, or feed a received signal from the antenna to the receiver.

Characteristic Impedance

One electrical property of a feed line is its *characteristic impedance*. In Chapter 5 we learned that resistance is an opposition to electric current. Impedance is another form of resistance to electric current. Impedance includes factors other than resistance, however.

The spacing between line conductors and the type of insulating material determines the characteristic impedance. Characteristic impedance is important because we want the feed line to take all the transmitter power and feed it to the antenna. For this to occur, the transmitter (source) must have the same impedance as its load (the feed line). In turn, the feed line must have the same impedance as its load (the antenna).

We can use special circuits called *matching devices* or *matching networks* if any of these impedances are different. *Network* just refers to a combination of inductors and capacitors that forms a special circuit. Still, careful selection of a feed line can minimize such matching problems.

Coaxial Cable

Several types of feed line are available for amateur use. The most common is **coaxial cable**. Called "coax" (pronounced kó-aks) for short, this feed line has one conductor inside the other. It's like a wire inside a flexible tube. The center conductor is surrounded by insulation, and the insulation is surrounded by a wire braid called the shield. The whole cable is then encased in a tough vinyl outer coating, which makes the cable weatherproof. See Figure 9-1. Coax comes in different sizes, with different electrical properties. Figure 9-1 also shows other types of coaxial cables used by amateurs.

The most common types of coax have either a 50-ohm or 72-ohm characteristic impedance. Coax designated RG-58, RG-8 and RG-213 are 50-ohm cables. Some coax designations may also include a suffix such as /U, A/U or B/U, or bear the label "polyfoam." Feed line of this type may be used with most antennas. Cables labeled RG-59 or RG-11 are 72-ohm lines. Many hams use these types to feed **dipole antennas**.

The impedance of a $^1/_2$-λ dipole far from other objects is about 73 ohms. Practical dipoles placed close to the earth, trees, buildings, etc, have an input impedance closer to 50 ohms. In any case, the small impedance mismatch caused by using 50 or 72-ohm cable as antenna feed line is unimportant.

In choosing the feed line for your installation, you'll have a trade-off between electrical characteristics and physical properties. The RG-58 and RG-59 types of cable are about $^1/_4$ inch in diameter, comparatively lightweight, and reasonably flexible. RG-8, RG-213 and RG-11 are about $^1/_2$ inch in diameter, nearly three times heavier, and considerably less flexible. As far as operation goes, RG-8, RG-213 and RG-11 will handle much more power than RG-58 and RG-59.

Any line that feeds an antenna absorbs a small amount of transmitter power. That power is lost, because it serves no useful purpose. (The lost power warms the feed line slightly.) The loss occurs because the wires are not perfect conductors, and the insulating material is not a perfect insulator. Signal loss also increases slightly when the SWR is greater than 1:1, so we try to keep the SWR below 2:1 if possible.

Better-quality coaxial cables have lower loss than poor-quality cables. More of the transmitter power is lost as heat in a poor-quality coaxial cable. You usually get better-quality coax if you stick with name brands. You can also look at the shield braid on the coax. Better-quality cables have more complete coverage of the center insulator. If you can easily see through the holes in the braid, you should probably select another cable.

The larger coax types, RG-8, RG-213 and RG-11, have less signal loss than the smaller types. If your feed line is less than 100 feet long, you probably won't notice the small additional signal loss, at least on the HF bands. This, combined with light weight and flexibility, is why many

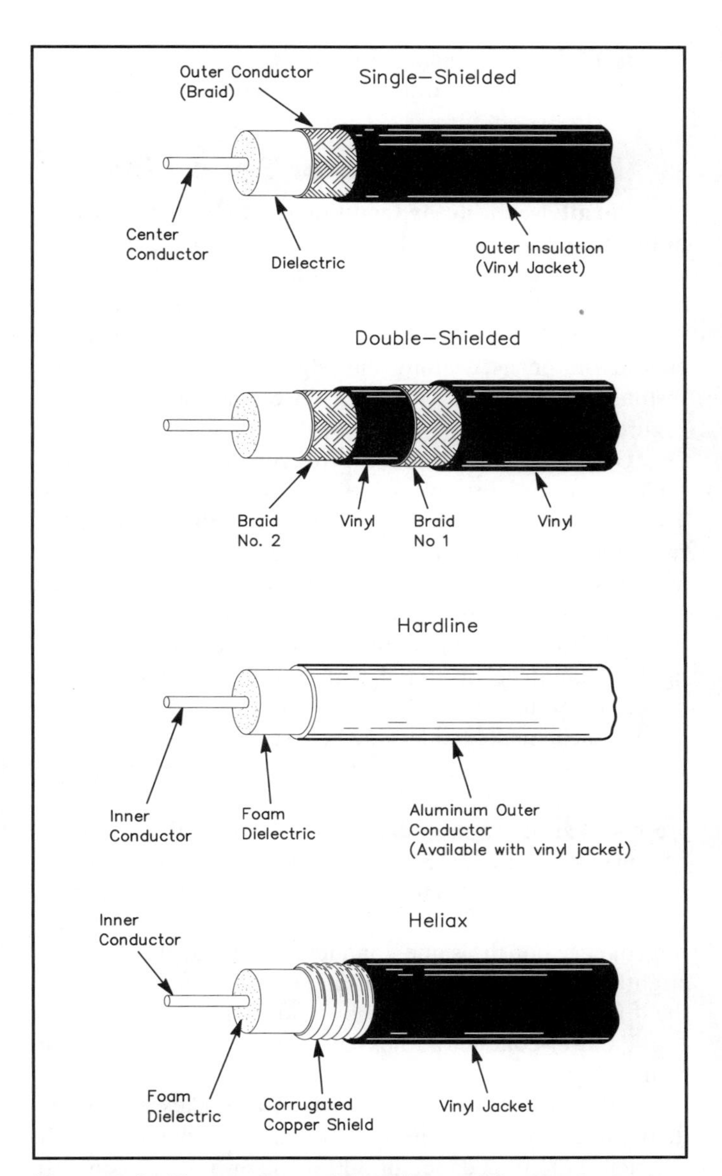

Figure 9-1—Types of coaxial cable used by amateurs. Abbreviated coax, it has a center conductor surrounded by insulation. The second conductor, called the shield, goes around that. Plastic insulation covers the entire cable.

HF operators find the smaller coax better suited to their needs. Also, the smaller feed line costs about half as much per foot as the larger types.

On the VHF/UHF bands, however, you will find the losses in RG-58 and RG-59 more noticeable, especially if your feed line is longer than about 50 feet. On these bands, most amateurs use higher-quality RG-213 coax or even lower-loss special coaxial cables. It is also important to use good-quality connectors at VHF and UHF.

Coaxial cable has several advantages as a feed line. It is readily available, and is resistant to weather. Most common amateur antennas have characteristic impedances of about 50 ohms. Coax can be buried in the ground if necessary. It can be bent, coiled and run next to metal with little effect. Its major drawback is its cost.

Parallel-Conductor Feed Line

Parallel-conductor feed line is another popular type of line for use below 30 MHz. Although it comes in several varieties, all consist of two wires held apart by a piece of insulation. The most familiar example of this feed line is the 300-ohm ribbon used for TV antennas. It has two parallel conductors encased along the edges of a strip of plastic insulation. We often call this kind of line *twin lead*. See Figure 9-2A.

Open-wire feed line is a third type of parallel-conductor line. It contains two wires separated by spacer rods. There is a plastic insulating rod every few inches along the feed line to maintain a uniform wire separation. The primary insulation is air. See Figure 9-2B. Often called **ladder line**, this type usually has a characteristic impedance between 450 and 600 ohms. The conductors can be bare wire, or they might be insulated with plastic. Ladder line can handle much higher power than twin lead.

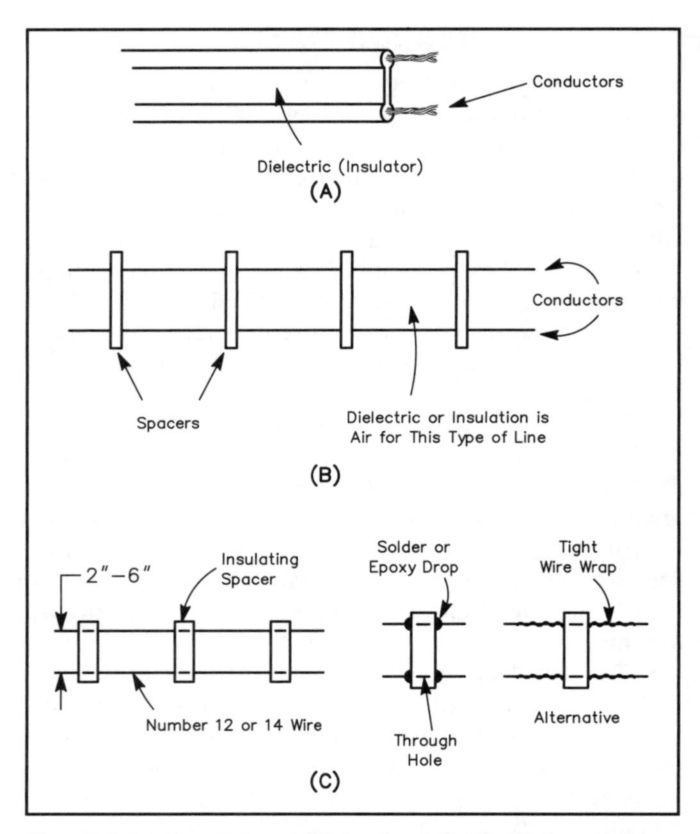

Figure 9-2—Parallel-conductor feed lines. At A, common 300-ohm TV twin lead. B shows a parallel-conductor feed line that uses air as the dielectric, with insulating spacers every few inches. Ladder line is its common name. C shows the construction of home-built ladder line.

You can make ladder line yourself. It is also available commercially, but may be difficult to find. A near equivalent is a cross between twin lead and ladder line. The wires are encased along the edges of a plastic ribbon, but the ribbon is not solid. Instead, it has punched rectangular holes along its length, leaving air as the primary insulation material. The common variety of this line has a 450-ohm impedance. For the same conductor spacing, this line has slightly more loss than ladder line. For most amateur work, the difference is negligible, however.

Parallel-conductor lines have some disadvantages. For example, they cannot be coiled or tied to metal drain pipes and gutters without adverse effects. This makes them more difficult to install properly. Another drawback is their characteristic impedance of 300 ohms or higher. You will need an **impedance-matching device** if you use any type of parallel-conductor feed line because it cannot be connected directly to most transmitters. Connect such a device between your transmitter and your feed line. If you use a matching device, even inexpensive TV twin lead can be used as your feed line. The need for a matching device is the main reason parallel-conductor lines are not used on VHF/UHF. Matching devices are more difficult to build for these frequencies, so most amateurs use coax instead.

The major advantage of ladder line is its very low loss. This means that for the same feed-line length, more of your transmitter power will get to your antenna. It can also tolerate a higher **standing-wave ratio (SWR)** than coax.

It is even possible to use a single wire from your rig to the antenna. We call this a *single-wire feed line*. With this feed arrangement, the feed line often becomes a part of the antenna. It can radiate outdoors, of course, but it can also radiate right in the shack. The impedance of this type of antenna system is seldom within the tuning range of most transmitters. You will need an **impedance-matching device** to connect between the rig and the feed line.

Many new amateurs worry needlessly about feed-line length. If you use coaxial cable between a rig and a dipole or vertical antenna, the cable can be any reasonable length. A good length to start with is simply the distance between the rig and the antenna! The feed line cannot be any shorter, and any extra length is probably unnecessary.

[Now study the questions in Chapter 13 numbered N9C01 through N9C10. Review this section if you have difficulty with any of the questions.]

[If you are studying for the Technician exam, you should also study question T9B13 and T9B14 in Chapter 14. Review this section if you have difficulty.]

IMPEDANCE-MATCHING DEVICES

Your transmitter won't operate very well if connected to a mismatched feed line. Let's say you want to use parallel-conductor feed lines on your dipoles. Your transmitter probably has an output circuit designed for a 50-ohm load. But that's not what the load will be with this antenna system. An **impedance-matching device** or network will provide the proper impedance correction. For some mismatches, a suitable network might contain only an inductor and a capacitor.

A *Transmatch* is a special type of matching device. Transmatches contain variable matching components (inductors and capacitors), and often a band switch. They offer the flexibility of matching a wide range of impedances over a wide frequency range. With a Transmatch, it is possible to use one antenna on several bands. For example, you might use a center-fed wire antenna with inexpensive 300-ohm twin lead. Each band will use its own Transmatch settings: one combination for 80, one for 40, one for 15, and one for 10 meters.

Connect the Transmatch between the antenna and the **SWR meter**, as Figure 9-3 shows. (An SWR meter measures the impedance mismatch on the feed line between the two pieces of equipment it is connected to.) Adjust the controls on your Transmatch for minimum SWR. Don't worry if you can't achieve a perfect match (1:1). Anything lower than 2:1 will work just fine.

Figure 9-4A shows the schematic diagram for a versatile Transmatch circuit. Part B shows a homemade Transmatch built from this circuit.

Baluns

A center-fed wire with open ends, such as a dipole, is a *balanced* antenna. In a balanced center-fed antenna, the current flowing into one half of the antenna is equal to the current in the other half. The two currents are also opposite in *phase*. (*Phase* refers to the relative positions of two points on a wave, or on two different waves at a particular instant of time. If the two currents are opposite in phase, one is in the positive half of the cycle when the other is in the negative half cycle.) You can think of a balanced antenna as one where neither side connects to ground. A balanced antenna is balanced with respect to ground.

If we feed the antenna with a parallel-conductor line, we keep balance throughout the system. Parallel-conductor feed line, with the same space between the two conductors, is **balanced**: Neither side connects to ground. You can connect this feed line directly to a dipole antenna.

If we feed a dipole at the center with coax, we upset the system balance. One side of the antenna connects to the coax inner, or center, conductor. The other side connects to the coax shield. The shield connects to ground, so coax is

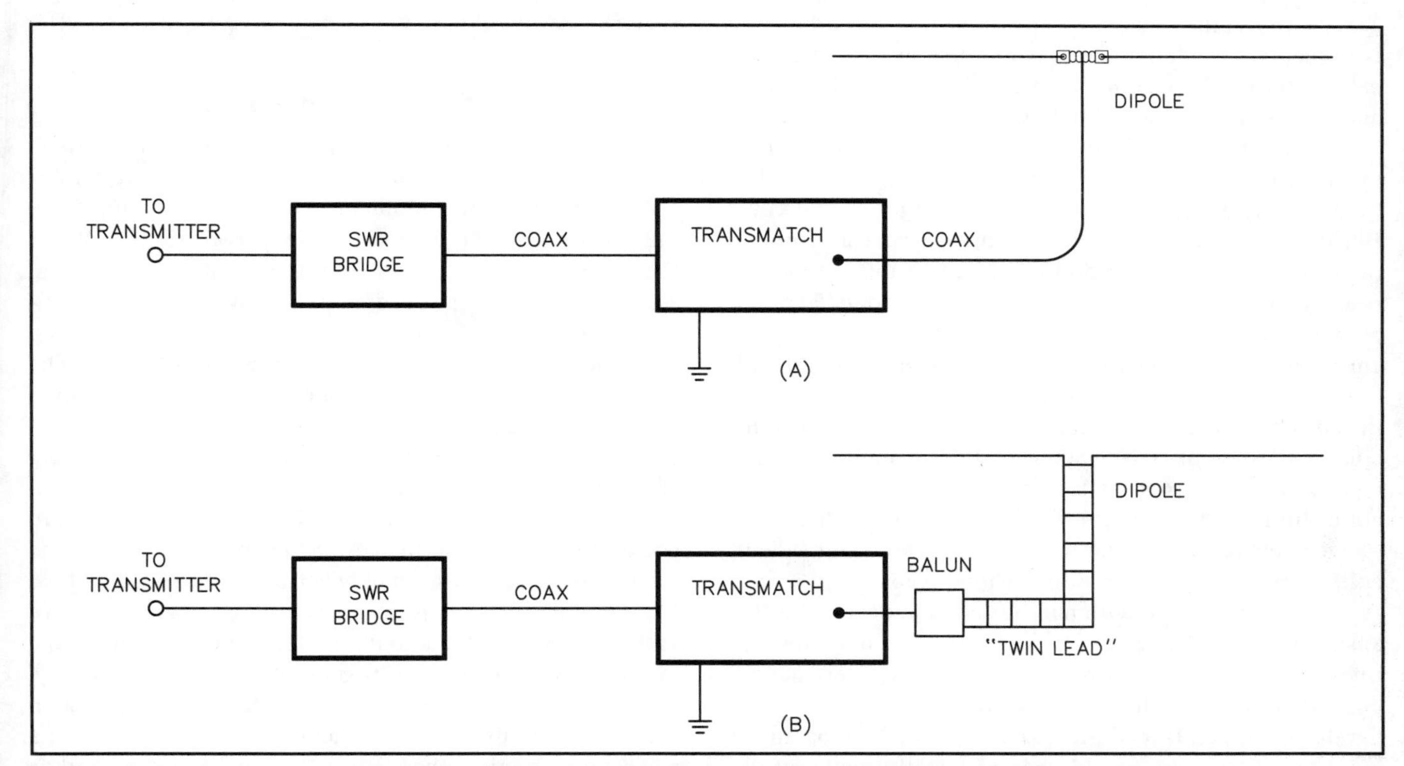

Figure 9-3—With a Transmatch you have flexibility in designing your antenna system. You can use the antenna on several bands, and the length isn't critical. You can use a dipole fed by coax (A) or twin lead (B). With a Transmatch, the dipole legs can be any length, although they should be as long as possible.

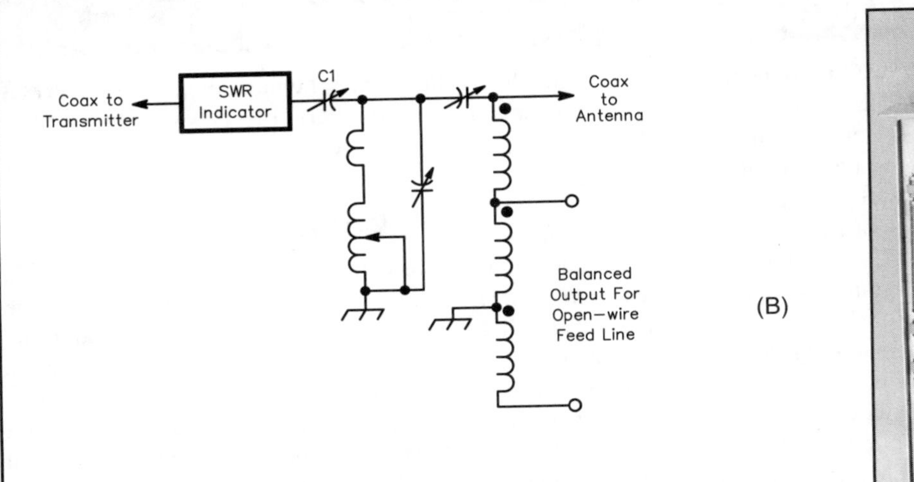

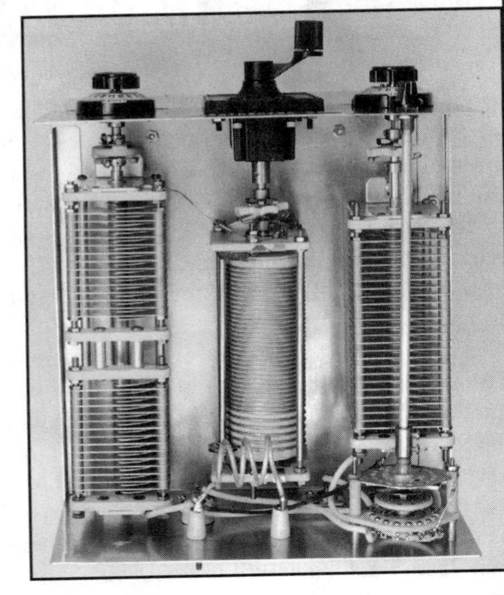

Figure 9-4—At A, the schematic diagram for a versatile Transmatch circuit. B shows a homemade Transmatch constructed from the circuit of A.

an *unbalanced line*. This unbalanced condition may allow some antenna current to flow down the outside of the coax braid from the antenna. This can lead to several antenna problems, and should be avoided.

You don't have to go searching for 72-ohm twin lead to have a balanced dipole, however. You can feed it with ordinary 300-ohm twin lead. (To do this, you'll also need a Transmatch at the transmitter end of the line.) Another way is to feed your antenna with coaxial line and use a device called a **balun**. Balun is a contraction for *bal*anced to *un*balanced. You install the balun at the antenna feed point.

Different types of baluns are available commercially, or you can make your own. A common type is a balun transformer, which uses wires wound on a toroidal core. Besides providing a balance, these baluns can transform an impedance, such as from 50 to 75 ohms. Another type is a bead balun. Several ferrite beads go over the outside of the coax, one after the other. The beads tend to choke off any RF current that might otherwise flow on the outside of the shield.

You can easily make another type of choke balun from the coax transmission line itself. At the antenna feed point, coil up 10 turns of coax into a roll about 6 inches in diameter. Tape the coax turns together. The inductance of the coiled turns tends to choke off RF currents on the shield. An older type of choke balun, made from a pair of air-wound coils, connects parallel conductor line to coax.

Feeding a dipole with parallel-conductor line doesn't ensure a balanced antenna. You must also consider how the line connects to the transmitter. Most transmitters have a coaxial-style connector. Suppose you are using 450-ohm parallel-conductor feed line. To feed such a balanced antenna system, you would need to install a balun at the transmitter end of the line (or use a special balanced matching network). Unless you use a balanced matching network, you need a balun at the transmitter end of the line if you are using any type of parallel-conductor feed line.

[Study questions in Chapter 13 numbered N9C11 and N9C12. Review this section if you have difficulty answering these questions.]

[**If you are studying for the Technician exam**, also turn to Chapter 14 and study questions T9B15 through T9B18. Review this section if you have any difficulty.]

Standing-Wave Ratio (SWR)

If an antenna system does not match the characteristic impedance of the transmitter, some of the transmitter energy is reflected from the antenna. The power traveling from the transmitter to the antenna is called *forward power*. When that power reaches the antenna in an unmatched system, some of the power is reflected back down the feed line toward the transmitter. Some of the power is also radiated from the antenna, which is what you want to happen. The power that returns to the transmitter from the antenna is called *reflected power*.

The forward power and the reflected power passing each other on the feed line cause voltage standing waves on the line. When this happens, the RF voltage and current are not uniform along the line. An **SWR meter** measures the relative impedance match between an antenna and its feed line. It does this by measuring voltage **standing-wave ratio**, the ratio of the maximum voltage on the line to the minimum voltage. (These two points will always be $^1/_4$ λ apart.) Lower SWR values mean a better impedance match exists between the transmitter and the antenna system. If a perfect match exists, the SWR is 1:1. Your SWR meter thus gives a relative measure of how well the antenna system impedance matches that of the transmitter.

SWR: What Does It Mean?

You already know that SWR is defined as the ratio of the maximum voltage to the minimum voltage in the standing wave:

$$SWR = \frac{E_{max}}{E_{min}}$$

An SWR of 1:1 means you have no reflected power. The transmission line is said to be "flat." If the load is completely resistive (no reactance), then the SWR can be calculated. Divide the line characteristic impedance by the load resistance, or vice versa. Use whichever gives a value greater than one:

$$SWR = \frac{Z_0}{R} \text{ or } SWR = \frac{R}{Z_0}$$

where:

Z_0 = characteristic impedance of the transmission line
R = load resistance (not reactance)

For example, if you feed a 100-ohm antenna with 50-ohm transmission line, the SWR is 100/50 or 2:1. Similarly, if the impedance of the antenna is 25 ohms the SWR is 50/25 or 2:1.

When a high SWR exists, losses in the feed line are increased. This is because of the multiple reflections from the antenna and transmitter. Each time the transmitted power has to travel up and down the feed line, a little more energy is lost as heat.

This effect is not so great as some people believe, however. Some line losses are less than 2 dB (such as for 100 feet of RG-213 or RG-58 cables up to about 30 MHz).

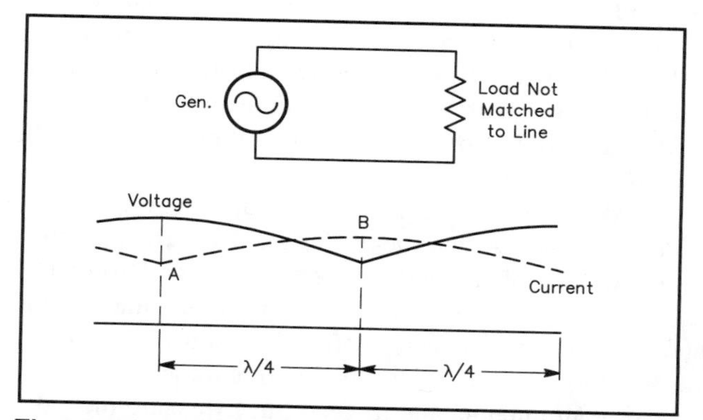

The standing-wave ratio is the ratio of the voltage amplitude at point A to the voltage amplitude at point B, or the ratio of the current amplitude at point B to the current amplitude of point A.

The SWR would have to be greater than 3:1 to add an extra decibel of loss because of the SWR. (A high SWR will cause the output power to drop drastically with many solid-state transceivers. The drop in power is caused by internal circuits that sense the high SWR and automatically reduce output power to protect the transceiver.)

A transmission line should be terminated in a resistance equal to its characteristic impedance. Then maximum power is delivered to the antenna, and transmission line losses are minimized. This would be an ideal condition. Such a perfect match is seldom realized in a practical antenna system.

Modern transmitters are designed to match 52-ohm coaxial lines and antennas. Most commercial antennas are designed to have nearly the same characteristic impedance *when properly adjusted*. So, if your SWR is higher than 2:1, it means your antenna is not adjusted properly for the frequency you are using. We adjust the antenna for minimum SWR (not always 1:1) somewhere in the middle of the band of interest. On other frequencies the SWR may be higher. If the antenna is assembled properly, an SWR of 2:1 or less is probably all right.

If you are using a matching device, you can probably adjust it so the SWR meter reads 1:1. The SWR on the transmission line between the tuner and the antenna will, however, be different. A matching device can cover up the mismatch, but it does not eliminate it.

SWR Meters

The most common **SWR meter** application is tuning an antenna to resonate on the frequency you want to use. (This discussion applies if you connect the feed line directly to the transmitter output, with no Transmatch.)

An SWR reading of 2:1 or less is quite acceptable. A reading of 4:1 or more is unacceptable. This means there is a serious impedance mismatch between your antenna and your feed line.

To use the SWR meter, you transmit through it. (You must have a license to operate a transmitter! What if your license has not arrived by the time you are ready to test your antenna? Just invite a licensed ham over to operate the transmitter.)

How you measure the SWR depends on your type of meter. Some SWR meters have a SENSITIVITY control and a FORWARD-REFLECTED switch. If so, the meter scale usually gives you a direct SWR reading. To use the meter, first put the switch in the FORWARD position. Then adjust the SENSITIVITY control and the transmitter power output until the meter reads full scale. Some meters have a mark on the meter face labeled SET or CAL. The meter pointer should rest on this mark. Next, set the selector switch to the REFLECTED position. Do this without readjusting the transmitter power or the meter SENSITIVITY control. Now the meter pointer shows you the SWR value.

Most SWR meters are designed for operation on the high-frequency (HF) bands. You may be able to use your SWR meter on the VHF bands if you are able to adjust the meter for a full-scale reading in the SET or CAL position. The readings may not be as accurate as the readings you obtain on the HF bands, but it may give an indication of the impedance match to your VHF antenna.

You can also measure SWR with a wattmeter, which measures RF power in watts. Your wattmeter may have meters to read both forward and reflected power. If not, it should have a switch or another way to change from forward to reflected power readings. To compute the SWR with a wattmeter, first note both the forward and reflected power for a given transmitter setting. Then consult a graph provided with the meter to find the corresponding SWR.

Most RF wattmeters are accurate over a limited frequency range. A wattmeter designed for HF operation (3 to 30 MHz) probably will not be accurate at VHF or UHF.

Find the **resonant frequency** of an antenna by connecting the meter between the feed line and your antenna. This technique will measure the relative impedance match between your antenna and its feed line. Measure the SWR at different frequencies across the band. Ideally, you will measure the lowest SWR at the center of the band, with higher readings at each end.

Sometimes it isn't practical to put your SWR meter at the antenna feed point, between the feed line and the antenna. Most hams just put the meter in their shack, at the transmitter, and make measurements there. You should realize that this is a compromise, however. You are measuring the relative impedance match between your transmitter and the *antenna system*, which includes the feed line.

If you are using a Transmatch with your antenna system, you should place the SWR meter between the transmitter and the Transmatch. The meter then indicates when you have adjusted the Transmatch to provide the best impedance match to your transmitter.

Sometimes your antenna may resonate far off frequency. See Figure 9-5. In this situation you will not get a "dip" in SWR readings with frequency. Instead, your readings will increase as you change frequency from one end of the band to the other. For example, you might read 2.5:1 at the low-frequency band edge, and the reading might increase across the band to 5:1 at the high-frequency end. This means antenna resonance is closer to the low-frequency end of the band than the high. It also means that resonance is below the low-frequency band edge. For a dipole or vertical antenna, this condition exists when the antenna is too long. Trimming the length will correct the problem.

Suppose the readings were 5:1 at the low-frequency end of the band and decreased across the band to 2.5:1 at the high end. Here, the antenna is too short. Adding to the antenna length will correct this problem. Adjusting the antenna length for resonance in this way is what we call *tuning the antenna*.

This method works for dipoles or vertical antennas. It does not show antenna resonance if you have a matching device between the SWR meter and the antenna. Nor does it show resonance for antenna systems that include a matching device at the antenna. An SWR meter *will* show when you have adjusted the matching device properly, however. Use the settings that give you the lowest SWR at your preferred operating frequency.

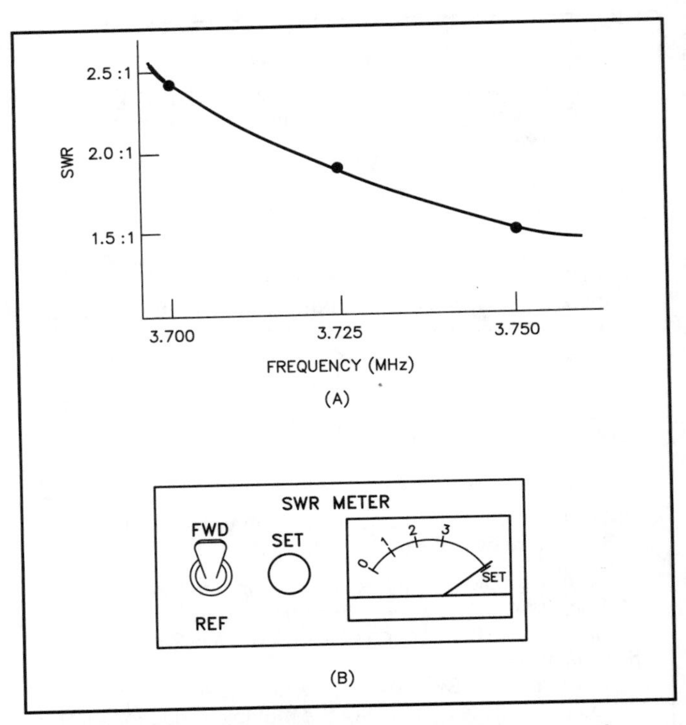

Figure 9-5—To determine if your antenna is cut to the right length, measure the SWR at different points in the band. Plot these values and draw a curve. Use the curve to see if the antenna is too long or too short. Adjusting the length will bring the lowest SWR to the desired frequency. Here the SWR is higher at the low end of the band. This antenna needs to be longer.

Finding Antenna Problems with an SWR Meter

Sometimes problems occur when you first install an antenna, or only after the weather batters your antenna for weeks, months or years.

It's handy to have an **SWR meter** or power meter to help diagnose antenna problems. This section tells how to interpret SWR meter readings to solve specific problems. This information applies to any type of antenna. We assume the antenna you're using normally provides a good match to your feed line at the measurement frequency.

One common problem is a loose connection where the feed line from your station attaches to the antenna wire. Splices or joints are another possible failure point. Your SWR meter will tell if the problem you're experiencing is a poor connection somewhere in the antenna. Observe the SWR reading. It should remain constant. If it is erratic, jumping markedly, chances are you have a loose connection. This problem is very easy to see on windy days.

If your SWR reading is unusually high, greater than 10:1 or so, you probably have a worse problem. *Caution:* Do not operate your transmitter with a very high SWR any longer than it takes to read the SWR! The problem could be an open connection or a short circuit. The most likely failure point is at the antenna feed point. The problem might be at the connector attaching your feed line to your transmitter. Carefully check your connections and your feed line for damage. You can also get unusually high SWR readings if the antenna is far from the correct length. This would happen if you try to operate your antenna on the wrong amateur band!

Most hams leave an SWR meter in the line all the time. Any sudden changes in the SWR mean you have a problem, such as a broken wire or bad connection.

[Now turn to Chapter 13 and study questions N4C01 through N4C11. Review this section if you have difficulty with any of these questions.]

[**If you are preparing for the Technician exam**, also study questions T9B09 through T9B12 in Chapter 14. Review this section if these questions give you any difficulty.]

PRACTICAL ANTENNAS

Hams use many different kinds of antennas. There is no one best kind. Beginners usually prefer simpler, less expensive types. Some hams with more experience have antenna systems that cost thousands of dollars. Others have antennas that use several acres of property!

The Half-Wave Dipole Antenna

Probably the most common amateur antenna is a wire cut to $\frac{1}{2}\lambda$ at the operating frequency. The feed line attaches across an insulator at the center of the wire. This is the **half-wave dipole**. We often refer to an antenna like this as a **dipole antenna**. (*Di* means two, so a dipole has two equal parts. A dipole could be a length other than $\frac{1}{2}\lambda$.) The total length of a half-wavelength dipole is $\frac{1}{2}\lambda$. The feed line connects to the center. This means that each side of this dipole is $\frac{1}{4}$-λ long.

Use Equation 9-2 to find the total length of a $\frac{1}{2}$-λ dipole for a specific frequency. Notice that the frequency is given in megahertz and the antenna length is in feet for this equation.

$$\frac{1}{2}\lambda \text{ (in feet)} = \frac{468}{\text{f (in MHz)}} \qquad \text{(Eq 9-2)}$$

Equation 9-2 gives values that are less than $\frac{1}{2}$ the free-space wavelength calculated by Equation 9-1. This equation accounts for various effects on the length of a real antenna. Equation 9-2 gives us the following approximate lengths for $\frac{1}{2}$-λ dipoles.

Wavelength	*Frequency*	*Length*
80 meters	3.725 MHz	125.6 feet
40 meters	7.125 MHz	66 feet
15 meters	21.125 MHz	22 feet
10 meters	28.150 MHz	16.6 feet
10 meters	28.475 MHz	16.4 feet
2 meters	146.0 MHz	3.25 feet
1.25 meters	223 MHz	2.1 feet = 25 inches

Equation 9-2 is only accurate for frequencies up to around 30 MHz. One reason for this frequency limitation is that the element diameter is a larger percentage of the wavelength at VHF and higher frequencies. Another reason is because of effects commonly known as *end effects*. So the values given here for 2 and 1.25 meters are only for comparison purposes. These lengths might serve as a starting point for building antennas for 2 meters or 1.25 meters, but they will probably be too long.

Figure 9-6A shows the construction of a basic $\frac{1}{2}$-λ dipole antenna. Parts B through D show enlarged views of how to attach the insulators. You can use just about any kind of copper or copper-clad steel wire for your dipole. Most hardware or electrical supply stores carry suitable wire.

House wire and stranded wire will stretch with time, so a heavy gauge copper-clad steel wire is best. This wire consists of a copper jacket over a steel core. Such construction provides the strength of steel combined with the excellent conducting properties of copper. You can sometimes find copper-clad steel wire at a radio store. This wire is used for electric fences to keep farm animals in their place, so another place to try is a farm supply store.

Remember, you want a good conductor for the antenna, but the wire must also be strong. The wire must support itself *and* the weight of the feed line connected at the center.

We use wire gauge to rate wire size. Larger gauge numbers represent smaller wire diameters. Conversely, smaller gauge numbers represent larger wire diameters. Although you can make a dipole antenna from almost any size wire, 12 or 14 gauge is usually best. Smaller-diameter wires may stretch or break easily.

Cut your dipole according to the dimension found by Equation 9-2, but leave a little extra length to wrap the ends around the insulators. You'll need a feed line to connect it to your transmitter. For the reasons mentioned earlier, the most popular feed line for use with dipole antennas is coaxial cable. When you shop for coax, look for some with a heavy braided shield. If possible, get good quality cable that has at least 95 percent shielding. If you

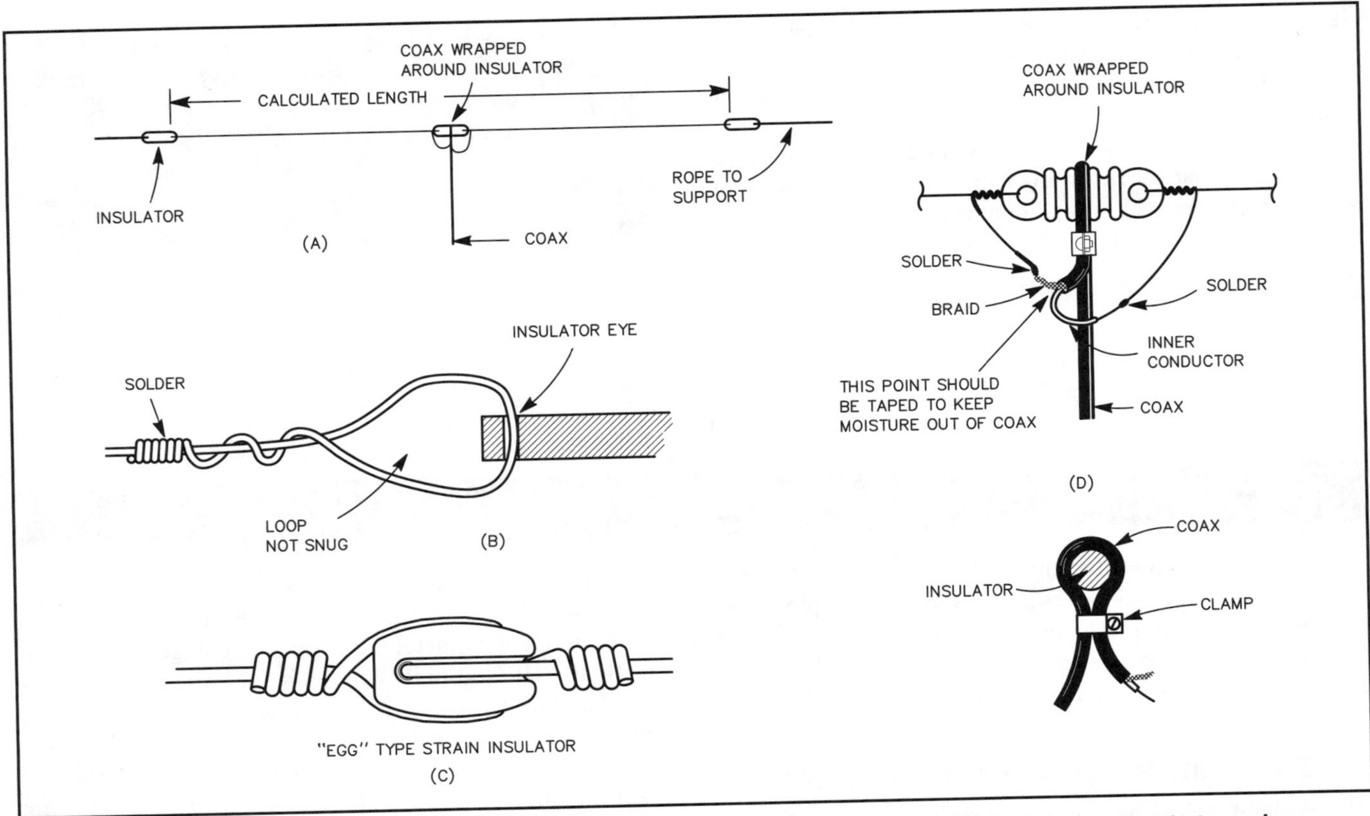

Figure 9-6—Simple half-wave dipole antenna construction. B and C show how to connect the wire ends to various insulator types. D shows the feed-line connection at the center.

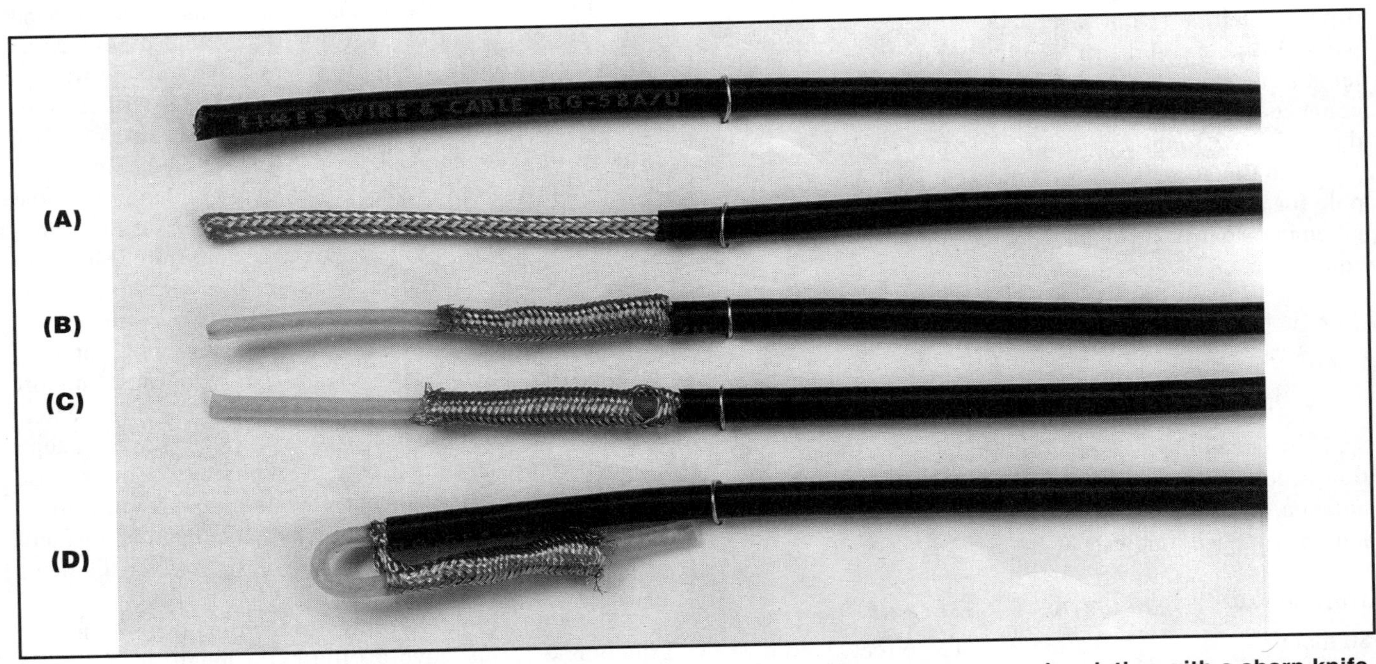

Figure 9-7—Preparing coaxial cable for connection to antenna wire. A—Remove the outer insulation with a sharp knife or wire stripper. If you nick the braid, start over. B—Push the braid in accordion fashion against the outer jacket. C—Spread the shield strands at the point where the outer insulation ends. D—Fish the center conductor through the opening in the braid. Now strip the center conductor insulation back far enough to make the connection and tin (flow solder onto) both center conductor and shield. Be careful not to use too much solder, which will make the conductors inflexible. Also be careful not to apply too much heat, or you will melt the insulation. A pair of pliers used as a heat sink will help. The outer jacket removed in step A can be slipped over the braid as an insulator, if necessary. Be sure to slide it onto the braid before soldering the leads to the antenna wires.

stick with name brand cable, you'll get a good quality feed line. Figure 9-7 shows the steps required to prepare the cable end for attachment to the antenna wires at the center insulator.

The final items you'll need for your dipole are three insulators. You can purchase them from your local radio or hardware store (Figure 9-8). You can also make your own insulators from plastic or Teflon blocks. See Figure 9-9. One insulator goes on each end and another holds the two wires together in the center. Figure 9-10 shows some examples of how the feed line can attach to the antenna wires at a center insulator.

Dipole antennas send radio energy best in a direction that is 90° to the antenna wire. For example, suppose you install a dipole antenna so the ends of the wire run in an east/west direction. Assuming it was well off the ground (preferably $1/2$ λ high), this antenna would send stronger signals in north and south directions. A dipole also sends radio energy straight up and straight down. Of course the dipole also sends some energy in directions off the ends of the wire, but these signals won't be as strong. So you will be able to contact stations to the east and west with this antenna, but you may find that the signals are stronger with stations to the north and south.

[Now turn to Chapter 13 and study questions N4B06, N4B07 and questions N9A01, N9A03, N9A04 and N9A07. Also study question N9B10. Review this section if you need to.]

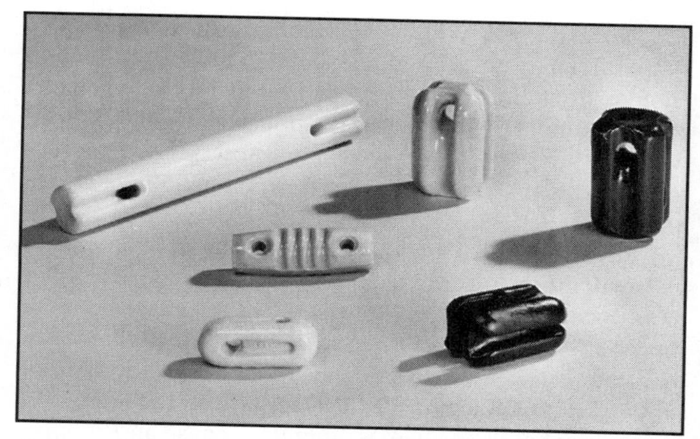

Figure 9-8—Various commercially made antenna insulators.

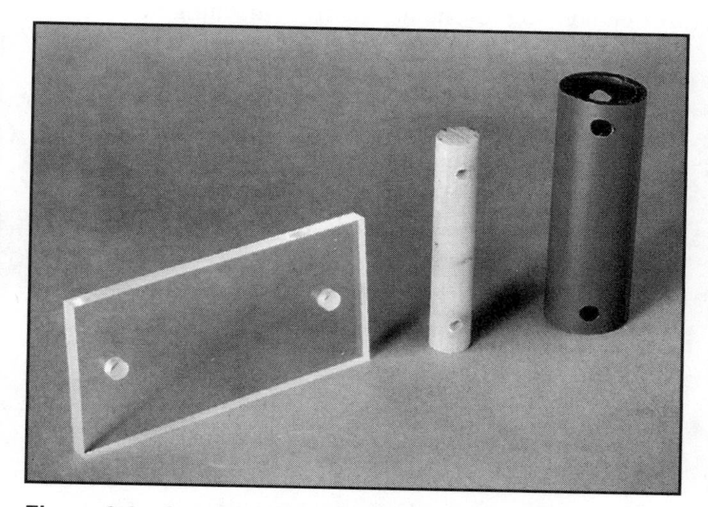

Figure 9-9—Some ideas for homemade antenna insulators.

Antenna Location

Once you have assembled your dipole, find a good place to put it. *Never* put your antenna or feed line under, or over the top of electrical power lines. If they ever come into contact with your antenna, you could be electrocuted. Avoid running your antenna parallel to power lines that come close to your station. Otherwise you may receive unwanted electrical noise. Sometimes power-line noise can cover up all but the strongest signals your receiver hears. You'll also want to avoid running your antenna too close to metal objects. These could be rain gutters, metal beams, metal siding, or even electrical wiring in the attic of your house. Metal objects tend to shield your antenna, reducing its capability.

The key to good dipole operation is height. How high? One wavelength above ground is good, and this ranges from about 35 feet on 10 meters to about 240 feet on 80 meters. On the 2-meter band, one wavelength is only about 7 feet. You should still try to install the antenna higher, to get it clear of buildings and trees. Of course very few people can get their antennas 240 feet in the air, so 40 to 60 feet is a good average height for an 80-meter dipole. Don't despair if you can get your antenna up only 20 feet or so, though. Low antennas can work well. Generally, the higher above ground and surrounding objects you can get your antenna, the greater the success you'll have. You'll find this to be true even if you can get only part of your antenna up high.

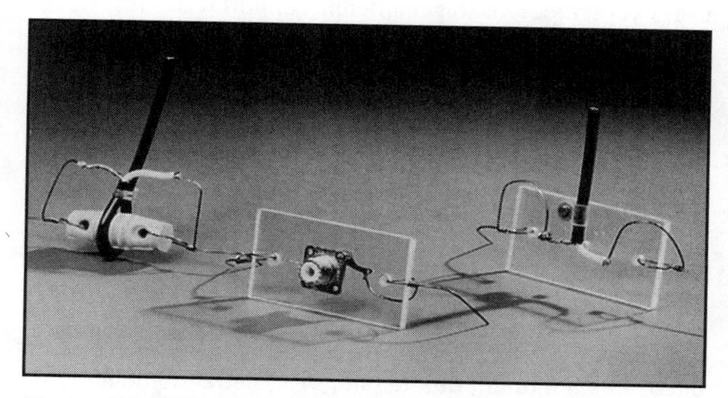

Figure 9-10—Some dipole center insulators have connectors for easy feed-line removal. Others have a direct solder connection to the feed line.

Normally you will support the dipole at both ends. The supports can be trees, buildings, poles or anything else high enough. Sometimes, however, there is just no way you can put your dipole high in the air at both ends. If you're faced with this problem, you have two reasonably good alternatives. You can support your dipole in the middle or at one end.

If you choose to support the antenna in the middle, both ends will droop toward the ground. This antenna, known as an *inverted-V dipole*, works best when the angle between the wires is no less than 90°. See Figure 9-11. If you use an inverted-V dipole, make sure the ends are high enough that no one can touch them. When you transmit, the high voltages present at the ends of a dipole can cause an *RF burn*. Yes, radio energy can burn your skin.

If you support your antenna at only one end, you'll have what is known as a *sloper*. This antenna also works well. As with the inverted-V dipole, be sure the low end is high enough to prevent anyone from touching it.

If you don't have the room to install a dipole in the standard form, don't be afraid to experiment a little. You can get away with bending the ends to fit your property, or even making a horizontal V-shaped antenna. Many hams have enjoyed countless hours of successful operating with antennas bent in a variety of shapes and angles.

On the 6 and 2-meter bands, dipoles for FM or packet operation work much better if they are installed vertically. Now you need only one support. The coax should come away from the antenna at a right angle for as far as possible, so it doesn't interfere with the radiation from the antenna. Dipoles are not often used on frequencies above the 2-meter band, as other, simpler, antennas work better. We describe one such antenna later in this chapter.

Antenna Installation

After you've built your antenna and chosen its location, how do you get it up? There are many schools of thought on putting up antennas. Can you support at least one end of your antenna on a mast, tower, building or in an easily climbed tree? If so, you have solved some of your problems. Unfortunately, this is not always the case. Hams use several methods to get antenna support ropes into trees. Most methods involve a weight attached to a rope or line. You might be able to tie a rope around a rock and throw it over the intended support. This method works for low antennas. Even a major league pitcher, however, would have

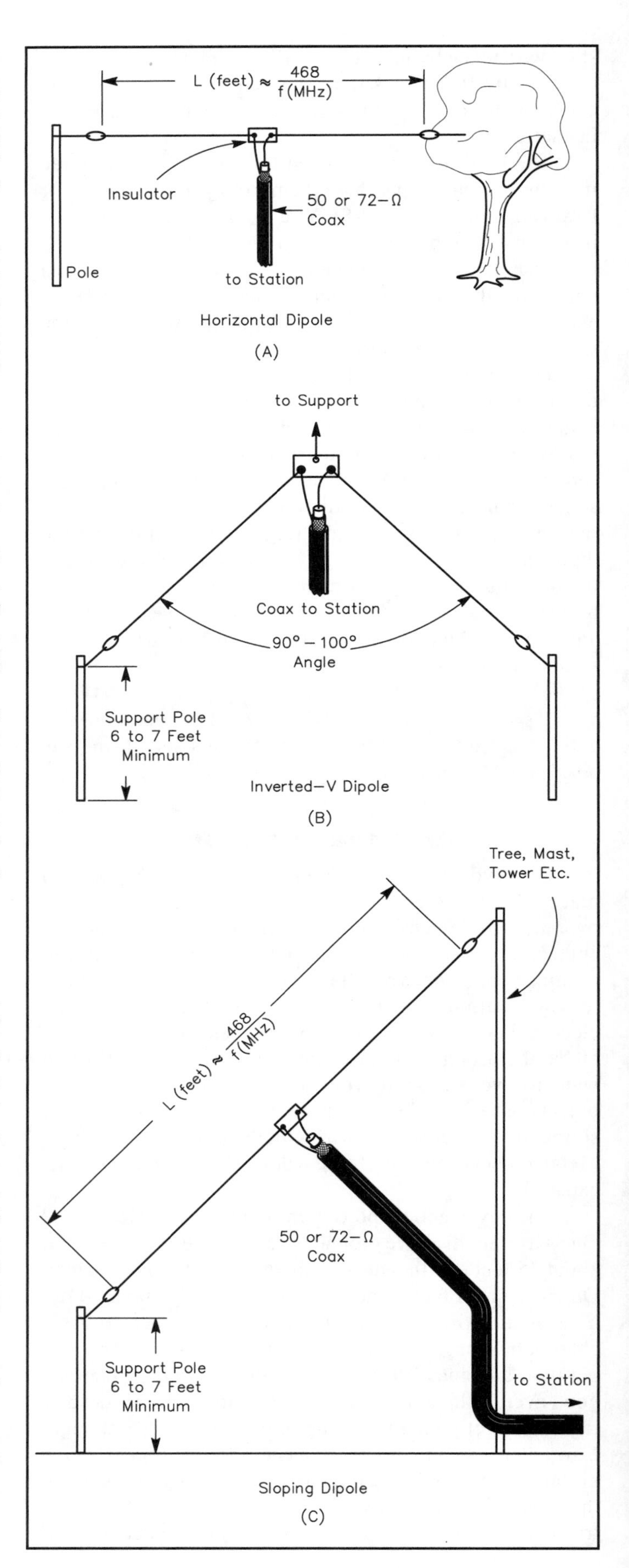

Figure 9-11—These are simple but effective wire antennas. A shows a horizontal dipole. The legs can be drooped to form an inverted-V dipole, as at B. C illustrates a sloping dipole (sloper). The feed line should come away from the sloper at a 90° angle for best results. If the supporting mast is metal, the antenna will have some directivity in the direction of the slope.

trouble getting an antenna much higher than 40 feet with this method.

A better method is to use a bow and arrow, a fishing rod or even a slingshot to launch the weight and rope. See Figures 9-12 and 9-13. You'll find that strong, lightweight fishing line is the best line to attach to the weight. (Lead fishing weights are a good choice.) Regular rope is too heavy to shoot any great distance. When you have successfully cleared the supporting tree, remove the weight. Then tie the support rope to the fishing line and reel it in.

If your first attempt doesn't go over the limb you were hoping for, try again. Don't just reel in the line, however. Let the weight down to the ground first and take it off the line. Then you can reel the line in without getting the weight tangled in the branches.

You can put antenna supports in trees 120 feet and higher with this method. As with any type of marksmanship, make sure all is clear downrange before shooting. Your neighbors will not appreciate stray arrows, sinkers or rocks falling in their yards! Be sure the line is strong enough to withstand the shock of shooting the weight. Always check the bow and arrow or slingshot to ensure they are in good working order.

When your support ropes are in place, attach them to the ends of the dipole and haul it up. Pull the dipole reasonably tight, but not so tight that it is under a lot of strain. Tie the ends off so they are out of reach of passersby. Be sure to leave enough rope so you can let the dipole down temporarily if necessary. Dacron rope is resistant to the sun's ultraviolet radiation and other weather effects, and is a good choice for an antenna-support rope. Nylon is strong, but slowly deteriorates in sunlight. Inexpensive polypropylene rope is a poor choice because it disintegrates rapidly when exposed to sunlight and weather.

Just one more step and your antenna installation is complete. After routing the coaxial cable to your station, cut it to length and install the proper connector for your rig. Usually this connector will be a PL-259, sometimes called a UHF connector. Figure 9-14 shows how to attach one of these fittings to RG-8 or RG-11 cable. Follow the step-by-step instructions exactly as illustrated and you should have no trouble. Be sure to place the coupling ring on the cable *before* you install the connector body! If you are using RG-58 or RG-59 cable, use an adapter to fit the cable to the connector. Figure 9-15 illustrates the steps for installing the connector with an adapter. The PL-259 is standard for most rigs. If you require another kind of connector, consult your radio instruction manual or *The ARRL Handbook for Radio Amateurs* for installation information.

Figure 9-12—There are many ways to get an antenna support rope into a tree. These hams use a bow and arrow to shoot a lightweight fishing line over the desired branch. Then they attach the support rope to the fishing line and pull it up into the tree.

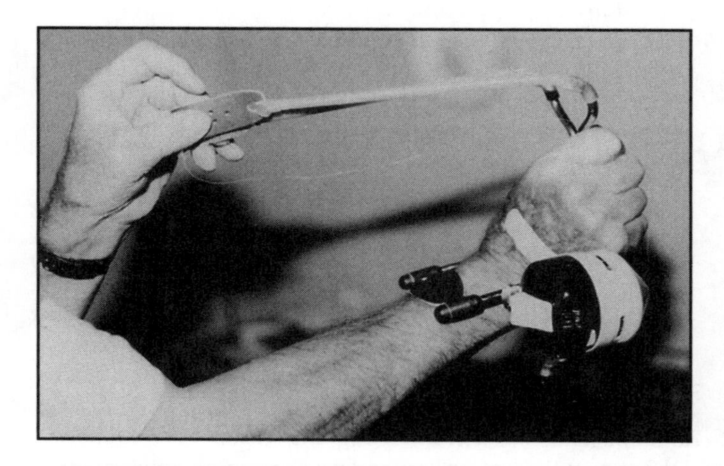

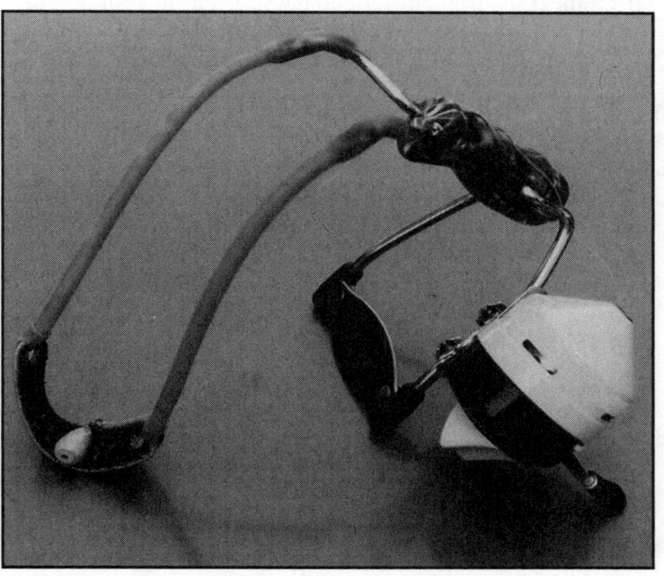

Figure 9-13—Another method for getting an antenna support into a tree. Small hose clamps attach a casting reel to the wrist bracket of a slingshot. Monofilament fishing line attached to a 1-ounce sinker is easily shot over almost any tree. Remove the sinker and rewind the line for repeated shots. When you find a suitable path through the tree, use the fishing line to pull a heavier line over the tree.

Tuning the Antenna

When you build an antenna, you cut it to the length given by an equation. This length is just a first approximation. Nearby trees, buildings or large metal objects and height above ground all affect the antenna resonant frequency. An SWR meter can help you determine if you should shorten or lengthen the antenna. The correct length provides the best impedance match for your transmitter.

The first step is to measure the SWR at the bottom, middle and top of the band. On 80 meters, for example, you would check the SWR at 3.626, 3.700 and 3.724 MHz. (A friend with a higher license class can help you check the SWR over a wider frequency range.) Graph the readings, as shown in Figure 9-16. You could be lucky—no further antenna adjustments may be necessary, depending on your transmitter.

Many tube-type transmitters include an output tuning network. They will usually operate fine with an SWR of 3:1 or less. Most solid-state transmitters (using all transistors and integrated circuits) do not include such an output tuning network. These no-tune radios begin to shut down—the power output drops off—with an SWR much higher than 1.5:1. In any event, most hams like to prune their antennas for the lowest SWR they can get at the center of the band. With a full-size dipole 30 or 40 feet high, your SWR should be less than 2:1. If you can get the SWR down to 1.5:1, great! It's not worth the time and effort to do any better than that.

If the SWR is lower at the low-frequency end of the band, your antenna is probably too long. Making the antenna *shorter* will *increase* the resonant frequency. Disconnect the transmitter and try shortening your antenna at each end. The amount to trim off depends on two things. First is which band the antenna is operating on, and, second, how much you want to change the resonant frequency. Let's say the antenna is cut for the 80-meter band. You'll probably need to cut 8 or 10 inches off each end to move the resonant frequency 50 kHz. You may have to trim only an inch or less for small frequency changes on the 10-meter band. Measure the SWR again (remember to recheck the calibration). If the SWR went down, keep shortening the antenna until the SWR at the center of the band is less than 2:1.

If the SWR is lower at the high-frequency end of the band, your antenna was probably too short to begin with. If so, you must add more wire until the SWR is acceptable. Making the antenna *longer* will *decrease* the resonant frequency. Before you solder more wire on the antenna ends, try attaching a 12-inch wire on each end. Use alligator clips, as Figure 9-17 shows. You don't need to move the insulators yet. Clip a wire on each end and again measure the SWR. Chances are the antenna will now be too long. You will need to shorten it a little at a time until the SWR is below 2:1.

Once you know how much wire you need to add, cut two pieces and solder them to the ends of the antenna. When you add wire, be sure to make a sound mechanical connection before soldering. Figure 9-18 shows how. Remember that these joints must bear the weight of the antenna and the feed line. After you solder the wire, reinstall the insulators at the antenna ends, past the solder connections.

If the SWR is very high, you may have a problem that can't be cured by simple tuning. A very high SWR may mean that your feed line is open or shorted. Perhaps a connection isn't making good electrical contact. It could also be that your antenna is touching metal. A metal mast, the

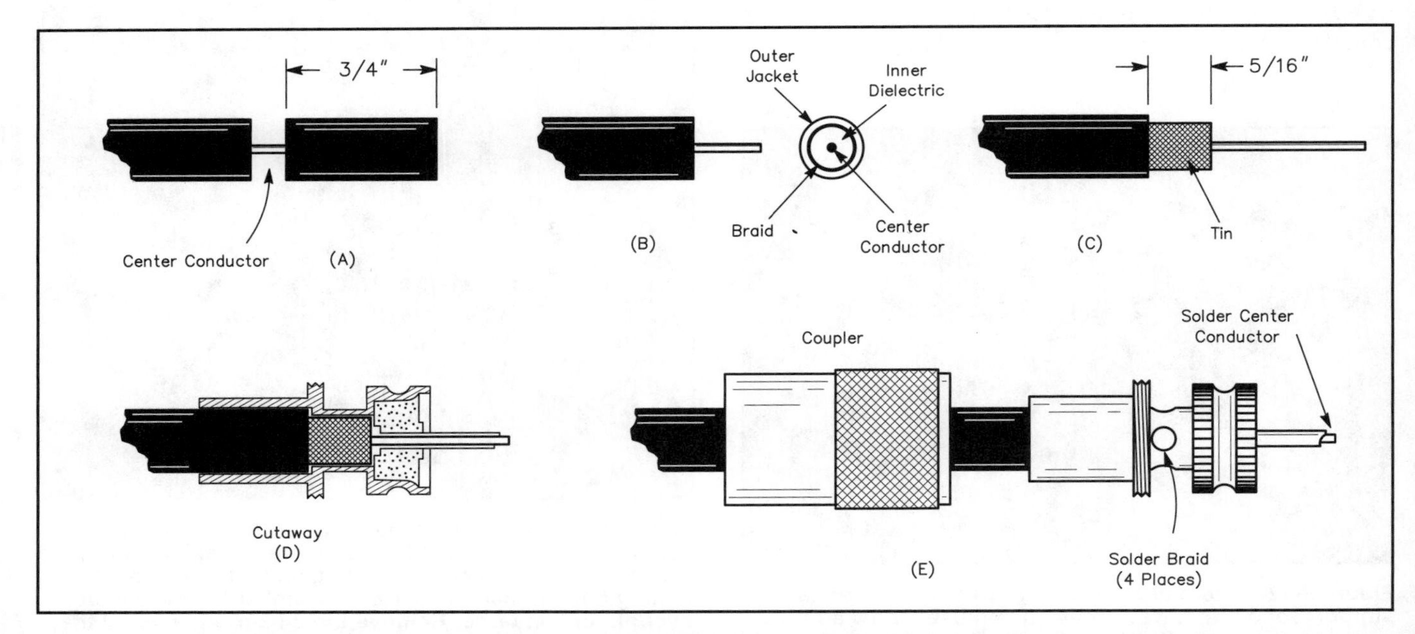

Figure 9-14—The PL-259 or UHF connector is almost universal for amateur HF work. It is also popular for equipment operating in the VHF range. Steps A through E illustrate how to install the connector properly. Despite its name, the UHF connector is rarely used on frequencies above 225 MHz.

rain gutter on your house or some other conductor would add considerable length to the antenna. If the SWR is very high, check all connections and feed lines, and be sure the antenna clears surrounding objects.

Now you have enough information to construct, install and adjust your dipole antenna. You'll need a separate dipole antenna for each band you expect to operate on. Sometimes a 40-meter dipole will also work on 15 meters, though. Check the SWR on 15 meters—you may have a two-for-one dipole!

Multiband Dipole Operation

The single-band dipole is fine if you operate on only one band. If you want to operate on more than one band, however, you could build and install a dipole for each band. What if you don't have supports for all these dipoles? Or what if you don't want to spend the money for extra coaxial cable? The popular and inexpensive multiband dipole will enable you to operate on more than one band with a single feed line.

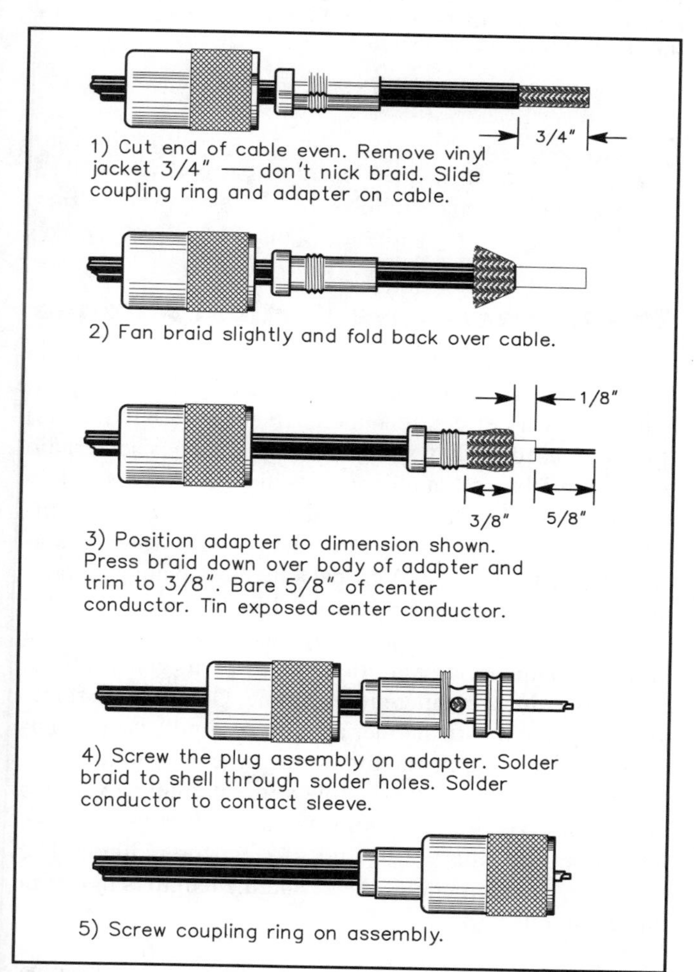

Figure 9-15—If you use RG-58 or RG-59 with a PL-259 connector, you should use an adapter, as shown here. Thanks to Amphenol Electronic Components, RF Division, Bunker Ramo Corp, for this information.

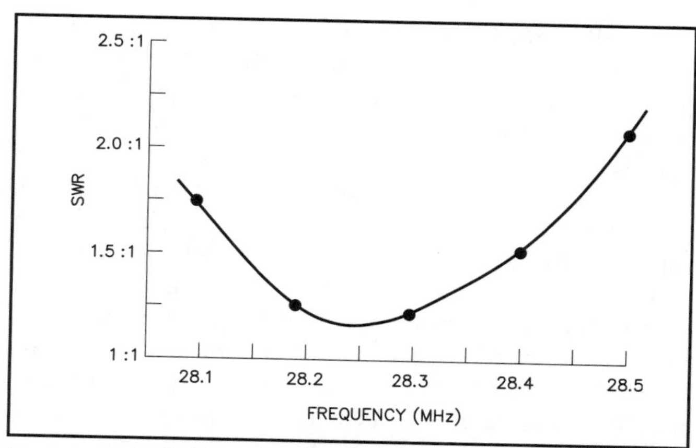

Figure 9-16—This graph shows how the SWR might vary across an amateur frequency band. The point of lowest SWR here is near the center of the band, so no further antenna-length adjustments are necessary.

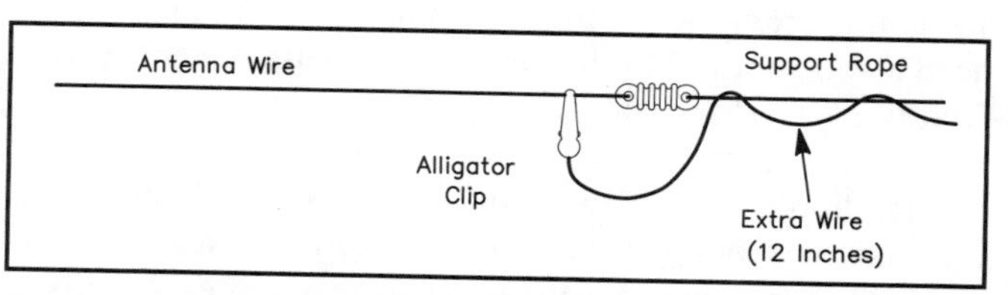

Figure 9-17—If your antenna is too short, attach an extra length of wire to each end with an alligator clip. Then shorten the extra length a little at a time until you get the correct length for the antenna. Finally, extend the length inside the insulators with a soldered connection. (See Figure 9-18.)

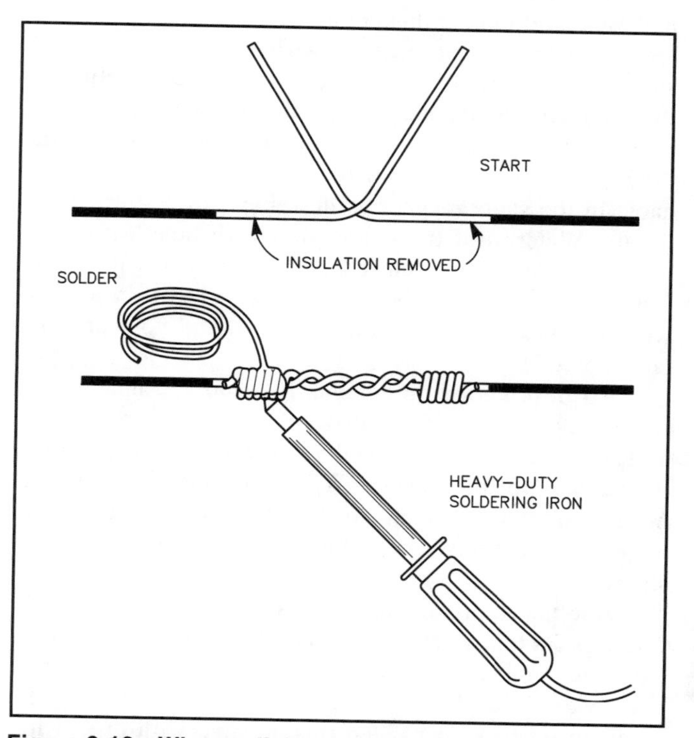

Figure 9-18—When splicing antenna wire, remember that the connection must be strong mechanically and good electrically.

A single-band dipole can be converted into a *multi-band antenna* without too much difficulty. All you need to do is connect two additional 1/4-λ wires for each additional band you want to use. Each added wire connects to the same feed line as the original dipole. The result is a single antenna system, fed with a single coaxial cable, that works on several different bands without adjustment. There is one potential problem with this antenna, though. The antenna will radiate signals on two or more bands simultaneously, so make sure your transmitter is adjusted properly. A poorly adjusted transmitter may produce harmonics of the desired output. If so, energy from your transmitter may show up on more than one band. The FCC takes a dim view of such operation!

Three-Band Dipole

You can build a three-band dipole for 80, 40 and 15 meters from ladder line. To build this antenna, you'll need a 100-foot roll of this line, three insulators and a coax feed line.

This antenna construction is similar to that for a regular dipole. Carefully remove the line from the spool and lay it on the ground. Take care to avoid twists and kinks in the wire. See Figure 9-19. At 33 feet, 6 inches from one end (X), cut *one* of the two wires. At 63 feet, 8 inches from the same end, cut the *other* wire. Remove the plastic spacers between these cuts, separating the open-wire line into two pieces. Measuring from the other end (Y), cut the shorter wire at 33 feet, 6 inches and the longer one at 63 feet, 9 inches. Figure 9-19A shows how the two antenna halves should look at this point.

Reverse the position of the two halves, as Figure 9-19B shows. Now prepare the wire ends for connection to the feed line. Sandpaper the protective coating off both wires at the ends. (X and Y identify these ends in the drawing.) Connect the wire and a piece of coaxial cable in the same manner as described for a single-band dipole. Waterproof the connection with tape. Spray the taped connection with clear lacquer or coat it with liquid rubber or silicone sealant for added protection against weather. Attach insulators to the ends, and your antenna is complete.

All information on antenna location and installation for a single-band dipole also applies to this multiband antenna. Use the same procedure described earlier to tune the antenna for the lowest SWR. Just remember that you must adjust the SWR on two bands. Adjust both the 80-meter part (the longer wire) and the 40/15-meter part (the shorter wire).

Another multiband antenna is shown in Figure 9-20. The legs for this antenna should each be 1/4 λ at the lowest frequency you want to use. In other words, if you want to operate on all bands, 80 through 10 meters, each leg should be about 63 feet long. Feed this antenna with open-wire feed line (either 300 or 450-ohm), or TV twin lead. It requires a Transmatch at the station end.

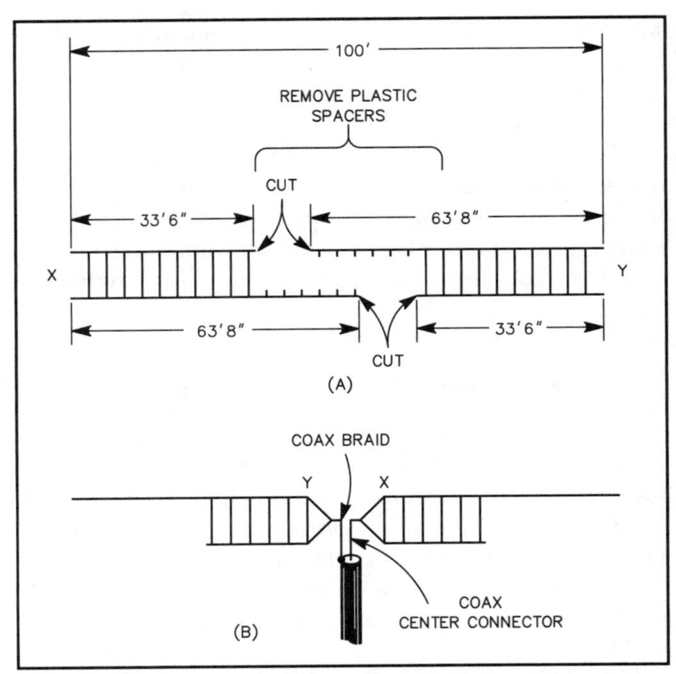

Figure 9-19—At A, cut the two lengths of ladder line as shown. Reverse the two sections to make the three-band dipole antenna, as shown at B.

[Before you go on to the next section turn to Chapter 13 and study questions N4B08, N4B09, N4B13 and questions N9A08 through N9A13. Review this section if you have difficulty with any of these questions.]

The Quarter-Wave Vertical Antenna

The **quarter-wavelength vertical antenna** is simple and popular. It requires only one element and one support, and can be very effective. On the HF bands (80-10 meters) it is often used for DX work. Vertical antennas send radio energy equally well in all compass directions. This is why we sometimes call them *nondirectional antennas* or *omni-directional antennas*. They also tend to concentrate the signals toward the horizon. Vertical antennas do not generally radiate strong signals straight up, like horizontal dipoles do. Because they concentrate signals toward the horizon, they have a *low-angle radiation pattern*. This gives vertical antennas **gain** as compared to a dipole. Gain always refers to a comparison with another antenna. A dipole is one common comparison antenna for stating antenna gain.

Figure 9-21 shows a simple vertical antenna you can make. This **vertical antenna** has a radiator that is 1/4-λ long. Use Equation 9-3 to find 1/4 λ for the radiator. The frequency is given in megahertz and the length is in feet in this equation.

$$\frac{1}{4}\lambda \text{ (in feet)} = \frac{234}{f \text{ (in MHz)}} \quad \text{(Eq 9-3)}$$

Equation 9-3 gives values that are less than 1/4 the free-space wavelength calculated by Equation 9-1. Like Equa-

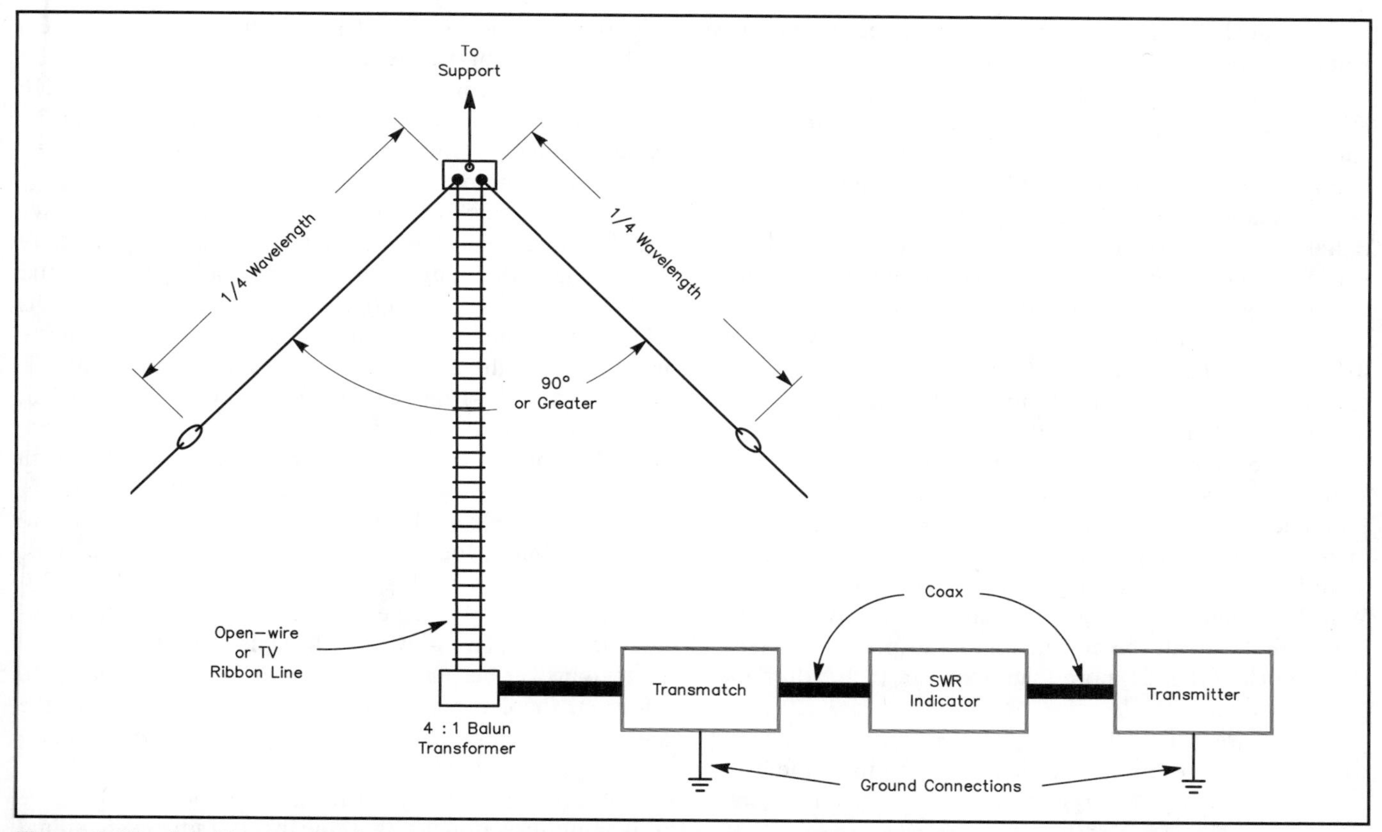

Figure 9-20—A half-wave dipole antenna can be made to work on several bands. Use twin-lead feed line and a Transmatch.

tion 9-2, this one accounts for various effects on the length of a real antenna. Equation 9-3 gives us the following approximate lengths for the radiator and each ground radial of a $^1/_4$-λ vertical.

Wavelength	*Frequency*	*Length*
80 meters	3.700 MHz	63.24 feet
40 meters	7.125 MHz	32.8 feet
15 meters	21.125 MHz	11.1 feet
10 meters	28.150 MHz	8.3 feet
10 meters	28.4 MHz	8.2 feet
6 meters	52.5 MHz	4.5 feet
2 meters	146.0 MHz	1.6 feet (19.25 inches)
1.25 meters	223.0 MHz	1.05 feet (12.6 inches)
70 cm	440.0 MHz	0.53 feet (6.4 inches)
23 cm	1282.5 MHz	0.18 feet (2.2 inches)

Equation 9-3 is only accurate for frequencies up to around 30 MHz. One reason for this frequency limitation is because the ratio of wavelength to element diameter is larger at VHF and higher frequencies. Another reason is because of effects commonly known as *end effects*. So the values given here for the VHF and UHF bands are only for comparison purposes. These lengths might serve as a starting point for building antennas for 2 meters, 1.25 meters, 70 cm or 23 cm, but they will probably be too long.

As with $^1/_4$-λ dipoles, the resonant frequency of a

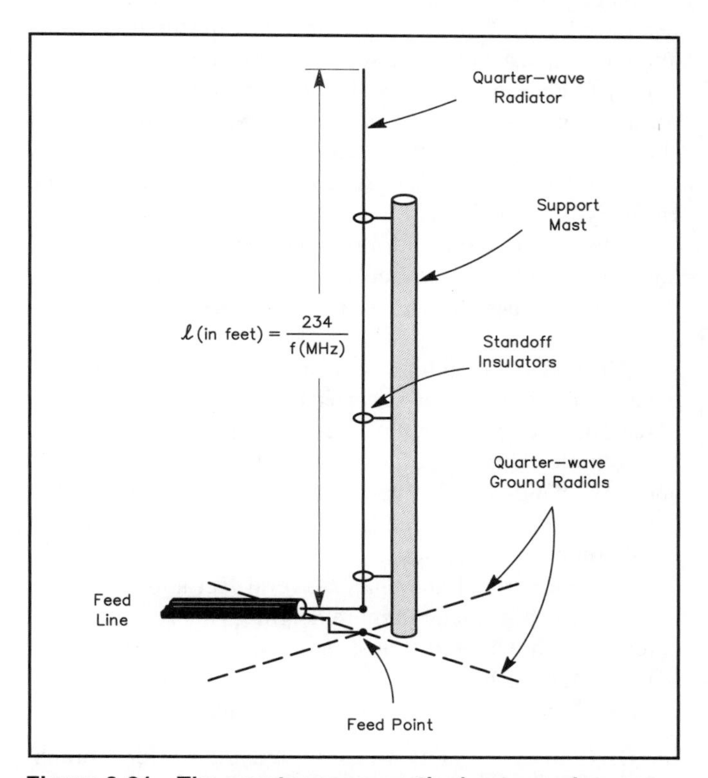

Figure 9-21—The quarter-wave vertical antenna has a center radiator and four or more radials spread out from the base. These radials form a ground plane.

$^1/_4$-λ vertical decreases as the length increases. Shorter antennas have higher resonant frequencies.

The $^1/_4$-λ vertical also has radials. For operation on 80 through 10 meters, the vertical may be at ground level, and the radials placed on the ground. The key to successful operation with a ground-mounted vertical antenna is as good radial system. The best radial system uses *many* ground radials. Lay them out like the spokes of a wheel, with the vertical at the center. Some hams have buried ground radial systems containing over 100 in-dividual wires.

Ideally, these wires would be $^1/_4$ λ long or more at the lowest operating frequency. With such a system, earth or ground losses will be negligible. When the antenna is mounted at ground level, radial length is not very critical, however. Studies show that with fewer radials you can use shorter lengths, but with a corresponding loss in antenna efficiency.[1,2] Some of your transmitter power does no more than warm the earth beneath your antenna. With 24 radials, there is no point in making them longer than about $^1/_8$ λ. With 16 radials, a length greater than 0.1 λ is unwarranted. Four radials should be considered an absolute minimum. Don't put the radials more than about an inch below the ground surface.

Compared with 120 radials of 0.4 λ, antenna efficiency with 24 radials is roughly 63%. For 16 radials, the efficiency is roughly 50%. So it pays to put in as many radials as you can.

If you place the vertical above ground, you reduce earth losses drastically. Here, the wires should be cut to $^1/_4$ λ for the band you plan to use. Above ground, you need only a few radials—two to five. If you install a multiband vertical antenna above ground, use separate ground radials for each band you plan to use. These lengths are more critical than for a ground-mounted vertical. For elevated verticals, you should have a minimum of two radials for each band. You can mount a vertical on a pipe driven into the ground, on the chimney or on a tower.

The radials at the bottom of a vertical antenna mounted above ground form a surface that acts like the ground under the antenna. These antennas are sometimes called *ground-plane antennas*.

Ground-plane antennas are popular mobile antennas because the car body can serve as the ground plane. You can place a magnetic-base whip antenna on the roof of your car, for example. This type of antenna is often used on the VHF and UHF bands for FM voice communication. Because it is a vertical antenna, it sends radio energy out equally well in all compass directions.

Once a vertical antenna is several feet above ground, there is little advantage in more height. (This assumes your antenna is above nearby obstructions.) For sky-wave signals, a height of 15 feet for the base is almost as good as 50. This isn't the case for a horizontal antenna, where height is important for working DX. Only if you can get the vertical up 2 or 3 wavelengths does the low-angle radiation begin to improve. Even then the improvement is only slight. At VHF and UHF, however, it pays to get the antenna up high. At these frequencies you want the antenna higher than even distant obstructions.

Most vertical antennas used at lower frequencies are $^1/_4$ λ long. For VHF and UHF, antennas are physically short enough that longer verticals may be used. A popular mobile antenna is a $^5/_8$-λ vertical, often called a "five-eighths whip." This antenna is popular because it may concentrate more of the radio energy toward the horizon than a $^1/_4$-λ vertical. Mounted on the roof of a car, a $^5/_8$-λ vertical may provide more **gain** than a $^1/_4$-λ vertical. Simply stated, "gain" means a concentration of transmitter power in some direction. A $^5/_8$-λ vertical concentrates the power toward the horizon. Naturally, this is the most useful direction, unless you want to talk to airplanes or satellites. At 222 MHz, a $^5/_8$-λ whip is only 28$^1/_2$ inches long.

Don't use any of the equations in this chapter to calculate the length of a $^5/_8$-λ whip. The equations won't give the correct answer because of a variety of antenna factors. In addition, there is an impedance-matching device at the antenna feed point. See *The ARRL Antenna Book* for complete construction details.

A $^5/_8$-λ vertical is great for mobile operation because it is omnidirectional. That means it radiates a signal equally well in all compass directions. This is especially useful for mobile operation because you change direction often. One minute you may be driving toward the repeater, and the next minute you may be driving away from it.

Vertical antennas that are $^1/_2$-λ long can be used without ground radials. This may sometimes be a definite advantage because it takes less wire and occupies less horizontal space. To find the size of a $^1/_2$-λ vertical antenna, double the lengths given in the table of $^1/_4$-λ verticals, above.

Commercially available vertical antennas need a coax feed line, usually with a PL-259 connector. Just as with the dipole antenna, you can use RG-8, RG-11 or RG-58 coax. The instructions that accompany the antenna should provide details for attaching both the feed line and the ground radials.

Some manufacturers offer *trap verticals*. Traps are tuned circuits that change the antenna electrical length. They allow the antenna to work on several bands, making it a *multiband antenna*. Some manufacturers even offer 20- to 30-foot-high vertical antennas that cover all HF bands. Figure 9-22 shows one such antenna.

Antennas for Hand-Held Transceivers

When you buy a new VHF hand-held transceiver, it will have a flexible rubber antenna commonly called a "rubber duck." This antenna is inexpensive, small, lightweight and difficult to break. On the other hand, it has some disadvantages: It is a compromise design that is

[1] J. O. Stanley, "Optimum Ground Systems for Vertical Antennas," *QST*, December 1976, pp 14-15.

[2] B. Edward, "Radial Systems for Ground-Mounted Vertical Antennas," *QST*, June 1985, pp 28-30.

inefficient and thus does not perform as well as larger antennas. Two better-performing antennas are the $^1/_4$ and $^5/_8$-λ telescoping types (see Figure 9-23).

[Now turn to Chapter 13 and study questions numbered N9A02, N9A05 and N9A06. Also study questions N9B08, N9B09 and N9B11. Review this section if you have difficulty with any of these questions.]

[If you are studying for the Technician exam, turn to Chapter 14 now. Study questions T9A10 through T9A12, and T9B03. Review this section if you have trouble answering any of these questions.]

Figure 9-23—The $^1/_4$-λ telescoping antenna is a good substitute for the rubber flex antenna that comes with most hand-held transceivers. *(photo courtesy Larsen Electronics)*

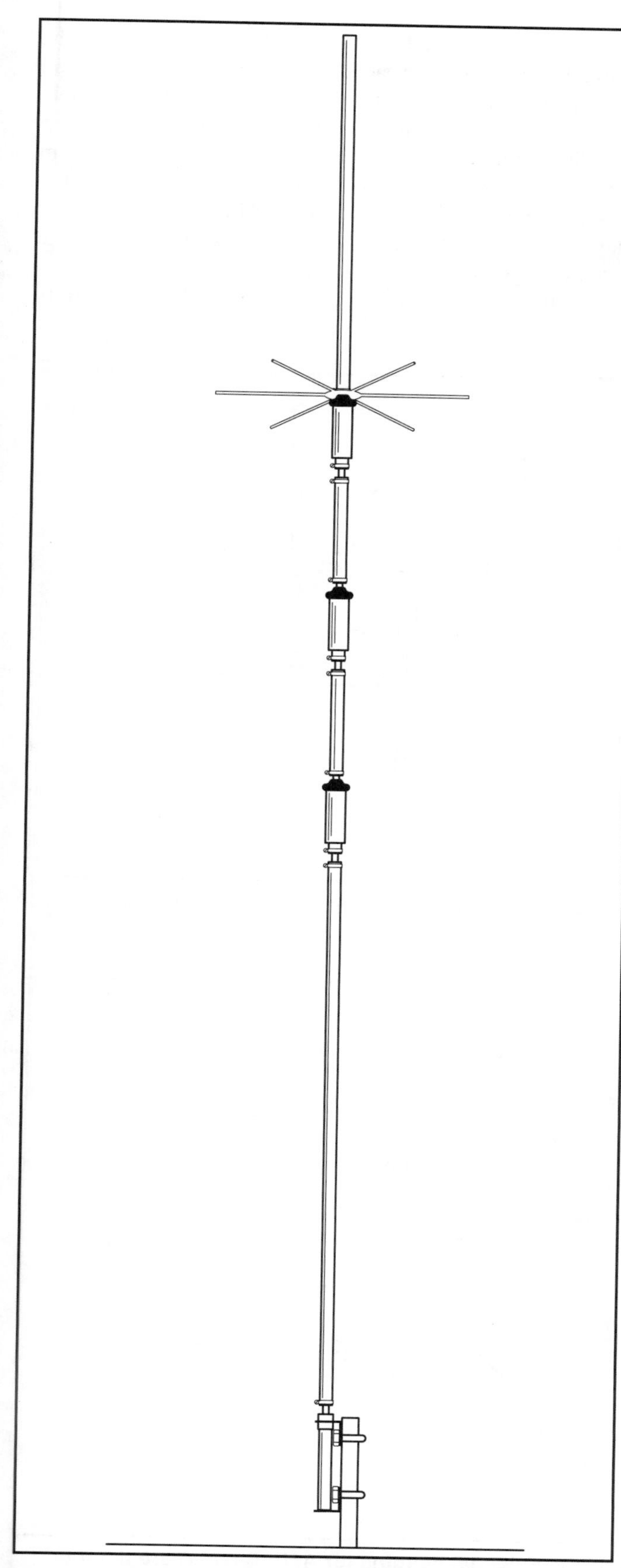

Figure 9-22—Commercial "trap vertical" antennas generally look something like this. These antennas operate on several bands.

Vertical Antennas for 146, 222 and 440 MHz

For FM and packet radio operation with nearby stations, the ease of construction and low cost of a $^1/_4$-λ vertical make it an ideal choice. Three different types of construction are shown in Figures 9-24 through 9-27; the choice of construction method depends on the materials available and the desired style of antenna mounting.

The 146-MHz model shown in Figure 9-24 uses a flat piece of sheet aluminum, to which radials are connected with machine screws. A 45° bend is made in each of the radials. This bend can be made with an ordinary bench vise. An SO-239 chassis connector is mounted at the center of the aluminum plate with the threaded part of the connector facing down. The vertical portion of the antenna is made of no. 12 copper wire soldered directly to the center pin of the SO-239 connector.

The 222-MHz version, Figure 9-25, uses a slightly different technique for mounting and sloping the radials. In this case the corners of the aluminum plate are bent down at a 45° angle with respect to the remainder of the plate. The four radials are held to the plate with machine screws, lock washers and nuts. A mounting tab is included in the design of this antenna as part of the aluminum base. A compression type of hose clamp could be used to secure the antenna to a mast. As with the 146-MHz version, the vertical portion of the antenna is soldered directly to the SO-239 connector.

A very simple method of construction, shown in Figures 9-26 and 9-27, requires nothing more than an SO-239 connector and some 4-40 hardware. A small loop formed at the inside end of each radial is used to attach the radial directly to the mounting holes of the coaxial connector. After the radial is fastened to the SO-239 with no. 4-40 hardware, a large soldering iron or propane torch is used to solder the radial and the mounting hardware to the coaxial connector. The radials are bent to a 45° angle and the vertical portion is soldered to the center pin to complete the antenna. The antenna can be mounted by passing the feed line through a mast of 3/4-inch ID plastic or aluminum tubing. A compression hose clamp can be used to secure the PL-259 connector, attached to the feed line, in the end of the mast. Dimensions for the 146, 222, and 440-MHz bands are given in Figure 9-26.

If these antennas are to be mounted outside it is

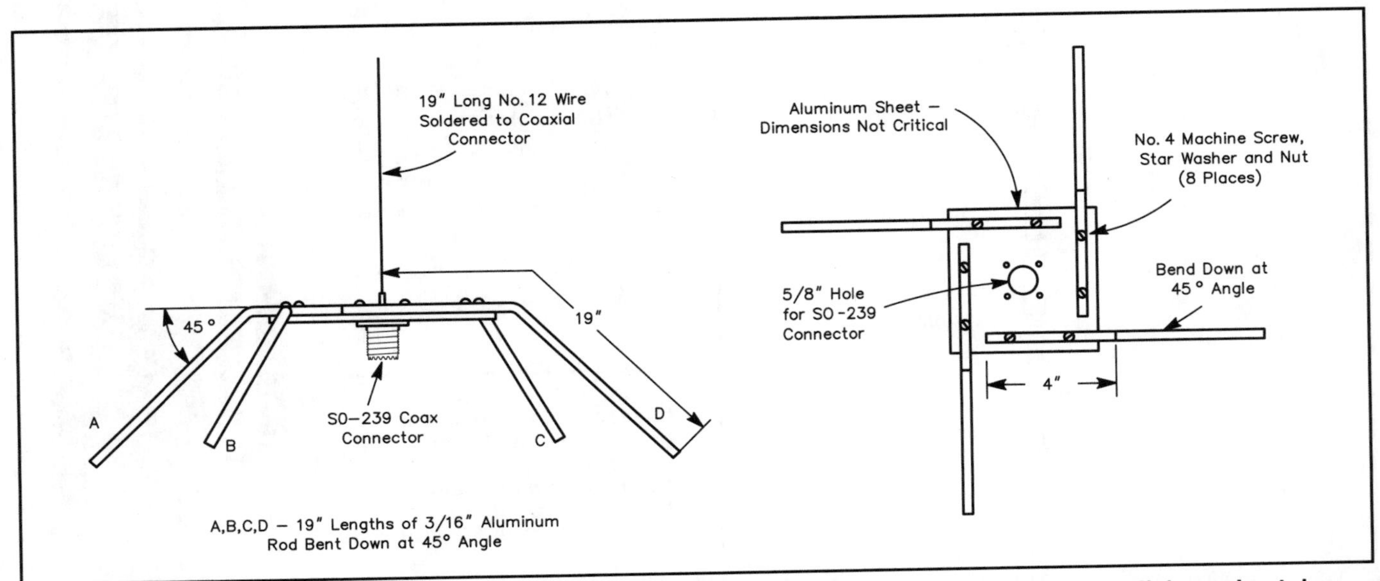

Figure 9-24—These drawings show the dimensions for the 146-MHz ground-plane antenna. The radials are bent down at a 45° angle.

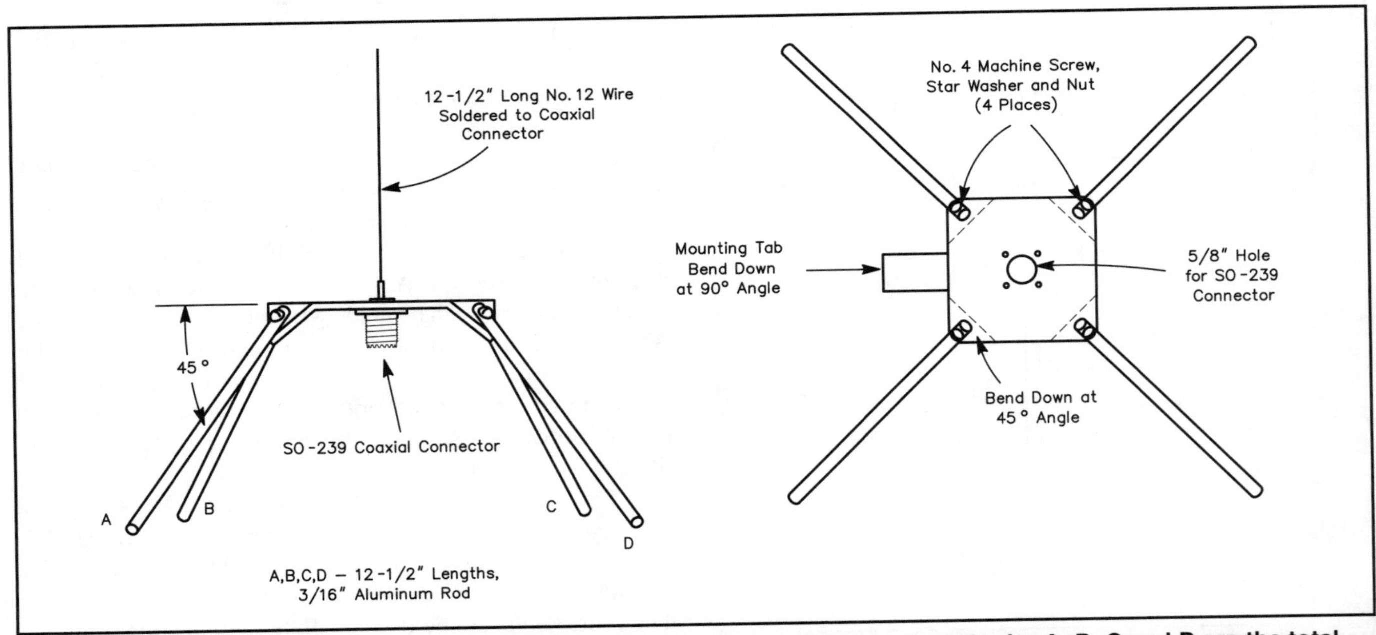

Figure 9-25—Dimensional information for the 222-MHz ground-plane antenna. Lengths for A, B, C and D are the total distances measured from the center of the SO-239 connector. The corners of the aluminum plate are bent down at a 45° angle rather than bending the aluminum rod as in the 146-MHz model. Either method is suitable for these antennas.

wise to apply a small amount of RTV sealant or similar material around the areas of the center pin of the connector to prevent the entry of water into the connector and coax line.

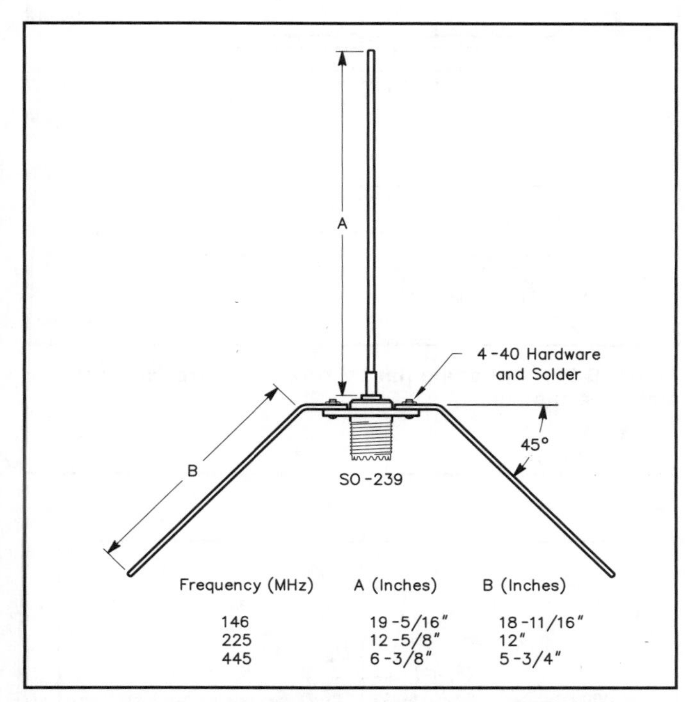

Frequency (MHz)	A (Inches)	B (Inches)
146	19-5/16"	18-11/16"
225	12-5/8"	12"
445	6-3/8"	5-3/4"

Figure 9-26—Simple ground-plane antenna for the 146, 222 and 440-MHz bands. The vertical element and radials are 3/32 or 1/16-in. brass welding rod. Although 3/32-in. rod is preferred for the 146-MHz antenna, no. 10 or 12 copper wire can also be used.

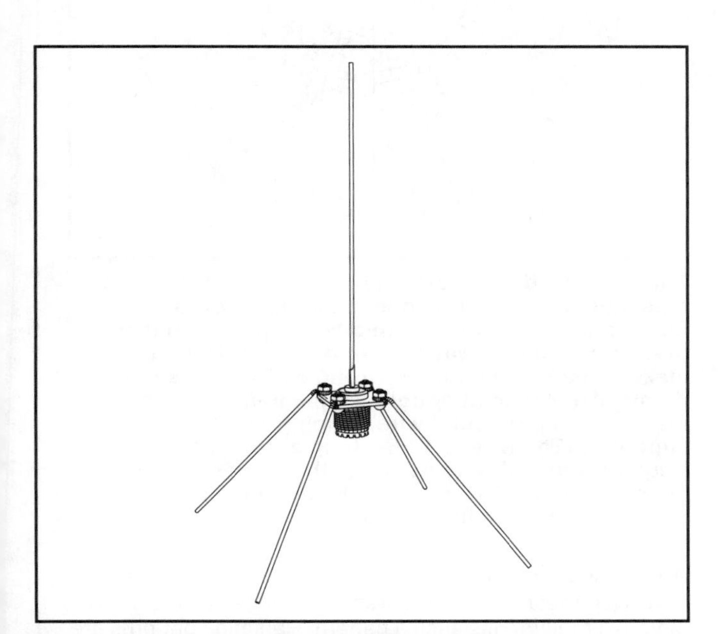

Figure 9-27—A 440-MHz ground-plane antenna constructed using only an SO-239 connector, no. 4-40 hardware and 1/16-in. brass welding rod.

A Simple Novice Vertical

Figure 9-28 shows a simple, inexpensive vertical antenna for 10 and 15 meters. The antenna requires very little space and it's great for working DX. A few materials make up the entire antenna.

12-foot piece of clean 2 × 2 pine from the local lumberyard
20 feet of flat four-wire rotator control cable
20 feet of regular TV twin lead
Several TV standoff insulators

Cut the twin lead into lengths of 11 feet 3 inches, and 8 feet 6 inches. Remove 1 inch of insulation from both wires at both ends on each twin-lead length. Wrap the two wires together securely at each end, as Figure 9-29 shows, and solder. Cover one end of each length of twin lead with electrical tape.

Mount the TV standoff insulators at regular intervals on opposite sides of the 2 × 2. These will support the two pieces of twin lead. Separate the control cable into two pieces of two-conductor cable. Do this by slitting the cable a small amount at one end, then pulling the two pieces apart like a zipper.

Carefully cut the rotator cable as Figure 9-30 shows. Separate the pieces between the center cuts to make four identical sets of two-conductor cable. These make up the radials for the antenna system. Strip about 1 inch of insulation from each wire on the evenly cut end. The unevenly cut ends will be away from the antenna.

Attach a suitable length of RG-58 coaxial cable to the antenna. Connect the coax center conductor to the two wires on the 2 × 2. Then attach the cable braid to all the radial ends soldered together. Figure 9-31 shows the antenna construction. Make all the connections waterproof.

Now install the 2 × 2. You can do this in different ways.

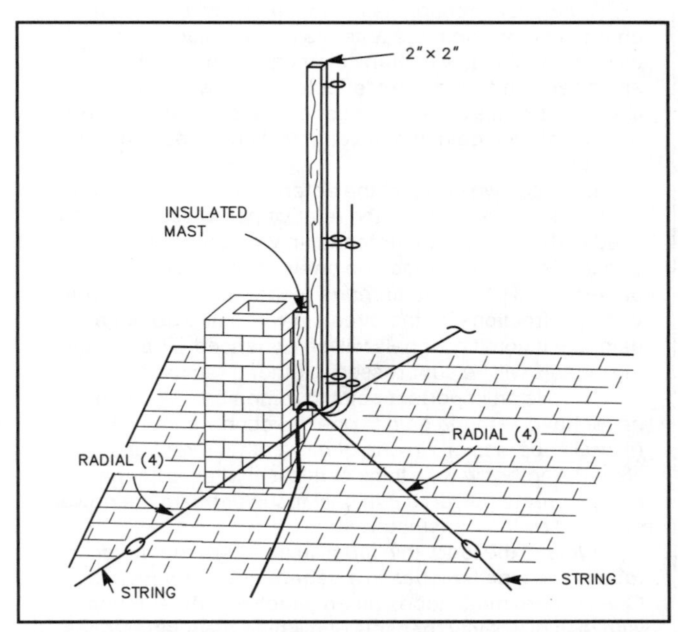

Figure 9-28—A simple Novice vertical antenna for use on 10 and 15 meters.

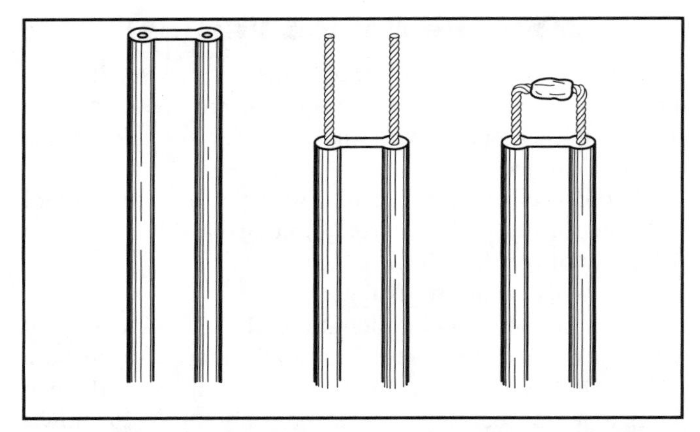

Figure 9-29—For each length of twin lead, strip back the insulation from both ends, connect the two wires, and solder them.

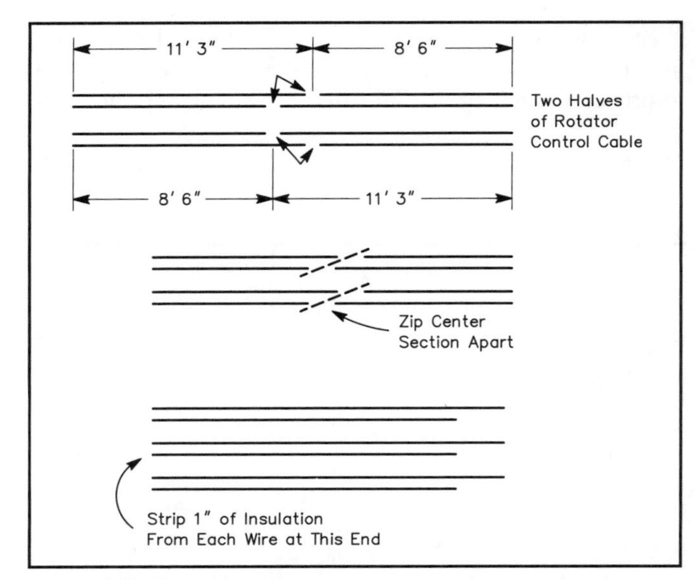

Figure 9-30—Cut some flat TV rotator-control cable to make the radials.

Antenna Radiation Patterns

Often one desirable antenna feature is **directivity**. Directivity means the ability to pick up signals from one direction, while suppressing signals from other unwanted directions. Going hand in hand with directivity is **gain**. Gain tells how much signal a given antenna will pick up as compared with that from another antenna, usually a dipole.

An antenna that has directivity should also have gain. These two antenna properties are useful not only for picking up or receiving radio signals, but also for transmitting them. An antenna that has gain will boost your transmitted energy in the favored direction while suppressing it in other directions.

When you mention gain and directivity, most amateurs envision large antenna arrays, made from aluminum tubing, with many elements. Simple wire antennas can also be very effective, however, as illustrated in this antenna radiation pattern. Such patterns reveal both the gain and the directivity of a specific antenna.

Let's say we connect the antenna to a transmitter and send. The pattern shows the relative power received at a fixed distance from the antenna, in various compass directions. If you connect the antenna to a receiver, the pattern shows how the antenna responds to signals from various directions. In the direction where the antenna has gain, the incoming signals will be enhanced. The incoming signals will be suppressed in other directions.

Here is an important point to remember. You can never have antenna gain in one direction without a loss (signal suppression) in one or more other directions. Never! Another way to think of this is that an antenna cannot create power. It can only focus or beam the power supplied by the transmitter.

We call the long, thin lobes in a pattern the *major lobes*. The smaller lobes in a pattern are *minor lobes*. One or more major lobes mean directivity. An antenna with less directivity than this one would have fatter lobes.

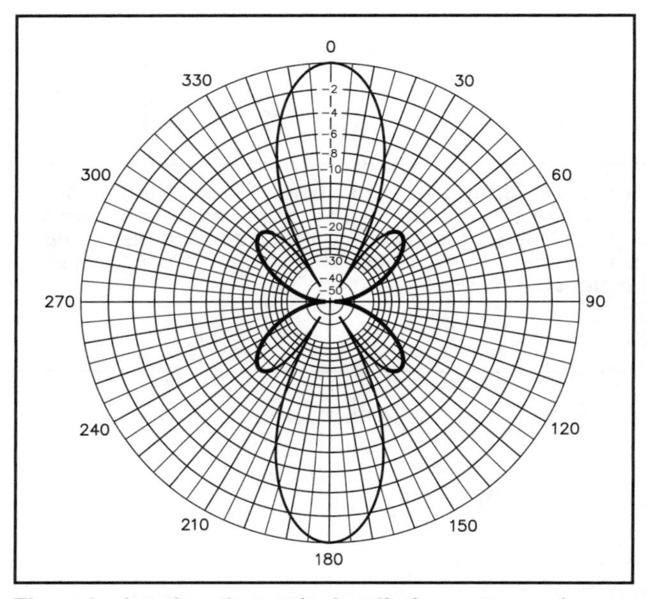

The calculated or theoretical radiation pattern of an extended double Zepp antenna. In its favored directions, this antenna exhibits roughly 2 decibels of gain over a half-wavelength dipole. This would make a 200-watt signal as strong as 317 watts into a dipole. An extended double Zepp antenna may be made with a horizontal wire hanging between two supports. (The wire is 1.28-λ long at the operating frequency and should be fed at the center with open-wire line.) The wire axis is along the 90/270-degree line shown in the chart.

An antenna with no directivity at all would have a pattern that is a perfect circle. A theoretical antenna called an isotropic radiator has such a pattern. Radiation patterns are a very useful tool in measuring antenna capability.

You could clamp it with U bolts to a TV mast, or hang it with a hook over a high branch. Or you might mount it on your house, at the side, near the roof peak. Mount the antenna as high as possible. Hang the radials at about a 45° angle away from the antenna base. For example, you could clamp the wood to your chimney with TV chimney-mount hardware. Let the radials follow the roof slope. Tie them off at the four corners. Unless you are surrounded by buildings or high hills, this antenna should perform well on 10 and 15 meters.

Random-Length Wire Antennas

If you can't install either a dipole or a vertical, you can still get on the air. Try a *random-length wire antenna*. See Figure 9-32. As the name implies, the antenna requires no specific length. As a rule of thumb, you should make your random-length antenna as long as possible. If you live in an apartment, you might use an antenna running along the ceiling in a few rooms. On the other hand, you may be able to string up a long length of wire outdoors. Use small wire (no. 22 to no. 28) if you want an antenna with low visibility. The only disadvantage of small wire is that it breaks easily. You may have to replace it often.

Random-length antennas are versatile: They can be used almost anywhere. But they do have one major disadvantage. Unlike the dipole and vertical, which can be fed directly from the transmitter through coaxial cable, a random-length antenna requires a matching device. The matching device is required because the antenna impedance is not likely to be 50 ohms.

Beam Antennas

Although generally impractical on 80 meters, and very large and expensive even on 40 meters, *directional antennas* often see use on 15 and 10 meters. The most

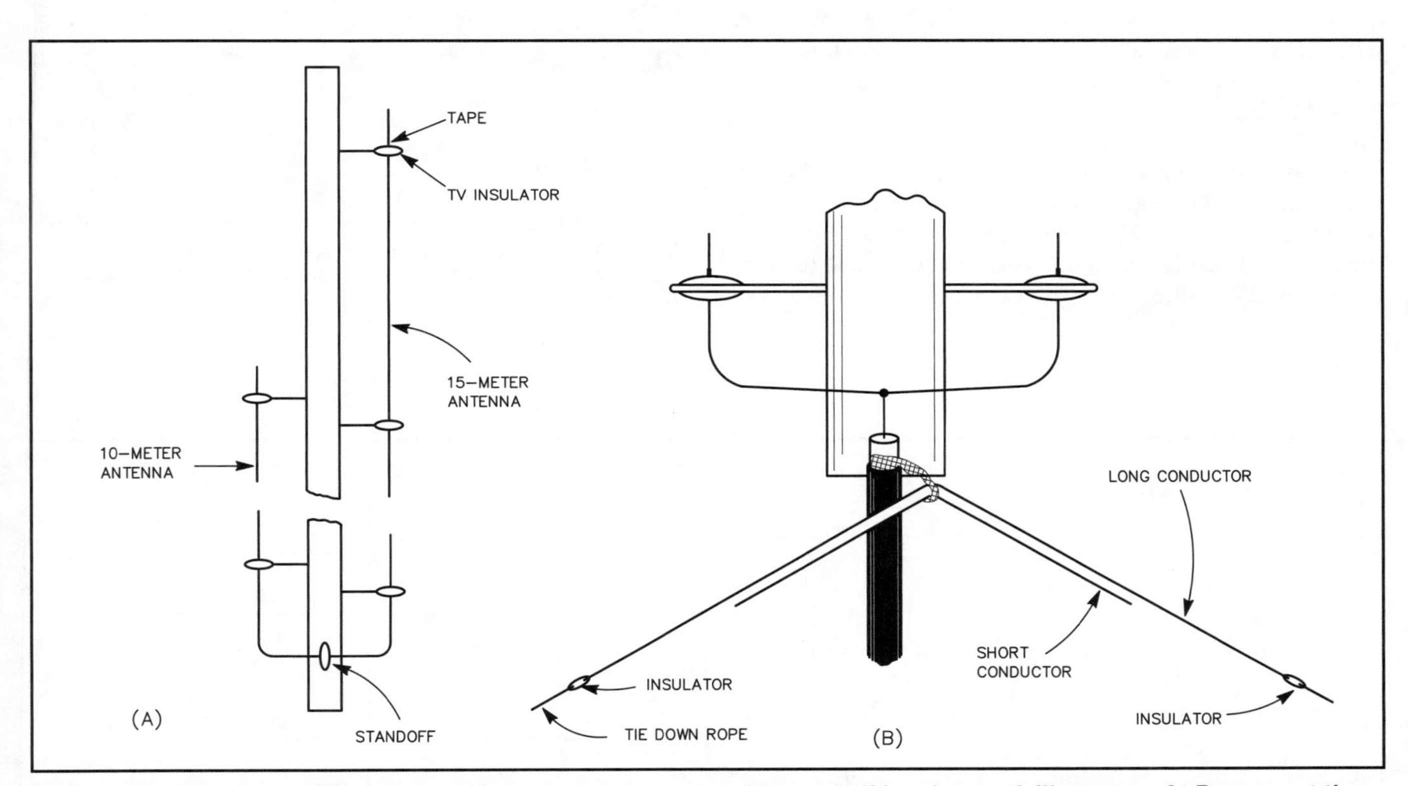

Figure 9-31—Attach the twin lead to the sides of the 2 × 2 with TV standoff insulators. A illustrates. At B, connect the feed line to both twin-lead lengths at the base.

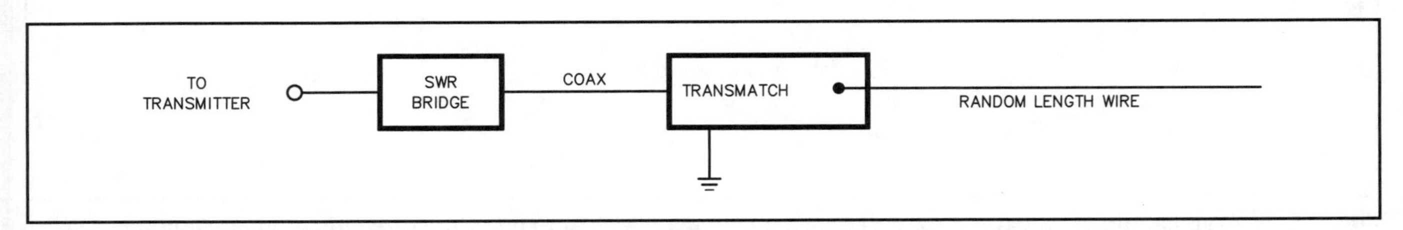

Figure 9-32—With a Transmatch, you can use a wire cut to a convenient random length as an antenna.

common directional antenna that amateurs use is the **Yagi antenna**, but there are other types, too.

Generally called **beam antennas**, these directional antennas have two important advantages over dipole and vertical systems. First, a beam antenna concentrates most of its transmitted signal in one compass direction. The antenna provides **gain** or *directivity* in the direction it is pointed. Gain makes your signal sound stronger to other operators, and their signals sound stronger to you.

The second important advantage is that the antenna reduces the strength of signals coming from directions other than where you point it. This increases your operating enjoyment by reducing the interference from stations in other directions.

A graph of an antenna's gain and directivity shows its *radiation pattern*. Figure 9-33 shows some of the Yagi beams at W1AW, the station located at ARRL Headquarters in Newington, Connecticut. Figure 9-34 shows the typical radiation pattern of a Yagi beam.

A Yagi beam antenna has several elements attached to a central *boom*, as Figure 9-35 shows. The elements are parallel to each other and are placed in a straight line along the boom. Although several factors affect the amount of gain of a Yagi antenna, *boom length* has the largest effect: The longer the boom, the higher the gain.

The feed line connects to only one element. We call this element the **driven element**. On a three-element Yagi like the one shown in Figure 9-35, the driven element is in the middle. The element at the front of the antenna (toward the favored direction) is a **director**. Behind the driven

Figure 9-33—Yagi beams at W1AW. A single beam sits atop the tower at left. Three stacked beams adorn the taller tower at the right.

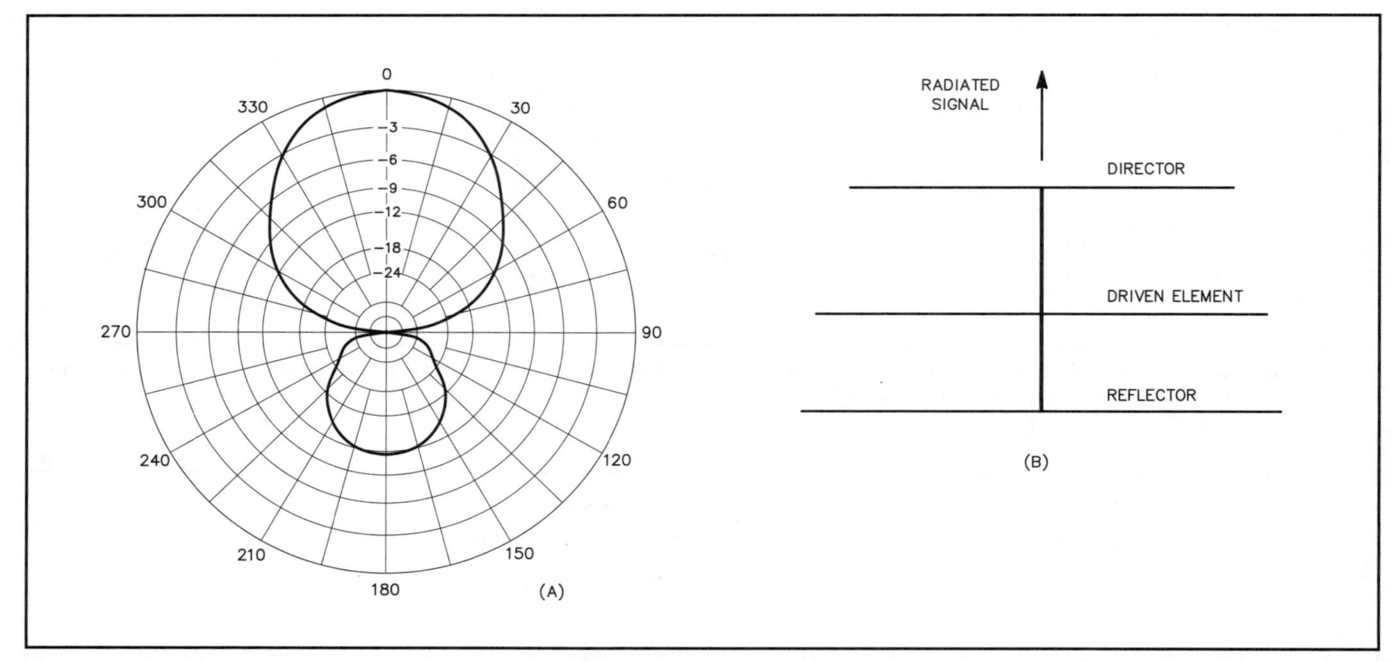

Figure 9-34—Typical radiation pattern for a Yagi beam antenna. The inset shows the direction of beam pointing. The transmitted signal is stronger in the forward direction than in others.

element is the **reflector** element. The driven element is about $^1/_2$ λ long at the antenna design frequency. The director is a bit shorter than $^1/_2$ λ, and the reflector a bit longer.

Yagi beams can have more than three elements. Seldom is there more than one reflector. Instead, the added elements are directors. A 4-element Yagi has a reflector, a driven element and two directors. Directors and reflectors are called *parasitic* elements, since they are not fed directly. Beams are sometimes called **parasitic beam antennas.**

Because beams are directional, you'll need something to turn them. A single-band beam for 10 or 15 meters can be mounted in the same manner as a TV antenna. You can use a TV mast, hardware and rotator. You could plan to buy a large triband beam for use on 10 and 15 meters as a Novice. It will cover 20 meters when you upgrade to a General or higher license class. For such a big antenna you'll need a heavy-duty mount and rotator. You can get good advice about the equipment you'll need from your instructor or from local hams.

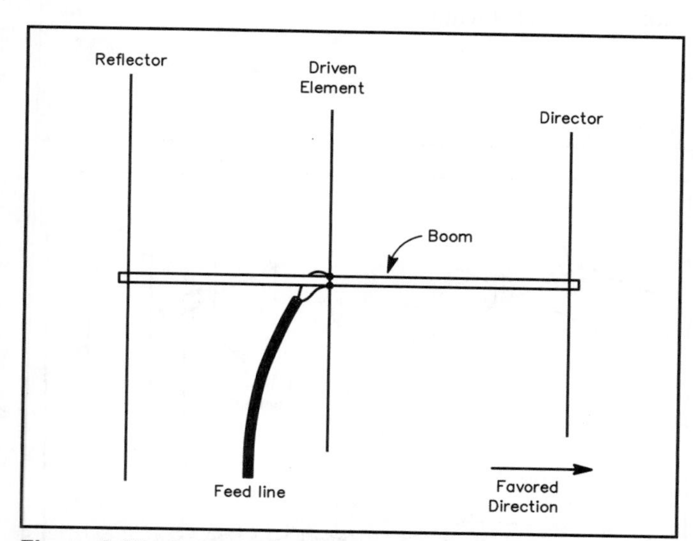

Figure 9-35—A three-element Yagi antenna has a director, a driven element and a reflector. A boom supports the elements.

[Now turn to Chapter 13 and study questions N9B01 through N9B07. Review this section if you have any difficulty answering these questions.]

Converting CB Antennas

It is usually a simple matter to convert a Citizen's Band (CB) antenna to work on the Novice 10-meter band. This is because the 11-meter Citizens Band at 27 MHz is close in frequency to the Novice 10-meter band. Since most CB antennas are $^1/_4$-λ verticals, the formula for the correct length is quite simple—234 divided by the frequency in megahertz (Equation 9-3). The antenna length is given in feet. For example, 234 divided by 27 MHz equals 8.67 feet or 8 feet, 8 inches.

To resonate (or tune) such an antenna on 10 meters, you shorten it to the required length. Using Equation 9-2, $^1/_2$ λ for 28.1 MHz works out to be 16 feet, 8 inches. The length for a $^1/_4$-λ vertical, using Equation 9-3, is 8 feet 4 inches. This means removing 8 inches from a dipole, or 4 inches from a full-sized vertical (not loaded with a coil). Before shortening any CB antenna, try the antenna on 10 meters. If your transmitter loads and tunes up, it may be a waste of time to prune the antenna. It may work well just as it is!

ANTENNA POLARIZATION

For VHF and UHF base-station operation, most amateurs use a vertical antenna or a beam antenna. For VHF and UHF FM and repeater operation, almost everyone uses vertical **polarization**.

Polarization refers to the electrical-field characteristic of a radio wave. You can think of it as how the antenna is positioned. An antenna that is parallel to the Earth's surface (like a dipole antenna) produces horizontally polarized radio waves. One that is perpendicular (at a 90° angle) to the Earth's surface, such as a $^{1}/_{4}$-λ vertical, produces vertically polarized waves.

For SSB and CW VHF/UHF work, almost everyone uses a beam with horizontal polarization. If you plan to do both VHF weak-signal and FM repeater work, you'll probably need separate antennas for each. This is because at VHF and UHF, signal strength suffers greatly if your antenna has different polarization than the station you're trying to work.

A beam antenna is impractical for VHF and UHF FM mobile operation. Most hams use a vertical of some kind. Mobile antennas are available in several varieties. Most mount on the automobile roof or trunk lid, but some even mount on a glass window.

Antenna Polarization

The signal sent from an amateur station depends on the antenna type and how it is oriented. A horizontal antenna, parallel to the Earth's surface (like a dipole), will produce a *horizontally polarized signal.* A Yagi antenna with horizontal elements will also produce a horizontally polarized signal. See Part A in the drawing.

A vertical antenna (perpendicular to the Earth's surface) will produce a *vertically polarized signal.* A Yagi with vertical elements also produces vertical polarization. See Part B in the drawing.

Most communications on the HF bands (80 through 10 meters) use horizontal polarization. Polarization on the HF bands is not critical, however. As a signal travels through the ionosphere, its polarization can change.

On the VHF/UHF bands, most FM communications use vertical polarization. Here, the polarization is important. The signals retain their polarization from transmitter to receiver. If you use a horizontal antenna, you will have difficulty working through a repeater with a vertical antenna.

Since the antennas on orbiting satellites are normally *circularly polarized*, earth-based antennas that are used for satellite communications should also be circularly polarized. One type of circularly polarized antenna is the crossed-Yagi, shown in Part C in the drawing.

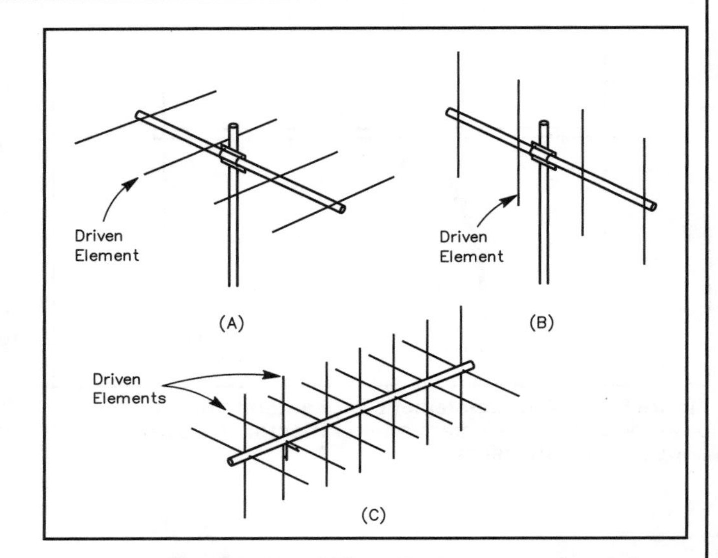

The plane of the elements in a Yagi antenna determines the transmitted-signal polarization. If the elements are horizontal, as shown at A, the signal will have horizontal polarization. Vertical mounting, shown at B, produces a vertically polarized wave. An antenna with elements that are both horizontal and vertical, shown at C, produce signals that are circularly polarized.

INSTALLATION SAFETY

Antenna work sometimes requires that someone climb up on a tower, into a tree, or onto the roof. Never work alone! Work slowly, thinking out each move before you make it. The person on the ladder, tower, tree or roof top should always wear a safety belt, and keep it securely anchored. Before each use, inspect the belt carefully for damage such as cuts or worn areas. The belt will make it much easier to work on the antenna and will also prevent an accidental fall. A hard hat and safety glasses are also important safety equipment.

Never try to climb a tower carrying tools or antenna components in your hands. Carry what you can with a tool belt, including a long rope leading back to the ground. Then use the rope to pull other needed objects up to your workplace after securing your safety belt! It is helpful (and safe) to tie strings or lightweight ropes to all tools. You can save much time in retrieving dropped tools if you tie them to the tower. This also reduces the chances of injuring a helper on the ground.

Helpers on the ground should never stand directly under the work being done. All ground helpers should wear hard hats and safety glasses for protection. Even a small tool can make quite a dent if it falls from 50 or 60 feet. A ground helper should always observe the tower work carefully. Have you ever wondered why electric utility crews seem to have someone on the ground "doing nothing"? Now you know that for safety's sake, a ground observer with no other duties is free to notice potential hazards. That person could save a life by shouting a warning.

[Now study the questions in Chapter 13 with numbers N4B03 through N4B05. Review this section if you have any problems.]

How to Live Long Enough to Upgrade

Keep antenna safety in mind when you're setting up your station. Antennas for the ham bands are often large, requiring care and attention to detail when installed. Here are two points to keep in mind when putting up your antenna.

1) Be sure your antenna materials and supports are strong—strong enough to withstand heavy winds without breaking.

2) Keep away from power lines!

If your antenna falls, it could damage your house, garage or property. If it falls into a power line, your house might end up like the unlucky CBer's shown in the photo. A windstorm knocked over his ground-plane antenna, sending it into a 34,500-V power line. The resulting fire damaged his house extensively.

Safety pays! Accidents are avoidable, if you use good sense.

Photo by Charles Stokes, WB4PVT

KEY WORDS

Balanced line—Feed line with neither conductor at ground potential, such as open-wire line or ladder line.

Cubical quad antenna—An antenna built with its elements in the shape of four-sided loops.

Delta loop antenna—A variation of the cubical quad with triangular elements.

Horizontally polarized wave—An electromagnetic wave with its electric lines of force parallel to the ground.

Major lobe—The shape or pattern of field strength that points in the direction of maximum radiated power from an antenna.

Parasitic beam antenna—Another name for the beam antenna.

Parasitic element—Part of a directive antenna that derives energy from mutual coupling with the driven element. Parasitic elements are not connected directly to the feed line.

Polarization—Describes the electrical-field characteristic of a radio wave. An antenna that is parallel to the surface of the Earth, such as a dipole, produces horizontally polarized waves. One that is perpendicular to the Earth's surface, such as a quarter-wave vertical, produces vertically polarized waves.

Unbalanced line—Feed line with one conductor at ground potential, such as coaxial cable.

Vertically polarized wave—A radio wave that has its electric lines of force perpendicular to the surface of the Earth.

9 Additional Antenna Information for the Technician Exam

FEED-LINE ATTENUATION

As we have seen, transmission lines can be constructed in a variety of forms. Both parallel-conductor feed line and coaxial cable can be divided into two classes: those in which the majority of the space between the conductors is air, and those in which the conductors are embedded in and separated by a solid plastic or foam insulation (dielectric).

Air-Insulated Feed Lines

Air-insulated feed lines have the lowest loss per unit length (usually expressed in dB/100 ft). Adding a solid dielectric between the conductors increases the losses in the feed line. The power loss causes heating of the dielectric. (*Dielectric* is the technical term for an insulating material.) As frequency increases, conductor and dielectric losses become greater for coaxial and parallel-conductor feed lines. An advantage of this type of transmission line is that it can be operated at a high SWR and still retain its low-loss properties.

The characteristic impedance of an air-insulated parallel-conductor line depends on the diameter of the wires used in the feed line and the spacing between them. The greater the spacing between the conductors, the higher the characteristic impedance of the feed line. The impedance decreases, however, as the size of the conductors increases. The characteristic impedance of a feed line is not affected by the length of the line.

Solid-Dielectric Feed Lines

Transmission lines in which the conductors are separated by a flexible dielectric have several advantages over air-dielectric line: They are less bulky, maintain more uniform spacing between conductors, are generally easier to install and are neater in appearance. Both parallel-conductor and coaxial lines are available with this type of insulation.

One disadvantage of these types of lines is that the power loss per unit length is greater than air-insulated lines because of the dielectric. As the frequency increases, the dielectric losses become greater. The power loss causes heating of the dielectric. Under conditions of high power or high SWR, the dielectric may actually melt and cause short circuits or arcing inside the line.

We mentioned TV-type parallel-conductor line, commonly called twin lead. Twin lead consists of two no. 20 wires that are molded into the edges of a polyethylene ribbon about a half-inch wide. The presence of the solid dielectric lowers the characteristic impedance of the line as compared to the same conductors in air.

The fact that part of the field between the conductors exists outside the solid dielectric leads to operating disadvantages. Dirt or moisture on the surface of the ribbon tends to change the characteristic impedance. Weather effects can be minimized, however, by coating the feed line with silicone grease or car wax. In any case, the changes in the impedance will not be very serious if the line is terminated in its characteristic impedance (Z_0). If there is a considerable standing-wave ratio, however, small changes in Z_0 may cause wide fluctuations of the input impedance.

Some hams try inexpensive lamp cord (also called *zip cord*). Although it may be acceptable for very short runs at low frequencies, such as 80 or 40 meters, zip cord is too lossy to work well at higher frequencies, such as 10 or 6 meters.

Parallel-conductor feed lines are **balanced lines** because neither conductor connects to ground. They can be connected directly to *balanced antennas*.

Coaxial Cable

The characteristic impedance of a coaxial line depends on the diameter of the center conductor, the dielectric constant of the insulation between conductors, the inside diameter of the shield braid and the distance between conductors. The characteristic impedance of coaxial cables increases for larger-diameter shield braids, but it decreases for larger-diameter center conductors.

The larger the diameter, the higher the power capability of the line because of the increased dielectric thickness and conductor size. In general, losses decrease as the cable diameter increases, because there is less power lost in the conductor.

Amateurs commonly use RG-8, RG-58, RG-174 and RG-213 coaxial cable. RG-8 and RG-213 are similar cables, and they have the least loss of the types listed here. RG-174 is only about 1/8 inch in diameter, and it has the highest loss of the cables listed here.

As the frequency increases, the conductor and dielectric losses become greater, causing more attenuation of the signal in the cable. Although the cost, size and weight are larger for RG-213, it is usually the best choice (of the cables mentioned) for a run of over 150 feet for frequencies up to 54 MHz (the amateur 6-meter band). Open-wire transmission line has the least loss of any feed line type commonly used by amateurs.

Of the common coaxial cables discussed here, RG-213 has the least loss. RG-58 has a bit more loss than RG-213, and RG-174 has the most loss of these cables. RG-174 is normally used for cables that connect sections of a transmitter or receiver, or for short interconnecting cables in a low-power system. Some amateurs use RG-174 cable as the feed line for a low-power portable station.

Extra cable length increases attenuation. When using coaxial cable, you should try to use a matched antenna and feed line. You should then be able to change feed line lengths without significantly affecting the antenna system. Then your feed line has only to be long enough to reach your antenna. A low SWR on the line means that the impedance "seen" by the transmitter will be about the same regardless of line length. You can cut off or shorten excess cable length to reduce attenuation of the signal caused by antenna-system loss. (This does not apply to multiple antennas in phased arrays or line sections used for impedance-matching purposes.)

Attenuation is not affected by the characteristic impedance of a matched line if the spacing between the conductors in the coaxial cable is a small fraction of a wavelength at the operating frequency.

The outside shield of coaxial cable is normally connected to ground. Coax, then, is an **unbalanced line**. When connecting coaxial *(unbalanced)* feed lines to *balanced antennas*, many hams prefer to use a *balun*.

Coaxial-cable connectors are an important part of a coaxial feed line. Your choice of connectors normally depends on the matching connectors on your radios. Most HF radios and many VHF radios use SO-239 connectors. The mating connector is called a PL-259. (These are military-type designations.) The PL-259 is sometimes called a *UHF connector,* although they are not the best choice for the UHF bands. Figure 9-36C shows the SO-239 connector and its mating PL-259.

Many VHF and UHF hand-held radios use BNC connectors. These connectors are designed for use with RG-58 coax. They produce a low-loss connection that is also weatherproof. Figure 9-36A shows a pair of BNC connectors. BNC connectors are well-suited for use with hand-held-radio antennas because they require only a quarter

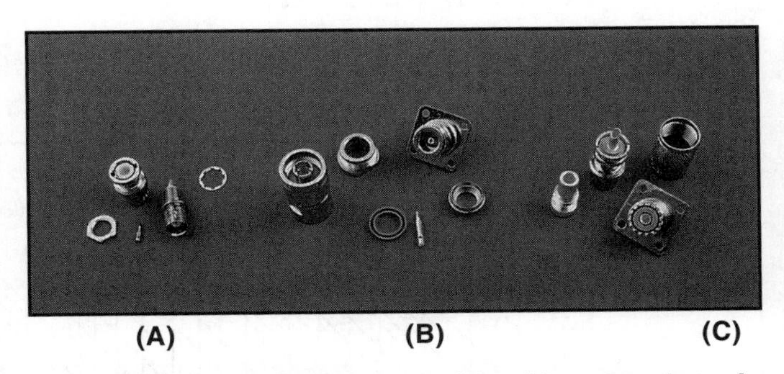

Figure 9-36—Some common coaxial-cable connectors. A shows a BNC connector pair. Many hand-held radios use BNC connectors. They are also popular when a weatherproof connector is needed for RG-58 sized cable. B shows a pair of N connectors. These are often used for UHF equipment because of their low loss. Part C shows an SO-239 chassis connector and its mating PL-259. Most HF equipment uses these connectors.

turn to install or remove, yet they lock securely in place.

Many amateurs prefer N-type connectors for UHF equipment. N connectors are low-loss, high-power connectors. They are designed for use with RG-213 or RG-8 coaxial cable. N connectors have the lowest loss of the common coaxial connectors, which is why they are often used at UHF. Figure 9-36B shows a set of N connectors.

It is a good idea to check your coaxial connectors on a regular basis. Be sure they are clean and tight to minimize their resistance. If you suspect a bad solder connection, you should resolder the joints.

[Turn to Chapter 14 and study questions T9C01 through T9C11 before continuing with this chapter. Review this section if needed.]

POLARIZATION OF ANTENNAS AND RADIO WAVES

An electromagnetic wave consists of moving electric and magnetic fields. Remember that a field is an invisible force of nature. We can't see radio waves, but we can show a representation of where the energy is in the electric and magnetic fields. We did this in Chapter 5 to show the magnetic flux around a coil, and in Chapter 6 to show the electric field in a capacitor. We can visualize a traveling radio wave as looking something like Figure 9-37. The lines of electric and magnetic force are at right angles (90°) to each other. They are also perpendicular (90°) to the direction of travel. These fields can have any position with respect to the Earth.

Horizontal Polarization

Polarization is defined by the direction of the electric lines of force in a radio wave. A wave with electric lines of force that are parallel to the surface of the Earth is a **horizontally polarized wave**.

The polarization of a radio wave as it leaves the antenna is determined by the orientation of the antenna. For example, a half-wavelength dipole parallel to the surface of the Earth transmits a horizontally polarized wave.

Vertical Polarization

The electric lines of force can also be perpendicular to the Earth. A wave with vertical electric field lines is a **vertically polarized wave**. A half-wavelength dipole perpendicular to the surface of the Earth transmits a vertically polarized wave. A mobile whip antenna is mounted vertically on a car. It transmits a wave that is vertically polarized.

The quarter-wavelength vertical antenna is a popular HF antenna because it provides low-angle radiation when a beam or dipole cannot be placed far enough above ground. Low-angle radiation refers to signals that travel closer to the horizon, rather than signals that are high above the horizon. Low-angle radiation is usually better when you are trying to contact distant stations. Vertical antennas of any length radiate vertically polarized waves.

Polarization is most important when installing antennas for VHF or UHF. Propagation at these frequencies is mostly line of sight. The polarization of a terrestrial (on the ground, as opposed to out in space) VHF or UHF signal does not change from transmitting antenna to receiving antenna. Best signal reception occurs when both transmitting and receiving stations use the same polarization. The polarization of an HF signal may change many times as it passes through the ionosphere. Antenna polarization at HF is not as important.

Most VHF/UHF FM and data communications is done with vertically polarized antennas. Vertically polarized antennas are more popular for repeaters and other

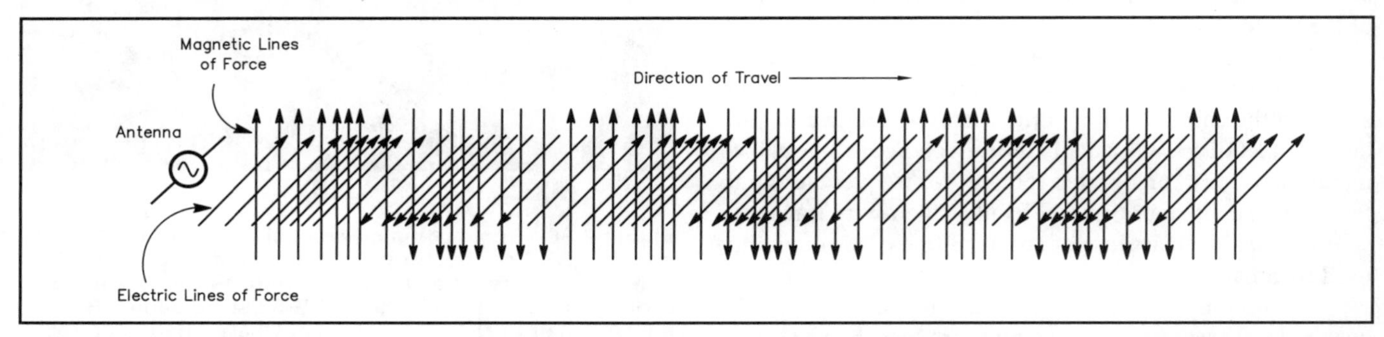

Figure 9-37—A horizontally polarized electromagnetic wave. The electrical lines of force are parallel to the ground. The fields are stronger where the arrows are closer together. The points of maximum field strength correspond to the points of maximum amplitude for the voltage producing the RF signal.

VHF/UHF FM communications, because the antennas used on cars are almost always verticals. Vertical antennas are also more useful for repeaters and home-station use on these bands because they are not directional. For long-distance FM operating, a vertically polarized beam is the best antenna. Data communication on the VHF/UHF bands is also mostly done with vertically polarized antennas. VHF/UHF CW and SSB operation, however, is done mostly with horizontally polarized antennas.

Yagi beam antennas can have either horizontal or vertical polarization. If the antenna elements are parallel to the Earth, or horizontal, the antenna will produce horizontally polarized waves.

If the elements are turned to be perpendicular to the Earth, the antenna will produce vertically polarized waves. Figure 9-38 shows three Yagi antennas. The antenna at A produces horizontally polarized waves and the one at B produces vertically polarized waves. Notice that the boom is horizontal in both cases.

Most man-made noise tends to be vertically polarized. Thus, a horizontally polarized antenna will receive less noise of this type than will a vertical antenna.

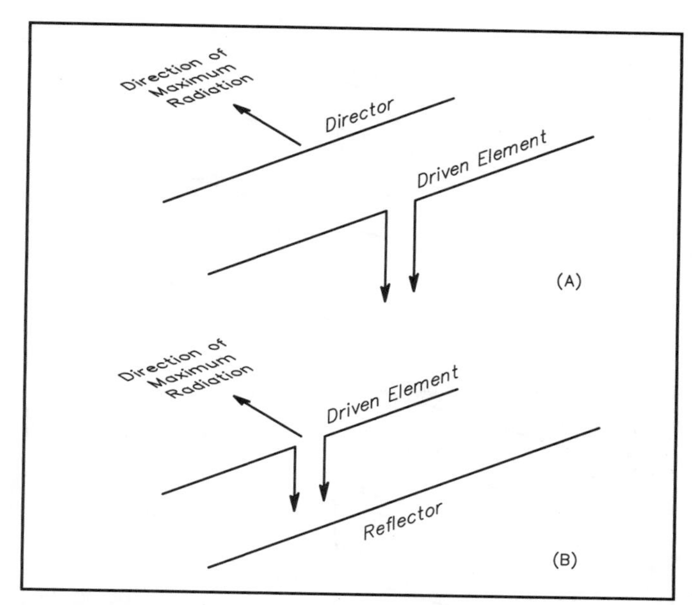

Figure 9-39—In a directional antenna, the reflector element is placed behind the driven element. The director goes in front of the driven element.

Circular Polarization

There is one more type of polarization: circular. Signals from orbiting satellites are circularly polarized, so the ground-based antennas that receive these signals should also be circularly polarized. Figure 9-38 Part C shows an example of a circularly polarized antenna.

[Turn to Chapter 14 and study questions T9B01 and T9B02, and T9B04 through T9B08. Review the material in this section as needed.]

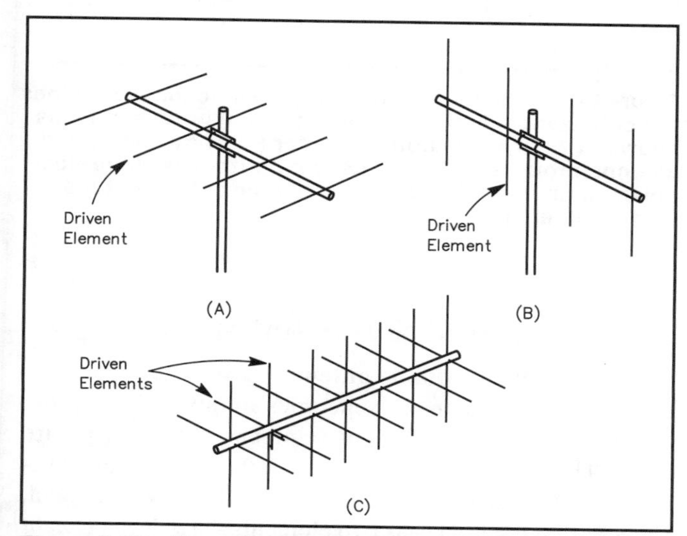

Figure 9-38—The orientation of the elements in a Yagi antenna determines the polarization of the transmitted wave. If the elements are horizontal as shown at A, the wave has horizontal polarization. Vertical mounting, shown at B, produces a vertically polarized wave. A circularly polarized antenna, used for space communications, is shown at C.

More About Beam Antennas

In most multiple-element antennas, the additional elements are not directly connected to the feed line. They receive power by mutual coupling from the *driven element*. The driven element is the element connected to the feed line. The additional elements then reradiate the power in the proper phase relationship. Proper phasing achieves *gain* or *directivity* over a simple half-wavelength dipole. These elements are called **parasitic elements**. Beam antennas that use parasitic elements, such as the Yagi, quad and delta loop, are sometimes called *parasitic beam antennas*.

There are two types of parasitic elements. A *director* is generally shorter than the driven element. A director is located at the front of the antenna. A *reflector* is generally longer than the driven element. A reflector is located at the back of the antenna. See Figure 9-39. The direction of maximum radiation from a parasitic beam antenna is from the reflector through the driven element to the director.

The term *major lobe* refers to the region of maximum radiation from a directional antenna. The major lobe is also sometimes referred to as the *main lobe*. Communication in different directions may be achieved by rotating the array in the *azimuthal*, or *horizontal, plane*, to point it in different compass directions.

Yagi Arrays

The Yagi arrays in Figure 9-40 are examples of antennas that make use of parasitic elements to produce a *unidirectional radiation pattern*. (This means most of the radio signal goes in one direction.) A Yagi antenna has at least two elements. One element is a driven element. The other elements are directors and/or reflectors. These

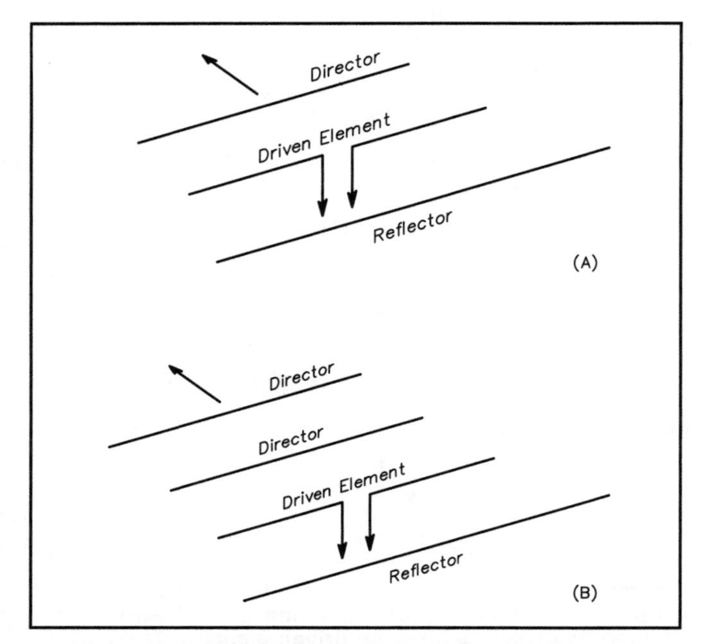

Figure 9-40—At A, a three-element Yagi. At B, a four-element Yagi with two directors.

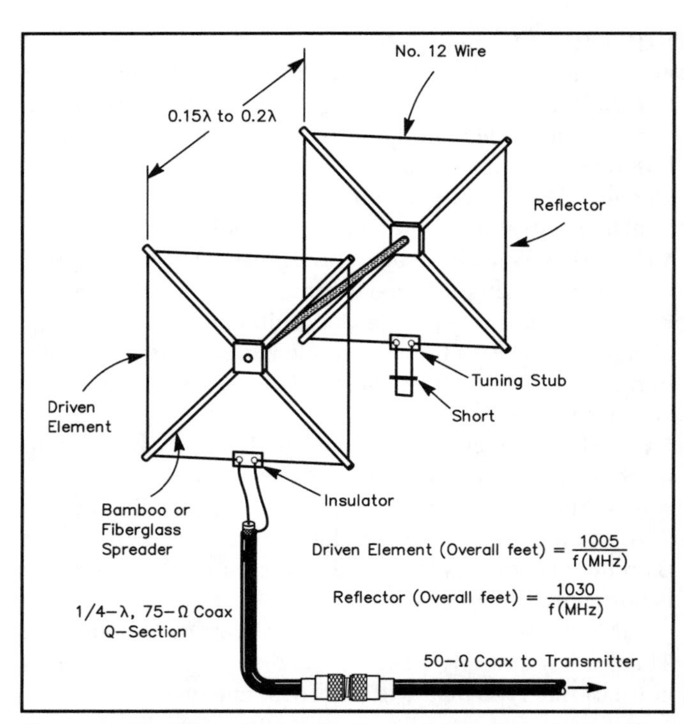

Figure 9-41—The cubical quad antenna. The total length of the driven element is about one wavelength. This antenna can be fed directly with coaxial cable.

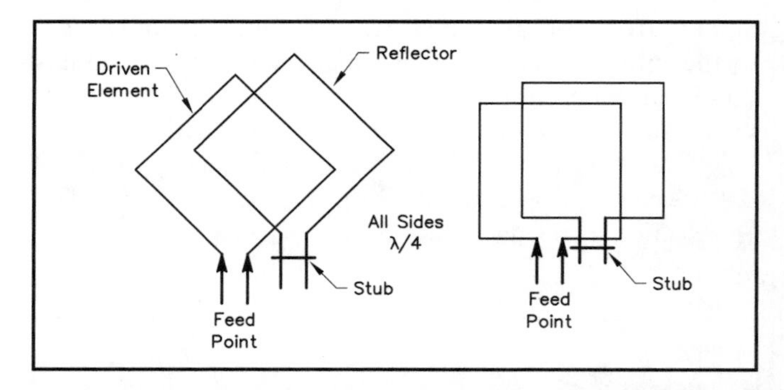

Figure 9-42—The feed point of a quad antenna determines the polarization. Fed in the middle of the bottom side (as shown at A) or at the bottom corner (as shown at B), the antenna produces a horizontally polarized wave. Vertical polarization is produced by feeding the antenna at the side in either case.

elements are usually parallel to each other and made of straight metal tubing. They line up with each other along the antenna *boom.*

Though typical HF antennas of this type have three elements, some may have six or more. Multiband Yagi antennas have many elements. Some elements work on some frequencies and others are used for different frequencies. The radiation pattern of the Yagi antenna is shown in Figure 9-34. This pattern indicates that the antenna will reject signals coming from the sides and back. It selects mainly those signals from a desired direction. There are several types of Yagi antennas.

The length of the driven element in the most common type of Yagi antenna is approximately an electrical half-wavelength. This means that Yagi antennas are most often used for the 20-meter band (14.0 to 14.35 MHz) and higher frequencies. Yagis for frequencies below 14 MHz are very large. They require special construction techniques, heavy-duty supporting towers and large rotators. Yagis for 40 meters are fairly common. Rotatable Yagis for 80 meters are few and far between. You can overcome the mechanical difficulties of large Yagis. Some amateurs build nonrotatable Yagi antennas for 40 and 80 meters, with wire elements supported on both ends.

In a three-element beam, the director is approximately 5% shorter than the driven element. The reflector is approximately 5% longer than the driven element. These lengths can vary considerably, however. The actual lengths depend on the spacing between elements and the diameter of the elements. Whether the elements are made from tapered or cylindrical tubing also makes a difference.

Cubical Quad Antennas

The **cubical quad antenna** also uses parasitic elements. This antenna is sometimes simply called a *quad.* The two elements of the quad antenna are usually wire loops. The total length of the wire in the driven element is approximately one electrical wavelength. A typical quad, shown in Figure 9-41, has two elements—a *driven element* and a *reflector.* A two-element quad could also use a driven element with a *director.* You can add more elements, such as a reflector and one or more directors. The radiation pattern of a typical quad is similar to that of the Yagi shown in Figure 9-34.

The elements of the quad are usually square. Each loop is about an electrical wavelength long. Each side of the square is about $^1/_4$-λ long.

The polarization of the signal from a quad antenna can be changed. Polarization is determined by where the feed point is located on the driven element. See Figure 9-42. If the feed point is located in the center of a horizontal side, parallel to the earth's surface, the transmitted wave will be horizontally polarized. When the antenna is fed in the center of a vertical side, the transmitted wave is vertically polarized. We can turn the antenna 45 degrees, so it looks like a diamond. When the antenna is fed at the bottom corner, the transmitted wave is horizontally polarized. If the antenna is fed at a side corner, the transmitted wave is vertically polarized.

Delta Loop Antennas

The **delta loop antenna**, shown in Figure 9-43, is similar to the quad. A delta loop antenna has triangular elements, rather than square. The total loop length is still one wavelength. Divide the total length by 3 to find the length of each side of the elements. The radiation pattern of a delta loop is similar to that of the quad and Yagi, shown in Figure 9-34.

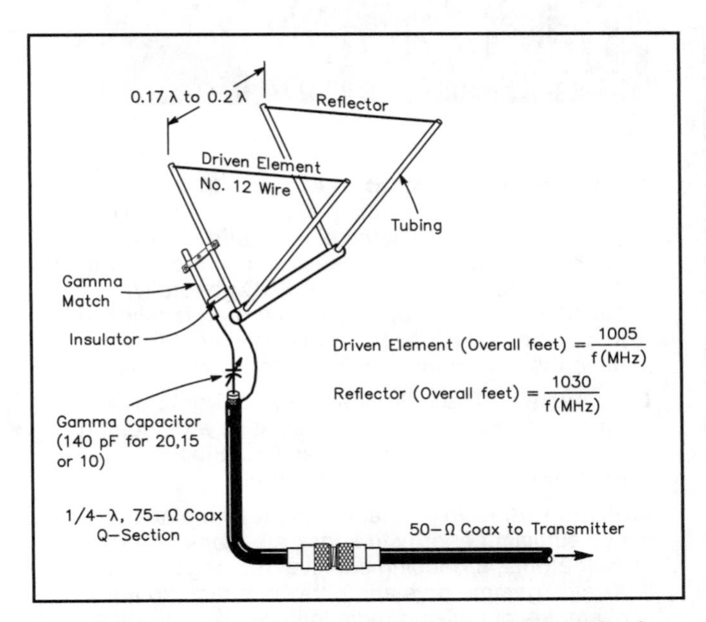

Figure 9-43—A delta loop antenna. The total length of each element can be found using the same equations as used for a quad antenna. The antenna is fed with a gamma match.

[Congratulations. This completes your study of antennas for your Technician exam. You will probably want to learn more about antennas later. After you pass the exam and have your license you'll be able to experiment with lots of antenna types. Now turn to Chapter 14 and study questions T9A01 through T9A09. Review this section if any of these questions give you difficulty.]

10 KEY WORDS

Controlled environment — Any area in which an RF signal may cause radiation exposure to people who are aware of the radiated electric and magnetic fields and who can exercise some control over their exposure to these fields. The FCC generally considers amateur operators and their families to be in a controlled RF exposure environment to determine the maximum permissible exposure levels.

Duty cycle — The ratio of average power to peak envelope power (PEP) expressed as a percentage. A lower duty cycle means less RF radiation exposure for the same PEP output.

Ionizing radiation — Electromagnetic radiation that has sufficient energy to knock electrons free from their atoms, producing positive and negative ions. X-rays, gamma rays and ultraviolet radiation are examples of ionizing radiation.

Mobile device — A radio transmitting device designed to be mounted in a vehicle. In Amateur Radio operation, the transmitter is usually activated by a push-to-talk (PTT) switch.

Nonionizing radiation — Electromagnetic radiation that does not have sufficient energy to knock electrons free from their atoms. Radio frequency (RF) radiation is nonionizing.

Portable device — A radio transmitting device designed to have a transmitting antenna that is generally within 20 centimeters of a human body.

RF burn — A burn produced by coming in contact with RF voltages.

RF radiation — Waves of electric and magnetic energy. Such electromagnetic radiation with frequencies as low as 3 kHz and as high as 300 GHz are considered to be part of the RF region.

RF safety — Preventing injury or illness to humans from the effects of radio-frequency energy.

Uncontrolled environment — Any area in which an RF signal may cause radiation exposure to people who may not be aware of the radiated electric and magnetic fields. The FCC generally considers members of the general public and an amateur's neighbors to be in an uncontrolled RF radiation exposure environment to determine the maximum permissible exposure levels.

The "antenna farm" at W1AW (ARRL Headquarters).

10 RF Environmental Safety Practices

FCC Rules now require all amateurs to meet certain maximum permissible exposure limits for RF radiation from their stations. In this chapter, you will learn how to evaluate your station to be sure you meet these important safety requirements.

RF RADIATION AND ELECTROMAGNETIC FIELD SAFETY

Amateur Radio is basically a safe activity. In recent years, however, there has been considerable discussion and concern about the possible hazards of electromagnetic radiation (EMR), including both RF energy and power-frequency (50-60 Hz) electromagnetic fields. FCC regulations set limits on the maximum permissible exposure (MPE) allowed from the operation of radio transmitters. These regulations do not take the place of RF-safety practices, however. This section deals with the topic of RF safety. See the sidebar, "FCC RF-Exposure Regulations," for information about the rules.

Extensive research on RF safety is underway in many countries. This section was prepared by members of the ARRL RF Safety Committee and coordinated by Dr Robert E. Gold, WBØKIZ. It summarizes what is now known and offers safety precautions based on the research to date.

All life on Earth has adapted to survive in an environment of weak, natural, low-frequency electromagnetic fields (in addition to the Earth's static geomagnetic field). Natural low-frequency EM fields come from two main sources: the sun, and thunderstorm activity. But in the last 100 years, man-made fields at much higher intensities and with a very different spectral distribution have altered this natural EM background in ways that are not yet fully understood. Much more research is needed to assess the biological effects of EMR.

Both RF and 60-Hz fields are classified as **nonionizing radiation** because the frequency is too low for there to be enough photon energy to ionize atoms. (**Ionizing radiation**, such as X-rays, gamma rays and even some ultraviolet radiation has enough energy to knock electrons loose from their atoms. When this happens, positive and negative ions are formed.) Still, at sufficiently high power densities, EMR poses certain health hazards. It has been known since the early days of radio that RF energy can cause injuries by heating body tissue. (Anyone who has ever touched an improperly grounded radio chassis or energized antenna and received an *RF burn* will agree that this type of injury can be quite painful.) In extreme cases, RF-induced heating in the eye can result in cataract formation and can even cause blindness. Excessive RF heating of the reproductive organs can cause sterility. Other serious health problems can also result from RF heating. These heat-related health hazards are called *thermal effects*. In addition, there is evidence that magnetic fields may produce biologic effects at energy levels too low to cause body heating. The proposition that these *athermal effects* may produce harmful health consequences has produced a great deal of research.

In addition to the ongoing research, much else has been done to address this issue. For example, FCC regulations set limits on exposure from radio transmitters. The American National Standards Institute and the National Council for Radiation Protection and Measurement, among others, have recommended voluntary guidelines to limit human exposure to RF energy. And the ARRL has established the RF Safety Committee, a committee of concerned medical doctors and scientists, serving voluntarily to monitor scientific research in the fields and to recommend safe practices for radio amateurs.

Thermal Effects of RF Energy

Body tissues that are subjected to very high levels of RF energy may suffer serious heat damage. These effects depend upon the frequency of the energy, the power density of the RF field that strikes the body, and even on factors such as the polarization of the wave.

At frequencies near the body's natural resonant frequency, RF energy is absorbed more efficiently, and maximum heating occurs. In adults, this frequency usually is about 35 MHz if the person is grounded, and about 70 MHz if the person's body is insulated from the ground. Also, body parts may be resonant; the adult head, for example is resonant around 400 MHz, while a baby's smaller head resonates near 700 MHz. Body size thus determines the frequency at which most RF energy is absorbed. As the frequency is increased above resonance, less RF heating generally occurs. However, additional longitudinal resonances occur at about 1 GHz near the body surface. *Specific absorption rate (SAR)* is a term that describes the rate at which RF energy is absorbed into the human body. Maximum permissible exposure (MPE) limits are based on whole-body SAR values. This helps explain why these safe exposure limits vary with frequency.

Nevertheless, thermal effects of RF energy should not be a major concern for most radio amateurs because of the relatively low RF power we normally use and the intermittent nature of most amateur transmissions. Amateurs spend more time listening than transmitting, and many amateur transmissions such as CW and SSB use low-duty-cycle modes. (With FM or RTTY, though, the RF is present continuously at its maximum level during each transmission.) In any event, it is rare for radio amateurs to be subjected to RF fields strong enough to produce thermal effects unless they are fairly close to an energized antenna or unshielded power amplifier. Specific suggestions for avoiding excessive exposure are offered later.

[Turn to Chapter 13 now and study questions N0A02, N0A04, N0A05, N0B06, N0E07 and N0E08. Review this section as needed.]

[**If you are studying for a Technician license exam**, turn to Chapter 14 and study questions T0A03, T0A11, T0B01, T0B02, T0B03 T0B04, T0B05, T0B08 and T0B09. Review this section if you are uncertain of the answer to any of these questions.]

Athermal Effects of EMR

Nonthermal effects of EMR may be of greater concern to most amateurs because they involve lower level energy fields. Research about possible health effects resulting from exposure to the lower level energy fields, the athermal effects, has been of two basic types: epidemiological research and laboratory research.

Scientists conduct laboratory research into biological mechanisms by which EMR may affect animals including humans. Epidemiologists look at the health patterns of large groups of people using statistical methods. These epidemiological studies have been inconclusive. By their basic design, these studies do not demonstrate cause and effect, nor do they postulate mechanisms of disease. Instead, epidemiologists look for associations between an environmental factor and an observed pattern of illness. For example, in the earliest research on malaria, epidemiologists observed the association between populations with high prevalence of the disease and the proximity of mosquito infested swamplands. It was left to the biological and medical scientists to isolate the organism causing malaria in the blood of those with the disease and identify the same organisms in the mosquito population.

In the case of athermal effects, some studies have identified a weak association between exposure to EMF at home or at work and various malignant conditions including leukemia and brain cancer. A larger number of equally well designed and performed studies, however, have found no association. A risk ratio of between 1.5 and 2.0 has been observed in positive studies (the number of observed cases of malignancy being 1.5 to 2.0 times the "expected" number in the population). Epidemiologists generally regard a risk ratio of 4.0 or greater to be indicative of a strong association between the cause and effect under study. For example, men who smoke one pack of cigarettes per day increase their risk for lung cancer tenfold compared to nonsmokers, and two packs per day increase the risk to more than 25 times the nonsmokers' risk.

Epidemiological research by itself is rarely conclusive, however. Epidemiology only identifies health patterns in groups—it does not ordinarily determine their cause. And there are often confounding factors: Most of us are exposed to many different environmental hazards that may affect our health in various ways. Moreover, not all studies of persons likely to be exposed to high levels of EMR have yielded the same results.

There has also been considerable laboratory research about the biological effects of EMR in recent years. For example, it has been shown that even fairly low levels of EMR can alter the human body's circadian rhythms, affect the manner in which cancer-fighting T lymphocytes function in the immune system, and alter the nature of the electrical and chemical signals communicated through the cell membrane and between cells, among other things.

Much of this research has focused on low-frequency magnetic fields, or on RF fields that are keyed, pulsed or modulated at a low audio frequency (often below 100 Hz). Several studies suggested that humans and animals can adapt to the presence of a steady RF carrier more readily than to an intermittent, keyed or modulated energy source. There is some evidence that while EMR may not directly cause cancer, it may sometimes combine with chemical agents to promote its growth or inhibit the work of the body's immune system.

None of the research to date conclusively proves that low-level EMR causes adverse health effects. Given the fact that there is a great deal of ongoing research to examine the health consequences of exposure to EMF, the American Physical Society (a national group of highly respected scientists) issued a statement in May 1995 based on its review of available data pertaining to the possible connections of cancer to 60-Hz EMF exposure. This report is exhaustive and should be reviewed by anyone with a serious interest in the field. Among its general conclusions were the following:

1. "The scientific literature and the reports of reviews by other panels show no consistent, significant link between cancer and powerline fields."
2. "No plausible biophysical mechanisms for the systematic initiation or promotion of cancer by these extremely weak 60-Hz fields has been identified."
3. "While it is impossible to prove that no deleterious health effects occur from exposure to any environmental factor, it is necessary to demonstrate a consistent, significant, and causal relationship before one can conclude that such effects do occur."

The APS study is limited to exposure to 60-Hz EMF. Amateurs will also be interested in exposure to EMF in the RF range. A 1995 publication entitled *Radio Frequency and ELF Electromagnetic Energies, A Handbook for Health Professionals* includes a chapter called "Biologic Effects of RF Fields." In it the authors state: "In conclusion, the data do not support the finding that exposure to RF fields is a causal agent for any type of cancer" (page 176). Later in the same chapter they write: "Although the data base has grown substantially over the past decades, much of the information concerning nonthermal effects is generally inconclusive, incomplete, and sometimes contradictory. Studies of human populations have not demonstrated any reliably effected end point" (page 186).

Readers may want to follow this topic as further studies are reported. Amateurs should be aware that exposure to RF and ELF (60 Hz) electromagnetic fields at all power levels and frequencies may not be completely safe. Prudent avoidance of any avoidable EMR is always a good idea. However, an Amateur Radio operator should not be fearful of using his or her equipment. If any risk does exist, it will almost surely fall well down on the list of causes that may be harmful to your health (on the other end of the list from your automobile).

Safe Exposure Levels

How much EM energy is safe? Scientists and regulators have devoted a great deal of effort to deciding upon safe RF-exposure limits. This is a very complex problem, involving difficult public health and economic considerations. The recommended safe levels have been revised downward several times in recent years—and not all scientific bodies agree on this question even today. An Institute of Electrical and Electronics Engineers (IEEE) standard for recommended EM exposure limits went into effect in 1991. It replaced a 1982 American National-Standards Institute (ANSI) standard that permitted somewhat higher exposure levels. The new IEEE standard was adopted by ANSI in 1992.

The IEEE standard recommends frequency-dependent and time-dependent maximum permissible exposure levels. Unlike earlier versions of the standard, the 1991 standard recommends different RF exposure limits in **controlled environments** (that is, where energy levels can be accurately determined and everyone on the premises is aware of the presence of EM fields) and in **uncontrolled environments** (where energy levels are not known or where some persons present may not be aware of the EM fields).

The graph in Figure 10-1 depicts the 1991 IEEE standard. It is necessarily a complex graph because the standards differ not only for controlled and uncontrolled environments but also for electric fields (E fields) and magnetic fields (H fields). Basically, the lowest E-field exposure limits occur at frequencies between 30 and 300 MHz. The lowest H-field exposure levels occur at 100-300 MHz. The ANSI standard sets the maximum E-field limits between 30 and 300 MHz at a power density of 1 mW/cm^2 (61.4 V/m) in controlled environments—but at one-fifth that level (0.2 mW/cm^2 or 27.5 V/m) in uncontrolled environments. The H-field limit drops to 1 mW/cm^2 (0.163 A/m) at 100-300 MHz in controlled environments and 0.2 mW/cm^2 (0.0728 A/m) in uncontrolled environments. Higher power densities are permitted at frequencies below 30 MHz (below 100 MHz for H fields) and above 300 MHz, based on the concept that the body will not be

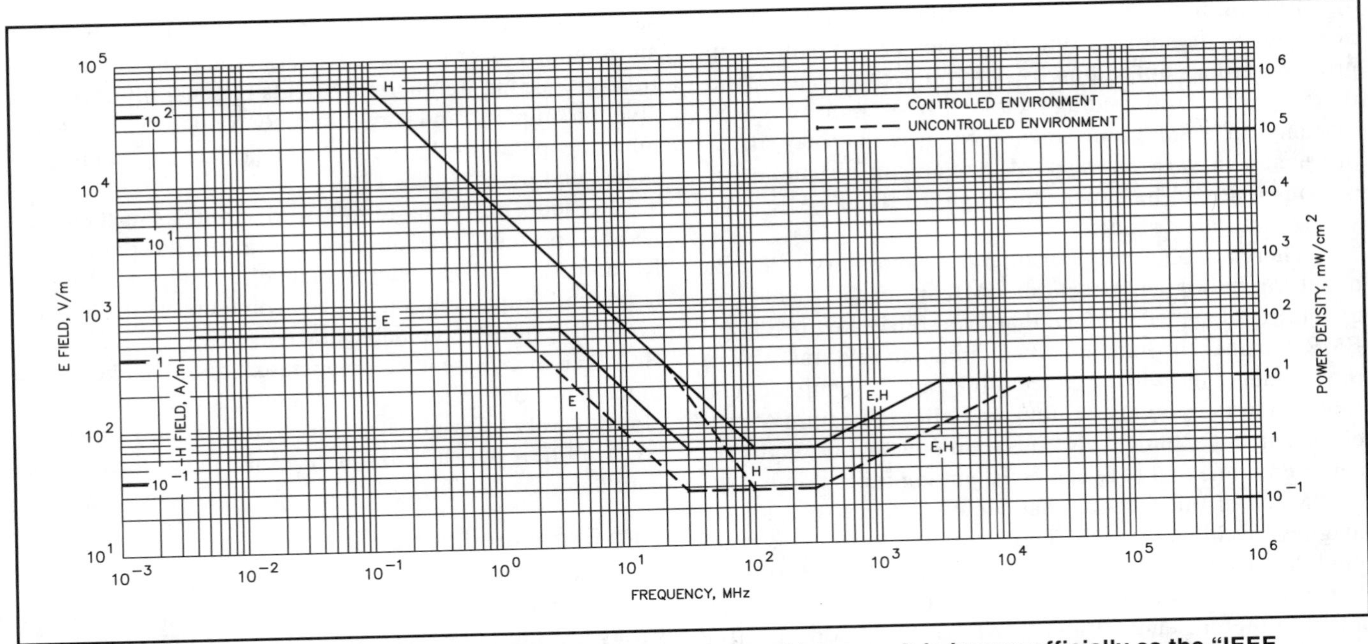

Figure 10-1 — The 1991 RF protection standard for body exposure of humans. It is known officially as the "IEEE Standard for Safety Levels with Respect to Human Exposure to Radio Frequency Electromagnetic Fields, 3 kHz to 300 GHz." (Note that the exposure levels set by this standard are not the same as the FCC maximum permissible exposure limits.)

resonant at those frequencies and will therefore absorb less energy.

In general, the 1991 IEEE standard requires averaging the power level over time periods ranging from 6 to 30 minutes for power-density calculations, depending on the frequency and other variables. The ANSI exposure limits for uncontrolled environments are lower than those for controlled environments, but to compensate for that the standard allows exposure levels in those environments to be averaged over much longer time periods (generally 30 minutes). This long averaging time means that an intermittently operating RF source (such as an Amateur Radio transmitter) will show a much lower power density than a continuous-duty station for a given power level and antenna configuration.

Time averaging is based on the concept that the human body can withstand a greater rate of body heating (and thus, a higher level of RF energy) for a short time than for a longer period. Time averaging may not be appropriate, however, when considering nonthermal effects of RF energy.

The IEEE standard excludes any transmitter with an output below 7 W because such low-power transmitters would not be able to produce significant whole-body heating. (Recent studies show that hand-held transceivers often produce power densities in excess of the IEEE standard within the head.)

There is disagreement within the scientific community about these RF exposure guidelines. The IEEE standard is still intended primarily to deal with thermal effects, not exposure to energy at lower levels. A small but significant number of researchers now believe athermal effects should also be taken into consideration. Several European countries and localities in the United States have adopted stricter standards than the recently updated IEEE standard.

Another national body in the United States, the National Council for Radiation Protection and Measurement (NCRP), has also adopted recommended exposure guidelines. NCRP urges a limit of 0.2 mW/cm^2 for non-occupational exposure in the 30-300 MHz range. The NCRP guideline differs from IEEE in two notable ways: It takes into account the effects of modulation on an RF carrier, and it does not exempt transmitters with outputs below 7 W.

[Turn to Chapter 13 and study questions N0B02 through N0B05. Review this section as needed.]

Cardiac Pacemakers and RF Safety

It is a widely held belief that cardiac pacemakers may be adversely affected in their function by exposure to electromagnetic fields. Amateurs with pacemakers may ask whether their operating might endanger themselves or visitors to their shacks who have a pacemaker. Because of this and similar concerns regarding other sources of electromagnetic fields, pacemaker manufacturers apply design methods that for the most part shield the pacemaker circuitry from even relatively high EM field strengths.

It is recommended that any amateur who has a pacemaker or is being considered for one discuss this matter with his or her physician. The physician will probably put the amateur into contact with the technical representative of the pacemaker manufacturer. These representatives are generally excellent resources and may have data from laboratory or "in the field" studies with pacemaker units of the type the amateur needs to know about.

One study examined the function of a modern (dual chamber) pacemaker in and around an Amateur Radio station. The pacemaker generator has circuits that receive and process electrical signals produced by the heart and also generate electrical signals that stimulate (pace) the heart. In one series of experiments the pacemaker was connected to a heart simulator. The system was placed on top of the cabinet of a 1-kW HF linear amplifier during SSB and CW operation. In addition, the system was placed in close proximity to several 1 to 5-W 2-meter hand-held transceivers. The test pacemaker connected to the heart simulator was also placed on the ground 9 meters below and 5 meters in front of a three-element Yagi HF antenna. No interference with pacemaker function was observed in this experimental system.

Although the possibility of interference cannot be entirely ruled out by these few observations, these tests represent more severe exposure to EM fields than would ordinarily be encountered by an amateur with an average amount of common sense. Of course prudence dictates that amateurs with pacemakers using hand-held VHF transceivers keep the antenna as far from the site of the implanted pacemaker generator as possible and use the lowest transmitter output required for adequate communication. For high power HF transmission, the antenna should be as far from the operating position as possible and all equipment should be properly grounded.

Low-Frequency Fields

Recently, much concern about EMR has focused on low-frequency energy rather than RF. Amateur Radio equipment can be a significant source of low-frequency magnetic fields, although there are many other sources of this kind of energy in the typical home. Magnetic fields can be measured relatively accurately with inexpensive 60-Hz dosimeters that are made by several manufacturers.

Table 10-1 shows typical magnetic field intensities of Amateur Radio equipment and various household items. Because these fields dissipate rapidly with distance, "prudent avoidance" would mean staying perhaps 12 to 18 inches away from most Amateur Radio equipment (and 24 inches from power supplies with 1-kW RF amplifiers) whenever the ac power is turned on. The old custom of leaning over a linear amplifier on a cold winter night to keep warm may not be the best idea!

There are currently no non-occupational US standards for exposure to low-frequency fields. Some epidemiological evidence, however, suggests that when the general level of 60-Hz fields exceeds 2 milligauss, there is an increased cancer risk in both domestic environments and industrial environments. Typical home environments (not close to appliances or power lines) are in the range of 0.1-0.5 milligauss.

Table 10-1

Typical 60-Hz Magnetic Fields Near Amateur Radio Equipment and AC-Powered Household Appliances

Item	*Field (milligauss)*	*Distance*
Electric blanket	30-90	Surface
Microwave oven	10-100	Surface
	1-10	12″
IBM personal computer	5-10	Atop monitor
	0-1	15″ from screen
Electric drill	500-2000	At handle
Hair dryer	200-2000	At handle
HF transceiver	10-100	Atop cabinet
	1-5	15″ from front
1-kW RF amplifier	80-1000	Atop cabinet
	1-25	15″ from front

(Source: measurements made by members of the ARRL RF Safety Committee)

Determining RF Power Density

Unfortunately, determining the power density of the RF fields generated by an amateur station is not as simple as measuring low-frequency magnetic fields. Although sophisticated instruments can be used to measure RF power densities quite accurately, they are costly and require frequent recalibration. Most amateurs don't have access to such equipment, and the inexpensive field-strength meters that we do have are not suitable for measuring RF power density. The best we can usually do is to estimate our own RF power density based on measurements made by others or, given sufficient computer programming skills, use computer modeling techniques. The FCC has prepared a bulletin, "OET Bulletin 65: Evaluating Compliance With FCC-Specified Guidelines for Human Exposure to Radio Frequency Radiation," that contains charts and tables that amateurs can use to estimate compliance with the rules.

Table 10-2 shows a sampling of measurements made at Amateur Radio stations by the Federal Communications Commission and the Environmental Protection Agency in 1990. As this table indicates, a good antenna well removed from inhabited areas poses no hazard under any of the various exposure guidelines. However, the FCC/EPA survey also indicates that amateurs must be careful about using indoor or attic-mounted antennas, mobile antennas, low directional arrays or any other antenna that is close to inhabited areas, especially when moderate to high power is used.

Ideally, before using any antenna that is in close proximity to an inhabited area, you should measure the RF power density. If that is not feasible, the next best option is to make the installation as safe as possible by observing the safety suggestions listed in Table 10-3.

It is also possible, of course, to calculate the probable power density near an antenna using simple equations. Such calculations have many pitfalls. For one, most of the situations in which the power density would be high enough to be of concern are in the near field. The boundary between the near field and the far field of an antenna is approximately several wavelengths from the antenna. In the near field, ground interactions and other variables produce power densities that cannot be determined by simple arithmetic. In the far field, conditions become

Table 10-2

Typical RF Field Strengths Near Amateur Radio Antennas

A sampling of values as measured by the Federal Communications Commission and Environmental Protection Agency, 1990

Antenna Type	*Freq (MHz)*	*Power (W)*	*E Field (V/m)*	*Location*
Dipole in attic	14.15	100	7-100	In home
Discone in attic	146.5	250	10-27	In home
Half sloper	21.5	1000	50	1 m from base
Dipole at 7-13 ft	7.14	120	8-150	1-2 m from earth
Vertical	3.8	800	180	0.5 m from base
5-element Yagi at 60 ft	21.2	1000	10-20	In shack
			14	12 m from base
3-element Yagi at 25 ft	28.5	425	8-12	12 m from base
Inverted V at 22-46 ft	7.23	1400	5-27	Below antenna
Vertical on roof	14.11	140	6-9	In house
			35-100	At antenna tuner
Whip on auto roof	146.5	100	22-75	2 m from antenna
			15-30	In vehicle
			90	Rear seat
5-element Yagi at 20 ft	50.1	500	37-50	10 m from antenna

Table 10-3

RF Awareness Guidelines

These guidelines were developed by the ARRL RF Safety Committee, based on the FCC/EPA measurements of Table 10-2 and other data.

- Although antennas on towers (well away from people) pose no exposure problem, make certain that the RF radiation is confined to the antennas' radiating elements themselves. Provide a single, good station ground (earth), and eliminate radiation from transmission lines. Use good coaxial cable, not open-wire lines or end-fed antennas that come directly into the transmitter area.
- No person should ever be near any transmitting antenna while it is in use. This is especially true for mobile or ground-mounted vertical antennas. Avoid transmitting with more than 25 W in a VHF mobile installation unless it is possible to first measure the RF fields inside the vehicle. At the 1-kW level, both HF and VHF directional antennas should be at least 35 ft above inhabited areas. Avoid using indoor and attic-mounted antennas if at all possible.
- Don't operate high-power amplifiers with the covers removed, especially at VHF/UHF.
- In the UHF/SHF region, never look into the open end of an activated length of waveguide or microwave feed-horn antenna or point it toward anyone. (If you do, you may be exposing your eyes to more than the maximum permissible exposure level of RF radiation.) Never point a high-gain, narrow-bandwidth antenna (a paraboloid, for instance) toward people. Use caution in aiming an EME (moonbounce) array toward the horizon; EME arrays may deliver an effective radiated power of 250,000 W or more.
- With hand-held transceivers, keep the antenna away from your head and use the lowest power possible to maintain communications. Use a separate microphone and hold the rig as far away from you as possible. This will reduce your exposure to the RF energy.
- Don't work on antennas that have RF power applied.
- Don't stand or sit close to a power supply or linear amplifier when the ac power is turned on. Stay at least 24 inches away from power transformers, electrical fans and other sources of high-level 60-Hz magnetic fields.

easier to predict with simple calculations. It is difficult to accurately evaluate the effects of RF radiation exposure in the near field. The boundary between the near field and the far field depends on the wavelength of the transmitted signal and the physical size and configuration of the antenna.

Computer antenna-modeling programs such as *MININEC* or other codes derived from *NEC* (Numerical Electromagnetics Code) are suitable for estimating RF magnetic and electric fields around amateur antenna systems. (See Chapter 2 of *The ARRL Antenna Book* for more information about *NEC* and *MININEC*.) And these too have limitations. Ground interactions must be considered in estimating near-field power densities. Also, computer modeling is not sophisticated enough to predict "hot spots" in the near field—places where the field intensity may be far higher than would be expected.

Intensely elevated but localized fields often can be detected by professional measuring instruments. These "hot spots" are often found near wiring in the shack and metal objects such as antenna masts or equipment cabinets. But even with the best instrumentation, these measurements may also be misleading in the near field.

One need not make precise measurements or model the exact antenna system, however, to develop some idea of the relative fields around an antenna. Computer modeling using close approximations of the geometry and power input of the antenna will generally suffice. Those who are familiar with *MININEC* can estimate their power densities by computer modeling, and those who have access to professional power-density meters can make useful measurements.

While our primary concern is ordinarily the intensity of the signal radiated by an antenna, we should also remember that there are other potential energy sources to be considered. You can also be exposed to RF radiation directly from a power amplifier if it is operated without proper shielding. Transmission lines may also radiate a significant amount of energy under some conditions.

Further RF Exposure Suggestions

Potential exposure situations should be taken seriously. Based on the FCC/EPA measurements and other data, the "RF awareness" guidelines of Table 10-3 were developed by the ARRL RF Safety Committee. A longer version of these guidelines, along with a complete list of references, appeared in a *QST* article by Ivan Shulman, MD, WC2S (see the list of RF Safety References at the end of this chapter).

In addition, *QST* carries information regarding the latest developments for RF safety precautions and regulations at the local and federal levels.

[Now turn to Chapter 13 and study questions N0A06, N0E04, N0E05, N0E06, N0E09 and N0E10. Review this section if you are uncertain of the answer to any of these questions.]

[**If you are studying for a Technician class license,** you should also turn to Chapter 14 and study questions T0A01, T0A08, T0A09, T0D02, T0D03 and T0E09. Review this section as needed.]

ADDITIONAL RF SAFETY INFORMATION FOR THE NOVICE LICENSE EXAM

Limiting RF Exposure

You must always take care to position your amateur antennas in a manner so they cannot harm you or anyone else. The simplest way to do this is to always install them high and in the clear, away from buildings or other locations where people might be close to them. To prevent **RF burns** you must be sure no one can touch the antenna while you are transmitting into it. (Some feed lines can also give RF burns, as can some circuits, such as the inductor and capacitors in an antenna tuner.) It doesn't matter what type of antenna it is, or how much power you are running. If you or someone else can touch the antenna, it is too close. The insulation on many hand-held radio antennas, along with the lower power of these radios, reduce the danger of RF burns. You still should not touch the antenna while transmitting.

As you learned in the previous section of this chapter, even if you can't touch the antenna, the RF fields can still produce some heating of your body tissue. You will learn more about how to estimate a safe distance to minimize any RF heating later in this chapter.

The combination of electric and magnetic waves of energy that are produced by your transmitter are called electromagnetic radiation. Electromagnetic radiation with frequencies between about 3 kHz and 300 GHz is called **radio frequency (RF) radiation**. (Higher-frequency electromagnetic waves, such as light, ultraviolet radiation and even gamma rays and X-rays are all above the RF range.)

The FCC has established limits for the maximum exposure that people can receive from transmitted electric and magnetic fields. You must ensure that people in and around your amateur station do not receive more than the maximum permissible exposure (MPE) from your station. FCC Rules

FCC RF-Exposure Regulations

FCC regulations control the amount of RF exposure that can result from your station's operation (§§97.13, 97.503, 1.1307 (b)(c)(d), 1.1310 and 2.1093). The regulations set limits on the maximum permissible exposure (MPE) allowed from operation of transmitters in all radio services. They also require that certain types of stations be evaluated to determine if they are in compliance with the MPEs specified in the rules. The FCC has also required that five questions on RF environmental safety practices be added to Novice, Technician and General class examinations.

These rules were announced on August 1, 1996. They were originally scheduled to go into effect on January 1, 1997, but the FCC responded to an ARRL petition and delayed the implementation date until January 1, 1998. This was done to give amateurs time to understand the rules, to conduct the required station evaluation and to make any changes necessary to be in compliance. The material presented here is the latest available at the time of printing. This discussion offers only an overview of this topic.

The Rules

Maximum Permissible Exposure (MPE)

The regulations control *exposure* to RF fields, not the *strength* of RF fields. There is no limit to how strong a field can be as long as no one is being exposed to it, although FCC regulations require that amateurs use the minimum necessary power at all times (§97.311 [a]). All radio stations must comply with the requirements for MPEs, even QRP stations running only a few watts or less. The MPEs vary with frequency, as shown in Table 10-4.

MPE limits are specified in maximum electric and magnetic fields for frequencies below 30 MHz, in power density for frequencies above 300 MHz and all three ways for frequencies from 30 to 300 MHz. For compliance purposes, all of these limits must be considered separately—if any one is exceeded, the station is not in compliance. For example, your 2-meter (146 MHz) station radiated electric field strength and power density may be less than the maximum allowed. If the radiated magnetic field strength exceeds that limit, however, your station does not meet the requirements.

Environments

In the latest rules, the FCC has defined two exposure environments—controlled and uncontrolled. A **controlled environment** is one in which the people who are being exposed are aware of that exposure and can take steps to minimize that exposure, if appropriate. In an **uncontrolled environment**, the people being exposed are not normally aware of the exposure. The uncontrolled environment limits are more stringent than the controlled environment limits.

Although the controlled environment is usually intended as an occupational environment, the FCC has determined that it generally applies to amateur operators and members of their immediate households. In most cases, controlled-environment limits can be applied to your home and property to which you can control physical access. The uncontrolled environment is intended for areas that are accessible by the general public, normally your neighbors' properties and the public sidewalk areas around your home. In either case, you can apply the more restrictive limits, if you choose.

Station Evaluations

The FCC requires that certain amateur stations be evaluated for compliance with the MPEs. This will help ensure a safe operating environment for amateurs, their families and neighbors. Although an amateur can have someone else do the evaluation, it is not difficult for hams to evaluate their own stations. FCC Bulletin 65 contains basic information about the regulations and a number of tables that show compliance distances for specific antennas and power levels. Generally, hams will use these tables to evaluate their stations. If they choose, however, they can do more extensive calculations, use a computer to model their antenna and exposure, or make actual measurements.

In most cases, hams will be able to use an FCC table that best describes their station's operation to determine the minimum compliance distance for their specific operation. Although such tables are not yet available from the FCC at the time of printing, we expect that they will show the compliance distances for uncontrolled environments for a particular type of antenna at a particular height. The power levels shown in a table would be average power levels, adjusted for the duty cycle of the operating mode being used, and operating on and off time. The data would be averaged over 6 minutes for controlled environments or 30 minutes for uncontrolled environments.

Categorical Exemptions

Some types of amateur stations do not need to be evaluated, but these stations must still comply with the MPE limits. The station licensee remains responsible for ensuring that the station meets these requirements.

The FCC has exempted these stations from the evaluation requirement because their output power, operating mode and frequency are such that they are presumed to be in compliance with the rules.

Amateur station licensees are not required to perform a routine RF environmental evaluation if the transmitter output PEP is less than or equal to the limits specified in Section 97.13(c) of the FCC Rules. (This section is printed on page 10-10.)

When the RF safety rules were first put into effect, in August 1996, any amateur station transmitting with more than 50 watts PEP was required to perform an RF environmental evaluation. In August 1997 the FCC released a change to the Rules that resulted in the table of transmitter output power levels listed in § 97.13(c). The ARRL/VEC believes questions N0C04, N0D04, N0D05 and T0D01 are no longer valid because of this change in the FCC Rules. While the ARRL/VEC will not use any of these questions on exams they coordinate, the VEC Question Pool

may require you to perform a routine RF radiation exposure evaluation to show that your station meets these requirements. You should have a basic understanding of some factors that can affect the strength of the RF fields that radiate from your antenna. This will help you understand the procedures you can follow to perform such an evaluation, and the steps you can take to minimize exposure.

Higher transmitter power will produce stronger radiated RF fields. So using the minimum power necessary to carry out your communications will minimize the exposure of anyone near your station. Reducing power may be a way of meeting the FCC MPE limits.

An antenna that is higher and farther away from people also reduces the strength of the radiated fields that anyone will be exposed to. Generally, if you can raise your antenna higher in the air or move it farther from your neighbor's property line you may reduce exposure.

A half-wavelength dipole antenna that is only 5 meters above the ground would generally create a stronger RF field on the ground beneath the antenna than other antennas

Committee was still considering whether to withdraw the questions from use when this second printing of *Now You're Talking!* went to the printer in early September, 1997. If you take an exam coordinated by a VEC other than the ARRL/VEC, your exam *may* included one (or more) of those questions. In that case, you should keep in mind that *any* transmitter with 50 watts PEP output or less, regardless of frequency, does not require an RF environmental evaluation. *Some* transmitters with more than 50 watts PEP output may require the evaluation to be performed. The power levels are now higher than 50 watts for stations operating on bands other than the 10, 6, 2 and 1.25-meter bands. For the latest news and information about the FCC RF radiation safety rules and the exam question pools visit the ARRL's website: **http://www.arrl.org/rfsafety/**

Hand-held radios and vehicle-mounted mobile radios that operate using a push-to-talk (PTT) button are also categorically exempt from performing the routine evaluation.

Table 10-4

(From §1.1310) Limits for Maximum Permissible Exposure (MPE)

(A) Limits for Occupational/Controlled Exposure

Frequency Range (MHz)	Electric Field Strength (V/m)	Magnetic Field (A/m)	Power Density (mW/cm²)	Averaging Time (minutes)
0.3-3.0	614	1.63	(100)*	6
3.0-30	1842/f	4.89/f	(900/f²)*	6
30-300	61.4	0.163	1.0	6
300-1500	——	——	f/300	6
1500-100,000	——	——	5	6
f = frequency in MHz		* = Plane-wave equivalent power density (see Note 1)		

(B) Limits for General Population/Uncontrolled Exposure

Frequency Range (MHz)	Electric Field Strength (V/m)	Magnetic Field (A/m)	Power Density (mW/cm²)	Averaging Time (minutes)
0.3-1.34	614	1.63	(100)*	30
1.34-30	824/f	2.19/f	(180/f²)*	30
30-300	27.5	0.073	0.2	30
300-1500	——	——	f/1500	30
1500-100,000	——	——	1.0	30
f = frequency in MHz		* = Plane-wave equivalent power density (see Note 1)		

Note 1: This means the equivalent far-field strength that would have the E or H-field component calculated or measured. It does not apply well in the near field of an antenna. The equivalent far-field power density can be found in the near or far regions from the relationships:

$P_d = |E_{total}|^2 / 3770 \text{ mW/cm}^2$ or from $P_d = |H_{total}|^2 \times 37.7 \text{ mW/cm}^2$

Correcting Problems

Most hams are already in compliance with the MPE requirements. Some amateurs, especially those using indoor antennas or high-power, high-duty-cycle modes such as a RTTY bulletin station and specialized stations for moonbounce operations and the like may need to make adjustments to their station or operation to be in compliance.

The FCC permits amateurs considerable flexibility in complying with these regulations. Hams can adjust their operating frequency, mode or power to comply with the MPE limits. They can also adjust their operating habits or control the direction their antenna is pointing. For example, if an amateur were to discover that the MPE limits had been exceeded for uncontrolled exposure after 28 minutes of transmitting, the FCC would consider it perfectly acceptable to take a 2-minute break after 28 minutes.

Ongoing Developments

A number of organizations and individuals, including the ARRL, have filed petitions seeking changes to these regulations. Any of these could result in significant changes to the rules. ARRL will announce any of these changes that could affect Amateur Radio. Check *QST,* W1AW bulletins and the RF Exposure News page on the ARRL Web site (**http://www.arrl.org/news/rfsafety**).

[After you have studied the text and table in this sidebar you should turn to Chapter 13 and check your understanding of the material by answering questions N0B01, N0B07, N0B09, N0C01 through N0C10 and questions N0D03 through N0D05. Review this sidebar if any of these questions give you difficulty.]

[**If you are studying for a Technician license**, you should also turn to Chapter 14 and study questions T0A02, T0A10, T0B07, T0D01, T0D04 and T0E11. Review this sidebar if you are uncertain of the answers to any of these questions.]

that are located much higher. For example, a large horizontal loop, a 3-element Yagi antenna or a 3-element Quad antenna generally have significantly more gain than a dipole. Yet at a height of 30 meters each of these antennas would produce a smaller RF field strength on the ground beneath the antenna than would the low dipole. As a general rule, place your antenna at least as high as necessary to ensure that you meet the FCC radiation exposure guidelines.

You can also use the radiation pattern of the antenna to your advantage in controlling exposure. For example, if you position your $1/2$-λ dipole antenna (with maximum radiation off the sides of the antenna and minimum radiation off the ends) so the ends are pointed at your neighbor's house (or your house) you will reduce the exposure. A beam antenna can have an even more dramatic effect on reducing the exposure. Simply do not point the antenna in the direction where people will most likely be located.

Even your choice of operating frequency can have an affect. Humans absorb less RF energy at some frequencies (and the MPE is higher at those frequencies). You can reduce exposure by selecting an operating frequency with a higher MPE.

Everything else being equal, some emission modes will result in less RF radiation exposure than others. For example, modes like RTTY or FM voice transmit at full power during the entire transmission. On CW, you transmit at full power during dots and dashes and at zero power during the space between these elements. A single-sideband (SSB) phone signal generally produces a lower radiation exposure than a CW or FM transmitter because the transmitter is at full power for only a small percentage of the time during a single transmission. The **duty cycle** of an emission is the ratio of the average power to the peak envelope power (PEP) of a transmission, expressed as a percentage. In effect, the concept of duty cycle takes into account the time a transmitter is operating at full power during a single transmission. An emission with a lower duty cycle produces less RF radiation exposure for the same PEP output.

Lower duty cycles, then, result in lower RF radiation exposures. That also means the antenna can be closer to people without exceeding their MPE limits. Compared to a 100% duty-cycle mode, people can be closer to your antenna if you are using a 50% duty-cycle mode.

Another step you can take to limit your RF radiation exposure is by reducing your actual transmitting time. The FCC regulations specify time averaged MPE limits. For a controlled RF exposure environment, the exposure is averaged over any 6-minute period. For an uncontrolled RF exposure environment, the exposure is averaged over any 30-minute period. So if your routine RF radiation exposure evaluation indicates that you might exceed the MPE limits for a controlled RF environment, reduce your actual transmit time during any 6-minute period. If you might exceed the MPE limits for an uncontrolled environment, reduce your actual transmit time during any 30-minute period, as requred to meet the RF exposure limits.

[You have been studying some very important information. Let's review what you have learned before going on to more new material. Turn to Chapter 13 and study questions N0A01, N0A03 and N0A07 through N0A11. Also study questions N0B08, N0D01, N0D11, N0E01 through N0E03 and N0E11 through N0E13. Review this section if any of those questions give you trouble.]

[**If you are studying for the Technician license exam**, then you should also turn to Chapter 14 and study questions T0B06 and T0B10 through T0B12. Review this section as needed.]

FCC Maximum Permissible Exposure (MPE) Limits

In the first part of this chapter, we described the 1991 IEEE standard, which ANSI adopted in 1992, as well as the National Council for Radiation Protection and Measurement (NCRP) recommended exposure guidelines. When the FCC adopted RF radiation protection rules in 1996, they chose a blend of these guidelines. Table 10-4 in the RF Exposure Regulations sidebar lists the FCC Maximum Permissible Exposure limits for both the controlled and uncontrolled radiation environments at various frequencies.

Amateurs normally look only to Part 97 for the exact Rules governing Amateur Radio. There are other Parts to the FCC Rules, though, and in the case of the RF radiation exposure limits, you will have to look in Part 1 to find the exact limits. If we look in Part 97 for information about the exposure limits we find:

> **§ 97.13 Restrictions on station location.**
>
> (c) Before causing or allowing an amateur station to transmit from any place where the operation of the station could cause human exposure to RF electromagnetic field levels in excess of those allowed under § 1.1310 of this chapter, the licensee is required to take certain actions.
>
> (1) The licensee must perform the routine RF environmental evaluation prescribed by § 1.1307(b) of this chapter, if the transmitter PEP exceeds the following limits:

Wavelength Band	Transmitter Power (watts)
MF	
160 m	500
HF	
80 m	500
75 m	500
40 m	500
30 m	425
20 m	225
17 m	125
15 m	100
12 m	75
10 m	50
VHF (all bands)	50
UHF	
70 cm	70
33 cm	150
23 cm	200
13 cm	250
SHF (all bands)	250
EHF (all bands)	250

> (2) If the routine environmental evaluation indicates that the RF electromagnetic fields could exceed the limits contained in § 1.1310 of this chapter in accessible areas, the licensee must take action to prevent human exposure to such RF electromagnetic fields. Further information on evaluating compliance with these limits can be found in the FCC's OET Bulletin 65, "Evaluating Compliance with FCC-Specified Guidelines for Human Exposure to Radio Frequency Electromagnetic Fields."

As mentioned in the last line, FCC's Office of

Engineering and Technology (OET) Bulletin 65 also lists the specific MPE limits and provides some information that will help amateurs evaluate their stations. Table 10-4 will help you determine the MPE limits that apply to your amateur operations. If you will only be using one or two bands, you will only have to perform a station evaluation for those bands. If you will be using a variety of bands over the HF, VHF and UHF ranges, you will have to perform an evaluation for each band you will be using. Don't worry about trying to memorize the table. You will have a copy of it included with your exam papers, so you only have to know how to read the table to find the information you need.

For example, if you will be using the Novice HF bands, your first step will be to look at Table 10-4 and locate the MPE limits for that frequency range. (Remember that the Novice HF bands fall in the range from 3 to 30 MHz.) Section A of that table shows that for a range of 3 to 30 MHz, the **controlled environment** MPE values are listed for electric field strength (measured in volts per meter — V/m), magnetic field strength (measured in amperes per meter — A/m) and as plane-wave equivalent power density (measured in milliwatts per square centimeter — mW/cm^2). Because the MPE values in the HF range change with frequency, you will notice three equations given there rather than a specific value. The electric field strength MPE, given in V/m, is equal to 1842 divided by the operating frequency, in MHz. The magnetic field strength MPE, given in A/m, is 4.89 divided by the operating frequency, in MHz. The plane-wave equivalent power density MPE, given in mW/cm^2, is 900 divided by the square of the operating frequency, in MHz.

Of course you will also be interested in the **uncontrolled environment** MPE limits, so you will have to look at Section B of Table 10-4 for those values. Here you will notice that the HF bands are included in the range from 1.34 to 30 MHz. The MPE values again change with frequency, so you find three equations. The electric field strength MPE, given in V/m, is equal to 824 divided by the operating frequency, in MHz. The magnetic field strength MPE, given in A/m, is 2.19 divided by the operating frequency, in MHz. The plane-wave equivalent power density MPE, given in mW/cm^2, is 180 divided by the square of the operating frequency, in MHz.

If you want to know the MPE limits at a specific operating frequency, you have some simple calculations to perform. Simply use the three equations listed in Table 10-4 Section A and the three in Section B. For example, the controlled environment plane-wave equivalent power density MPE limit on 3.7 MHz is:

HF controlled Plane - Wave Equivalent Power Density MPE

$$= \frac{900}{(f\ (MHz))^2} \qquad \text{(Eq 10-1)}$$

HF controlled Plane - Wave Equivalent Power Density MPE

$$= \frac{900}{3.7^2} = \frac{900}{13.69} = 65.7 \frac{mW}{cm^2}$$

This value is called the *plane-wave equivalent power density* or the *equivalent far-field power density* because power density values are only meaningful in an antenna's far-field radiation zone. You will learn more about the near field and far-field zones of an antenna shortly.

Suppose you wanted to know the MPE limits for an uncontrolled environment at 28.4 MHz. Again, just perform the calculations indicated in Table 10-4 Section B for the HF range.

HF Uncontrolled Plane-Wave Equivalent Power Density MPE

$$= \frac{180}{(f\ (MHz))^2} \qquad \text{(Eq 10-2)}$$

HF Uncontrolled Plane-Wave Equivalent Power Density MPE

$$= \frac{180}{28.4^2} = \frac{180}{806.56} = 0.223 \frac{mW}{cm^2}$$

For some frequency ranges, the MPE limits are given as constant values. For example, over the VHF (30 to 300 MHz) range, the controlled environment MPE limits are 61.4 V/m for the electric field and 0.163 A/m for the magnetic field. The power density is 1.0 mW/cm^2. Look at Section B of Table 10-4 to find the uncontrolled environment limits over the VHF range (30 to 300 MHz). Here you will notice that the electric field MPE is 27.5 V/m and the magnetic field MPE is 0.073 A/m. The power density is 0.2 mW/cm^2. These same values apply to the amateur 2-meter (146-MHz) band as well as the 1.25-meter (222-MHz) band.

In the UHF (300 to 1500 MHz) and SHF (1500 to 100,000 MHz) ranges, the FCC only specifies MPE limits in power densities. This makes it easier for you to determine if your station meets the safe exposure limits because you only have to consider this one value, rather than electric field, magnetic field and power density values. Suppose you wanted to know the controlled environment MPE limit for the 70-cm (440-MHz) band or the Novice 1270-MHz band. Look at Section A of Table 10-4 and find the 300 to 1500-MHz value. This is another one that varies with frequency, so you are given an equation: divide the operating frequency in megahertz by 300.

$$\text{UHF Controlled Power Density MPE} = \frac{f\ (MHz)}{300} \qquad \text{(Eq 10-3)}$$

$$\text{UHF Controlled Power Density MPE} = \frac{1270}{300} = 4.2 \frac{mW}{cm^2}$$

To find the MPE limits for uncontrolled environments over this frequency range, you will have to look at Section B of Table 10-4. Here you can find the maximum power density by dividing the operating frequency by

$$\text{UHF Uncontrolled Power Density MPE} = \frac{f(MHz)}{1500} \qquad \text{(Eq 10-4)}$$

[Now it is time for a trip to Chapter 13 to study questions N0B10 through N0B14 and questions N0C11 through N0C14. Review this section as needed.]

[If you are studying for a Technician license exam, you should also turn to Chapter 14 and study questions T0C01 through T0C09. Review this section if any of those questions give you trouble.]

Determining the Field Strengths Around Your Antenna

You have just learned that the FCC maximum permissible exposure limits are given in terms of electric and magnetic field strengths. So how do you determine if the transmitted signal from your station is within these RF exposure limits? You must analyze, measure or otherwise determine your transmitted field strengths and power density. This means there are a number of ways you can perform the required routine RF radiation evaluation.

One way to do this is by making direct measurements of the electric and magnetic field strengths around your antenna while transmitting a signal. If you happen to have a calibrated field-strength meter with a calibrated field-strength sensor, you can make accurate measurements. Unfortunately, such calibrated meters are expensive and not normally found in an amateur's tool box. The relative-field-strength meters many amateurs use are not accurate enough to make this type of measurement.

Even if you do have access to a laboratory-grade calibrated field-strength meter, you must be aware of factors that can upset your readings. Reflections from ground and nearby conductors (power lines, other antennas, house wiring, etc) can easily confuse field strength readings. For example, the measuring probe and the person making the measurement can interact with the antenna radiation if they are in the near-field zone. In addition, you must know the frequency response of the test equipment and probes, and use them only within the appropriate range. Even the orientation of the test probe with respect to the test antenna polarization is important.

A wide-bandwidth instrument used to measure RF fields is calibrated over a wide frequency range and responds instantly to any signal within that range. The nice thing about a wide-bandwidth instrument is that it requires no tuning over its entire operating range. A narrow-bandwidth instrument, on the other hand, may be able to cover a wide frequency range, but would have a bandwidth of perhaps only a few kilohertz at any instant. You have to tune the instrument to the particular frequency of interest before making your measurements.

Why should we be concerned with the separation between the source antenna and the field-strength meter, which has its own receiving antenna? One important reason is that if you place a receiving antenna very close to an antenna when you measure the field strength, *mutual coupling* between the two antennas may actually alter the radiation from the antenna you are trying to measure.

This sort of mutual coupling can occur in the region very close to the antenna under test. This region is called the *reactive near-field* region. The term "reactive" refers to the fact that the mutual impedance between the transmitting and receiving antennas can be either capacitive or inductive in nature. The reactive near field is sometimes called the "induction field," meaning that the magnetic field usually is predominant over the electric field in this region. The transmitting antenna acts as though it were a rather large, lumped-constant inductor or capacitor, storing energy in the reactive near field rather than propagating it into space. For simple wire antennas, the reactive near field is considered to be within about a half wavelength from an antenna's radiating center.

The strength of the reactive near field decreases in a complicated fashion as you increase the distance from the antenna. Beyond the reactive near field, the antenna's radiated field is divided into two other regions: the *radiating near field* and the *radiating far field*. Nearly any metal object or other conductor that is located within the radiating near field can alter the radiation pattern of the antenna. Conductors such as telephone wiring or aluminum siding on a building that is located in the radiating near field will interact with the theoretical electric and magnetic fields to add or subtract intensity. This results in areas of varying field strength. Although you have measured the fields in the general area around your antenna and found that your station meets the MPE limits, there may still be "hot spots" or areas of higher field strengths within that region. In the near field of an antenna, the field strength varies in a way that depends on the type of antenna and other nearby objects as you move farther away from the antenna.

Because the boundary between the fields is rather "fuzzy," experts debate where one field begins and another leaves off, but the boundary between the radiating near and far fields for a full-sized dipole or Yagi antenna is generally accepted as:

$$D \approx \frac{2L^2}{\lambda} \qquad \text{(Eq 10-5)}$$

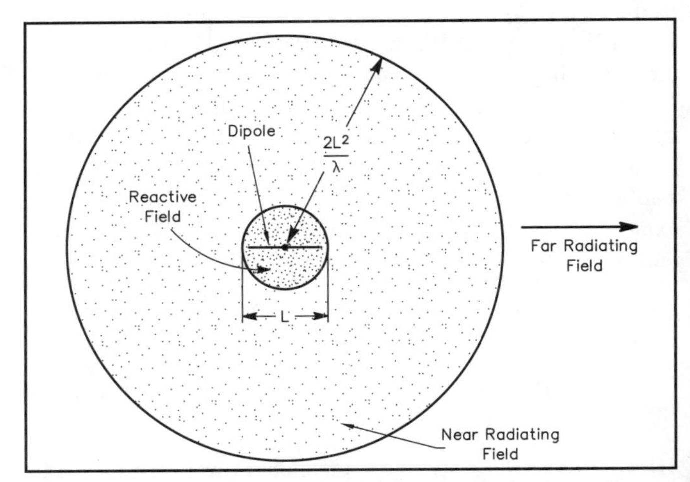

Figure 10-2 — This drawing illustrates the reactive near field, the radiating near field and the far field around a half-wavelength dipole antenna.

where L is the largest dimension of the physical antenna, expressed in the same units of measurement as the wavelength, λ. Remember, this is a rule of thumb, and this equation does not exactly define the near/far field boundary. Figure 10-2 shows the three fields surrounding of a simple wire antenna.

The radiating far-field forms the traveling electromagnetic waves. Far-field radiation is distinguished by a number of characteristics: The power density is proportional to the inverse square of the distance. (That means if you double the distance from the antenna, the power density will be one fourth as strong.) The electric and magnetic fields are perpendicular to each other in the wave front, and are in phase. The total energy is equally divided between the electric and magnetic fields. Beyond several wavelengths from the antenna, these are the only fields we need to consider.

In the far field there is a simple relationship between the electric (E) field and the magnetic (H) field of an electromagnetic wave. The strength of the E field divided by the strength of the H field is a constant 377 ohms, sometimes called the *intrinsic impedance of free space*.

$$\frac{E}{H} = \text{Intrinsic Impedance of Free Space} = 377\ \Omega \quad \text{(Eq 10-6)}$$

where:

E = electric field strength in volts per meter
H = magnetic field strength in amperes per meter

This relationship is only true in the far field of the antenna. In the near field the relationship is not this simple. This calculation may prove useful to you as you analyze your station for compliance with the FCC MPE limits. If you know the E or H field strength at some point in the far field then you can calculate the other value at that same point.

If you don't have the necessary measuring equipment, you can perform some calculations to determine the electric and magnetic field strengths from your station. This is normally done using computer programs, which take into account the gain and directivity of an antenna. There are a variety of programs available. The most reliable ones are based on the Numerical Electromagnetic Code (NEC) program that uses the Method of Moments analysis. (MININEC is a variation of the original program shortened to run on personal computers. Newer versions, such as NEC-2 and NEC-4 will run on modern PCs.) While these are powerful analysis tools, you must be careful to accurately model the antenna and all conductors that might be within a few wavelengths of the antenna for reliable results.

The FCC does not require you to keep any records of your routine RF radiation exposure evaluation. It is a good idea to keep them, however. They may prove useful if the FCC would ask for documentation to substantiate (prove) that an evaluation has been performed. The FCC will ask you to demonstrate that you have read and understood the FCC Rules about RF-radiation exposure by indicating that understanding on FCC Form 610 when you apply for your license.

[Turn to Chapter 13 now and study questions N0C15, N0D02 and N0D06 through N0D10. Review this section as needed.]

[If you are also preparing for a Technician class license exam, you should turn to Chapter 14 now and study questions T0A04 through T0A07. Also study questions T0E07 and T0E08.

Routine Evaluation by Tables

There is one more method of performing a routine RF radiation exposure evaluation that you should be aware of. In fact, this is probably the method most amateurs will use. It involves the use of some tables created to help you determine safe exposure distances from your antennas. You simply look through the tables to find the station and antenna configuration that is most like your station and then read the minimum recommended distances for controlled and uncontrolled environments. The tables are generated by using a computer analysis and modeling of various station configurations.

The FCC has promised to publish such tables in a supplement to their Office of Engineering and Technology (OET) Bulletin 65. Amateurs can use the information in that bulletin and supplement to evaluate their stations. The FCC was still working on that bulletin when this book went to press in September 1997. The ARRL is also working on a book that will include tables from a large number of such computer calculations.

The VEC Question Pool Committee wanted to be sure you knew how to read tables like these and evaluate your station in this simple, straightforward manner, so they created some tables and included questions. These tables were created using relatively simple computer calculations, so they may not be as accurate as those being created for OET Bulletin 65 or the ARRL book. They will serve as a good first approximation, however, so you can actually use them to evaluate your station until you gain access to tables calculated by more detailed computer analysis.

Table 10-5 shows the set of tables created for the question pools. When you are looking for a configuration that matches your station you must be careful to read all the notes at the beginning of each section, so you are sure to select the proper table. Be sure you read the appropriate column for either the controlled or uncontrolled environment distances. Also be careful to select the appropriate power level. If there is no listing for the power you normally use, then select the next higher value. This will recommend a farther distance for safety, and will provide a higher margin for error.

Are you ready to try a couple of examples? Let's find the minimum safe distance in an uncontrolled environment for a station using a half-wavelength dipole antenna on 3.5 MHz, if the transmitter produces 100 watts continuous output. Section C gives the tables for half-wavelength dipole antennas, and the first block of data is for 3.5-MHz operation. Find the row for 100-W operation, and read across to learn that in an uncontrolled environment everyone should be a minimum of 1.5 feet from any part of the antenna.

Table 10-5

RF Exposure Limits

Section A.

Estimated distances to meet RF power density guidelines in the main beam of a typical 3-element "triband" Yagi for the 14, 21 and 28 MHz Amateur Radio bands. Calculations include the EPA ground reflection factor of 2.56.

Frequency: 14 MHz
Antenna gain: 6.5 dBi
Controlled limit: 4.59 mW/cm²
Uncontrolled limit: 0.92 mW/cm²

Transmitter power (watts)	Distance to controlled limit	Distance to uncontrolled limit
100	4.6′	10.3′
500	10.3′	23.1′
1000	14.6′	32.7′
1500	17.9′	40′

Frequency: 21 MHz
Antenna gain: 7 dBi
Controlled limit: 2.04 mW/cm²
Uncontrolled limit: 0.408 mW/cm²

Transmitter power (watts)	Distance to controlled limit	Distance to uncontrolled limit
100	7.3′	16.4′
500	16.4′	36.7′
1000	23.2′	51.9′
1500	28.4′	63.6′

Frequency: 28 MHz
Antenna gain: 8 dBi
Controlled limit: 1.15 mW/cm²
Uncontrolled limit: 0.23 mW/cm²

Transmitter power (watts)	Distance to controlled limit	Distance to uncontrolled limit
100	11′	24.5′
500	24.5′	54.9′
1000	34.7′	77.6′
1500	42.5′	95.1′

Section B.

Estimated distances to meet RF power density guidelines with an omni-directional HF quarter-wave vertical or ground-plane antenna (estimated gain, 1 dBi). Calculations include the EPA ground reflection factor of 2.56.

Frequency: 3.5 MHz
Estimated antenna gain: 1 dBi
Controlled limit: 73.5 mW/cm²
Uncontrolled limit: 14.7 mW/cm²

Transmitter power (watts)	Distance to controlled limit	Distance to uncontrolled limit
100	0.6′	1.4′
500	1.4′	3.1′
1000	1.9′	4.3′
1500	2.4′	5.3′

Frequency: 7 MHz
Estimated antenna gain: 1 dBi
Controlled limit: 18.37 mW/cm²
Uncontrolled limit: 3.67 mW/cm²

Transmitter power (watts)	Distance to controlled limit	Distance to uncontrolled limit
100	1.2′	2.7′
500	2.7′	6.1′
1000	3.9′	8.7′
1500	4.7′	10.6′

Frequency: 14 MHz
Estimated antenna gain: 1 dBi
Controlled limit: 4.59 mW/cm²
Uncontrolled limit: 0.918 mW/cm²

Transmitter power (watts)	Distance to controlled limit	Distance to uncontrolled limit
100	2.5′	5.5′
500	5.5′	12.3′
1000	7.8′	17.3′
1500	9.5′	21.2′

Frequency: 21 MHz
Estimated antenna gain: 1 dBi
Controlled limit: 2.04 mW/cm²
Uncontrolled limit: 0.408 mW/cm²

Transmitter power (watts)	Distance to controlled limit	Distance to uncontrolled limit
100	3.7′	8.2′
500	8.2′	18.4′
1000	11.6′	26′
1500	14.2′	31.9′

Frequency: 28 MHz
Estimated antenna gain: 1 dBi
Controlled limit: 1.15 mW/cm²
Uncontrolled limit: 0.23 mW/cm²

Transmitter power (watts)	Distance to controlled limit	Distance to uncontrolled limit
100	4.9′	11′
500	11′	24.5′
1000	15.5′	34.7′
1500	19′	42.5′

Section C.

Estimated distances to meet RF power density guidelines with a horizontal half-wave dipole antenna (estimated gain, 2 dBi). Calculations include the EPA ground reflection factor of 2.56.

Frequency: 3.5 MHz
Estimated antenna gain: 2 dBi
Controlled limit: 73.5 mW/cm²
Uncontrolled limit: 14.7 mW/cm²

Transmitter power (watts)	Distance to controlled limit	Distance to uncontrolled limit
100	0.7′	1.5′
500	1.5′	3.4′
1000	2.2′	4.9′
1500	2.7′	6′

Frequency: 7 MHz
Estimated antenna gain: 2 dBi
Controlled limit: 18.37 mW/cm²
Uncontrolled limit: 3.67 mW/cm²

Transmitter power (watts)	Distance to controlled limit	Distance to uncontrolled limit
100	1.4′	3.1′
500	3.1′	6.9′
1000	4.3′	9.7′
1500	5.3′	11.9′

Frequency: 14 MHz
Estimated antenna gain: 2 dBi
Controlled limit: 4.59 mW/cm²
Uncontrolled limit: 0.918 mW/cm²

Transmitter power (watts)	Distance to controlled limit	Distance to uncontrolled limit
100	2.8′	6.2′
500	6.2′	13.8′
1000	8.7′	19.5′
1500	10.7′	23.8′

Frequency: 21 MHz
Estimated antenna gain: 2 dBi
Controlled limit: 2.04 mW/cm²
Uncontrolled limit: 0.408 mW/cm²

Transmitter power (watts)	Distance to controlled limit	Distance to uncontrolled limit
100	4.1′	9.2′
500	9.2′	20.6′
1000	13′	29.2′
1500	16′	35.7′

Frequency: 28 MHz
Estimated antenna gain: 2 dBi
Controlled limit: 1.15 mW/cm²
Uncontrolled limit: 0.23 mW/cm²

Transmitter power (watts)	Distance to controlled limit	Distance to uncontrolled limit
100	5.5′	12.3′
500	12.3′	27.5′
1000	17.4′	38.9′
1500	21.3′	47.7′

Section D.

Estimated distances to meet RF power density guidelines with a VHF quarter-wave ground-plane or mobile whip antenna (estimated gain, 1 dBi). Calculations include the EPA ground reflection factor of 2.56.

Frequency: 146 MHz
Estimated antenna gain: 1 dBi
Controlled limit: 1 mW/cm²
Uncontrolled limit: 0.2 mW/cm²

Transmitter power (watts)	Distance to controlled limit	Distance to uncontrolled limit
10	1.7′	3.7′
50	3.7′	8.3′
150	6.4′	14.4′

Table 10-5 (continued)

Section E.

Estimated distances to meet RF power density guidelines in the main beam of a UHF 5/8-wavelength ground-plane or mobile whip antenna (estimated gain, 4 dBi). Calculations include the EPA ground reflection factor of 2.56.

Frequency: 446 MHz
Estimated antenna gain: 4 dBi
Controlled limit: 1.49 mW/cm²
Uncontrolled limit: 0.3 mW/cm²

Transmitter power (watts)	Distance to controlled limit	Distance to uncontrolled limit
10	1.9′	4.3′
50	4.3′	9.6′
150	7.5′	16.7′

Section F.

Estimated distances to meet RF power density guidelines in the main beam of a 17-element Yagi on a five-wavelength boom designed for weak signal communications on the 144 MHz Amateur Radio band (estimated gain, 16.8 dBi). Calculations include the EPA ground reflection factor of 2.56.

Frequency: 144 MHz
Estimated antenna gain: 16.8 dBi
Controlled limit: 1 mW/cm²
Uncontrolled limit: 0.2 mW/cm²

Transmitter power (watts)	Distance to controlled limit	Distance to uncontrolled limit
10	10.2′	22.9′
100	32.4′	72.4′
500	72.4′	162′
1500	125.5′	280.6′

Section G.

Estimated distances to meet RF power density guidelines in the main beam of an array of eight 17-element Yagis with five-wavelength booms designed for earth-moon-earth ("moonbounce") communications on the 144 MHz Amateur Radio band (estimated gain, 24 dBi). Calculations include the EPA ground reflection factor of 2.56.

Frequency: 144 MHz
Estimated antenna gain: 24 dBi
Controlled limit: 1 mW/cm²
Uncontrolled limit: 0.2 mW/cm²

Transmitter power (watts)	Distance to controlled limit	Distance to uncontrolled limit
150	90.9′	203.3′
500	166′	371.1′
1500	287.4′	642.7′

Suppose your station operates on 28 MHz and you use a quarter-wavelength vertical antenna. What is the minimum safe distance if your transmitter produces 100-W continuous output? This time you will have to look in Section B of the table, and then find the data block for 28-MHz operation. Again, find the row for 100-W operation and read across to the controlled environment value. In this case no one who is aware of the RF (such as you or members of your family) should be able to get closer than 4.9 feet while the station is being operated for a full 6 minutes.

Take a look at Section A of Table 10-5. This section is for a typical three-element triband Yagi antenna. What is the safe exposure distance from this antenna when you are operating on 21 MHz with a 100-W continuous output transmitter? The second data block in this section is for 21-MHz operation, so go there and read the 100-W row. The minimum controlled environment distance is 7.3 feet and the minimum uncontrolled environment distance is 16.4 feet. These distances assume you are in the main beam of the antenna. In other words, the antenna is pointed directly at you. The radiated field strengths will be less in any direction off the main beam, so you could be closer to the antenna if you are off the side or back of the beam.

Suppose you mounted your three-element HF triband beam on top of a 30-foot tower. That should mean the installation is completely safe at the 100-W power level, even for a person standing directly under the antenna. Take a look around before you reach that conclusion, however. For example, if your neighbor's house is less than 16.4 feet from the antenna, and the second (or third) floor reaches the same height as your antenna, then it is possible that someone in your neighbor's house could get too close to your antenna.

Of course you have many ways to ensure that no one receives more than the maximum permissible exposure of RF radiation from your station. In this example you might simply take steps to ensure that you can't transmit with your antenna pointed directly at your neighbor's house. You can also reduce operating power, select an operating mode with a lower duty cycle or even limit your transmit time during any 30-minute averaging period.

You also have the option of restricting access to any areas of high RF radiation levels. This might mean putting up a fence around your property to ensure that no one can wander too close to your antennas while you are transmitting.

[Congratulations! You have now studied all the RF safety material for your Novice license exam. You should turn to Chapter 13 and study questions N0C16 through N0C27. Also study questions N0E14 through N0E17. Review this section if you don't understand the answers to any of these questions.]

[**If you are studying for a Technician license**, you should also turn to Chapter 14 and study questions T0C13 through T0C22. Also study questions T0D12 through T0D23, T0E10 and T0E12 through T0E28. Review this section as needed.]

ADDITIONAL RF SAFETY PRACTICES FOR THE TECHNICIAN EXAM

Using Graphs to Evaluate RF Radiation Exposure

It is possible to create graphs of field strength or power density based on computer analysis or other calculations. Figure 10-3 shows one such graph. This figure represents a beam antenna, such as a Yagi, that you might use with your amateur station. Some people might find it easier to read such graphs than search through the data in a table. Each antenna type requires its own graph, so you still may have to search through many drawings to find the one that best describes your station. Graphs such as these may be included in the FCC OET Bulletin 65. All this information will help you perform the most accurate possible evaluation of your station's RF radiation field strengths.

The power density of Figure 10-3 represents the signal in the main beam of this antenna. It is expressed for various levels of effective radiated power (ERP). ERP takes the antenna gain into account. For example, if you are using an antenna with 10-dBd of gain, and your transmitter produces 100 watts PEP output, then you would use the 1000-W ERP line. If you use only 10-watts PEP output with this antenna then you would use the 100-W ERP line.

Suppose you want to know the power density at a point 10 meters from your antenna when you have 1000-W ERP. Point 1 on this graph conveniently locates the 10-meter distance on the 1000-W ERP line of the graph. Now look to the axis along the left edge of the graph and read the power density. If you judged the value to be about 0.35 mW/cm^2 you would be pretty close.

Of course your evaluation is not complete at this point. Now you will have to determine the MPE limits for controlled and uncontrolled environments at your operating frequency. For a signal in the VHF range (30 to 300 MHz), the controlled environment power density limit is 1.0 mW/cm^2, so the power density at 10 meters is below this limit. For an uncontrolled environment, however, the power density limit is 0.2 mW/cm^2, so you will have to increase the distance to meet this limit.

To find the distance for this uncontrolled environment limit you should find 0.2 mW/cm^2 on the power density axis and look across to the right until you come to the 1000-W ERP line. You should come to point 5 on this graph. Now look down to the distance axis, and you should estimate that at about 15 meters you will meet the uncontrolled limit.

As you can see, a graph like this one can be quite helpful in evaluating the RF radiation exposure from your station at various distances and ERP levels.

[You should turn to Chapter 14 now and study questions T0D05 through T0D11. If you have difficulty with any of these questions, you should review this section.]

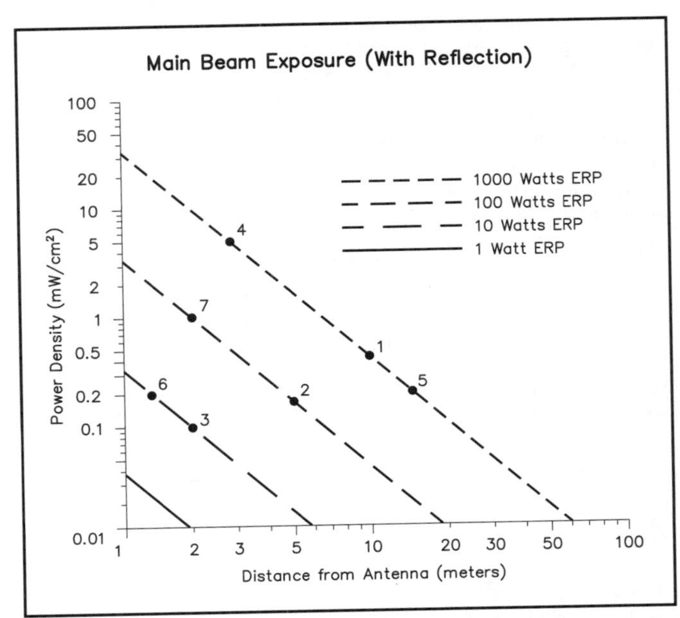

Figure 10-3 — Using computer analysis or other calculations, it is possible to create a graphical display of the field strengths and power density for various antennas and transmitter power levels. This graph represents the power density in the main beam of an antenna such as a Yagi. Various effective radiated power (ERP) levels are given. ERP takes the antenna gain into account.

General Safety Recommendations

There are a few additional RF safety points that you should be aware of when operating your Amateur Radio station. This section includes some general guidelines to help keep you and anyone near your station safe while you are operating your station.

Hand-held radios are very popular for VHF and UHF operation, especially with FM repeaters. They transmit with less than 7-watts of power, which is generally considered safe. Because the radios are designed to be operated with an antenna that is within 20 centimeters of your body, they are classified as **portable devices** by the FCC. Some special considerations are in order to ensure safe operation. This is especially true because hand-held radios generally place the antenna close to your head. Try to position the radio so the antenna is as far from your head (and especially your eyes) while transmitting. An external speaker microphone can be helpful. See Figure 10-4.

A mobile transceiver is one that is designed to be mounted in a vehicle. If the mobile transceiver uses a push-to-talk (PTT) switch to activate the transmitter, then it is exempt from the requirement to perform a routine RF radiation exposure evaluation. Mobile operations also require some special considerations. For example, you

Figure 10-4 — Keep the antenna of a hand-held radio as far from your head (especially your eyes) as possible while transmitting. A remote microphone allows you to hold the radio farther away.

should try to mount the antenna in the center of the metal roof of your vehicle, if possible. This will use the metal body of the vehicle as an RF shield to protect people inside the car. Glass-mounted antennas can result in higher exposure levels, as can antennas mounted on a trunk lid or front fender. Glass does not form an RF shield.

Although mobile transceivers usually transmit with higher power levels than hand-held radios, the mobile unit often produces less RF radiation exposure. This is because an antenna mounted on a metal vehicle roof is generally well shielded from vehicle occupants. The duty cycle of such operation is generally low.

Don't operate RF power amplifiers or transmitters with the covers or shielding removed. This practice helps you avoid both electric shock hazards and RF safety hazards. A safety interlock prevents the gear from being turned on accidentally while the shielding is off. (If your equipment does not have such a safety interlock you should take other steps to ensure that the amplifier cannot be turned on accidentally.) This is especially important for VHF and UHF equipment. When reassembling transmitting equipment, replace all the screws that hold the RF compartment shielding in place. Tighten all the screws securely before applying power to the equipment.

Another area you should pay attention to is the feed line connecting your transmitter to your antenna. If you are using poor-quality coax, with shielding that is inadequate, or if there are other causes leading to signals radiating from your feed line, you should consider corrective measures. (Use only good-quality coax and be sure your connectors are installed properly.) Improper grounding can also lead to a condition known as RF in the shack. This is especially a problem with stations installed in the second or third floor (or higher) where the ground lead begins to act more like an antenna. If you notice that your SWR reading changes as you touch your equipment, or if you feel a tingling sensation in your fingers when you touch the radio or microphone, these may be indicators of RF in the shack. You will have to take some steps to correct these conditions to ensure a safe operating environment.

If you are installing a repeater or other transmitter in a location that includes antennas and transmitters operating in other services, you must be aware that the total site installation must meet the FCC RF radiation MPE limits. This means your signal is only one part of the total RF radiation from that location. You will probably have to cooperate with the licensees for the other transmitters to determine the total exposure.

[Congratulations! You have now studied all the RF safety material for your Technician class Amateur Radio license. Turn to Chapter 14 now and study questions T0C10 through T0C12 and questions T0E01 through T0E06. Review this section if you have any problems. If you have studied the material in this book in the order it was presented, then you have also completed your study of all the material for your license exam. You will want to review any areas that gave you particular difficulty before going for your exam. Good luck with the test!]

RF SAFETY REFERENCES

IEEE Standard for Safety Levels with Respect to Human Exposure to Radio Frequency Electromagnetic Fields, 3 kHz to 300 GHz, IEEE Standard C95.1-1991, Institute of Electrical and Electronics Engineers, New York, 1992.

For an unbiased assessment of ELF hazards, read the series in *Science*, Vol 249 beginning 9/7/90 (p 1096), continuing 9/21/90 (p 1378), and ending 10/5/90 (p 23). Also see *Science*, Vol 258, p 1724 (1992). You can find *Science* in any large library.

An excellent and timely document is available on the Internet by an anonymous FTP from: **rtfm.mit.edu/pub/usenet-by-group/news.answers powerlines-cancer-faq/part1 and part2.**

The Environmental Protection Agency publishes a free consumer-level booklet entitled, "EMF in Your Environment," document 402-R-92-008, dated December 1992. Look for the nearest office of the EPA in your phone book.

W. R. Adey, "Tissue Interactions with Nonionizing Electromagnetic Fields," *Physiology Review*, 1981; 61:435-514.

W. R. Adey, "Cell Membranes: The Electromagnetic Environment and Cancer Promotion," *Neurochemical Research*, 1988; 13:671-677.

W. R. Adey, "Electromagnetic Fields, Cell Membrane Amplification, and Cancer Promotion," in B. W. Wilson, R. G. Stevens, and L. E. Anderson, *Extremely Low Frequency Electromagnetic Fields: The Question of Cancer* (Columbus, OH: Batelle Press, 1989), pp 211-249.

W. R. Adey, "Electromagnetic Fields and the Essence of Living Systems," Plenary Lecture, 23rd General Assembly, International Union of Radio Sciences (URSI), Prague, 1990; in J. Bach Andersen, Ed., *Modern Radio Science* (Oxford: Oxford Univ Press), pp 1-36.

Q. Balzano, O. Garay and K. Siwiak, "The Near Field of Dipole Antennas, Part I: Theory," *IEEE Transactions on Vehicular Technology (VT) 30*, p 161, Nov 1981. Also "Part II; Experimental Results," same issue, p 175.

R. F. Cleveland and T. W. Athey, "Specific Absorption Rate (SAR) in Models of the Human Head Exposed to Hand-Held UHF Portable Radios," *Bio-electromagnetics*, 1989; 10:173-186.

R. F. Cleveland, E. D. Mantiply and T. L. West, "Measurements of Environmental Electromagnetic Fields Created by Amateur Radio Stations," presented at the 13th annual meeting of the Bioelectromagnetics Society, Salt Lake City, Utah, Jun 1991.

R. L. Davis and S. Milham, "Altered Immune Status in Aluminum Reduction Plant Workers," *American Journal of Industrial Medicine*, 1990; 131:763-769.

F. C. Garland, et al, "Incidence of Leukemia in Occupations with Potential Electromagnetic Field Exposure in United States Navy Personnel," *American Journal of Epidemiology*, 1990; 132:293-303.

A. W. Guy and C. K. Chou, "Thermograph Determination of SAR in Human Models Exposed to UHF Mobile Antenna Fields," Paper F-6, Third Annual Conference, Bioelectromagnetics Society, Washington, DC, Aug 9-12, 1981.

C. C. Johnson and M. R. Spitz, "Childhood Nervous System Tumors: An Assessment of Risk Associated with Paternal Occupations Involving Use, Repair or Manufacture of Electrical and Electronic Equipment," *International Journal of Epidemiology*, 1989; 18:756-762.

D. L. Lambdin, "An Investigation of Energy Densities in the Vicinity of Vehicles with Mobile Communications Equipment and Near a Hand-Held Walkie Talkie," *EPA Report ORP/EAD 79-2*, Mar, 1979.

D. B. Lyle, P. Schechter, W. R. Adey and R. L. Lundak, "Suppression of T-Lymphocyte Cytotoxicity Following Exposure to Sinusoidally Amplitude Modulated Fields," *Bioelectromagnetics*, 1983; 4:281-292.

G. M. Matanoski et al, "Cancer Incidence in New York Telephone Workers," *Proc Annual Review, Research on Biological Effects of 50/60 Hz Fields*, U.S. Dept of Energy, Office of Energy Storage and Distribution, Portland, OR, 1989.

D. I. McRee, "A Technical Review of the Biological Effects of Non-Ionizing Radiation," Office of Science and Technology Policy, Washington, DC, 1978.

G. E. Myers, "ELF Hazard Facts" *Amateur Radio News Service Bulletin*, Alliance, OH, Apr 1994.

S. Milham, "Mortality from Leukemia in Workers Exposed to Electromagnetic Fields," *New England Journal of Medicine*, 1982; 307:249.

S. Milham, "Increased Mortality in Amateur Radio Operators due to Lymphatic and Hematopoietic Malignancies," *American Journal of Epidemiology*, 1988; 127:50-54.

W. W. Mumford, "Heat Stress Due to RF Radiation," *Proc IEEE,* 57, 1969, pp 171-178.

W. Overbeck, "Electromagnetic Fields and Your Health," *QST*, Apr 1994, pp 56-59.

S. Preston-Martin et al, "Risk Factors for Gliomas and Meningiomas in Males in Los Angeles County," *Cancer Research*, 1989; 49:6137-6143.

D. A. Savitz et al, "Case-Control Study of Childhood Cancer and Exposure to 60-Hz Magnetic Fields," *American Journal of Epidemiology*, 1988; 128:21-38.

D. A. Savitz et al, "Magnetic Field Exposure from Electric Appliances and Childhood Cancer," *American Journal of Epidemiology*, 1990; 131:763-773.

I. Shulman, "Is Amateur Radio Hazardous to Our Health?" *QST*, Oct 1989, pp 31-34.

R. J. Spiegel, "The Thermal Response of a Human in the Near-Zone of a Resonant Thin-Wire Antenna," *IEEE Transactions on Microwave Theory and Technology* (MTT) 30(2), pp 177-185, Feb 1982.

B. Springfield and R. Ely, "The Tower Shield," *QST*, Sep 1976, p 26.

T. L. Thomas et al, "Brain Tumor Mortality Risk among Men with Electrical and Electronic Jobs: A Case-Controlled Study," *Journal of National Cancer Institute*, 1987; 79:223-237.

N. Wertheimer and E. Leeper, "Electrical Wiring Configurations and Childhood Cancer," *American Journal of Epidemiology*, 1979; 109:273-284.

N. Wertheimer and E. Leeper, "Adult Cancer Related to Electrical Wires Near the Home," *International Journal of Epidemiology*, 1982; 11:345-355.

"Safety Levels with Respect to Human Exposure to Radio Frequency Electromagnetic Fields (300 kHz to 100 GHz)," *ANSI C95.1-1991* (New York: IEEE-American National Standards Institute).

"Biological Effects and Exposure Criteria for Radiofrequency Electromagnetic Fields," *NCRP Report No. 86* (Bethesda, MD: National Council on Radiation Protection and Measurements, 1986).

US Congress, Office of Technology Assessments, "Electric and Magnetic Fields—Background Paper," *OTA-BP-E53* (Washington, DC: US Government Printing Office), 1989.

11 KEY WORDS

A1A emission — The FCC emission type designator that describes international Morse code telegraphy (CW) communications without audio modulation of the carrier.

Bandwidth — A range of associated frequencies (measured in hertz). For example, bandwidth is used to describe the range of frequencies in the radio spectrum that a radio transmission occupies.

Code key — A device used as a switch to generate Morse code. Also called a **hand key**, a **straight key** or a **telegraph key**.

Code-practice oscillator — A device that produces an audio tone, used for learning the code.

CW (continuous wave) — The FCC emission type that describes international Morse code telegraphy communication without audio modulation of the carrier. Hams usually produce Morse code signals by interrupting the continuous-wave signal from a transmitter to form the dots and dashes.

Dash — The long sound used in Morse code. Pronounce this as "dah" when verbally sounding Morse code characters. Dashes are three times longer than dots.

Dot — The short sound used in Morse code. Pronounce this as "dit" when verbally sounding Morse code characters if the dot comes at the end of the character. If the dot comes at the beginning or in the middle of the character, pronounce it as "di."

Emission — RF signals transmitted from a radio station.

Emission privileges — Permissible types of transmitted signals.

Fist — The unique rhythm of an individual amateur's Morse code sending.

Hand key — Another name for a **code key**.

MCW — The FCC emission type that describes international Morse code telegraphy communication with audio modulation of the carrier.

Phone — Voice communications.

Q signals — Three-letter symbols beginning with "Q," used in amateur CW work to save time and for better comprehension.

Straight key — Another name for a **code key**.

Telegraph key — Another name for a **code key**.

Traffic net — An on-the-air meeting of amateurs, for the purpose of relaying messages.

W1AW — The headquarters station of the American Radio Relay League. This station is a memorial to the League's cofounder, Hiram Percy Maxim. The station provides daily on-the-air code practice and bulletins of interest to hams.

11 Morse Code —The Amateur's International Language

How can one ham who speaks only *English communicate with another ham who speaks* no *English? Using Morse code, of course! This chapter gives you some helpful hints and study suggestions for learning the skill of communicating by Morse code.*

To qualify for a Novice license, you must pass a written exam about FCC Rules, electronics theory and basic amateur operating practices. In addition, you will have to show your ability to communicate using the international Morse code. The code exam is given at 5 words per minute (wpm).

To earn the Technician license, you do not have to pass a Morse code exam. That license does not include privileges on the amateur high-frequency bands, however. Those bands provide direct worldwide communications. As a Technician class licensee, you can earn those privileges, however. To join the excitement of communicating with other Amateur Radio operators around the world, you simply pass the 5-wpm Morse code exam.

This book won't teach you the Morse code. In fact, you won't even find a copy of the Morse code printed in this book. That's because Morse code is best learned by sound, not by sight. We don't want to confuse you with printed dots and dashes. That will only slow you down later. (If you really want to see a printed copy of the Morse code dots and dashes, you can find it in *The ARRL Handbook* and *Morse Code: the Essential Language*, both published by the ARRL. Most dictionaries also include a copy of the Morse code, as do many encyclopedias.)

As we said, the code is best learned by listening to sounds, and for that you'll need a method that teaches you the Morse code characters by producing the sounds for you. ARRL's *Your Introduction to Morse Code* is available as a set of two cassette tapes or two audio CDs. There are several good computer programs that will teach you the code and give you unlimited practice without repeating the same text. *Your Introduction to Morse Code* teaches you all of the

Morse code characters required by the FCC. Practice text at 5 wpm ensures that you are prepared for the 5-wpm code exam. ARRL's *Increasing Your Code Speed* series will help you increase your code speed right up to 22 wpm to help you prepare for the 20-wpm Amateur Extra code exam! There are sets to help you go from 5 to 10 wpm, 10 to 15 wpm and 15 to 22 wpm. (All are available as a set of two cassette tapes, and they are being produced on audio CDs as new master recordings are produced. Check with the ARRL Publications Sales staff about the availability of CDs for the sets you want.)

The ARRL also offers two computer programs to teach Morse code. For the IBM PC and compatible computers, the ARRL sells the GGTE *Morse Tutor* program, which is an excellent Morse-code training program. You can tell the computer what code speed you want it to use for the characters and words it sends. *Morse Tutor* drills you on individual characters and combinations of characters. You can practice with a mixture of random characters or with words. The program also includes a random QSO generator, which produces an unlimited variety of Amateur-Radio-type conversations. These practice QSOs are very similar to the code exams.

The ARRL also sells the GGTE *Morse Tutor Gold*. If you have a SoundBlaster or compatible sound card in your computer, or you may want to send Morse code from an ASCII text file, you will want the Gold version of the program! It also includes all the features of the regular *Morse Tutor*. This version also includes some additional text analysis and reporting features.

The audio CDs, cassette tapes and both computer programs are available directly from ARRL Headquarters or from the many dealers who sell ARRL publications. See the advertising section at the back of this book for an ARRL publications order form.

CODE IS FUN!

Using the code is an exciting way to communicate. Many long-time hams beam with pride when they proclaim that they don't even own a microphone! You are fluent in another complete language when you know the code. You can chat with hams from all around the world using this common language. With the practice you will gain by making on-the-air contacts, your speed will increase quickly. With more advanced licenses, you'll be operating on more portions of the bands, increasing your realm of contacts.

Amateur Radio operators must know the international Morse code if their license permits them to transmit on frequencies below 30 MHz. An international treaty sets this rule, and the United States follows the treaty. Because of this, the Federal Communications Commission requires you to pass a code test to earn your Novice license.

There's a lot more to the code than just satisfying the terms of a treaty, however. Morse code goes back to the very beginning of radio, and is still one of the most effective radio-communication methods. We send Morse code by interrupting the **continuous-wave** signal generated by a transmitter, and so we call it **CW** for short.

For one thing, it takes far less power to establish reliable communications with CW than it does with voice (**phone**). On phone, we sometimes need high power and elaborate antennas to communicate with distant stations, or *DX* in the ham's lingo. On CW, less power and more modest stations will provide the same contacts. Finally, there is great satisfaction in being able to communicate using Morse code. This is similar to the satisfaction you might feel from using any acquired skill.

Conserve Time and Spectrum Space

Another advantage of CW over phone is its very narrow **bandwidth**. Morse code makes efficient use of spectrum space. The group of frequencies where hams operate — the *ham bands* — are narrow portions of the whole spectrum. Many stations use the bands, and because they are so crowded, interference is sometimes a problem. A CW signal occupies only about one-tenth the bandwidth of a phone signal. This means as many as 10 CW signals can fit into the space taken up by one phone signal.

Over the years, radiotelegraph operators have developed a vocabulary of three-letter **Q signals**, which other radio-telegraphers throughout the world understand. For example, the Q signal *QRM* means "you are being interfered with." Just imagine how hard it would be to communicate that thought to someone who didn't understand one word of English. (Chapter 2 includes a list of common Q signals.)

Another advantage to using Q signals is speed. It's much faster to send three letters than to spell out each word. That's why you'll use these Q signals even when you're chatting by code with another English-speaking ham. Speed of transmission is also the reason radiotelegraphers use a code "shorthand." For example, to acknowledge that you heard what was transmitted to you, send the letter R. This means "I have received your transmission okay."

Standard Q signals and shorthand abbreviations reduce the total time necessary to send a message. When radio conditions are poor and signals are marginal, a short message is much more likely to get through.

Many hams prefer to use CW in **traffic nets**. (A net is a regular on-the-air meeting of a group of hams.) Using

A Family of Hams

The Dalton family of Independence, Missouri, is a special group of hams. David (NØLOG) and Elaine (NØNLF) began to study for Amateur Radio licenses in 1989. Two of their children, Sarah (10 at the time) and Daniel (then 8) also became interested when they saw mom and dad studying. Both children passed their Novice exams in January 1990.

Encouraged by the excitement of the Novice bands and the interest of other hams in their local radio club, both children upgraded to Technician in May. Less than a year after first being licensed, Daniel (NØMAT) and Sarah (NØMAS) upgraded to General. David has upgraded to Amateur Extra and Elaine to General.

Sarah and Daniel have become very excited about ham radio. They have taken their hand-held 2-meter radios to school and demonstrated them to their classes, they attend club meetings and enjoy participating in nets. Daniel took his hand-held radio along when he went out selling popcorn for his Cub Scout Pack and on other Webelos Scout activities. Now he takes it on Boy Scout campouts. Both children carry their radios while on their newspaper routes. This may be Elaine's favorite aspect of ham radio — keeping in touch with her busy family. The family has helped their local radio club provide public-service communications at several local events.

James, another member of the Dalton family, began to study Morse code at the age of 6, and learned over half the Morse code characters after only a few weeks of studying. James (KBØJPG) passed his Novice license before he completed the first grade and has since upgraded to Technician Plus! Jonathan, the sixth member of the Dalton family received his Technician license (KBØPQH) just a few days after his eighth birthday.

these nets, hams send messages across the country for just about anyone. When they send messages using CW, there is no confusion about the spelling of names such as Lee, Lea or Leigh.

CW: Sometimes the Only Choice

When WA6INJ's jeep went over a cliff in a February snowstorm, he was able to call for help using his mobile rig. This worked well at first, but as the search for him continued, he became unable to speak. The nearly frozen man managed to tap out Morse code signals with his microphone push-to-talk button. That was all his rescuers had to work with to locate him. Morse code saved this ham's life.

For some types of transmissions, CW (also called type **A1A emission** by the FCC) is the only available choice. As a Novice, CW is the only **emission privilege** you may use on the 80, 40 and 15-meter amateur bands. (One exception to this rule is if you are in Alaska, and involved in some emergency communications. In that case you may use single sideband [J3E or R3E] on 5167.5 kHz.) CW is the only mode that may be used on *all* amateur frequencies.

Some hams like to bounce VHF and UHF signals off

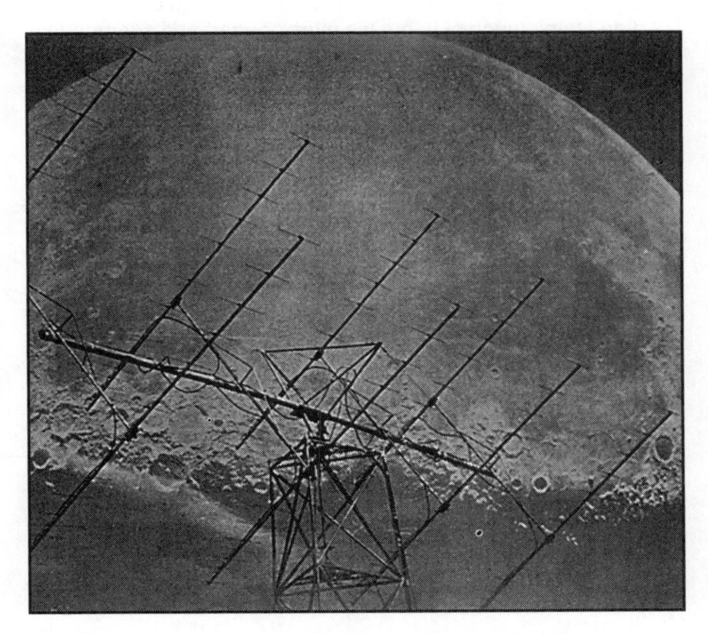

Figure 11-1 — Even the moon isn't immune from ever-ambitious hams, and code is the most efficient type of signal to bounce off its surface and be reflected back to Earth.

the surface of the moon to another ham station on the Earth. Because of its efficiency, hams use CW in most of this *moonbounce* work. They could use voice signals, but this increases the power and antenna-gain requirements quite a bit.

On some frequencies, amateurs communicate by bouncing their signals off an auroral curtain in the northern sky. (Stations in the Southern Hemisphere would use an auroral curtain in the southern sky.) Phone signals become so distorted in the process of reflecting off an auroral curtain that they are difficult or impossible to understand. CW is the most effective way to communicate using signals bounced off an aurora.

You will feel a special thrill and a warm satisfaction when you use the Morse code to communicate with someone. This feeling comes partly from sending messages to another part of the world without regard to language barriers. Sending and receiving Morse code is a skill that helps set amateurs apart from other people. It provides a common bond between amateurs worldwide.

Morse Code From the Heart

The power of modern medicine kept one ham alive; the power of Morse code kept him in touch with loved ones.

Reprinted from July 1990 QST.

Have you ever felt Morse code rather than heard it? Although I've been a licensed amateur for 34 years and have had many wonderful experiences in this hobby, none can compare to the one I'll never forget. It happened on Monday, August 26, 1985.

My husband Ralph, W8LCU, went to the hospital the Friday before for an ECG and heart catheterization. He was told that an immediate quadruple bypass was needed, and he could not leave the hospital. Arrangements were made for the following Monday. Ralph had always been healthy and never showed an inkling of heart problems, so this came as quite a shock to us. The doctors told us that after his surgery he would not be speaking or doing much else, until perhaps the next day or until he was off the respirator, heart pump and everything else that goes with this type of surgery.

Over the weekend we had many things to discuss, and Amateur Radio never entered my mind.

Monday evening, in the intensive care unit (ICU) after surgery, I held Ralph's hand. He began tapping on my palm. I didn't think much of it, but his eyes opened and seemed to be telling me something. He moved his fingers to my wrist and suddenly, as clear as if I were hearing CW on the radio, came the letters P A (tears coming down my face) I N (pause) I S H E L L. "Pain is hell."

Before I could say or do anything, four nurses were in the room. I told them what Ralph had just said to me in Morse code. I think they were ready to have me taken to the mental ward, but they gave him a shot. Then they stood by in disbelief as he tapped out a short message to our son and daughter, and a final I LUV U before drifting off to sleep.

Meanwhile, in another hospital across the state, a friend of ours, Vanessa, KA8THR, was back in her room after major surgery. Her first words to her husband, Chuck, N8EOJ, were "How is Ralph doing?" Sure enough, through the wonders of CW, hand-held transceivers, repeaters and

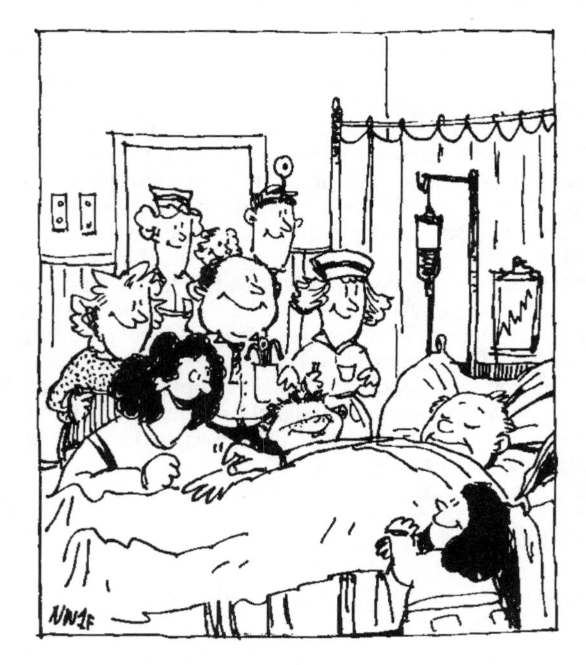

many hams along the way, relays of Ralph's progress were sent to N8EOJ all day long. When Ralph tapped CW to me, within moments hams all across the state knew all was well.

The ICU and other units were buzzing for days as doctors came into Ralph's room to find out about this "new" form of communication. They told Ralph they thought more people should know Morse code, themselves included. Of course, Ralph told them that with a little study and determination, anyone can do it.

Needless to say, CW will always be important to me, either in an emergency or just for plain fun.

Ralph recovered from his surgery, and I, while I was an Advanced at the time, have gone on to earn my Amateur Extra license. — *Donna Burch, W8QOY*

GETTING ACQUAINTED WITH THE CODE

The basic element of a Morse code character is a **dot**. The length of the dot determines how long a **dash** should be. The dot length also determines the length of the spaces between elements, characters and words. Figure 11-2 shows the proper timing of each piece. Notice that a dash is three times as long as a dot. The time between dots and dashes in a character is equal to the length of a dot. The time between letters in a word is equal to three dot lengths and the space between words is seven dot lengths.

The lengths of Morse code characters are not all the same, of course. Samuel Finley Breese Morse (1791-1872) developed the system of dots and dashes in 1838. He assigned the shortest combinations to the most-used letters in plain-language text. The letter E has the shortest sound, because it is the most-used letter. T and I are the next most-used characters, and they are also short. Character lengths get longer for letters used less often.

An analysis of English plain-language text shows that the average word (including the space after the word) is 50 units long. By a unit we mean the time of a single dot or space between the parts of a character. The word PARIS is 50 units long, so we use it as a standard word to check code speed accurately. For example, to transmit at 5 words per minute (wpm), adjust your code-speed timing to send PARIS five times in one minute. To transmit at 10 wpm, adjust your timing to send PARIS 10 times in one minute.

As you can see, the correct dot length (and the length of dashes and spaces) changes for each code speed. As a result, the characters sound different when the speed changes. This leads to problems for a person learning the code. Also, at slower speeds, the characters seem long and drawn out. The slow pace encourages students to count dots and dashes, and to learn the code through this counting method.

Learning the code by counting dots and dashes introduces an extra translation in your brain. (Learning the code by memorizing Morse-code-character dot/dash patterns from a printed copy introduces a similar extra translation.) That extra translation may seem okay at first. As you try to increase your speed, however, you will soon find out what a problem it is. You won't be able to count the dots and dashes and then make the translation to a character fast enough!

People learning the Morse code with either of these methods often reach a learning plateau at about 10 words per minute. This is just below the 13-wpm speed required to upgrade to the General license. That frustration (I just can't copy faster than 10 wpm!) is overcome by other methods, however.

Learning Morse Code

Many studies have been done, and various techniques tried, to teach Morse code. The method that has met with the most success is called the *Farnsworth method*. That is the method used in the ARRL package, *Your Introduction to Morse Code*. With this technique, we send each character at a faster speed (we use an 18-wpm character speed at ARRL). At speeds in this range, the characters — and even some short words — begin to take on a distinctive rhythmic pattern.

With this faster character speed, we use longer spaces between characters and words to slow the overall code speed. *Your Introduction to Morse Code* starts with code sent at an overall speed of 5 wpm. (You can still measure this timing by using the word PARIS, as described earlier.) Once you learn the character sounds, and can copy at 5 wpm, it will be easy to increase your speed. Just decrease

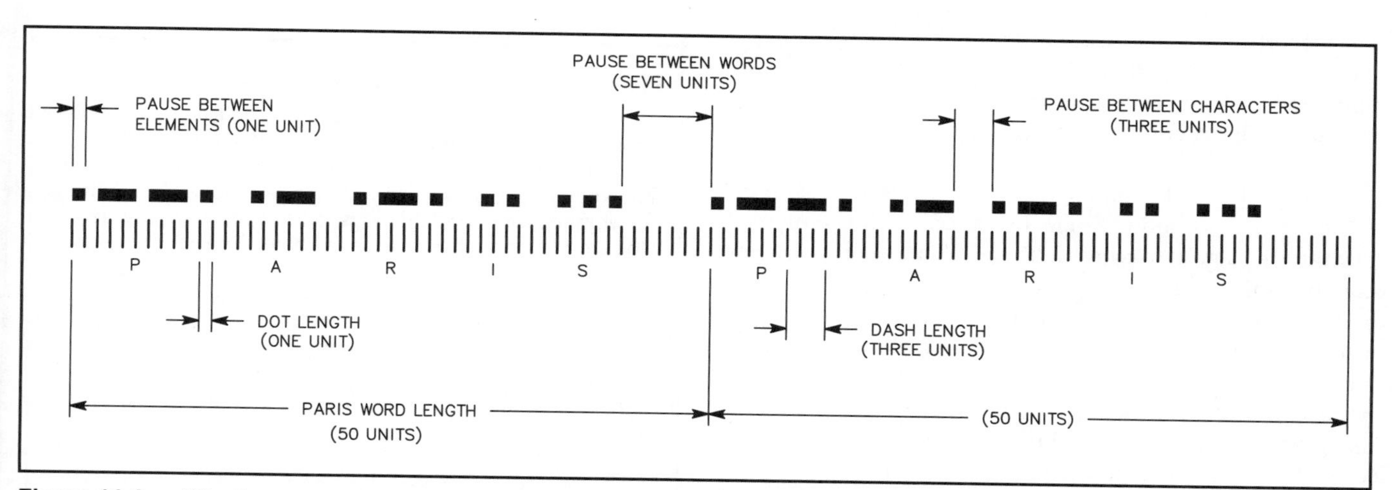

Figure 11-2 — Whether you are a beginner or an expert, good sending depends on maintaining proper time ratios or "weight" between the dots, dashes and spaces, as shown.

the spaces between letters and words, and your code speed increases without changing the rhythmic pattern of the characters. If you use a computer program to learn the code, or another method, we recommend this same technique, with the code characters sent at 18 to 20 wpm.

Learn to recognize that rhythmic pattern, and associate it directly with the character. You'll learn the code in the shortest possible time, and it will be much easier to increase your code speed. Decreasing the space between characters and words provides a natural progression to increase your code speed. The ARRL *Increasing Your Code Speed* series offers code practice in steps from 5 to 10 wpm, 10 to 15 wpm and 15 to 22 wpm using this technique.

Morse code is a communication method that depends on sounds. To understand the communication you must hear the sounds and interpret their meaning. This is why most code-teaching methods repeat the sounds for you to listen to and associate the characters with the sounds. It is also why you will not find a copy of the Morse code dots and dashes printed in this book. Audio CDs, cassette tapes, computer programs and even classroom or individual practice with a **code-practice oscillator** and a **code key** all rely on sounds to teach you the Morse code characters. (A code key is sometimes called a **hand key**, **straight key** or a **telegraph key**.)

There are many computer programs for teaching Morse code, and some of them may be very helpful. The ARRL sells the GGTE programs *Morse Tutor* and *Morse Tutor Gold* for the IBM PC and compatibles. Both are excellent teaching programs.

Beware of programs written in BASIC. Because of the way BASIC operates, the timing is often not quite right. Also beware of programs written "to teach myself Morse code." While such writers may be excellent computer programmers, they are often unaware of the finer points of code training. Many people are using the Microsoft Windows operating environment on their IBM-compatible computers. Many Morse code programs do not run properly under multitasking software because of timing variations. If you are using Windows, be sure it is set to devote 100% of the time to the software. A few Windows programs have come out recently. The shareware program *Nu-Morse* seems to work well, as does *Morse University II*, which was being marketed by AEA shortly before they went out of business. If you can find a copy of this program for sale it would be a good investment.

Several other techniques have also been successful for teaching Morse code. Some of these methods involve memorizing a printed copy of the code. There are even a few commercial packages that picture the dots and dashes of the code characters in various "creative" patterns to help you remember them. You can learn Morse code by following any of these methods. Most people will learn faster, and will be able to increase their code speed easier, however, by using a "listening" method. For those who seem unable to learn the Morse code using the audio CDs or tapes in *Your Introduction to Morse Code* or a computer program, one of these "printed" methods may prove helpful. Save that as a last resort, however. Practice faithfully with *Your Introduction to Morse Code* or a computer program every day for at least three to four weeks. Then, if you have not learned many of the characters, you may want to try one of the visual, or printed, methods.

The 26 letters of the alphabet and the numbers 0 through 9 each have a different sound. There are also different sounds for the period, comma, question mark, double dash (=), fraction bar (/) and some procedural signals that hams use. The + sign, which hams call $\overline{\text{AR}}$, means "end of message." $\overline{\text{SK}}$ means "end of work," or "end of contact." Hams sometimes refer to the double dash as $\overline{\text{BT}}$ and the fraction bar as $\overline{\text{DN}}$. These two-letter combinations are written with a line over the letters to indicate that two letters are sent as one character to form these symbols. That's a total of 43 character sounds that you will have to learn for your 5-wpm Morse code exam. You'll learn the sounds of all these characters as you practice with your code cassettes or computer program.

Morse Code: The Essential Language, by L. Peter Carron, Jr, W3DKV (published by the ARRL) contains other suggestions for learning Morse code. That book also describes the history of Morse code. It includes several stories about lives saved because of emergency messages transmitted over the radio.

Sounding the Code

Some people find it helpful to say the sounds of Morse code characters, especially when they are first learning the code. Instead of saying the names of the Morse code elements, dot and dash, we use the sounds "dit" and "dah." If the dot is at the beginning or in the middle of the character, we sound it out as "di" instead of "dit."

Listen to the difference between the sounds you make saying the word "dit" and saying "dah." If you can tell the difference between those sounds, you have all the ability

Morse Code is the Universal Language

Clark J. Evans, Sr., WA4DLL and Lee Paulet, KB4FBX were operating the Amateur Radio display station at the Museum of Science and Industry in Tampa, Florida. A tour guide came into the station with two Japanese boys. When the woman introduced the boys to the operators, she said, "You won't be able to talk to these boys. They don't speak English at all." She also mentioned that one of the boys was a ham radio operator in Japan. That was the door opener.

Lee whistled CQ CQ CQ. That got the boy's attention, and the youngster was soon whistling back. Lee told the guide what the boy was saying. He was 13 years old from a town 40 miles north of Tokyo. He had been in the USA for two weeks.

The tour guide said, "You can't be communicating with him!" Lee's answer was, "Oh yes, we hams can. Morse code is a universal language!"

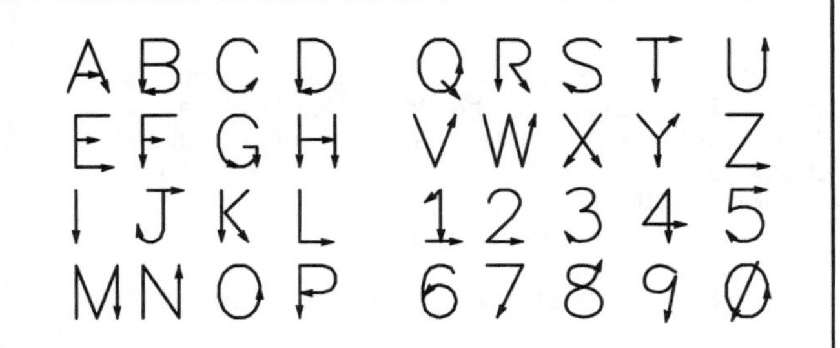

Figure 11-3 — A method of hand-printing letters and numbers with minimum effort and maximum speed.

you need to learn the code. Being able to receive Morse code is really nothing more than being able to recognize a sound. Try it yourself. Say "didah." Now say "didahdit." Can you hear the difference? Congratulations! You now know the sounds for the letters A and R, and are on your way to learning the Morse code.

Using this method, you hear the sound "didah" and associate that sound with the letter A. With practice, you'll learn all the sounds and associate them with the correct letters, numbers and punctuation.

Learning to Write

To learn the code, you train your hand to write a certain letter, number or punctuation mark whenever you hear a specific sound. You are forming a habit through your practice. After all, forming a habit is nothing more than doing something the same way time after time. Eventually, whenever you want to do that thing, you automatically do it the same way; it has become a habit.

You will need lots of practice writing the specific characters each time you hear a sound. Eventually you'll copy the code without thinking about it. In other words, you'll respond automatically to the sound by writing the corresponding character.

As you copy code sent at faster speeds, you may find that your ability to write the letters limits you. Practice writing the characters as quickly as possible. If you normally print, look for ways to avoid retracing lines. Don't allow yourself to be sloppy, though, because you may not be able to read your writing later.

Many people find that script, or cursive writing is faster than printing. Experiment with different writing methods, and find one that works for you. Then practice writing with that method so you don't have to think about forming the characters when you hear the sounds.

Figure 11-3 shows a systematic method of printing numbers and the letters of the alphabet. This system requires a minimum of pencil movement or retraced lines. You may find this technique helpful to increase your writing speed. Whatever method you choose, practice so it becomes second nature.

Some Study Suggestions

The secret to easy and painless mastery of the Morse code is regular practice. Set aside two 15 to 30-minute periods every day to practice the code. If you try longer sessions, you may become over tired, and you will not learn as quickly. Likewise, if you only practice every other day or even less often, you will tend to forget more between practice sessions. It's a good idea to work your practice sessions into your daily routine. For instance, practice first thing in the morning and before dinner. Daily practice gives quick results.

Learn the sound of each letter. Morse code character elements sound like dits and dahs, so that's what we call them. Each character has its own pattern of dits and dahs, so learn to associate the sound of that pattern with the character. Don't try to remember how many dots and dashes make up each character. Practice until you automatically recognize each Morse code character.

Feel free to review. You're learning a new way of communicating, by using the Morse code. If you are having trouble with a particular character, spend some extra time with it. If you are using ARRL's *Your Introduction to Morse Code*, replay the problem CD track or rewind the tape and play that section again. If you are using one of the computer programs, spend some extra time drilling on the problem character. After you've listened to the practice on one character two or three times, however, go on to the next one. You will get more practice with the problem character later, because you will be constantly reviewing all of the characters learned so far. You can even come back to the problem character again in a later practice session. Be sure to listen to the practice for at least 3 characters during each practice session, however.

If you hear a Morse code character you don't immediately know, just draw a short line on your copy paper. Then get ready for the next letter. If you ignore your mistakes now, you'll make fewer of them. If you sit there worrying about the letter you missed, you'll miss a lot more! You shouldn't expect to copy perfectly while you are learning the code. You will get better with more practice.

When you think you've recognized a word after copying a few letters, concentrate all the more on the actual code sent. If you try to anticipate what comes next, your guess may turn out to be wrong. When that happens, you'll probably get confused and miss the next few letters as well. Write each letter just after it is sent. With more practice, you'll learn to "copy behind," hearing and writing whole words at one time. For now, when you are just learning the code, concentrate on writing each character as it is sent. This helps reinforce the association of a sound and a character in your mind.

One trick that some people use while learning the code is to whistle or hum the code while walking or driving. You can also say the sounds di, dit and dah to sound out the characters. Send the words on street signs, billboards and store windows. This extra practice may be just the help you need to master the code!

Morse code is a language. Eventually you'll begin to

Assembling a Code-Practice Oscillator

It isn't difficult to construct a code-practice oscillator. A complete oscillator that mounts on a small piece of wood is shown in Figure A. Figure B shows all the parts for this project laid out ready for assembly. The circuit board for this project can be ordered from FAR Circuits, 18 N. 640 Field Court, Dundee, IL 60118-9269. A complete parts kit, including circuit board, is available from the Hoosier Lakes Amateur Radio Club, PO Box 981, Warsaw, IN 46581-0981 and from Jade Products, PO Box 368, East Hampstead, NH 03826-0368. Phone orders: 800-JADE-PRO (800-523-3776). E-mail to: **jadepro@jadeprod.com** or check out their World Wide Web page at: **http://www.jadeprod.com/**

Contact these vendors for the latest pricing.

Please read all instructions carefully before mounting any parts. Check the parts-placement diagram for the location of each part.

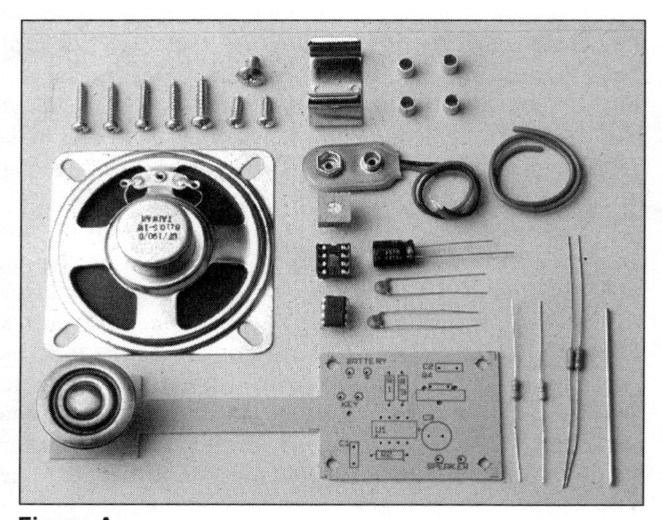

Figure A

Figure B

☐ Check each box as that part is installed and soldered.

Quantity	*Description*	*Radio Shack Part Number*	*Component Number*	*Used in Step Number*
Capacitors				
☐ 1	0.01-µF	272-131	C1	3
☐ 1	0.01-µF	272-131	C2	6
☐ 1	220-µF, 35-V electrolytic	272-1029	C3	5
Resistors				
☐ 1	10-kilohm, 1/4 W (brown-black-orange stripes)	272-1335	R2	4
☐ 1	47-kilohm, 1/4 W (yellow-violet-orange stripes)	272-1342	R3	8
☐ 1	10-kilohm, 1/4 W (brown-black-orange stripes)	272-1335	R1	9
☐ 1	47-ohm, 1/4 W (yellow-violet-black stripes)	271-009	R5	11
Miscellaneous				
☐ 1	100-kilohm potentiometer	271-284	R4	7
☐ 1	8-pin IC socket	276-1995		2
☐ 1	7555 CMOS Timer IC (or 555 Timer IC)	276-1718	U1	13
☐ 1	Loudspeaker — 2-inch, 8-ohm	40-245	LS1	11
☐ 1	Six to 10 inches of insulated wire, about 18 or 22 gauge			11
☐ 1	9-V battery connector	270-325		10
☐ 1	9-V battery	23-553	BT1	10
☐ 1	U-shaped battery holder	270-326		12
☐ 1	Brass rod, 2 inches long, approximately 18 gauge (about the diameter of a coat hanger). Available at hobby shops.			
☐ 4	1/4-inch spacers	64-3024		12
☐ 1	2 × 4 × 1/2-inch piece of wood for base			12
☐ 5	No. 6 wood screws, 3/4-inch long			12
☐ 2	No. 6 wood screws, 3/8-inch long			12
☐ 1	Five-lug tie point, used to mount speaker (optional)	274-688		11

Assembly instructions

☐ **Check each box as that step is completed.**

☐ **Step 1: Attach the brass rod. Check the parts-placement diagram (Figure E) for location.**

Clean the brass rod with sandpaper or steel wool. Bend one end of the rod slightly less than 90 degrees. Lay the circuit board on the table with foil side up. Place the hooked end of the brass rod over the large hole near the handle (see Figure D). Make sure the rod extends out over the handle area. Solder the rod to the board on the foil side. The end of the brass rod should not extend past the marked oval on the handle. This is your contact point. If it does extend beyond this point, cut the rod off just before the end of the oval.

☐ **Step 2: Solder the IC socket to the board.**

The socket for the IC is placed on the component (non-foil) side of the board first. Do not plug the IC into the socket now. After all the other parts are soldered to the board you will be instructed to plug the IC into the socket (Step 13). Identify the notched end of the socket. Insert the socket into the circuit board. Turn the board over and gently spread the pins on the socket so they make contact with the foil side of the board. Solder the socket in place.

☐ **Step 3: Place C1 (0.01-µF capacitor) on the component side of the board.**

Thread the wire leads on C1 through the holes on the board. (See Figure F.) Solder the wires onto the foil side of the board. Cut the extra wire off above the solder joint.

☐ **Step 4: Place R2 (10-kilohm resistor) on the component side of the board.**

Prepare resistors for mounting by bending each lead (wire) of the resistor to approximately a 90° angle. (See Figure G.) Insert the leads into the board holes and bend them over to hold the resistor in place. Solder the leads to the foil and trim them close to the foil.

☐ **Step 5: Place C3 (220-µF, 35-volt electrolytic capacitor) on the component side of the board.**

This capacitor has a plus (+) side and a negative (–) side. The (–) side is placed on the board facing away from the handle. (Notice the + sign printed on the circuit board at this location.) Insert the capacitor leads into the circuit board holes, solder them in place and trim off the extra wire.

☐ **Step 6: Place C2 (0.01-µF capacitor) on the component side of the board.**

Thread the wire leads from C2 through the holes on the circuit board. (See Step 3 and Figure F.) Solder the wires onto the board. Cut the extra wire off above the solder joint.

10 kΩ R1
47 kΩ R3
100 kΩ Tone R4
KEY
NC
8 7 6 5
0.01 µF C2
9 V BATTERY
V_{cc} DISCHARGE THRESHOLD CONTROL VOLTAGE
Timer
U1 555 or 7555
GROUND TRIGGER OUTPUT RESET
1 2 3 4
C3 R5
+ 220 µF 35 V
47 Ω
Speaker
R2 10 kΩ
C1 0.01 µF

Figure C—Schematic diagram of a code-practice oscillator.

BRASS ROD SOLDERED IN POSITION

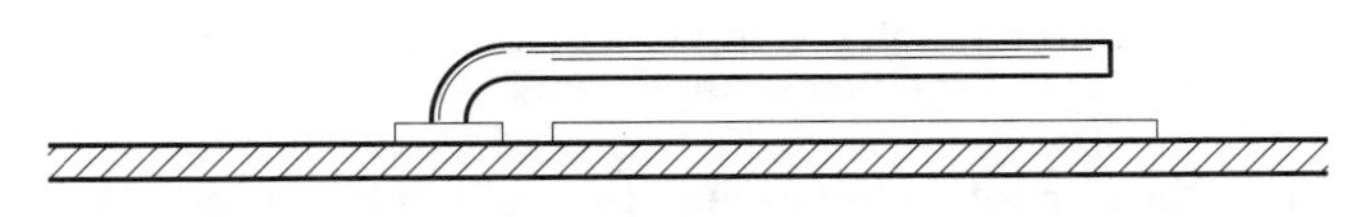

CIRCUIT–BOARD HANDLE

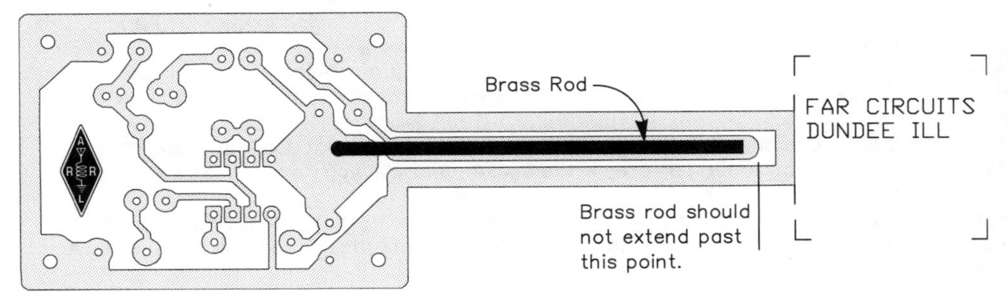

Figure D—Solder the brass rod in position on the foil side of the PC board. You can also use this figure as a circuit-board etching pattern if you want to make your own circuit board, since the pattern is printed full size.

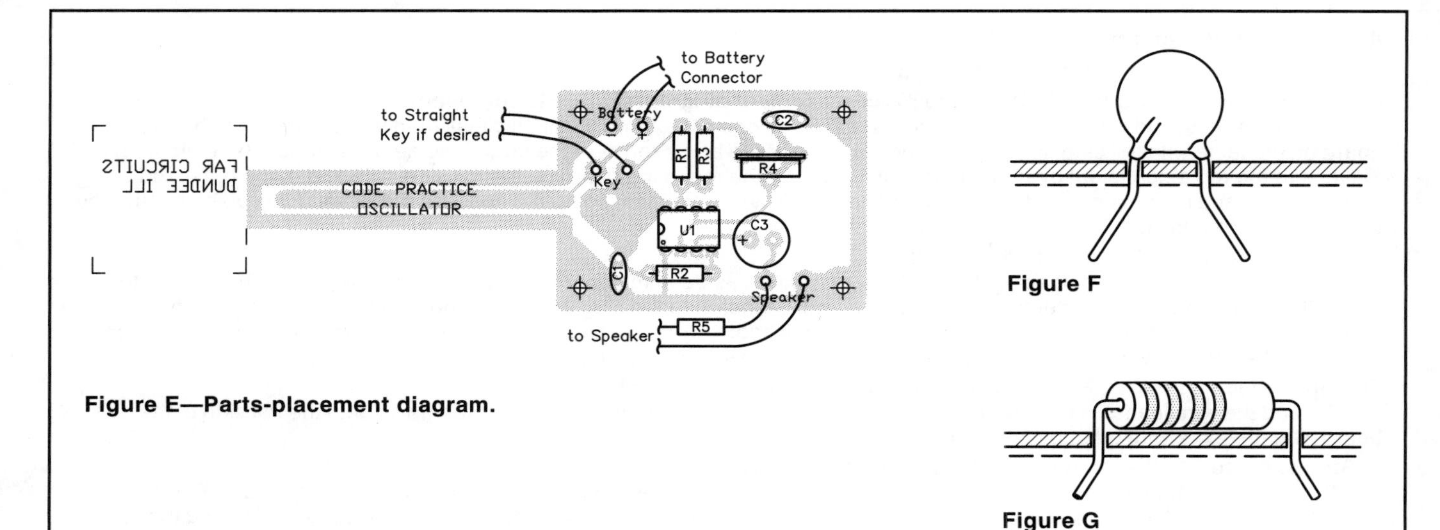

Figure E—Parts-placement diagram.

Figure F

Figure G

☐ **Step 7: Place R4 (100-kilohm potentiometer) on the component side of the circuit board.**
This component has three pins. All three pins must be plugged into the holes on the board. (It fits only one way.) Solder them in place and cut off any excess lead length.

☐ **Step 8: Place R3 (47-kilohm resistor) on the component side of the circuit board.**
Bend the wires on the resistor to plug it into the board. (See Step 4 and Figure G.) Plug the resistor into the board, spread the wires and solder it in place. Trim off the extra wire lengths.

☐ **Step 9: Place R1 (10-kilohm resistor) on the component side of the board.**
Bend the wires on the resistor to plug it into the board. (See Step 4 and Figure G.) Plug the resistor into the board, spread the wires and solder it in place. Trim the excess wire lengths.

☐ **Step 10: Hook up the battery connector leads.**
The battery connector consists of two wires, one red and one black, attached to a snap-on cap. Remove $1/4$ inch of plastic insulation from the end of both wires. The black wire is neagative and the red wire is positive. The positive and negative battery connections are marked on the component side of the board. Be sure the red wire goes in the hole marked "+" and the black wire goes in the hole marked "–". Solder the wires in place and trim any excess length close to the solder joint.

☐ **Step 11: Hook up the speaker.**
If you are using the tie lug to hold the speaker in place, solder the speaker lugs to tie-point lugs on each side of the center post. (If you are not using the tie lug then you will solder the wires directly to the speaker lugs.) Cut the speaker wire into two equal lengths. Remove $1/4$ inch of plastic insulation from each end of both wires. Solder one end of each wire to one of the tie-point solder lugs below the speaker terminals. Solder one end of R5 (47-ohm resistor) to the circuit board as shown in Figure E. Solder one wire to the other end of this resistor and the second speaker wire to the circuit board.

☐ **Step 11: Attach the circuit board to the wood base.**
Place the completed circuit board on the wood. Trace through the four corner holes with a pencil. Take the circuit board off the wood and lay it aside. Place the spacers on the wood, standing upright. Carefully put the circuit board on top of the spacers. Put the $3/4$-inch screws through the holes in the circuit board and through the spacers and screw them into the board until snug. Be sure not to overtighten the screws, or you may crack the circuit board. Attach the speaker to the end of the board opposite the handle with a $3/8$-inch screw through the tie-point mounting hole, or mount the speaker to the board using two screws and holes in the outside edge of the speaker's metal frame. Attach the U-shaped metal battery holder to the wooden base with a $3/8$-inch screw.

☐ **Step 13: Plug the integrated circuit (IC) into the socket, being careful to position it so the notch or dot on one end of the IC is toward the handle.**
CAUTION — The static electricity from your body could destroy the IC. Before touching the IC, be sure you have discharged any static that may be built up on your body. While sitting at your table or workbench, touch a metal pipe or other large metal object for a few seconds. Carefully remove the IC from its foam padding. Hold it by the black body and avoid touching the wires. Plug it into the socket, being sure that the notched end of the IC is facing toward the handle. The notch on the IC should line up with the notch on the socket.

Attach the battery to the snap-on battery connector and place it in the U-shaped battery holder. This unit uses electricity only when the telegraph key handle is pushed down. No ON/OFF switch is necessary, and you may leave the battery connected at all times.

You're done! The oscillator should produce a tone when you press the key. If your oscillator does not work, check all your connections carefully. Make sure the IC is positioned correctly in the socket, and that you have a fresh battery. If it still doesn't work check all your solder connections.

Once you have the oscillator working, you're ready to use it to practice Morse code. If you are studying with a friend, you can use the oscillator to send code to each other. If you are studying alone, tape record your sending and play it back later. Can you copy what you sent? How would it sound on the air? Good luck!

recognize common syllables and words. With practice you will know many complete words, and won't even listen to the individual letters. When you become this familiar with the code, it really starts to be fun!

Don't be discouraged if you don't seem to be breaking any speed records. Some people have an ear for code and can learn the entire code in a week or less. Others require a month or more to learn it. Be patient, continue to practice and you will reach your goal.

To pass the Novice and Technician Plus code examination, you must be able to understand a plain-language message sent at 5 wpm. As you know, in the English language some words are just one or two letters long, others 10 or more letters long. To standardize the code test, the FCC defines a word as a group of five letters. Numbers and punctuation marks normally count as two characters for this purpose. You will receive 25 characters in one minute for 5-wpm code.

COMFORTABLE SENDING

There's more to the code than just learning to receive it; you'll also have to learn to send it. To accomplish this, you'll need a **telegraph key** and a **code-practice oscillator**. You can get these items at most electronics-parts stores, or you can build your own simple oscillator. The "Assembling A Code-Practice Oscillator" sidebar gives you step-by-step instructions for building one simple oscillator that even includes a key for you to practice with. The key with this project is not the best key you could use, but it will allow you to do some practice sending. You may want to consider adding a real telegraph key to the project — the circuit board has two holes to connect wires to a **straight key**.

Many experienced amateurs prefer to use an electronic keyer to send Morse code. An electronic keyer produces properly timed dots and dashes, because it uses one circuit to produce dots and another circuit to produce dashes. In general, it is probably better to learn to send Morse code with a hand key at first. Some students may have good success with a keyer. Commercial keyers range from simple, basic units to full-featured Morse code machines. While they are comparatively expensive, some of the full-featured machines offer features that are quite helpful to a beginner. The keyers shown in Figure 11-4 can send random-character code practice at any speed you desire.

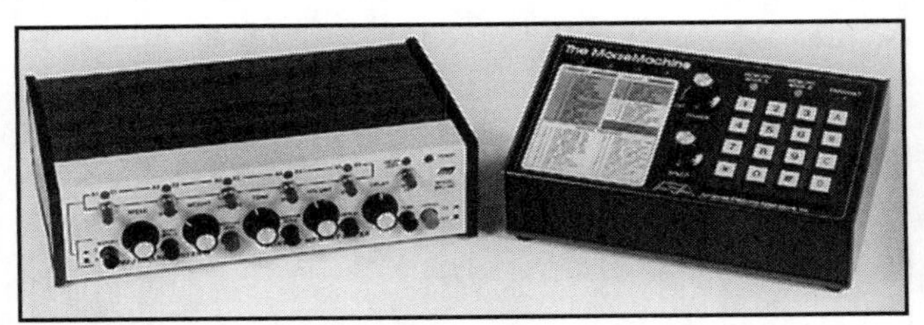

Figure 11-4 — Electronic keyers shown here can store messages in memory to be sent at the touch of a button and also offer various training features to help you learn Morse code and increase your speed. The MFJ-486 keyer sends 5-character random-letter groups, random 1 to 8-character groups and plain text in the format of an on-the-air Amateur Radio contact (QSO). In addition to sending random characters, the AEA MM-3 can connect to a computer to send text from a keyboard or text file. The MM-3 also contains a program that allows you to "contact" other stations in the keyer and exchange QSO information.

Figure 11-5 shows two different standard straight key models (one in the foreground and one to the left). There is a semiautomatic "bug" in the background and a popular Bencher paddle used with modern electronic keyers on the right. You'll want to obtain some type of code key to practice with before you're ready for your Novice code test. You'll need the key for some of your on-the-air operating after the license arrives from the FCC as well! Most new hams start with an inexpensive straight key.

Just as with receiving, it is important that you be comfortable when sending. It helps to rest your arm on the table, letting your wrist and hand do all the work. Grasp the key lightly with your fingertips. Don't grip it tightly. If you do, you'll soon discover a few muscles you didn't know you had, and each one will ache. With a light grasp, you'll be able to send for long periods without fatigue. See Figure 11-6.

Another important part of sending code is the proper adjustment of your telegraph key. There are only two adjustments that you will normally have to make on a straight key, but you'll find they are very important. Figure 11-7 illustrates these adjustments.

The first adjustment is the spacing between the contacts, which determines the distance the key knob must move to send a letter. Adjust the contacts so the knob moves about the thickness of a dime (1/16 inch). Try it. If you're not satisfied with this setting, try a wider space. If that doesn't do it, try reducing the space. Eventually, you'll find a spacing that works best for you. Don't be surprised if your feelings change from time to time, however, especially when your sending speed increases.

The second adjustment you must make is the spring tension that keeps the contacts apart. Just as there is no "correct" contact spacing, there is no correct tension adjustment for everyone. You will find, however, that adjusting the spacing may also require a tension adjustment. Adjust the tension to provide what you think is the best "feel" when sending.

Some straight keys have ball bearing pivot points on either side of the crossarms. These normally need no adjust-

Figure 11-5 — Older than radio itself, code still reigns as the most efficient and effective communications mode; many hams use it almost exclusively. A modern straight key, the device most beginners use, is in the foreground.

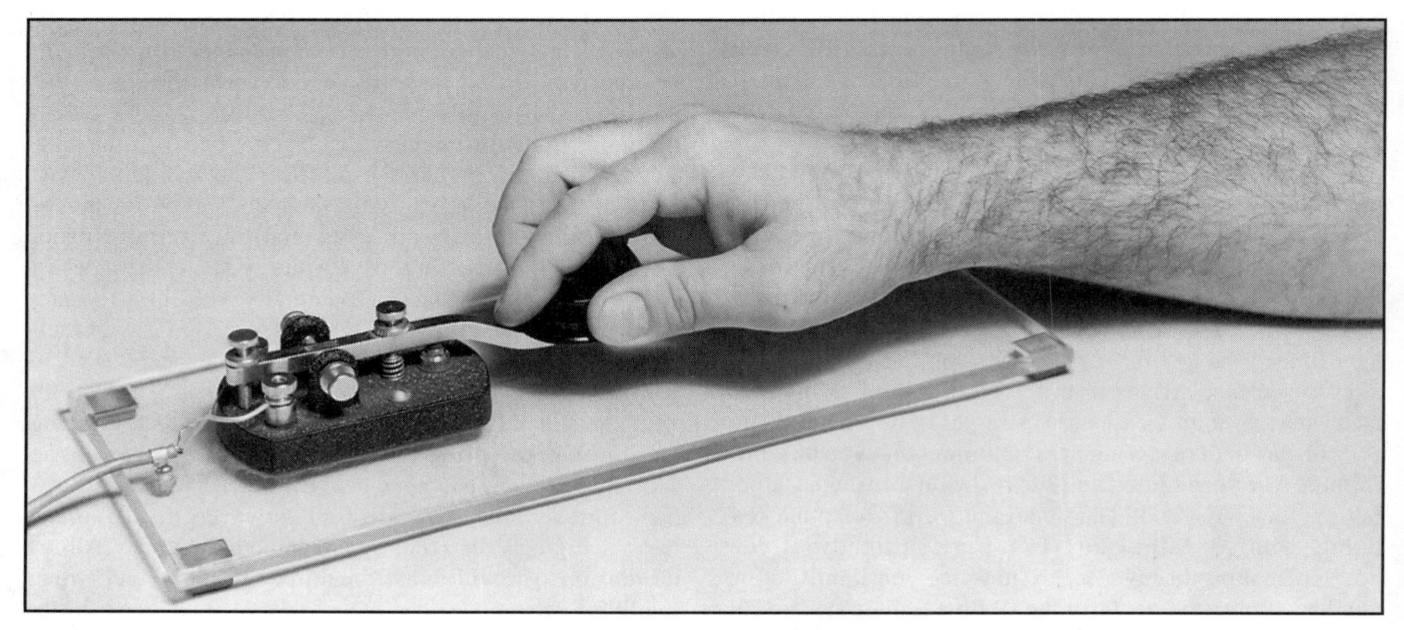

Figure 11-6 — Proper forearm support, with the wrist off the table, a gentle grip and a smooth up-and-down motion make for clean, effortless sending.

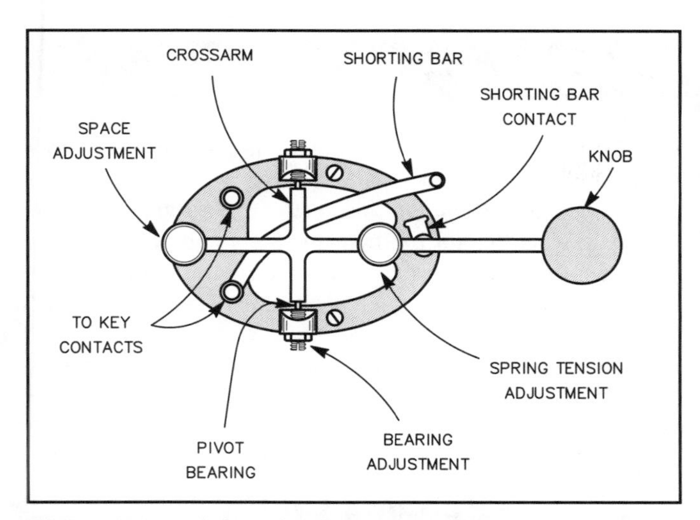

Figure 11-7 — A few simple adjustments to suit your style allow you to send for hours without fatigue. The contact spacing and spring tension should be set to provide the most comfortable feel.

ment, but you should be sure the crossarms move freely in these pivots. If the bearings are too tight the key will bind or stick. If they are too loose there will be excessive play in the bearings. The side screws also adjust the crossarms from side to side so the contact points line up.

All of these adjustment screws have lock nuts, so be sure you loosen them before you make any adjustments. Tighten the lock nuts securely after making the adjustment.

Some better-quality keys have a shorting bar like the one shown in Figure 11-7. This shorting bar can be used to close the key contacts for transmitter tuning or adjustments.

You'll probably want to fasten the key to a piece of wood or other heavy weight to prevent it from sliding around as you send. You might even want to fasten the key directly to the table. It's best to experiment with different positions before permanently attaching the key to any surface, however.

Some operators use a board that extends under their forearm and allows their arm to hold the key in position. This way the key can be moved aside for storage, or to clean off the table for other activities. A piece of 1/4-inch Plexiglas about 12 inches long works well for this type of key base. This technique also allows you to reposition the key to find the most comfortable sending position.

Learning to send good code, like learning to receive, requires practice. A good way to start is to send along with your code practice. Try to duplicate the sounds as much as possible.

Always remember that you're trying to send a complete sound, not a series of dots and dashes. With that in mind, take a moment to think about the sound you're trying to send. It consists of dits, dahs and pauses (or spaces). The key to good sending, then, is timing. If you're not convinced, listen to some code on the Novice bands. What transmissions would you prefer to receive? Why? That's right — because they have good timing.

Each person develops a unique rhythm in his or her sending. It's almost impossible to describe what makes one person's rhythm different from another person's. No two people send code exactly the same way, though. This unique rhythm, or **fist** is like the signature of the operator sending the code. Some experienced CW operators can even identify their friends without hearing call signs.

Some people send code that is very easy and enjoyable to copy. Others are not so good, and require a lot of concentration to understand. Since your fist is your on-the-air signature, try to make it as easy to read as possible. Learn to send code that is easy to copy. Then you'll have a fist you can be proud of. It can make a big difference in your success as an amateur operator.

One of the best ways to learn to send and receive code is to work with another person. That person may be another member of your own family, a friend who is also studying the code or a licensed ham. If you're attending an organized class, you may be able to get together with another student several times a week.

If you must work alone, make good use of your tape recorder. Try recording your sending. After waiting a day or two, try receiving what you sent. The wait between sending and receiving will help prevent you from writing down the message from memory. Not only will this procedure provide practice in receiving, but it will also let you hear exactly how your sending sounds. If your timing is off, you'll hear it. If you're having trouble with a specific letter or number or punctuation mark, you'll soon know about it. If you can't understand your own sending, neither will anyone else!

Regardless of whether you're sending or receiving, the key to your success with Morse code is regular practice. After you've learned all the characters you will want to continue your regular practice to gain confidence and increase your speed.

If you can, listen to actual contacts between hams. Make use of the code-practice material transmitted daily by **W1AW**, the ARRL station heard nationwide from Newington, Connecticut. Table 11-1 lists the code-practice schedule for W1AW. When we change from Standard Time to Daylight Saving Time, the code-practice is sent at the same local times.

THE CODE TEST

To pass the Novice or Technician Plus code test, you must demonstrate your ability to send and receive Morse code at 5 words per minute. Your Novice test examiners will give you the code test. The FCC specifies that the exam message should be at least 5 minutes long. The message must contain all letters of the alphabet, all numbers, period, comma, question mark and fraction bar (/). The exam message must also include some common procedural signals. These are the + sign ($\overline{\text{AR}}$), the = sign ($\overline{\text{BT}}$) and $\overline{\text{SK}}$.

The Volunteer Examiners giving you the test have several options in deciding if you pass the exam. If you have one minute of solid (perfect) copy out of a five-minute test, you pass. The examiners may also give you a 10-question fill-in-the-blank or multiple-choice test about the exam content. A passing grade on such a written exam is 70%. The examiners are simply certifying that you have proven your ability to send and receive code at 5 words per minute. Yes, they may even ask you to send some code, although that is not necessary, and isn't a standard part of most exams.

Examination Nerves

If you have prepared properly, the code exam should be easy — a pleasant experience, not an agonizing one. The key to a successful exam is to relax. One of the biggest reasons people fail the code examination is because they become nervous. You can be confident of passing, however, because you have studied hard and are properly prepared.

The exam isn't *that* important. You can always take the test again soon. It won't kill you even if you don't pass. But you are going to pass! You can copy 5 wpm at home with only a couple of errors every few minutes. The code test will be even easier.

Think positively: "I will pass!" Relax. Before you know it, you'll be grinning from ear to ear as you hear the words, "You've passed!"

Heading On Up

Once you obtain your Novice ticket, you'll soon set your sights on a Technician Plus or General license. The ARRL's *Increasing Your Code Speed* series of code practice will help you increase your code speed. There is practice to take you from 5 to 10 wpm, from 10 to 15 wpm and from 15 to 22 wpm. Each set includes either two audio CDs or two cassette tapes, which provide practice with words, sentences, random code groups, related text and QSO practice. The space between characters decreases gradually to increase the overall code speed. You'll be copying faster code before you know it! W1AW also provides practice at a variety of speeds. Contact the ARRL, 225 Main Street, Newington, CT 06111 for the practice sets you'll need to reach your desired code speed.

Amateur Radio Excitement

Seven-year-old Luke Ward of Alexandria, Virginia, and his dad Keith saw an article about how to convert a small AM radio to receive shortwave broadcasts. After completing the project over a weekend, they were able to hear shortwave stations from all around the world! When Luke asked about the strange sounds they often heard, his dad explained that they were Morse code messages being sent by Amateur Radio operators. (Keith had been a ham when he was a boy, but was no longer licensed.) When Luke learned that there was no minimum age to become a ham, he and his dad began studying for their Novice exams. They also bought a used shortwave receiver and became avid shortwave listeners (SWLs). Luke and Keith also joined the Mt Vernon (VA) Amateur Radio Club.

About six weeks later, in August 1990, Luke and Keith passed their Novice exams. The next morning they left for a vacation trip to Cape Cod, and stopped at ARRL Headquarters in Newington, Connecticut, on the way. Luke and his parents were thrilled to see the W1AW facilities, many of which they had read about in ARRL books. Luke was in third grade that fall when the tickets arrived from the FCC, and Luke (now KO4IQ) and Keith (now KO4IO) were soon on the air. Luke's first contact was with VE1VAZ in Nova Scotia, Canada! Within a month Luke had 60 CW contacts on 80, 40 and 15 meters. Luke and his dad have been competing for most states and countries worked.

Luke wasted no time in studying for his Technician Plus license, and he and Keith passed their Technician Plus exams in October 1990, about six weeks before Luke's eighth birthday.

"Learning Morse code took a lot of practice and so was harder than learning the other stuff necessary to pass the written exams," says Luke, who was already technically inclined before he got involved with ham radio. "Besides ham radio, I enjoy BASIC computer programming. Math and PE are my favorite school subjects," he adds. Luke also enjoys soccer and Scouts.

After their 10-meter antenna was tuned, Luke tried SSB phone, and came to enjoy that as much as CW. His first phone contact was with CU1AC on Santa Maria Island in the Azores (a chain of islands in the North Atlantic that belong to Portugal)! "Eighty meters used to be my favorite band, but now I like 10 meters best because of the DX. I really like to get QSL cards, especially from foreign countries. My CW QSOs helped improve my code speed. I think I am about ready to take my General exam, once I finish studying the radio theory part. I want a General ticket so that I can use the extra band space, and I hope I can do moonbounce QSOs with it, too." (Luke did pass his General exam in February 1991, at the age of eight. He and Keith have since upgraded to Amateur Extra.)

Table 11-1

W1AW schedule

Pacific	Mtn	Cent	East	Sun	Mon	Tue	Wed	Thu	Fri	Sat
6 am	7 am	8 am	9 am					Fast Code	Slow Code	
7 am	8 am	9 am	10 am					Code Bulletin		
8 am	9 am	10 am	11 am					Teleprinter Bulletin		
9 am	10 am	11 am	noon							
10 am	11 am	noon	1 pm		**Visiting Operator Time**					
11 am	noon	1 pm	2 pm							
noon	1 pm	2 pm	3 pm							
1 pm	2 pm	3 pm	4 pm	Slow Code	Fast Code	Slow Code	Fast Code	Slow Code	Fast Code	Slow Code
2 pm	3 pm	4 pm	5 pm	Code Bulletin						
3 pm	4 pm	5 pm	6 pm	Teleprinter Bulletin						
4 pm	5 pm	6 pm	7 pm	Fast Code	Slow Code	Fast Code	Slow Code	Fast Code	Slow Code	Fast Code
5 pm	6 pm	7 pm	8 pm	Code Bulletin						
6 pm	7 pm	8 pm	9 pm	Teleprinter Bulletin						
6^{45} pm	7^{45} pm	8^{45} pm	9^{45} pm	Voice Bulletin						
7 pm	8 pm	9 pm	10 pm	Slow Code	Fast Code	Slow Code	Fast Code	Slow Code	Fast Code	Slow Code
8 pm	9 pm	10 pm	11 pm	Code Bulletin						
9 pm	10 pm	11 pm	Mdnte	Teleprinter Bulletin						
9^{45} pm	10^{45} pm	11^{45} pm	12^{45} am	Voice Bulletin						

W1AW's schedule is at the same local time throughout the year. The schedule according to your local time will change if your local time does not have seasonal adjustments that are made at the same time as North American time changes between standard time and daylight time. From the first Sunday in April to the last Sunday in October, UTC = Eastern Time + 4 hours. For the rest of the year, UTC = Eastern Time + 5 hours.

• ***Morse code transmissions:***

Frequencies are 1.818, 3.5815, 7.0475, 14.0475, 18.0975, 21.0675, 28.0675 and 147.555 MHz.

Slow Code = practice sent at 5, 7½, 10, 13 and 15 wpm.

Fast Code = practice sent at 35, 30, 25, 20, 15, 13 and 10 wpm.

Code practice text is from the pages of *QST*. The source is given at the beginning of each practice session and alternate speeds within each session. For example, "Text is from July 1992 *QST*, pages 9 and 81," indicates that the plain text is from the article on page 9 and mixed number/letter groups are from page 81.

Code bulletins are sent at 18 wpm.

W1AW qualifying runs are sent on the same frequencies as the Morse code transmissions. West Coast qualifying runs are transmitted on approximately 3.590 MHz by W6OWP, with K6YR as an alternate. At the beginning of each code practice session, the schedule for the next qualifying run is presented. Underline one minute of the highest speed you copied, certify that your copy was made without aid, and send it to ARRL for grading. Please include your name, call sign (if any) and complete mailing address. Send a 9×12-inch SASE for a certificate, or a business-size SASE for an endorsement.

• ***Teleprinter transmissions:***

Frequencies are 3.625, 7.095, 14.095, 18.1025, 21.095, 28.095 and 147.555 MHz.

Bulletins are sent at 45.45-baud Baudot and 100-baud AMTOR, FEC Mode B. 110-baud ASCII will be sent only as time allows.

On Tuesdays and Saturdays at 6:30 PM Eastern Time, Keplerian elements for many amateur satellites are sent on the regular teleprinter frequencies.

• ***Voice transmissions:***

Frequencies are 1.855, 3.99, 7.29, 14.29, 18.16, 21.39, 28.59 and 147.555 MHz.

• ***Miscellanea:***

On Fridays, UTC, a DX bulletin replaces the regular bulletins.

W1AW is open to visitors during normal operating hours: from 1 PM until 1 AM on Mondays, 9 AM until 1 AM Tuesday through Friday, from 1 PM to 1 AM on Saturdays, and from 3:30 PM to 1 AM on Sundays. FCC licensed amateurs may operate the station from 1 to 4 PM Monday through Saturday. Be sure to bring your current FCC amateur license or a photocopy.

In a communication emergency, monitor W1AW for special bulletins as follows: voice on the hour, teleprinter at 15 minutes past the hour, and CW on the half hour.

Headquarters and W1AW are closed on New Year's Day, President's Day, Good Friday, Memorial Day, Independence Day, Labor Day, Thanksgiving and the following Friday, and Christmas Day.

12 KEY WORDS

Editor's note: No exam questions are covered in this chapter. The key words are often used when describing equipment. They are included to help you learn them.

Amplitude modulation (AM)—A method of combining an information signal and an RF (radio-frequency) carrier. In double-sideband voice AM transmission, we use the voice information to vary (modulate) the amplitude of an RF carrier. Shortwave broadcast stations use this type of AM, as do stations in the Standard Broadcast Band (540-1600 kHz). Few amateurs use double-sideband voice AM, but a variation, known as **single sideband**, is very popular.

Bandwidth—The range of frequencies that will pass through a given filter. Also the range of frequencies that a radio transmission occupies.

Beat-frequency oscillator (BFO)—A receiver circuit that provides a signal to the detector. The BFO signal mixes with the incoming signal to produce an audio tone for CW reception. A BFO is needed to copy CW and SSB signals.

Continuous wave (CW)—Morse code telegraphy.

Frequency modulation (FM)—A method of combining an information signal with a radio signal (see **Amplitude modulation**). As you might suspect, we use voice or data to vary the frequency of the transmitted signal. FM broadcast stations and most professional communications (police, fire, taxi) use FM. VHF/UHF FM voice is the most popular amateur mode.

General-coverage receiver—A receiver used to listen to a wide range of frequencies. Most general-coverage receivers tune from frequencies below the standard-broadcast band to at least 30 MHz. These frequencies include the shortwave-broadcast bands and the amateur bands from 160 to 10 meters.

Ham-bands-only receiver—A receiver designed to cover only the bands used by amateurs. Usually refers to the bands from 80 to 10 meters, sometimes including 160 meters.

Lower sideband (LSB)—The common single-sideband operating mode on the 160, 80 and 40-meter amateur bands.

Multimode transceiver—Transceiver capable of SSB, CW and FM operation.

Offset—The 300 to 1000-Hz difference in CW transmitting and receiving frequencies in a transceiver. For a repeater, offset refers to the difference between its transmitting and receiving frequencies.

Receiver—A device that converts radio waves into signals we can hear or see.

Receiver incremental tuning (RIT)—A transceiver control that allows for a slight change in the receiver frequency without changing the transmitter frequency. Some manufacturers call this a clarifier (CLAR) control.

Rig—The radio amateur's term for a transmitter, receiver or transceiver.

Selectivity—The ability of a receiver to separate two closely spaced signals.

Sensitivity—The ability of a receiver to detect weak signals.

Shack—The room where an Amateur Radio operator keeps his or her station equipment.

Single sideband (SSB)—A common mode of voice operation on the amateur bands. SSB is a form of **amplitude modulation**.

Ticket—A common name for an Amateur Radio license.

Transceiver—A radio transmitter and receiver combined in one unit.

Transmitter—A device that produces radio-frequency signals.

Upper sideband (USB)—The common single-sideband operating mode on the 20, 17, 15, 12 and 10-meter HF amateur bands, and all the VHF and UHF bands.

12 Selecting Your Equipment

You may be able to operate a club station or another amateur's station when your license is issued, but eventually you will want a station of your own. By shopping carefully, you can find just the right equipment!

Having an amateur license and no station is a little like having a driver's license but no car. Without a station, your **ticket** (a ham's name for an Amateur Radio license) is just another piece of paper. You can probably arrange to use a friend's equipment or operate from a club station when your license first arrives. Very soon, you'll want to have your own station.

Most hams look with pride at their operating position. When hams meet, conversation always turns to the station. Whenever hams visit each other, the **shack** is usually the first stop on the tour. Many amateurs who develop friendships on the air eventually swap pictures of their stations. It's no wonder, really. Hams devote many hours to their hobby, and they spend most of these hours in their shacks. The shack is where a ham operates on the air, repairs equipment, makes improvements and experiments with new projects.

As a newcomer to Amateur Radio, you may find the wide assortment of equipment, antennas and available accessories confusing at first. You'll wonder, "Why is this antenna better than that one?" or "What features does this **transceiver** offer? Do I really need them all?"

You have to decide what your goals are. Take a look at your available resources (how much money you have to spend). Do a little research and then choose the equipment that best suits your needs. Actually, selecting your station equipment can be very easy, if you know what you want it to do. The information in this chapter will help you select a radio that will provide you with many hours of enjoyable operating.

To help you select station equipment, the ARRL publishes a collection of Product Reviews from *QST*, our monthly membership journal. *The ARRL Radio Buyer's Sourcebooks* include dozens of in-depth reviews and technical information about HF, VHF and UHF radios and accessories. Sections of this book also explain equipment features and specifications that are important to enjoyable operating in more detail than we can cover them in this chapter.

Look Before You Leap

Before you go out and purchase a room full of equipment, find out how big the room is! Try to get some idea of how much space you will have for your shack. Chapter 4 has more detailed information about where to locate your station. Some hams use a corner of a bedroom or den, and some use a fold-out shelf in a closet. Available space may be an important consideration in what equipment you purchase.

When you know how much space you have, you're ready to select your equipment. There are many factors to consider. "Specs" (specifications) are very important, but don't choose your equipment on that basis alone. Some **rigs** are technical marvels but very difficult and frustrating for a new ham to operate. Other gear may look great but have a lot of technical problems. The most important consideration is what works for you. Choose equipment that will be enjoyable and comfortable to operate.

There are many different places to look for equipment. Check with local hams to see what they use. Find out what they like about certain pieces of gear and what problems they have had. Learn from their experience. They are as proud of their shacks as you will be of yours. When you are deciding what you want, there is no substitute for sitting down and listening to a rig. (For the name of a local radio club, contact ARRL Headquarters.)

Another place to look is your local radio store. Check the telephone book for ham radio retailers in your area. Most large metropolitan areas have at least one. At the store, you can see and compare several of the newest pieces of equipment. In addition, most ham stores have antennas set up so you can listen to equipment you might buy, even if you don't have your license yet.

The popular ham magazines carry many ads in each issue. When you look through any issue of *QST*, you can find ads from many manufacturers. Local dealers and large mail-order companies also advertise in *QST*. Many of the ads list features and specifications, and will give you a rough idea of what the equipment costs.

QST has an advertising acceptance policy that can help protect you. If a piece of equipment doesn't live up to its manufacturer's claims, you won't see it advertised in *QST*. It's a good idea to study the ads and have some equipment in mind when you visit the local ham store.

Should You Build Your Own Equipment?

There was a time when almost all hams built their own rigs. The cost was small, and the operators took great pride in their work. Many hams still build some of their own equipment, either from parts or as kits. We hope you will try building at least some part of your station as well. The satisfaction of being able to say, "I built it myself!" is a joy you will never forget.

Most beginners won't want to build a full-featured transmitter, receiver or transceiver. The electronics have gotten much more complicated over the years, and this complexity can frustrate an inexperienced builder. If you have a limited ham radio budget, buy a used rig.

VHF AND UHF EQUIPMENT

Whether you are working toward the Novice or the Technician license, you have a wide variety of VHF and UHF equipment to choose from. Many transceivers for these bands operate only in the **frequency modulation (FM)** mode. Other gear can be used only for SSB and CW work. For both FM and SSB/CW operation, you need a third type: the multimode transceiver, like the one in Figure 12-1. You have to know what modes and bands you'd like to try before you can choose a rig. Two and three-band rigs are becoming more and more popular.

Also consider where you'll use the equipment. Some VHF/UHF equipment is best for home station operation. Other, more compact, units work well for mobile operation. Portable, hand-held transceivers can be used anywhere.

Although manufacturers sometimes design equipment for a certain application, you can use it wherever you want. If you like, you can use a hand-held transceiver in a car, or a mobile rig at your home station. There are, however, some

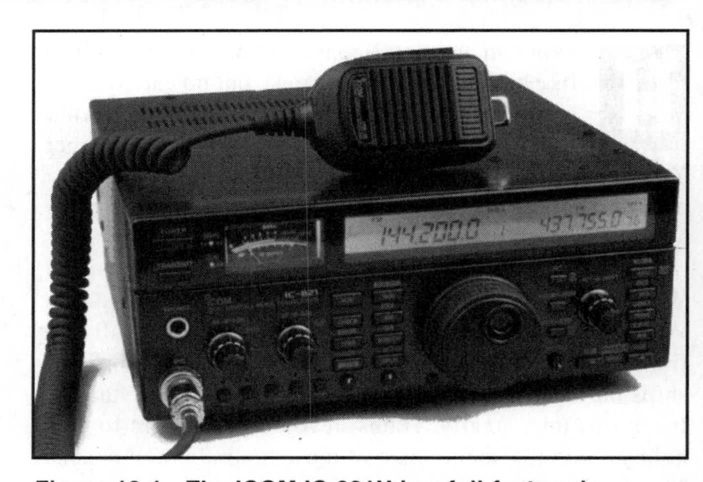

Figure 12-1—The ICOM IC-821H is a full-featured VHF/UHF multimode transceiver. It operates on both 2 meters and 70 cm, making it ideal for use with ham satellites as well as FM repeaters. In addition, it offers sideband and CW modes.

Figure 12-2—As with other types of transceivers, the trend in mobile radios like this one is to pack more and more features into a smaller and smaller package. The Standard C5900DA covers three bands, 6 and 2 meters and 70 cm, and receives a wide range of VHF and UHF frequencies. Among its other features is a detachable front panel that makes mounting easy in any type of vehicle.

Figure 12-3—The Kenwood TS-790A is a multiband, multimode transceiver that operates on the 144 and 440-MHz bands. Operation on the 1240-MHz band is possible with an optional module.

trade-offs to consider when you want to use a rig at home and in your car or portable.

Base-Station and Mobile Equipment

FM base stations often consist of a multimode transceiver or a mobile transceiver. Multimode base-station transceivers often have built-in ac power supplies and are larger than FM-only rigs.

Mobile rigs are often smaller than base-station transceivers, and they operate from a 13.8-V automobile electrical system. See Figure 12-2. You'll need an external, accessory power supply to use one at home.

FM mobile transceivers usually have all the features found in base-station equipment. Power output for mobile and base-station equipment is typically in the 10 to 50-watt range. The rig's physical size and the power supply are the most important things to consider. Other than size and power requirements, there is no particular advantage to either a base-station unit or a mobile unit for FM work. Many amateurs buy a mobile transceiver and external power supply, and use the same radio at home and in the car.

Transverters

There are fewer radios available for VHF/UHF SSB and CW work than for FM. Much SSB and CW operation is done with a *linear transverter* used with a regular 10-meter HF transceiver. A transverter converts signals from one band to another. For example, a 222 to 28-MHz transverter receives a signal at 222.1 MHz and converts it to 28.1 MHz for reception on an HF receiver. Likewise, the transverter accepts a 28.1-MHz signal from your HF transmitter and converts it for transmission on 222.1 MHz.

The transverter has no external controls. The transceiver operates exactly like it does on HF; the only difference is that you're transceiving on 222 MHz instead of 28. Most transverters have 10 to 25 watts RF output. Transverters are available for all VHF/UHF bands, as well as several microwave (above 1000 MHz) bands.

Multiband and Multimode Transceivers

Multiband transceivers combine more than one VHF/UHF band in one box. Optional modules for other bands may be available. For example, the unit shown in Figure 12-3 comes supplied with modules for 144 to 148 and 420 to 450 MHz. A separate 1270 to 1300-MHz module is available at extra cost. This rig has the additional ability to transmit on one band and receive on another, at the same time. Amateur Radio satellites require users to transmit and receive on separate bands, which is easy if you have two rigs. With transceivers like this one, you need only one rig.

The transceiver shown earlier in Figure 12-1 is also a **multimode transceiver**. It can be used on FM and SSB voice, and also on CW. Not all multiband rigs are multimode, and some multimode rigs operate on only one band.

Hand-Held Transceivers

Hand-held VHF or UHF FM transceivers (often called H-Ts, an abbreviation for "Handie-Talkie," a Motorola trademark) are very popular. Such equipment is the most compact of the lot. See Figure 12-4. H-Ts are self-contained stations. They use battery power and have a built-in antenna, speaker and microphone.

Most hand-held units can be used with an external 12-V supply and several types of antennas that are more efficient than the "rubber duck." Some offer accessory external speakers and microphones, which are handy

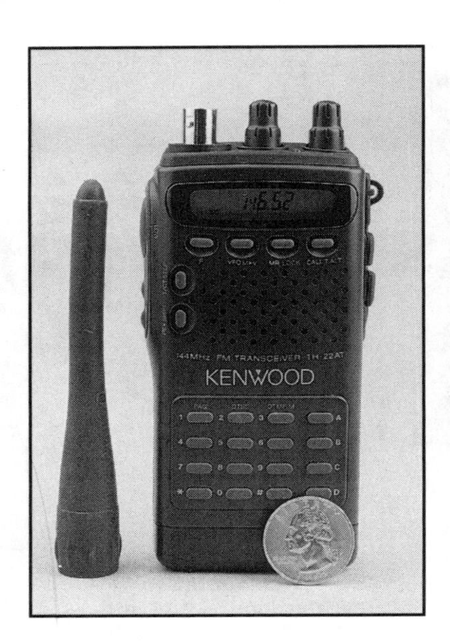

Figure 12-4—The Kenwood TH-22AT hand-held transceiver is only 4½ inches high, but does not lack for features. These include: wideband receive (136-174 MHz), 40 memory channels and wireless cloning capability—you can automatically transfer stored data such as memory channels from one TH-22AT to another.

for mobile or home-station use.

To reduce size and conserve battery power, H-Ts usually have lower power output than mobile or base-station rigs. Although a typical H-T has 1 or 2 watts output, a high-power model may produce 5-7 watts. Optional higher-power *battery packs* provide an increase in power and capacity.

FM Transceiver Features

VHF and UHF FM transceivers have many features not always found on HF rigs. While many of these features are not essential, they are useful. All FM transceivers have a *squelch* control that quiets the receiver when no signal is present. Squelch lets you leave your transceiver on without having to listen to noise when the frequency is not in use.

All newer transceivers use a synthesizer to control the operating frequency, while many older ones use VFOs or crystals. Synthesized and most VFO-type transceivers can operate on any frequency in the band. Crystal-controlled radios have a front-panel control to allow you to switch between two or more repeater or simplex channels. If you add the cost of crystals (about $15 per channel) to the cost of a second-hand crystal-controlled transceiver, the total cost may be more than that of a synthesized transceiver. For that reason, very few crystal-controlled transceivers are in use today.

Synthesized transceivers offer almost unlimited frequency flexibility. Going from simplex to repeater operation is as easy as flipping a switch. The transceiver comes factory programmed for the proper repeater frequency split for the band. You dial up the repeater output frequency and the transceiver automatically switches to the repeater input frequency when you transmit.

Nearly all newer VHF transceivers have *memories* to store your favorite frequencies. You store them in the memory and recall them as needed. For example, memory 1 may be for the local 145.45-MHz repeater, memory 2 for the 146.52-MHz simplex frequency, and so on.

Nearly all VHF transceivers also have automatic *scanning* up and down the band, or among the channels stored in memory. This feature is handy if you're traveling and want to find an active repeater.

There are other FM transceiver features, sometimes standard and sometimes available as options. *Tone pads* for autopatch use and *tone-burst* or *subaudible-tone generators* for repeaters that require such tones for access are standard on most new FM transceivers. (See Chapter 2 for more information on using tone pads and subaudible tones.) An *autopatch* provides a connection between the amateur station and the telephone line. A tone pad is a 12 or 16-button keypad used to generate audio tones similar to those used on TouchTone telephone systems. Some transceivers feature *wideband receivers* that let you listen to signals outside the amateur bands. These transceivers are useful for public-service work during emergencies.

HF EQUIPMENT

There are many different types of HF transceivers to choose from. You probably won't want to buy a top-of-the-line rig, one with all the "bells and whistles," until you've had some experience with a less complicated model. For one thing, HF rigs can cost several thousand dollars, even if purchased used. For another, they can be more difficult to learn to operate. Once you've become comfortable with the features of a decent-quality rig, you can always trade up to one that has all the latest features, if you wish to do so.

New or Used?

As with any purchase, you're safer buying a new piece of radio equipment than a used one. Fresh out of the box, it will have the manual or manuals, all the accessories and, perhaps best of all, the manufacturer's warrantee. If something goes wrong during the warrantee period, part or all the cost of any needed repairs will be covered. In addition, you'll know that the radio didn't have coffee spilled on it, and that it has not been abused in some way.

On the other hand, buying a good used transceiver is a great way to get on the air without investing a large sum of money. There is a healthy market in used gear that will ensure that you'll find a transceiver that's right for you and your pocketbook.

Before transceivers, there were separate transmitters

Used Equipment

What used rig should you buy? The answer to that question depends largely on available resources and personal preferences. Older issues of amateur magazines can provide a lot of information on equipment you may see on the used market. The New Products and Product Review columns in *QST*, and *The ARRL Radio Buyer's Sourcebooks* are a good start. Many local libraries carry back issues of *QST*. Local hams may also have some *QST*s. If you can't find the issues you need, you can purchase many of them from ARRL HQ. In addition, several other product guides and product-review anthologies are available. They summarize features and performance of recently manufactured equipment.

Don't just rush out and buy the first piece of equipment you can find. Buying used radio equipment is much like buying a used car: You really should "kick the tires" a bit. Examine the equipment closely (as well as you can). Don't get snowed by salesmanship. A couple of old adages apply: "You (generally) get what you pay for" and *caveat emptor*—buyer beware. Perhaps we should add a third: "All things come to those who wait." Have patience!

Surplus Equipment

There's still some WW II government surplus equipment on the market, and you may also see some newer surplus gear. Generally speaking, a beginner who doesn't have an Elmer should stay away from this equipment.

Much of the military equipment is big and heavy. You might also have to find some special connectors or build a power supply to use it. Some units produce large amounts of television interference (TVI), and that's one thing we can all do without!

Many units that appear to be bargains will require quite a bit of work to get them on the air. Many hams enjoy using surplus gear, but it's not cost-effective for the beginner. You might try a surplus rig later, when you have gained some experience and want to do some tinkering.

Tube Availability

You might wonder about the practicality of buying tube-type equipment. After all, modern equipment uses transistors and integrated circuits, and many companies have stopped making tubes. Tubes are still available, but they are becoming expensive. A few electronic supply houses still stock tubes. Many amateurs have plenty of tubes (often free for the asking) in their "junk boxes." Some excellent transceivers made in the 1970s used tubes in the power-amplifier stages. These tubes usually are readily available.

Where Can I Find Used Gear?

There are many different sources of used ham gear. You can buy gear from equipment retailers and local hams. You may see what you want listed in the pages of ham magazines (see the *QST* Ham Ads) or used-equipment flyers. Other sources are auctions, flea markets, hamfests and even garage sales.

If you belong to an Amateur Radio club, ask your instructor and other club members. They can often provide you with some leads and other useful information about buying used gear. The club members can help you select a particular unit or a complete station. They can also help you test it to ensure it works properly. They will certainly provide some helpful hints on connecting all the pieces.

One of the safest ways to purchase used equipment is to buy it from a local retailer. Often the dealer sets aside a section of the ham shop for used equipment. When you buy from such a source, there's usually some sort of warranty (30 days is typical) on the equipment. If a problem comes up within that time, you can bring it back to the retailer for repairs. Ask about the warranty before you buy.

Many dealers route the used equipment they've received through their service shop. This ensures that the gear is working properly before they place it on the used-equipment shelf. Some dealers will even allow you to operate the equipment before you buy it. This is the best way to determine the rig's capabilities and condition.

When you buy used gear from an individual, you generally have no guarantees. Ask to see the unit in operation. You might even wish to take along a more knowledgeable ham who can help you make the decision. A local ham can be your best source of information on used equipment.

You may already know what you want or what you need. The hard part is deciding which of those used rigs provides the best value for your hard-earned dollars. An "old-timer" can be an invaluable source of information. Chances are, he or she may have at one time owned a similar piece of equipment. With the seller's permission, your ham friend could operate the equipment and note any potential problems—your future headaches!

Local hamfests, flea markets and auctions are some of the better opportunities to see a quantity of used equipment. For the careful buyer, they may also be the source of some excellent equipment. But you have to know what you're looking for. What are the capabilities of the equipment, its current market value, cost of repairs and availability of repair parts?

At a hamfest or flea market, you may not be able to test equipment before you buy it. A thorough visual inspection by an experienced eye will generally suffice if the price is right, however. Again, help from an experienced amateur is a wise choice. You can usually tell a lot from the external appearance of equipment. If it looks physically abused, chances are it has been treated badly on the air as well.

There are some simple precautions you should take if you're buying gear from another ham through the mail. Try to be sure you'll have the right to return the equipment if you don't like what you receive. Shipment by truck freight with the right of inspection permits you to examine the package contents before you accept delivery. If you don't like what you receive, simply refuse delivery. You may be asked to pay by bank check or money order, rather than personal check.

There is one other very important point to keep in mind when you buy used equipment from any source: Be sure you get an owner's manual with the radio! Some hams may even have the service or shop manual. A shop manual can be a valuable addition, because it generally has more complete service procedures and troubleshooting guidelines.

If you are not getting at least an owner's manual with the radio, be cautious about making the purchase. Manuals for some pieces of equipment are available from the manufacturer. Several companies sell manuals for used equipment; see Table 12-1.

Table 12-1

Sources of Equipment Manuals

Alinco Electronics, Inc
438 Amapola Ave, #130
Torrance CA 90501
Tel 310-618-8616
(Alinco only)

Azden Corp
147 New Hyde Park Rd
Franklin Square NY 11010
Tel 516-328-7501
(Azden only)

Brock Publications
PO Box 5004
Oceanside CA 92052
Tel 619-757-0372
(Swan, Cubic, Siliconix)

Collins S-Line: See Surplus Sales of Nebraska and W7FG

Cubic Communications: See Brock Publications

R. L. Drake Co
230 Industrial Way
Franklin OH 45005
Tel 513-746-6990
(Drake only)

HI Manuals, Inc†
PO Box 802Q
Council Bluffs IA 51502
(large selection of older equipment)

Howard W. Sams Co.
Photofact Division
26476 Waterfront Pkwy East Dr
Indianapolis IN 46214-2041
Tel 800-428-7267

ICOM America, Inc
PO Box C-90029
2380 116th Ave NE
Bellevue, WA 98009-9029
Tel 206-454-7619
(ICOM only)

Kenwood East Coast Transistor Parts, Inc
2 Marlborough Rd
W Hempstead NY 11552
Tel 800-637-0388 (orders only)
(Kenwood only)

Kenwood Pacific Coast Parts Distributors, Inc
15024 Staff Ct
Gardena CA 90248
Tel 800-421-5080 (outside CA)
800-262-1312 (inside CA)
(Kenwood only)

Land Air Communications
95-15 108th St
Richmond Hill, NY 11419
Tel 718-847-3090
(Kenwood, ICOM and Yaesu, as well as some older equipment)

The Manual Man††
27 Walling St
Sayreville NJ 08872
Tel 908-238-8964
(large selection of older equipment)

Siliconix: See Brock Publications

Standard Amateur Radio Products
PO Box 48480
Niles IL 60714
Tel 773-763-0081

Surplus Sales of Nebraska
1502 Jones St
Omaha NE 68102
Tel 402-346-4750
(Collins and others)

Swan Manuals: See Brock Publications

Ten-Tec, Inc
1185 Dolly Parton Pkwy
Sevierville, TN 37862
Tel 423-453-7172
(Ten-Tec only)

W7FG Vintage Manuals
3300 Wayside Dr
Bartlesville OK 74006
Tel 800-807-6146
(Most Collins, Drake, Hallicrafters, Hammarlund, Heathkit, Johnson, National, Swan and others)

Yaesu USA
17210 Edwards Rd
Cerritos, CA 90701
310-404-2700

†Order their catalog ($3) to see if they have the manual you need.

††Send two First-Class stamps for catalog.

and receivers. Still available on the used market, these "separates" required special wiring between the two pieces of gear, along with a transmit-receive (TR) switch to switch the antenna between the two units. Nearly all the rigs sold today, new or used, are transceivers, although some hams use a separate "outboard" receiver for added convenience.

Examples of several types of new and used ham transceivers are shown in Figure 12-5.

Equipment Features

When you decide what kind of HF gear you want, you'll have to examine specific features. The next few sections describe some of the things to look for.

Frequency Display

Of all the controls on a rig, you use the frequency control most often. The frequency display includes the knobs, dials, gears and readout. Make certain that the mechanism works freely, the dial markings and frequency readout are understandable, and you feel comfortable operating it.

Two different types of tuning-dial mechanisms are available on recently manufactured transceivers: digital and analog. Most manufacturers now use digital displays. Digital displays are generally accurate and easy to use. Some transceivers have a control that allows the user to change the tuning rate, making it easy to scan the bands slowly or quickly. The analog circular dial (Figure 12-6) is more common on older and home-built equipment. The dial is usually calibrated in kilohertz. A fixed pointer indicates where the dial is tuned.

Choose the dial mechanism you are most comfortable reading and tuning. Make sure the knob operates smoothly. Avoid radios that feel sloppy and slip or skip—that's part of

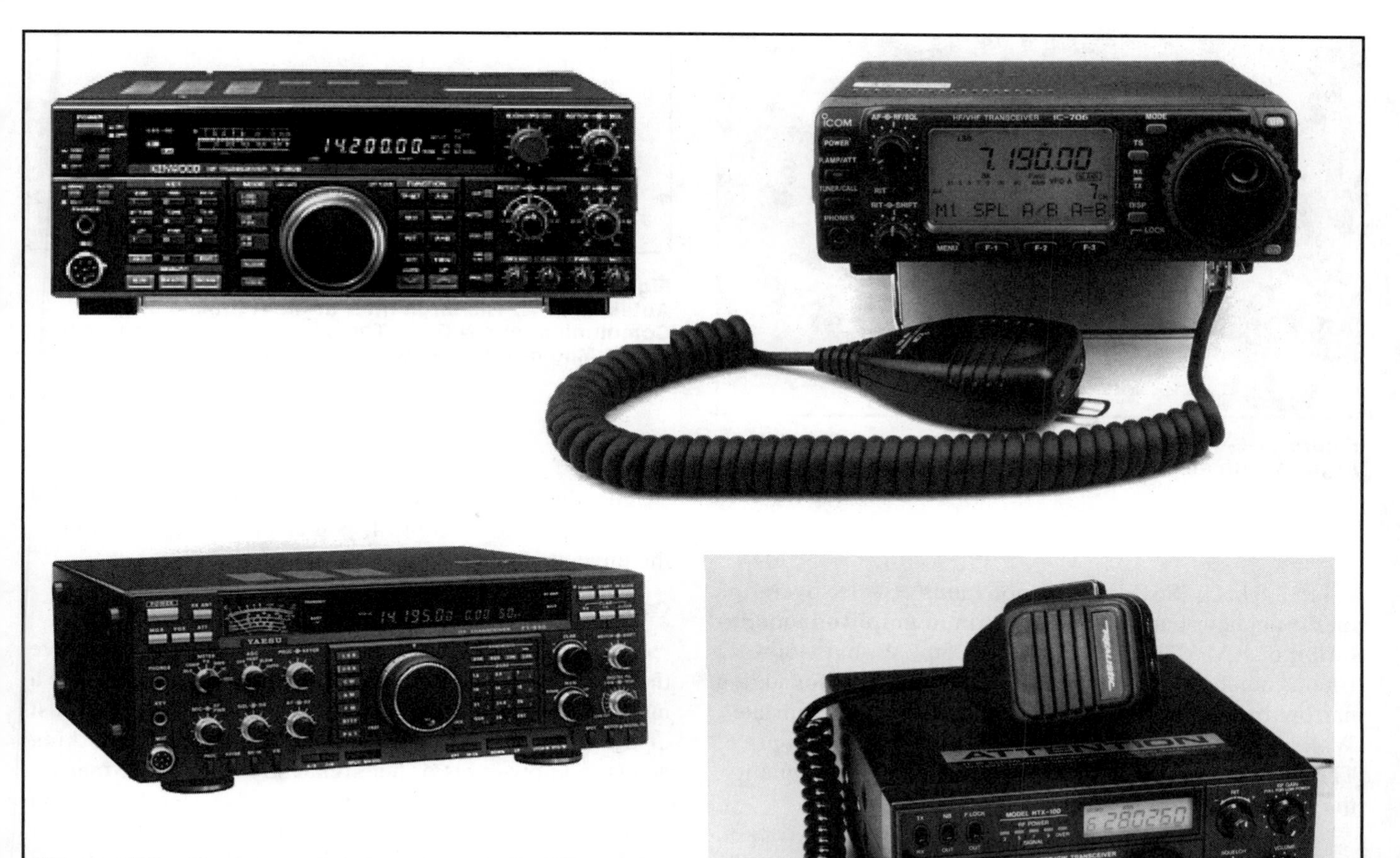

Figure 12-5—These transceivers are typical of the gear available today in the new and used market. (Clockwise from upper left) The Kenwood TS-450S is a compact transceiver whose many useful features include 100 memory channels, 9-band operation, general-coverage receiver and optional built-in antenna tuner. The ICOM IC-706 MF/HF/VHF transceiver is an ideal mobile radio because it operates on all bands from 160 through 2 meters, and all this capability fits in a package no larger than some single-band VHF rigs. The Radio Shack HTX-100 is a small, inexpensive 10-meter transceiver. The Yaesu FT-990, somewhat larger than the other rigs pictured here, has a long list of standard features, including full coverage of the 160 through 10-meter bands.

your comfort and accuracy.

Receiver Selectivity and Sensitivity

The receiver is the most critical part of the station. In this section, *receiver* applies to either a separate receiver or the receiver section of a transceiver.

There are two important receiver specifications you should know about: selectivity and sensitivity. **Selectivity** is the ability of a receiver to separate two closely spaced signals. This determines how well you can receive one signal that is very close to another. **Sensitivity** is the ability of a receiver to detect weak signals. Selectivity is more important than sensitivity. The ability to isolate the signal you are receiving from all the others nearby directly affects how much you will enjoy your time on the air.

We use **bandwidth** to measure selectivity. Bandwidth is nothing more than how wide a range of frequencies you hear with the receiver tuned to one frequency. With a 6-kHz bandwidth, you can hear signals as far as 3 kHz above and below where you are tuned.

If you can't hear signals more than 1200 Hz above or below where your receiver is tuned, your receiver has a bandwidth of 2400 Hz or 2.4 kHz. Narrower bandwidth means better selectivity. Narrow bandwidth makes it easier to copy one signal when another is close in frequency.

Special filters built into modern equipment determine the selectivity. Generally, these filters contain quartz crystals arranged to provide a specific selectivity. Some receivers come with several filters. You can choose the filter that gives the best reception.

Look for a selectivity of 500 Hz or less for **CW** operation. (CW is short for continuous wave, or Morse code telegraphy.) Receivers designed for **single-sideband (SSB)** voice operation come with a standard filter selectivity of

Figure 12-6—An example of a transceiver, the Ten-Tec Argosy, with an analog circular dial.

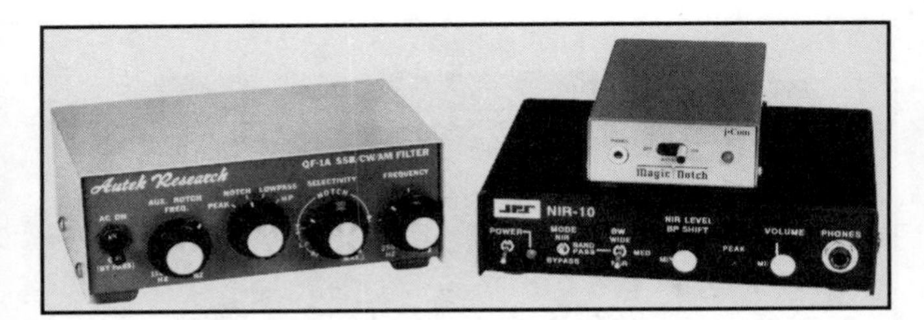

Figure 12-7—Several types of audio filters. At left is the Autek QF-1-A. The large filter at the right is a JPS Communications NIR-10. The smaller unit above it is the j-COM Magic Notch, a switched-capacitor notch filter.

2.4 to 2.8 kHz. (SSB is a common mode of voice operation on the amateur bands. SSB is a form of **amplitude modulation** or **AM**.) This filter is usable on CW, but a 500-Hz filter is much better. If the receiver has provision for adding narrow-bandwidth accessory filters, you can buy them later. While selectivity is very important, it comes with a price! The more filters a receiver has and the better their quality, the more the receiver or transceiver will cost.

If you're interested in operating RTTY, packet, AMTOR, or any other "digital" mode, you can also make use of a 500-Hz filter *if the receiver allows it to be used in FSK mode* (see Chapter 2 for more information on these modes). Some receivers, including those in modern transceivers, only allow use of the SSB filter when operating these modes. A few minutes spent listening to the receiver and reading the manual will tell you what filters are available when the different modes are selected.

To receive HF single-sideband (SSB), CW, RTTY, AMTOR or packet signals, your receiver must have a **beat-frequency oscillator (BFO)**. Almost all HF transceivers do, but many VHF/UHF transceivers don't. On VHF and UHF, RTTY and packet signals are usually sent with FM rigs, so you won't need a BFO for those modes on those bands.

An *audio filter* (Figure 12-7) is an inexpensive way to add selectivity to a receiver that has only an SSB filter. An audio filter doesn't work against interference as well as a crystal filter, but it may mean the difference between maintaining a QSO or losing one.

Audio filters are simple to construct (see *The ARRL Handbook for Radio Amateurs*). If you don't want to build a filter, you can purchase one of the many available commercial units. Audio filters are easy to use. They plug into the receiver headphone or speaker jack, and you plug your headphones or a speaker into the filter.

Another selectivity feature available on some receivers is a notch filter. This filter can be used to cut out, or notch, a specific frequency from within the received bandwidth. A notch filter is handy when you're trying to receive a signal that is very close in frequency to another signal. By adjusting the notch control, you can effectively eliminate the unwanted signal.

Other Receiver Features

You may find other features on receivers that improve their performance or make them easier to use. Most have a meter that shows the strength of the received signal. Most of the time, the first thing the operator you are working wants to know is his or her signal strength. These meters, called *S meters*, will tell you. They are also useful when you are comparing two antennas. Or, you can use the S meter when you are trying to rotate a directional antenna for maximum signal strength.

Some older receivers have a *crystal calibrator*. They produce a signal every 25 or 100 kHz. You can use the calibrator to make sure the dial or display is accurate. Most equipment with a digital display does not have or need a crystal calibrator.

Besides the filters already mentioned, some receivers have *noise blankers*. They are mostly useful for filtering ignition noise from cars and trucks. Some blankers can help reduce power-line noise as well. Some receivers have filters between the intermediate stages. These may be called variable IF filters, or the receiver may have a control labeled PASSBAND TUNING or VARIABLE BANDWIDTH TUNING. An RF gain control is sometimes helpful when interference is heavy.

Receiver incremental tuning (RIT) allows you to shift the receiver frequency over a limited range without affecting the transmitter frequency. Although transceivers theoretically transmit and receive on exactly the same frequency, there may actually be a slight frequency difference, or **offset**. This offset can vary from transceiver to transceiver. RIT lets you retune the receiver slightly. This feature also enables you to compensate if the station you are working has a different offset or is "drifting" (changing frequency slowly with time—a common occurrence with some older rigs.

Most HF transceivers now feature **general-coverage receivers**. Unlike **ham-bands-only receivers**, a general-coverage receiver lets you listen to shortwave broadcast stations and utility stations transmitting RTTY, facsimile, Morse code and voice.

All these features can add to your operating enjoyment. They may not be necessities if you are on a tight budget, however. Frequency coverage, tuning mechanism, frequency resolution, selectivity and sensitivity are the most important things to look for.

The Transmitter

Almost all transceivers offer voice operation. Voice capability is important if you want to operate voice or data on the 10-meter Novice/Tech Plus band (most data modes use the voice part of the radio).

Power output varies from rig to rig. Most transceivers produce about 100 watts output, which is less than the 200 watts output that Novices and Tech Plus licensees are allowed to use. This may sound bad, but it isn't. Most of the time, if 100 watts won't make the contact, neither will 200. One-hundred watts will even drive a high-power linear amplifier, if you decide to get one when you upgrade to General or a higher license. Figure 12-8 shows a simple but effective operating position.

A shortcoming of most rigs made before 1980 is their inability to operate on the 12, 17, and 30-meter bands. When you upgrade to the General or higher class license, you may want to use those bands. You'll also be able to use much larger portions of the 80, 40, 15 and 10-meter bands, plus the 160 and 20-meter bands. The rig you buy may cover bands you can't use right now, but they'll be there, waiting for you when you upgrade. If you later decide to sell the rig, they're sure to make it more valuable than a rig that doesn't have them.

A typical HF transmitter has an output power of between 80 and 200 watts. It operates **upper sideband (USB)**, **lower sideband (LSB)** and CW. The transmitter will cover the 80, 40, 30, 20, 17, 15, 12 and 10-meter bands. Some also include the 160-meter band, and older transceivers will not cover the 30, 17 and 12-meter bands.

Modern transceivers or transmitters with vacuum-tube final amplifiers develop more than 100 watts PEP output. The workhorse of the final-amplifier-tube family is the 6146. The Kenwood TS-830S is a popular transceiver that uses 6146 tubes in its final amplifier. It should be available at reasonable prices on the used market.

Tube-type transmitters use two basic types of keying circuits: cathode keying and grid-block keying. If you use a hand key, the voltage polarity at the key jack is of little consequence. If you use an electronic keyer with a transistor switch in the output circuit, polarity must be considered.

Keyers that use relays with floating (ungrounded) contacts in the output circuit are more flexible since key-jack voltage polarity is unimportant. But relay contacts can stick and the trend is to use transistorized output stages in modern keyers.

Cathode keying presents a positive voltage to the transmitter key jack. In cathode keying, the key line opens and closes the cathode circuit of one or more tubes in the transmitter. When the key is up, there is an open circuit in the tube cathode. This effectively shuts off the tube so it draws no plate or screen current; thus, no RF output.

Grid-block keying uses a negative voltage applied to the grid(s) of the keyed tube(s). This voltage, usually at a reduced amplitude, is also present at the transmitter key jack. With the key open, the bias voltage cuts off the tube. Closing the key removes or reduces the bias and the tube conducts.

As a rule, older transmitters used cathode keying while more modern tube equipment uses grid-block keying. Fully transistorized rigs may use either positive or negative polarity at the key jack.

Before you connect a keyer to any rig, you should check the key-jack voltage and polarity. Make sure your keyer works with the rig. Check your equipment manual. It should tell you the type of keying circuit used in your transmitter, and the key-jack voltage. If you built the keyer, the construction article should tell you the appropriate keying voltage.

There are ways to change the keyer output circuit. You can use a transistor or optoisolator, but these techniques are

Figure 12-8—Computers have become an integral part of most ham stations, and KA7JAW's is no exception. The operator, Ernie, was 92 when this photo was taken. *(Photo courtesy Lynn D. Baker, K7LUH)*

SIDEBANDS

There are two kinds of **single sideband: upper sideband (USB)** and **lower sideband (LSB).** Hams usually use LSB on 160, 80 and 40 meters, and USB on 20, 17, 15, 12 and 10 meters. Upper sideband is also used on VHF and UHF. Some radios automatically switch to the correct sideband for the operating frequency. Others have separate USB and LSB positions on the mode switch. You will always use the USB position for 10-meter and VHF/UHF SSB operation.

beyond the scope of this book. A more-experienced amateur may be able to help you if you face this problem. Also, *The ARRL Handbook for Radio Amateurs* contains more information.

Other Transmitter Features

Microphone

To talk on SSB, you will need to add a microphone to your transceiver. The microphone connects to the microphone jack on your radio.

Individual voices and microphones have different characteristics. Your rig should have a microphone gain control you can adjust for a clean signal. Most SSB transmitters have an automatic level control (ALC) meter to help you determine the correct microphone-gain setting. See your equipment manual for instructions on adjusting everything to the right levels.

Speech Processor

Speech processing increases the average power of a single-sideband signal. Used properly, a speech processor can greatly improve the readability of a signal. Misused, however, it can severely degrade the audio quality and make the signal more difficult to understand.

Almost all SSB transmitters and transceivers built since 1980 have speech processors as standard equipment. There is also a variety of accessory speech processing equipment for transmitters without built-in processing.

On some rigs, there is no adjustment for the speech processor; there is just an on/off control. Other gear has one or more variable controls to set the speech processing level. Always check your equipment instruction manual for information on setup and operation. When you think you've got everything adjusted properly, ask other amateurs for on-the-air checks.

Voice-Operated Switch (VOX)

Most SSB transceivers and transmitters have a voice-operated switch (VOX). The VOX switches the rig into transmit automatically when you speak into the microphone, then back to receive when you stop talking. VOX is handy because it allows you to listen during pauses and lets you keep both hands free.

There are usually three VOX controls: gain, delay and anti-VOX. *VOX gain* sets the sensitivity. Adjust this control so the VOX keys the transmitter when you speak in a normal voice.

VOX delay sets the interval between when you stop talking and when the transceiver switches back to receive. Most hams set this control so the rig switches after a short pause in speech.

Anti-VOX works with VOX gain to keep receiver audio from keying the VOX. If improperly set, speaker audio can key the transmitter. You could set the VOX gain so speaker audio wouldn't key the transmitter, but if you did, you'd have to shout to activate the VOX. Anti-VOX circuitry allows the VOX to ignore audio from the speaker, yet respond when you speak into the microphone.

For More Information

We've tried to give you a few guidelines in this chapter. There are lots of rigs out there, both new and used. Take your time and try to get as much information as you can. Talk to other hams and read the Product Reviews in *QST*. Advertisements in *QST* will tell you what's available. Remember that everyone has an opinion, though. What they like might not be what you like.

Most of all, don't worry! If you buy a used rig and you don't like it, you can probably sell it for almost as much as you paid for it. Start small and simple. Ham radio is fun; part of the fun is dreaming about new equipment and upgrading your station as you upgrade your license.

13 Novice (Element 2) Question Pool—With Answers

Study the questions in this chapter as directed in the previous chapters, to review your understanding of the material for your exam.

Don't Start Here!

This chapter contains the complete question pool for the Element 2 exam. Element 2 is the written part of the Novice exam. You must pass this written exam to earn any Amateur Radio license. To obtain a Novice license you must also pass the Element 1A 5-wpm Morse code exam.

The Element 2 exam is also part of the Technician license exam. The Technician exam also includes Element 3A. Chapter 14 contains the entire Element 3A question pool, with answers. The Technician license does not require a Morse code test, but by passing the Element 1A 5-wpm code exam you can earn a Technician Plus license.

Before you read the questions and answers printed in this chapter, be sure to read the text in the previous chap-

Table 13-1

Novice Exam Content

Subelement	*Topic*	*Number of Questions*
N1	Commission's Rules	10
N2	Operating Procedures	2
N3	Radio-Wave Propagation	1
N4	Amateur Radio Practices	4
N5	Electrical Principles	4
N6	Circuit Components	2
N7	Practical Circuits	2
N8	Signals and Emissions	2
N9	Antennas and Feed Lines	3
N0	RF Safety	5

ters. Use these questions as review exercises, when the text tells you to study them. Don't try to memorize all the questions and answers. This book was carefully written and prepared to guide you step-by-step as you learn about Amateur Radio. By understanding the electronics principles and Amateur Radio concepts in this book, you will enjoy our hobby more. You will also better appreciate the privileges granted by an Amateur Radio license.

This question pool, released by the Volunteer Examiner Coordinators' Question Pool Committee, will be used on exams beginning July 1, 1997. The pool is scheduled to be used until June 30, 2001. Changes to FCC Rules and other factors may result in earlier revisions. Such changes will be announced in *QST* and other Amateur Radio publications. Normally, the Question Pool Committee will simply withdraw specific questions from the question pool when such Rules changes occur.

How Many Questions?

Your Element 2 exam will consist of 35 of these questions, selected by the Volunteer Examiners giving you the test, or by the Volunteer Examiner Coordinator that coordinates the test session you attend. All VECs and examiners must use the questions, answers and distracters (incorrect answers) printed here.

This question pool contains more than 10 times the number of questions necessary to make up an exam. This ensures that the examiners have sufficient questions to choose from when they make up an exam.

Who Gives the Test?

All Amateur Radio license exams are given by teams of three or more Volunteer Examiners (VEs). Each of the examiners is accredited by a Volunteer Examiner Coordinator (VEC) to give exams under their program. A VEC is an organization that has entered into an agreement with the FCC to coordinate the efforts of VEs. The VEC reviews the paperwork for each exam session, and then forwards the information to the FCC.

Question Pool Format

The question pool is divided into 10 sections, called subelements. (A subelement is a portion of the exam element, in this case Element 2.) The FCC specifies how many questions from each section must be on your test. For example, there must be 10 questions from the Commission's Rules section, Subelement N1. There must also be two questions from the Operating Procedures section, Subelement N2. Four of the questions on your exam must come from the Electrical Principles section, Subelement N5, and so on. Table 13-1 summarizes the number of questions from each subelement that make up an Element 2 exam. The number of questions to be used from each subelement appears at the beginning of that subelement in the question pool, too.

The subelements are broken into smaller groups, and the VEC Question Pool Committee intends for one question to come from each of the smaller groups. This is not an FCC requirement, however, so your examiners may not select one question from each smaller group when they select questions for your exam.

There is a list of topics printed in bold type at the beginning of each small group. This list represents the syllabus, or study guide, topics for that section.

The small groups are listed alphabetically within each subelement. For example, since there are 10 exam questions from Subelement N1, these sections are labeled N1 through N1J. Two exam questions come from the Circuit Components subelement, so that subelement has sections labeled N6A and N6B.

The question numbers used in the question pool relate to the syllabus or study guide printed at the end of the Introduction chapter. The syllabus is an outline of topics covered by the exam.. Each question number begins with an N. This indicates the question is from the Novice question pool. Next is a number to indicate which subelement the question is from. These numbers will range from 1 to 0. Following this number is a letter to indicate which group the question is from in that subelement. Each question number ends with a two-digit number to specify its position in the set. So question number N1A01 is the first question in the A group of the first subelement. Question N5D09 is the ninth question in the D group of the fifth subelement.

The rest of this chapter contains the entire Element 2 question pool. We have printed the answer key to these questions along the edge of the page. There is a line to indicate where you should fold the page under to hide the answer key while you study. After making your best effort to answer the questions, you can look at the answers to check your understanding. We also have included page references along with the answers. These page numbers indicate where you will find the text discussion related to each question. If you have any problems with a question, refer to the page listed for that question. You may have to study beyond the listed page number to review all the related material. We also have included references to sections of Part 97 for the Commission's Rules Subelement, N1. This is to help you identify the particular rule citations for these questions. The complete text of the FCC Rules in Part 97 is included in the ARRL publication, *The FCC Rule Book*.

The Question Pool Committee included all the drawings for the question pool on one page, with two additional pages for tables related to the RF radiation safety questions in subelement N0. We placed a copy of these pages at the beginning of the question pool, and also placed the individual figures with the questions that reference them.

Good luck with your studies.

Element 2 (Novice)

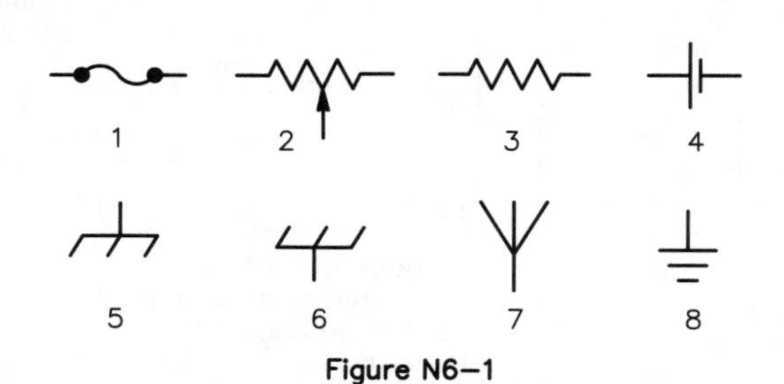

Figure N6–1

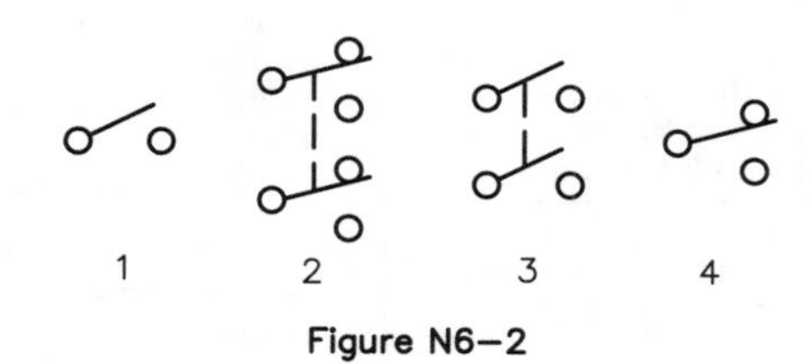

Figure N6–2

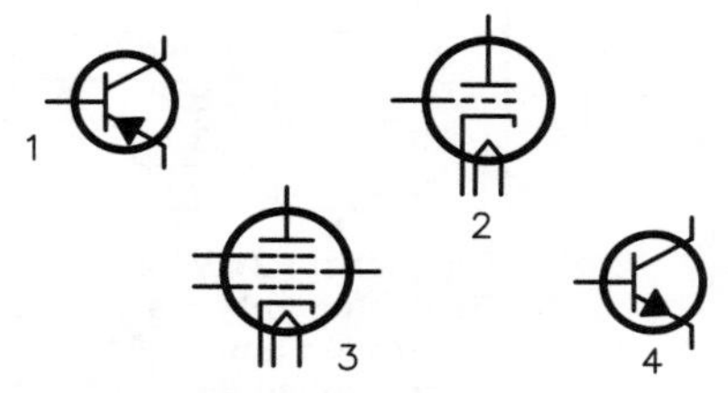

Figure N6–3

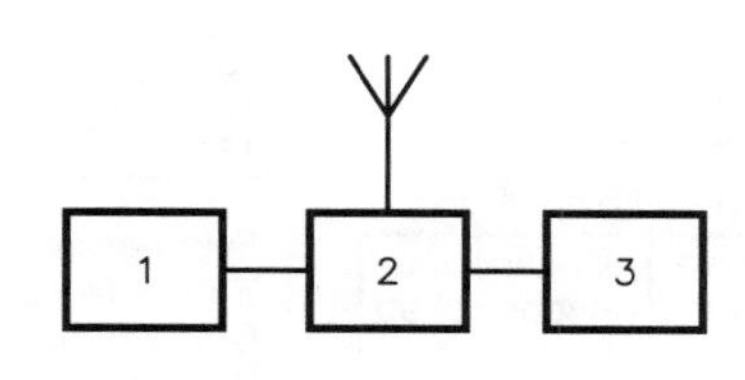

Figure N7–1

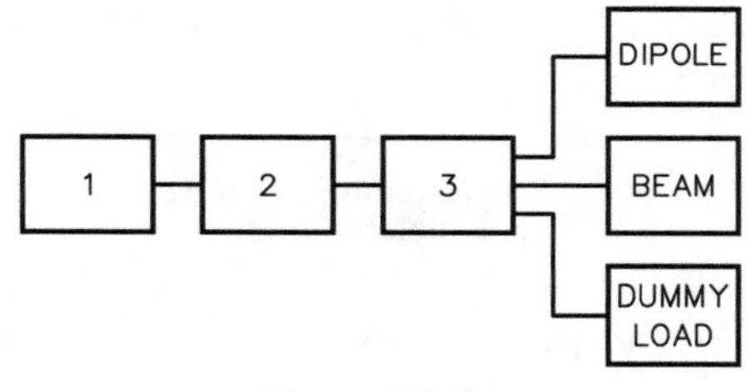

Figure N7–2

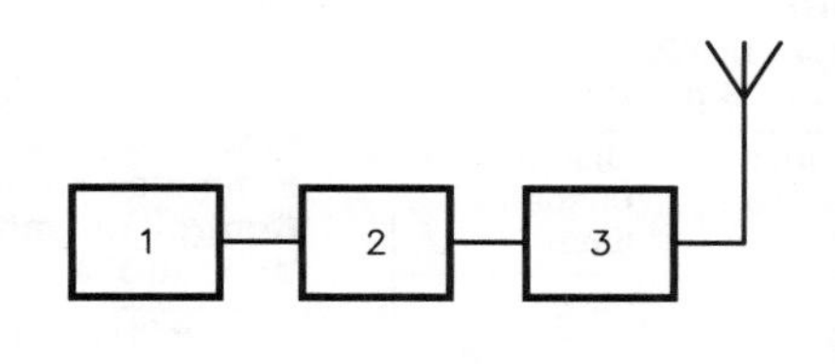

Figure N7–3

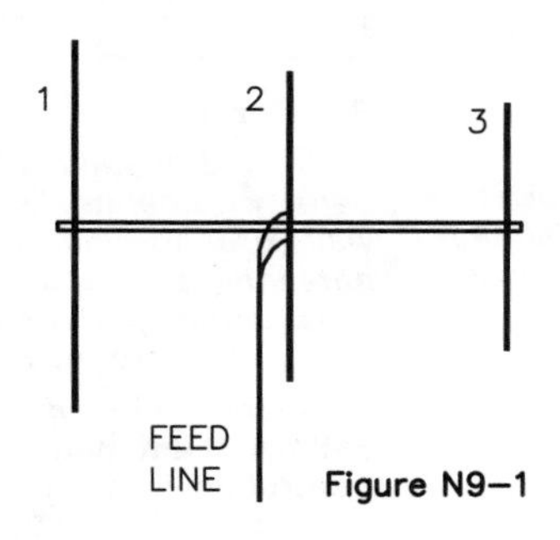

Figure N9–1

As Prepared by the
Question Pool Committee
for all examinations
administered
July 1, 1997 through June 30, 2001

Table NT0-1

RF Exposure Limits

Section A.

Estimated distances to meet RF power density guidelines in the main beam of a typical 3-element "triband" Yagi for the 14, 21 and 28 MHz Amateur Radio bands. Calculations include the EPA ground reflection factor of 2.56.

Frequency: 14 MHz
Antenna gain: 6.5 dBi
Controlled limit: 4.59 mW/cm²
Uncontrolled limit: 0.92 mW/cm²

Transmitter power (watts)	Distance to controlled limit	Distance to uncontrolled limit
100	4.6′	10.3′
500	10.3′	23.1′
1000	14.6′	32.7′
1500	17.9′	40′

Frequency: 21 MHz
Antenna gain: 7 dBi
Controlled limit: 2.04 mW/cm²
Uncontrolled limit: 0.408 mW/cm²

Transmitter power (watts)	Distance to controlled limit	Distance to uncontrolled limit
100	7.3′	16.4′
500	16.4′	36.7′
1000	23.2′	51.9′
1500	28.4′	63.6′

Frequency: 28 MHz
Antenna gain: 8 dBi
Controlled limit: 1.15 mW/cm²
Uncontrolled limit: 0.23 mW/cm²

Transmitter power (watts)	Distance to controlled limit	Distance to uncontrolled limit
100	11′	24.5′
500	24.5′	54.9′
1000	34.7′	77.6′
1500	42.5′	95.1′

Section B.

Estimated distances to meet RF power density guidelines with an omni-directional HF quarter-wave vertical or ground-plane antenna (estimated gain, 1 dBi). Calculations include the EPA ground reflection factor of 2.56.

Frequency: 3.5 MHz
Estimated antenna gain: 1 dBi
Controlled limit: 73.5 mW/cm²
Uncontrolled limit: 14.7 mW/cm²

Transmitter power (watts)	Distance to controlled limit	Distance to uncontrolled limit
100	0.6′	1.4′
500	1.4′	3.1′
1000	1.9′	4.3′
1500	2.4′	5.3′

Frequency: 7 MHz
Estimated antenna gain: 1 dBi
Controlled limit: 18.37 mW/cm²
Uncontrolled limit: 3.67 mW/cm²

Transmitter power (watts)	Distance to controlled limit	Distance to uncontrolled limit
100	1.2′	2.7′
500	2.7′	6.1′
1000	3.9′	8.7′
1500	4.7′	10.6′

Frequency: 14 MHz
Estimated antenna gain: 1 dBi
Controlled limit: 4.59 mW/cm²
Uncontrolled limit: 0.918 mW/cm²

Transmitter power (watts)	Distance to controlled limit	Distance to uncontrolled limit
100	2.5′	5.5′
500	5.5′	12.3′
1000	7.8′	17.3′
1500	9.5′	21.2′

Frequency: 21 MHz
Estimated antenna gain: 1 dBi
Controlled limit: 2.04 mW/cm²
Uncontrolled limit: 0.408 mW/cm²

Transmitter power (watts)	Distance to controlled limit	Distance to uncontrolled limit
100	3.7′	8.2′
500	8.2′	18.4′
1000	11.6′	26′
1500	14.2′	31.9′

Frequency: 28 MHz
Estimated antenna gain: 1 dBi
Controlled limit: 1.15 mW/cm²
Uncontrolled limit: 0.23 mW/cm²

Transmitter power (watts)	Distance to controlled limit	Distance to uncontrolled limit
100	4.9′	11′
500	11′	24.5′
1000	15.5′	34.7′
1500	19′	42.5′

Section C.

Estimated distances to meet RF power density guidelines with a horizontal half-wave dipole antenna (estimated gain, 2 dBi). Calculations include the EPA ground reflection factor of 2.56.

Frequency: 3.5 MHz
Estimated antenna gain: 2 dBi
Controlled limit: 73.5 mW/cm²
Uncontrolled limit: 14.7 mW/cm²

Transmitter power (watts)	Distance to controlled limit	Distance to uncontrolled limit
100	0.7′	1.5′
500	1.5′	3.4′
1000	2.2′	4.9′
1500	2.7′	6′

Frequency: 7 MHz
Estimated antenna gain: 2 dBi
Controlled limit: 18.37 mW/cm²
Uncontrolled limit: 3.67 mW/cm²

Transmitter power (watts)	Distance to controlled limit	Distance to uncontrolled limit
100	1.4′	3.1′
500	3.1′	6.9′
1000	4.3′	9.7′
1500	5.3′	11.9′

Frequency: 14 MHz
Estimated antenna gain: 2 dBi
Controlled limit: 4.59 mW/cm²
Uncontrolled limit: 0.918 mW/cm²

Transmitter power (watts)	Distance to controlled limit	Distance to uncontrolled limit
100	2.8′	6.2′
500	6.2′	13.8′
1000	8.7′	19.5′
1500	10.7′	23.8′

Frequency: 21 MHz
Estimated antenna gain: 2 dBi
Controlled limit: 2.04 mW/cm²
Uncontrolled limit: 0.408 mW/cm²

Transmitter power (watts)	Distance to controlled limit	Distance to uncontrolled limit
100	4.1′	9.2′
500	9.2′	20.6′
1000	13′	29.2′
1500	16′	35.7′

Frequency: 28 MHz
Estimated antenna gain: 2 dBi
Controlled limit: 1.15 mW/cm²
Uncontrolled limit: 0.23 mW/cm²

Transmitter power (watts)	Distance to controlled limit	Distance to uncontrolled limit
100	5.5′	12.3′
500	12.3′	27.5′
1000	17.4′	38.9′
1500	21.3′	47.7′

Section D.

Estimated distances to meet RF power density guidelines with a VHF quarter-wave ground-plane or mobile whip antenna (estimated gain, 1 dBi). Calculations include the EPA ground reflection factor of 2.56.

Frequency: 146 MHz
Estimated antenna gain: 1 dBi
Controlled limit: 1 mW/cm²
Uncontrolled limit: 0.2 mW/cm²

Transmitter power (watts)	Distance to controlled limit	Distance to uncontrolled limit
10	1.7′	3.7′
50	3.7′	8.3′
150	6.4′	14.4′

Section E.

Estimated distances to meet RF power density guidelines in the main beam of a UHF 5/8 wavelength ground-plane or mobile whip antenna (estimated gain, 4 dBi). Calculations include the EPA ground reflection factor of 2.56.

Frequency: 446 MHz
Estimated antenna gain: 4 dBi
Controlled limit: 1.49 mW/cm²
Uncontrolled limit: 0.3 mW/cm²

Transmitter power (watts)	Distance to controlled limit	Distance to uncontrolled limit
10	1.9′	4.3′
50	4.3′	9.6′
150	7.5′	16.7′

Section F.

Estimated distances to meet RF power density guidelines in the main beam of a 17-element Yagi on a five-wavelength boom designed for weak signal communications on the 144 MHz Amateur Radio band (estimated gain, 16.8 dBi). Calculations include the EPA ground reflection factor of 2.56.

Frequency: 144 MHz
Estimated antenna gain: 16.8 dBi
Controlled limit: 1 mW/cm²
Uncontrolled limit: 0.2 mW/cm²

Transmitter power (watts)	Distance to controlled limit	Distance to uncontrolled limit
10	10.2′	22.9′
100	32.4′	72.4′
500	72.4′	162′
1500	125.5′	280.6′

Section G.

Estimated distances to meet RF power density guidelines in the main beam of an array of eight 17-element Yagis with five-wavelength booms designed for earth-moon-earth ("moonbounce") communications on the 144 MHz Amateur Radio band (estimated gain, 24 dBi). Calculations include the EPA ground reflection factor of 2.56.

Frequency: 144 MHz
Estimated antenna gain: 24 dBi
Controlled limit: 1 mW/cm²
Uncontrolled limit: 0.2 mW/cm²

Transmitter power (watts)	Distance to controlled limit	Distance to uncontrolled limit
150	90.9′	203.3′
500	166′	371.1′
1500	287.4′	642.7′

RF Exposure Limits

(A) Limits for Occupational/Controlled Exposure

Frequency Range (MHz)	Electric Field Strength (V/m)	Magnetic Field (A/m)	Power Density (mW/cm²)	Averaging Time (minutes)
0.3-3.0	614	1.63	(100)*	6
3.0-30	1842/f	4.89/f	(900/f²)*	6
30-300	61.4	0.163	1.0	6
300-1500	—	—	f/300	6
1500-100,000	—	—	5	6

f = frequency in MHz * = Plane-wave equivalent power density

(B) Limits for General Population/Uncontrolled Exposure

Frequency Range (MHz)	Electric Field Strength (V/m)	Magnetic Field (A/m)	Power Density (mW/cm²)	Averaging Time (minutes)
0.3-1.34	614	1.63	(100)*	30
1.34-30	824/f	2.19/f	(180/f²)*	30
30-300	27.5	0.073	0.2	30
300-1500	—	—	f/1500	30
1500-100,000	—	—	1.0	30

f = frequency in MHz * = Plane-wave equivalent power density

Figure NT0-1

As prepared by the Question Pool Committee for all examinations administered July 1, 1997 through June 30, 2001

Answer Key

Page Numbers tell you where to look in this book for more information.

Element 2 (Novice Class) Question Pool

Subelement N1

Numbers in [square brackets] indicate sections in FCC Part 97, the Amateur Radio Rules.

Subelement N1 — Commission's Rules [10 Exam Questions — 10 Groups]

N1A Basis and purpose of amateur service and definitions

N1A01
(A)
[97]
Page 1-3

N1A01
What document contains the rules and regulations for the amateur service in the US?
A. Part 97 of Title 47 CFR (Code of Federal Regulations)
B. The Communications Act of 1934 (as amended)
C. The Radio Amateur's Handbook
D. The minutes of the International Telecommunication Unionmeetings

N1A02
(B)
[97]
Page 1-3

N1A02
Who makes and enforces the rules and regulations of the amateur service in the US?
A. The Congress of the United States
B. The Federal Communications Commission (FCC)
C. The Volunteer Examiner Coordinators (VECs)
D. The Federal Bureau of Investigation (FBI)

N1A03
(A)
[97]
Page 1-3

N1A03
Which three topics are part of the rules and regulations of the amateur service?
A. Station operation standards, technical standards, emergency communications
B. Notices of Violation, common operating procedures, antenna lengths
C. Frequency band plans, repeater locations, Ohm's Law
D. Station construction standards, FCC approved radios, FCC approved antennas

N1A04
(D)
[97]
Page 1-4

N1A04
Which of these topics is NOT part of the rules and regulations of the amateur service?
A. Qualifying examination systems
B. Technical standards
C. Providing emergency communications
D. Station construction standards

N1A05
What are three reasons that the amateur service exists?
A. To recognize the value of emergency communications, advance the radio art, and improve communication and technical skills
B. To learn about business communications, increase testing by trained technicians, and improve amateur communications
C. To preserve old radio techniques, maintain a pool of people familiar with early tube-type equipment, and improve tube radios
D. To improve patriotism, preserve nationalism, and promote world peace radio schematic drawings, and increase the pool of electrical drafting people

N1A05
(A)
[97.1]
Page 1-3

N1A06
What are two of the five purposes for the amateur service?
A. To protect historical radio data, and help the public understand radio history
B. To help foreign countries improve communication and technical skills, and encourage visits from foreign hams
C. To modernize radio schematic drawings, and increase the pool of electrical drafting people
D. To increase the number of trained radio operators and electronics experts, and improve international goodwill

N1A06
(D)
[97.1]
Page 1-3

N1A07
What is the definition of an amateur operator?
A. A person who has not received any training in radio operations
B. A person to whom the FCC has granted a license in the amateur service
C. A person who has very little practice operating a radio station
D. A person who is in training to become the control operator of a radio station

N1A07
(B)
[97.3a1]
Page 1-5

N1A08
What is the definition of the amateur service?
A. A private radio service used for profit and public benefit
B. A public radio service for US citizens that requires no exam
C. A personal radio service used for self-training, communication, and technical studies
D. A private radio service used for self-training of radio announcers and technicians

N1A08
(C)
[97.3a4]
Page 1-5

N1A09
What is the definition of an amateur station?
A. A station in a public radio service used for radiocommunications
B. A station using radiocommunications for a commercial purpose
C. A station using equipment for training new radiocommunications operators
D. A station in the Amateur Radio service used for radiocommunications

N1A09
(D)
[97.3a5]
Page 1-5

N1A10
What is the definition of a control operator of an amateur station?
A. Anyone who operates the controls of the station
B. Anyone who is responsible for the station's equipment
C. Any licensed amateur operator who is responsible for the station's transmissions
D. The amateur operator with the highest class of license who is near the controls of the station

N1A10
(C)
[97.3a12]
Page 1-6

N1A11
(C)
[97.509a]
Page 1-12

N1A11
What is a Volunteer Examiner (VE)?
A. A certified instructor who volunteers to examine amateur teaching manuals
B. An FCC employee who accredits volunteers to administer amateur license exams
C. An amateur, accredited by one or more VECs, who volunteers to administer amateur license exams
D. An amateur, registered with the Electronic Industries Association, who volunteers to examine amateur station equipment

N1B Station/Operator license; classes of US amateur licenses, including basic differences and privileges of the various license classes

N1B01
(D)
[97.5a]
Page 1-5

N1B01
Which of the following must you have an amateur license to do?
A. Transmit on public-service frequencies
B. Retransmit shortwave broadcasts
C. Repair broadcast station equipment
D. Transmit on amateur service frequencies

N1B02
(B)
[97.5a]
Page 1-5

N1B02
What does an amateur license allow you to control?
A. A shortwave-broadcast station's transmissions
B. An amateur station's transmissions
C. Non-commercial FM broadcast transmissions
D. Any type of transmitter, as long as it is used for non-commercial transmissions

N1B03
(C)
[97.5a]
Page 1-5

N1B03
Which of the following is required before you can operate an amateur station in the US?
A. You must hold an FCC operator's training permit for a licensed radio station
B. You must submit an FCC Form 610 together with a license examination fee
C. The FCC must grant you an amateur operator/primary station license
D. The FCC must issue you a Certificate of Successful Completion of Amateur Training

N1B04
(B)
[97.5d]
Page 1-5

N1B04
Where does a US amateur license allow you to operate?
A. Anywhere in the world
B. Wherever the amateur service is regulated by the FCC
C. Within 50 km of your primary station location
D. Only at the mailing address printed on your license

N1B05
(C)
[97.5d]
Page 1-7

N1B05
How many transmitters may a Novice licensee control at the same time?
A. Only one
B. No more than two
C. Any number
D. Any number, as long as they are transmitting in different bands

N1B06
(A)
[97.9a]
Page 1-5

N1B06
What must happen before you are allowed to operate an amateur station?
A. The FCC database must show that you have been granted an amateur license
B. You must have written authorization from the FCC
C. You must have written authorization from a Volunteer Examiner Coordinator
D. You must have a copy of the FCC Rules, Part 97, at your station location

N1B07
Which one of the following does NOT allow a person to control a US amateur station?
A. An operator/primary station license from the FCC
B. A reciprocal permit from the FCC for alien amateur licensee
C. An amateur service license from the United Nations Secretary of Communications
D. An amateur service license from the Government of Canada, if it is held by a Canadian citizen

N1B07
(C)
[97.7]
Page 1-6

N1B08
What is the FCC's full name for an amateur station license?
A. Restricted operating permit
B. General radiotelephone operator license
C. Amateur operator/primary station license
D. Amateur telegraphers radio station permit

N1B08
(C)
[97.5b1]
Page 1-5

N1B09
What document indicates your amateur station call sign?
A. Your operator/primary station license
B. The FCC's rules and regulations (Part 97)
C. None; you may choose any call sign you want
D. FCC Form 610, Application for Amateur License

N1B09
(A)
[97.3a11]
Page 1-5

N1B10
What are the six US amateur operator license classes?
A. Beginner, Novice, Communicator, General, Advanced, Expert
B. Novice, Technician, Technician Plus, General, Advanced, Expert
C. Communicator, Novice, Digital, Technician, General, Amateur Extra
D. Novice, Technician, Technician Plus, General, Advanced, Amateur Extra

N1B10
(D)
[97.9a]
Page 1-8

N1B11
What does the FCC consider to be the first two classes of US amateur operator licenses (one or the other of which most new amateurs initially hold)?
A. Novice and Technician
B. CB and Communicator
C. Novice and General
D. CB and Novice

N1B11
(A)
[97.9]
Page 1-8

N1B12
Which of the following would NOT be a new privilege if you upgraded your license class beyond the Novice level?
A. More operating frequencies
B. Higher transmitting power
C. Authority to prepare amateur license exams
D. Authority to send third-party messages

N1B12
(D)
[97.115, 97.301, 97.305, 97.507a]
Page 1-8

N1B13
Which US amateur license has no Morse code requirements?
A. Amateur Extra
B. Advanced
C. General
D. Technician

N1B13
(D)
[97.501e]
Page 1-8

N1C Novice control operator frequency privileges

N1C01
(B)
[97.301e]
Page 1-9

N1C01
What are the frequency limits of the 80-meter Novice band?
A. 3500 - 4000 kHz
B. 3675 - 3725 kHz
C. 7100 - 7150 kHz
D. 7000 - 7300 kHz

N1C02
(C)
[97.301e]
Page 1-9

N1C02
What are the frequency limits of the 40-meter Novice band in ITU Region 2?
A. 3500 - 4000 kHz
B. 3700 - 3750 kHz
C. 7100 - 7150 kHz
D. 7000 - 7300 kHz

N1C03
(A)
[97.301e]
Page 1-9

N1C03
What are the frequency limits of the 15-meter Novice band?
A. 21.100 - 21.200 MHz
B. 21.000 - 21.450 MHz
C. 28.000 - 29.700 MHz
D. 28.100 - 28.200 MHz

N1C04
(C)
[97.301e]
Page 1-9

N1C04
What are the frequency limits of the 10-meter Novice band?
A. 28.000 - 28.500 MHz
B. 28.100 - 29.500 MHz
C. 28.100 - 28.500 MHz
D. 29.100 - 29.500 MHz

N1C05
(B)
[97.301f]
Page 1-9

N1C05
What are the frequency limits of the 1.25-meter Novice band in ITU Region 2?
A. 225.0 - 230.5 MHz
B. 222.0 - 225.0 MHz
C. 224.1 - 225.1 MHz
D. 220.0 - 226.0 MHz

N1C06
(C)
[97.301f]
Page 1-9

N1C06
What are the frequency limits of the 23-centimeter Novice band?
A. 1260-1270 MHz
B. 1240-1300 MHz
C. 1270-1295 MHz
D. 1240-1246 MHz

N1C07
(A)
[97.301e]
Page 1-9

N1C07
If you are operating on 3710 kHz, in what amateur band are you operating?
A. 80 meters
B. 40 meters
C. 15 meters
D. 10 meters

N1C08
If you are operating on 7135 kHz, in what amateur band are you operating?
A. 80 meters
B. 40 meters
C. 15 meters
D. 10 meters

N1C08
(B)
[97.301e]
Page 1-9

N1C09
If you are operating on 21.165 MHz, in what amateur band are you operating?
A. 80 meters
B. 40 meters
C. 15 meters
D. 10 meters

N1C09
(C)
[97.301e]
Page 1-9

N1C10
If you are operating on 28.400 MHz, in what amateur band are you operating?
A. 80 meters
B. 40 meters
C. 15 meters
D. 10 meters

N1C10
(D)
[97.301e]
Page 1-9

N1C11
If you are operating on 223.50 MHz, in what amateur band are you operating?
A. 15 meters
B. 10 meters
C. 2 meters
D. 1.25 meters

N1C11
(D)
[97.301f]
Page 1-9

N1D Novice eligibility, exam elements, mailing addresses, US call-sign assignment and life of license

N1D01
Who can become an amateur licensee in the US?
A. Anyone except a representative of a foreign government
B. Only a citizen of the United States
C. Anyone except an employee of the US government
D. Anyone

N1D01
(A)
[97.5b1]
Page 1-8

N1D02
What age must you be to hold an amateur license?
A. 14 years or older
B. 18 years or older
C. 70 years or younger
D. There are no age limits

N1D02
(D)
[97.5b1]
Page 1-8

N1D03
What minimum examinations must you pass for a Novice amateur license?
A. A written exam, Element 1(A); and a 5-WPM code exam, Element 2(A)
B. A 5-WPM code exam, Element 1(A); and a written exam, Element 3(A)
C. A 5-WPM code exam, Element 1(A); and a written exam, Element 2
D. A written exam, Element 2; and a 5 WPM code exam, Element 4

N1D03
(C)
[97.501f]
Page 1-12

N1D04
(B)
[97.23]
Page 1-13

N1D04
Why must an amateur operator have a current US postal mailing address?
A. So the FCC has a record of the location of each amateur station
B. To follow the FCC rules and so the licensee can receive mail from the FCC
C. Because all US amateurs must be US residents
D. So the FCC can publish a call-sign directory

N1D05
(D)
[97.29]
Page 1-6

N1D05
What can you do to replace your license document if it is lost, mutilated or destroyed?
A. Nothing; the FCC does not replace license documents
B. Send a change of address request to the FCC using a current FCC Form 610
C. Retake all examination elements for your license
D. Ask the FCC for a replacement, explaining what happened to the original

N1D06
(B)
[97.23b]
Page 1-6

N1D06
What must you do to notify the FCC if your mailing address changes?
A. Fill out an FCC Form 610 using your new address, attach a copy of your license, and mail it to your local FCC Field Office
B. Fill out an FCC Form 610 using your new address, attach a copy of your license, and mail it to the FCC office in Gettysburg,PA
C. Call your local FCC Field Office and give them your new address over the phone
D. Call the FCC office in Gettysburg, PA, and give them your new address over the phone

N1D07
(C)
[No Part 97 Reference]
Page 1-16

N1D07
Which of the following call signs is a valid US amateur call?
A. UZ4FWD
B. KBL7766
C. KA1TMJ
D. VE3BKJ

N1D08
(B)
[No Part 97 Reference]
Page 1-16

N1D08
What letters must be used for the first letter in US amateur call signs?
A. K, N, U and W
B. A, K, N and W
C. A, B, C and D
D. A, N, V and W

N1D09
(D)
[No Part 97 Reference]
Page 1-16

N1D09
What numbers are normally used in US amateur call signs?
A. Any two-digit number, 10 through 99
B. Any two-digit number, 22 through 45
C. A single digit, 1 though 9
D. A single digit, 0 through 9

N1D10
(C)
[97.25]
Page 1-6

N1D10
For how many years is a new amateur license normally issued?
A. 2
B. 5
C. 10
D. 15

N1D11
How soon before the expiration date of your license should you send the FCC a completed Form 610 for a renewal?
A. No more than 90 days
B. No more than 30 days
C. Within 6 to 9 months
D. Within 6 months to a year

N1D11
(A)
[97.21a3i]
Page 1-6

N1D12
How soon after you pass the elements required for your first Amateur Radio license may you transmit?
A. Immediately
B. 30 days after the test date
C. As soon as the FCC grants you a license
D. As soon as you receive your license from the FCC

N1D12
(C)
[97.5a]
Page 1-5

N1E Novice control operator emission privileges

N1E01
What emission types are Novice control operators allowed to use in the 80-meter band?
A. CW only
B. Data only
C. RTTY only
D. Phone only

N1E01
(A)
[97.305, 97.307f9]
Page 1-10

N1E02
What emission types are Novice control operators allowed to use in the 40-meter band?
A. CW only
B. Data only
C. RTTY only
D. Phone only

N1E02
(A)
[97.305, 97.307f9]
Page 1-10

N1E03
What emission types are Novice control operators allowed to use in the 15-meter band?
A. CW only
B. Data only
C. RTTY only
D. Phone only

N1E03
(A)
[97.305, 97.307f9]
Page 1-10

N1E04
What emission types are Novice control operators allowed to use from 3675 to 3725 kHz?
A. Phone only
B. Image only
C. Data only
D. CW only

N1E04
(D)
[97.305, 97.307f9]
Page 1-10

N1E05
What emission types are Novice control operators allowed to use from 7100 to 7150 kHz in ITU Region 2?
A. CW and data
B. Phone
C. Data only
D. CW only

N1E05
(D)
[97.305, 97.307f9]
Page 1-10

N1E06
(D)
[97.305, 97.307f9]
Page 1-10

N1E06
What emission types are Novice control operators allowed to use on frequencies from 21.1 to 21.2 MHz?
A. CW and data
B. CW and phone
C. Data only
D. CW only

N1E07
(C)
[97.305]
Page 1-10

N1E07
What emission types are Novice control operators allowed to use on frequencies from 28.1 to 28.3 MHz?
A. All authorized amateur emission privileges
B. Data or phone
C. CW, RTTY and data
D. CW and phone

N1E08
(C)
[97.305, 97.307f10]
Page 1-10

N1E08
What emission types are Novice control operators allowed to use on frequencies from 28.3 to 28.5 MHz?
A. All authorized amateur emission privileges
B. CW and data
C. CW and single-sideband phone
D. Data and phone

N1E09
(D)
[97.305]
Page 1-10

N1E09
What emission types are Novice control operators allowed to use on the amateur 1.25-meter band in ITU Region 2?
A. Only CW and phone
B. Only CW and data
C. Only data and phone
D. All amateur emission privileges authorized for use on the band

N1E10
(D)
[97.305]
Page 1-10

N1E10
What emission types are Novice control operators allowed to use on the amateur 23-centimeter band?
A. Only data and phone
B. Only CW and data
C. Only CW and phone
D. All amateur emission privileges authorized for use on the band

N1E11
(D)
[97.305, 97.307f10]
Page 1-10

N1E11
On what HF frequencies may Novice control operators use single-sideband (SSB) phone?
A. 3700 - 3750 kHz
B. 7100 - 7150 kHz
C. 21100 - 21200 kHz
D. 28300 - 28500 kHz

N1E12
(C)
[97.305]
Page 1-10

N1E12
On which of the following frequencies may Novice control operators in ITU Region 2 use FM phone?
A. 28.3 - 28.5 MHz
B. 144.0 - 148.0 MHz
C. 222 - 225 MHz
D. 1240 - 1270 MHz

N1E13
On what frequencies in the 10-meter band may Novice control operators use RTTY?
A. 28.0 - 28.3 MHz
B. 28.1 - 28.3 MHz
C. 28.0 - 29.3 MHz
D. 29.1 - 29.3 MHz

N1E13
(B)
[97.301e, 97.305]
Page 1-10

N1E14
On what frequencies in the 10-meter band may Novice control operators use data emissions?
A. 28.0 - 28.3 MHz
B. 28.1 - 28.3 MHz
C. 28.0 - 29.3 MHz
D. 29.1 - 29.3 MHz

N1E14
(B)
[97.301e, 97.305]
Page 1-10

N1F Transmitter power on Novice sub-bands and digital communications [limited to concepts only]

N1F01
What amount of transmitter power must amateur stations use at all times?
A. 25 watts PEP output
B. 250 watts PEP output
C. 1500 watts PEP output
D. The minimum legal power necessary to communicate

N1F01
(D)
[97.313a]
Page 1-10

N1F02
What is the most transmitter power an amateur station may use on 3710 kHz?
A. 5 watts PEP output
B. 25 watts PEP output
C. 200 watts PEP output
D. 1500 watts PEP output

N1F02
(C)
[97.313c1]
Page 1-10

N1F03
What is the most transmitter power an amateur station may use on 7120 kHz?
A. 5 watts PEP output
B. 25 watts PEP output
C. 200 watts PEP output
D. 1500 watts PEP output

N1F03
(C)
[97.313c1]
Page 1-10

N1F04
What is the most transmitter power an amateur station may use on 21.150 MHz?
A. 5 watts PEP output
B. 25 watts PEP output
C. 200 watts PEP output
D. 1500 watts PEP output

N1F04
(C)
[97.313c1]
Page 1-10

N1F05
What is the most transmitter power a Novice station may use on 28.450 MHz?
A. 5 watts PEP output
B. 25 watts PEP output
C. 200 watts PEP output
D. 1500 watts PEP output

N1F05
(C)
[97.313c2]
Page 1-10

N1F06
(C)
[97.313c2]
Page 1-10

N1F06
What is the most transmitter power a Novice station may use on the 10-meter band?
A. 5 watts PEP output
B. 25 watts PEP output
C. 200 watts PEP output
D. 1500 watts PEP output

N1F07
(B)
[97.313d]
Page 1-10

N1F07
What is the most transmitter power a Novice station may use on the 1.25-meter band?
A. 5 watts PEP output
B. 25 watts PEP output
C. 200 watts PEP output
D. 1500 watts PEP output

N1F08
(A)
[97.313e]
Page 1-10

N1F08
What is the most transmitter power a Novice station may use on the 23-centimeter band?
A. 5 watts PEP output
B. 25 watts PEP output
C. 200 watts PEP output
D. 1500 watts PEP output

N1F09
(A)
[97.313c]
Page 1-10

N1F09
On which band(s) may a Novice station use up to 200 watts PEP output power?
A. 80, 40, 15, and 10 meters
B. 80, 40, 20, and 10 meters
C. 1.25 meters
D. 23 centimeters

N1F10
(C)
[97.313d]
Page 1-10

N1F10
On which band(s) must a Novice station use no more than 25 watts PEP output power?
A. 80, 40, 15, and 10 meters
B. 80, 40, 20, and 10 meters
C. 1.25 meters
D. 23 centimeters

N1F11
(D)
[97.313e]
Page 1-10

N1F11
On which band(s) must a Novice station use no more than 5 watts PEP output power?
A. 80, 40, 15, and 10 meters
B. 80, 40, 20, and 10 meters
C. 1.25 meters
D. 23 centimeters

N1F12
(D)
[97.313a]
Page 1-10

N1F12
If you make contact with another station and your signal is extremely strong and perfectly readable, what adjustment should you make to your transmitter?
A. Turn on your speech processor
B. Reduce your SWR
C. Don't make any changes, otherwise you may lose contact
D. Turn down your power output to the minimum necessary

N1F13
(C)
[97.3c2]
Page 1-10

N1F13
What name does the FCC use for telemetry, telecommand or computer communications emissions?
A. CW
B. Image
C. Data
D. RTTY

N1F14
What name does the FCC use for narrow-band direct-printing telegraphy emissions?
A. CW
B. Image
C. Data
D. RTTY

N1F14
(D)
[97.3c7]
Page 1-10

N1G Responsibility of licensee, control operator requirements

N1G01
What is the FCC's name for the person responsible for the transmissions from an amateur station?
A. Auxiliary operator
B. Operations coordinator
C. Third-party operator
D. Control operator

N1G01
(D)
[97.3a12]
Page 1-6

N1G02
Who is responsible for the proper operation of an amateur station?
A. Only the control operator
B. Only the station licensee
C. Both the control operator and the station licensee
D. The person who owns the station equipment

N1G02
(C)
[97.103a]
Page 1-6

N1G03
If you transmit from another amateur's station, who is responsible for its proper operation?
A. Both of you
B. The other amateur (the station licensee)
C. You, the control operator
D. The station licensee, unless the station records show that you were the control operator at the time

N1G03
(A)
[97.103a]
Page 1-7

N1G04
What is your responsibility as a station licensee?
A. You must allow another amateur to operate your station upon request
B. You must be present whenever the station is operated
C. You must notify the FCC if another amateur acts as the control operator
D. You are responsible for the proper operation of the station in accordance with the FCC rules

N1G04
(D)
[97.103a]
Page 1-6

N1G05
Who may be the control operator of an amateur station?
A. Any person over 21 years of age
B. Any person over 21 years of age with a General class license or higher
C. Any licensed amateur chosen by the station licensee
D. Any licensed amateur with a Technician class license or higher

N1G05
(C)
[97.103b]
Page 1-6

N1G06
If another amateur transmits from your station, which of these is NOT true?
A. You must first give permission for the other amateur to use your station
B. You must keep the call sign of the other amateur, together with the time and date of transmissions, in your station log
C. The FCC will think that you are the station's control operator unless your station records show that you were not
D. Both of you are equally responsible for the proper operation of the station

N1G06
(B)
[97.103b]
Page 1-7

N1G07
(A)
[97.105b,
97.119d]
Page 1-6

N1G07
If you let another amateur with a higher class license than yours control your station, what operating privileges are allowed?
A. Any privileges allowed by the higher license, as long as proper identification procedures are followed
B. Only the privileges allowed by your license
C. All the emission privileges of the higher license, but only the frequency privileges of your license
D. All the frequency privileges of the higher license, but only the emission privileges of your license

N1G08
(B)
[97.105b]
Page 1-7

N1G08
If you are the control operator at the station of another amateur who has a higher class license than yours, what operating privileges are you allowed?
A. Any privileges allowed by the higher license
B. Only the privileges allowed by your license
C. All the emission privileges of the higher license, but only the frequency privileges of your license
D. All the frequency privileges of the higher license, but only the emission privileges of your license

N1G09
(C)
[97.7]
Page 1-7

N1G09
When must an amateur station have a control operator?
A. Only when training another amateur
B. Whenever the station receiver is operated
C. Whenever the station is transmitting
D. A control operator is not needed

N1G10
(A)
[97.109b]
Page 1-7

N1G10
When a Novice station is transmitting, where must its control operator be?
A. At the station's control point
B. Anywhere in the same building as the transmitter
C. At the station's entrance, to control entry to the room
D. Anywhere within 50 km of the station location

N1G11
(B)
[97.109b]
Page 1-7

N1G11
Why can't unlicensed persons in your family transmit using your amateur station if they are alone with your equipment?
A. They must not use your equipment without your permission
B. They must be licensed before they are allowed to be control operators
C. They must first know how to use the right abbreviations and Q signals
D. They must first know the right frequencies and emissions for transmitting

N1H Station identification, points of communication and operation, and business communications

N1H01
(C)
[97.119a]
Page 1-18

N1H01
How often must an amateur station be identified?
A. At the beginning of a contact and at least every ten minutes after that
B. At least once during each transmission
C. At least every ten minutes during and at the end of a contact
D. At the beginning and end of each transmission

N1H02
What do you transmit to identify your amateur station?
A. Your "handle"
B. Your call sign
C. Your first name and your location
D. Your full name

N1H02
(B)
[97.119a]
Page 1-18

N1H03
What identification, if any, is required when two amateur stations begin communications?
A. No identification is required
B. One of the stations must give both stations' call signs
C. Each station must transmit its own call sign
D. Both stations must transmit both call signs

N1H03
(A)
[97.119a]
Page 1-18

N1H04
What identification, if any, is required when two amateur stations end communications?
A. No identification is required
B. One of the stations must transmit both stations' call signs
C. Each station must transmit its own call sign
D. Both stations must transmit both call signs

N1H04
(C)
[97.119a]
Page 1-18

N1H05
Besides normal identification, what else must a US station do when sending third-party communications internationally?
A. The US station must transmit its own call sign at the beginning of each communication, and at least every ten minutes after that
B. The US station must transmit both call signs at the end of each communication
C. The US station must transmit its own call sign at the beginning of each communication, and at least every five minutes after that
D. Each station must transmit its own call sign at the end of each transmission, and at least every five minutes after that

N1H05
(B)
[97.115c]
Page 1-18

N1H06
What is the longest period of time an amateur station can operate without transmitting its call sign?
A. 5 minutes
B. 10 minutes
C. 15 minutes
D. 30 minutes

N1H06
(B)
[97.119a]
Page 1-18

N1H07
With which non-amateur stations is a US amateur station allowed to communicate?
A. No non-amateur stations
B. All non-amateur stations
C. Only those authorized by the FCC
D. Only those who use international Morse code

N1H07
(C)
[97.111]
Page 1-18

N1H08
Under what conditions are amateur stations allowed to communicate with stations operating in other radio services?
A. Never; amateur stations are only permitted to communicate with other amateur stations
B. When authorized by the FCC or in an emergency
C. When communicating with stations in the Citizens Radio Service
D. When a commercial broadcast station is using Amateur Radio frequencies for news gathering during a natural disaster

N1H08
(B)
[97.113a3]
Page 1-18

N1H09
(D)
[97.5a]
Page 1-5

N1H09
When may you operate your amateur station somewhere in the US besides the address listed on your license?
A. Only during times of emergency
B. Only after giving proper notice to the FCC
C. During an emergency or an FCC-approved emergency practice
D. Whenever you want to

N1H10
(B)
[97.113a2]
Page 1-20

N1H10
If you work for a taxi service, under what conditions might you use your amateur station to tell taxi drivers where to pick up customers?
A. Only when you first obtain the proper FCC commercial endorsement for your license
B. Never, because this is clearly a business communication
C. Only between the hours of 6:00 PM and 6:00 AM local time
D. Only if the taxi driver is also a licensed Amateur Radio operator

N1H11
(A)
[97.113a]
Page 1-21

N1H11
Do the FCC Rules allow you to buy and sell amateur station equipment using amateur communications?
A. Yes, provided you do not do so on a regular basis
B. No, because this is clearly a business communication
C. Yes, provided you do this only once per calendar year
D. No, unless you collect all taxes and report them to the IRS

N1H12
(C)
[97.11]
Page 1-5

N1H12
When may you operate your amateur station aboard a commercial aircraft?
A. At any time
B. Only while the aircraft is not in flight
C. Only with the pilot's specific permission and not while the aircraft is operating under Instrument Flight Rules
D. Only if you have written permission from the commercial airline company and not during takeoff and landing

N1I International and space communications, authorized and prohibited transmissions

N1I01
(D)
[97.111a1]
Page 1-18

N1I01
When are you allowed to communicate with an amateur in a foreign country?
A. Only when the foreign amateur uses English
B. Only when you have permission from the FCC
C. Only when a third-party agreement exists between the US and the foreign country
D. At any time, unless it is not allowed by either government

N1I02
(C)
[97.3a38]
Page 1-22

N1I02
What is an amateur space station?
A. An amateur station operated on an unused frequency
B. An amateur station awaiting its new call letters from the FCC
C. An amateur station located more than 50 kilometers above the Earth's surface
D. An amateur station that communicates with space shuttles

N1I03
Who may be the licensee of an amateur space station?
A. An amateur holding an Amateur Extra class operator license
B. Any licensed amateur operator
C. Anyone designated by the commander of the spacecraft
D. No one unless specifically authorized by the government

N1I03
(B)
[97.207a]
Page 1-22

N1I04
When may someone be paid to transmit messages from an amateur station?
A. Only if he or she works for a public service agency such as the Red Cross
B. Under no circumstances
C. Only if he or she reports all such payments to the IRS
D. Only if the operator is a classroom teacher or works for a club station and special requirements are met

N1I04
(D)
[97.113c,d]
Page 1-21

N1I05
When is an amateur allowed to broadcast information to the general public?
A. Never
B. Only when the operator is being paid
C. Only when broadcasts last less than 1 hour
D. Only when broadcasts last longer than 15 minutes

N1I05
(A)
[97.113b]
Page 1-18

N1I06
When is an amateur station permitted to transmit music?
A. Never, except incidental music during authorized rebroadcasts of space shuttle communications
B. Only if the transmitted music produces no spurious emissions
C. Only if it is used to jam an illegal transmission
D. Only if it is above 1280 MHz, and the music is a live performance

N1I06
(A)
[97.113a4, 97.113e]
Page 1-21

N1I07
When is the use of codes or ciphers allowed to hide the meaning of an amateur message?
A. Only during contests
B. Only during nationally declared emergencies
C. Never, except when special requirements are met
D. Only on frequencies above 1280 MHz

N1I07
(C)
[97.113a4]
Page 1-21

N1I08
What is the definition of third-party communications?
A. A message sent between two amateur stations for someone else
B. Public service communications for a political party
C. Any messages sent by amateur stations
D. A three-minute transmission to another amateur

N1I08
(A)
[97.3a44]
Page 1-19

N1I09
What is a "third party" in amateur communications?
A. An amateur station that breaks in to talk
B. A person who is sent a message by amateur communications other than a control operator who handles the message
C. A shortwave listener who monitors amateur communications
D. An unlicensed control operator

N1I09
(B)
[97.3a44]
Page 1-19

N1I10
(A)
[97.115a2]
Page 1-19

N1I10
If you are allowing a non-amateur friend to use your station to talk to someone in the US, and a foreign station breaks in to talk to your friend, what should you do?
A. Have your friend wait until you find out if the US has a third-party agreement with the foreign station's government
B. Stop all discussions and quickly sign off
C. Since you can talk to any foreign amateurs, your friend may keep talking as long as you are the control operator
D. Report the incident to the foreign amateur's government

N1I11
(D)
[97.115a2]
Page 1-19

N1I11
When are you allowed to transmit a message to a station in a foreign country for a third party?
A. Anytime
B. Never
C. Anytime, unless there is a third-party agreement between the US and the foreign government
D. If there is a third-party agreement with the US government, or if the third party is eligible to be the control operator

N1I12
(D)
[97.405a]
Page 1-22

N1I12
If you hear a voice distress signal on a frequency outside of your license privileges, what are you allowed to do to help the station in distress?
A. You are NOT allowed to help because the frequency of the signal is outside your privileges
B. You are allowed to help only if you keep your signals within the nearest frequency band of your privileges
C. You are allowed to help on a frequency outside your privileges only if you use international Morse code
D. You are allowed to help on a frequency outside your privileges in any way possible

N1I13
(C)
[97.403]
Page 1-22

N1I13
When may you use your amateur station to transmit an "SOS" or "MAYDAY"?
A. Never
B. Only at specific times (at 15 and 30 minutes after the hour)
C. In a life- or property-threatening emergency
D. When the National Weather Service has announced a severe weather watch

N1I14
(B)
[97.405a]
Page 1-22

N1I14
When may you send a distress signal on any frequency?
A. Never
B. In a life- or property-threatening emergency
C. Only at specific times (at 15 and 30 minutes after the hour)
D. When the National Weather Service has announced a severe weather watch

N1J False signals or unidentified communications and malicious interference

N1J01
(A)
[97.113a4]
Page 1-22

N1J01
When may false or deceptive amateur signals or communications be transmitted?
A. Never
B. When operating a beacon transmitter in a "fox hunt" exercise
C. When playing a harmless "practical joke"
D. When you need to hide the meaning of a message for secrecy

N1J02
If an amateur pretends there is an emergency and transmits the word "MAYDAY," what is this called?
A. A traditional greeting in May
B. An emergency test transmission
C. False or deceptive signals
D. Nothing special; "MAYDAY" has no meaning in an emergency

N1J02
(C)
[97.113a4]
Page 1-22

N1J03
What is a transmission called that disturbs other communications?
A. Interrupted CW
B. Harmful interference
C. Transponder signals
D. Unidentified transmissions

N1J03
(B)
[97.3a22]
Page 1-21

N1J04
If you are operating FM phone on the 23-cm band and learn that you are interfering with a radiolocation station outside the US, what must you do?
A. Stop operating or take steps to eliminate this harmful interference
B. Nothing, because this band is allocated exclusively to the amateur service
C. Establish contact with the radiolocation station and ask them to change frequency
D. Change to CW mode, because this would not likely cause interference

N1J04
(A)
[97.303h]
Page 1-22

N1J05
Why is transmitting on a police frequency as a "joke" called harmful interference that deserves a large penalty?
A. It annoys everyone who listens
B. It blocks police calls that might be an emergency and interrupts police communications
C. It is in bad taste to communicate with non-amateurs, even as a joke
D. It is poor amateur practice to transmit outside the amateur bands

N1J05
(B)
[97.3a22]
Page 1-22

N1J06
When may you deliberately interfere with another station's communications?
A. Only if the station is operating illegally
B. Only if the station begins transmitting on a frequency you are using
C. Never
D. You may expect, and cause, deliberate interference because it can't be helped during crowded band conditions

N1J06
(C)
[97.101d]
Page 1-21

N1J07
When may an amateur transmit unidentified communications?
A. Only for brief tests not meant as messages
B. Only if it does not interfere with others
C. Never, except transmissions from a space station or to control a model craft
D. Only for two-way or third-party communications

N1J07
(C)
[97.119a]
Page 1-18

N1J08
What is an amateur communication called that does not have the required station identification?
A. Unidentified communications or signals
B. Reluctance modulation
C. Test emission
D. Tactical communication

N1J08
(A)
[97.119a]
Page 1-18

N1J09
(D)
[97.119a]
Page 1-18

N1J09
If you answer someone on the air and then complete your communication without giving your call sign, what type of communication have you just conducted?
A. Test transmission
B. Tactical signal
C. Packet communication
D. Unidentified communication

N1J10
(C)
[97.119a]
Page 1-18

N1J10
If an amateur transmits to test access to a repeater without giving any station identification, what type of communication is this called?
A. A test emission; no identification is required
B. An illegal unmodulated transmission
C. An illegal unidentified transmission
D. A non-communication; no voice is transmitted

N1J11
(B)
[97.3a22]
Page 1-22

N1J11
If an amateur repeatedly transmits on a frequency already occupied by a group of amateurs in a net operation, what type of interference is this called?
A. Break-in interference
B. Harmful or malicious interference
C. Incidental interference
D. Intermittent interference

Subelement N2

Subelement N2 — Operating Procedures [2 Exam Questions — 2 Groups]

N2A Preparing to transmit; choosing a frequency for tune-up, operating or emergencies; Morse code; RST signal reports; Q signals; voice communications and phonetics

N2A01
(A)
Page 2-3

N2A01
What should you do before you transmit on any frequency?
A. Listen to make sure others are not using the frequency
B. Listen to make sure that someone will be able to hear you
C. Check your antenna for resonance at the selected frequency
D. Make sure the SWR on your antenna feed line is high enough

N2A02
(C)
Page 2-7

N2A02
What is one way to shorten transmitter tune-up time on the air to reduce interference?
A. Use a random wire antenna
B. Tune up on 40 meters, then switch to the desired band
C. Tune the transmitter into a dummy load
D. Use twin lead instead of coax-cable feed lines

N2A03
(D)
Page 2-12

N2A03
If you are in contact with another station and you hear an emergency call for help on your frequency, what should you do?
A. Tell the calling station that the frequency is in use
B. Direct the calling station to the nearest emergency net frequency
C. Call your local Civil Preparedness Office and inform them of the emergency
D. Stop your QSO immediately and take the emergency call

N2A04
What is the correct way to call CQ when using Morse code?
A. Send the letters "CQ" three times, followed by "DE," followed by your call sign sent once
B. Send the letters "CQ" three times, followed by "DE," followed by your call sign sent three times
C. Send the letters "CQ" ten times, followed by "DE," followed by your call sign sent twice
D. Send the letters "CQ" over and over until a station answers

N2A04
(B)
Page 2-3

N2A05
How should you answer a Morse code CQ call?
A. Send your call sign four times
B. Send the other station's call sign twice, followed by "DE," followed by your call sign twice
C. Send the other station's call sign once, followed by "DE," followed by your call sign four times
D. Send your call sign followed by your name, station location and a signal report

N2A05
(B)
Page 2-3

N2A06
At what speed should a Morse code CQ call be transmitted?
A. Only speeds below five WPM
B. The highest speed your keyer will operate
C. Any speed at which you can reliably receive
D. The highest speed at which you can control the keyer

N2A06
(C)
Page 2-3

N2A07
What is the meaning of the procedural signal "CQ"?
A. "Call on the quarter hour"
B. "New antenna is being tested" (no station should answer)
C. "Only the called station should transmit"
D. "Calling any station"

N2A07
(D)
Page 2-3

N2A08
What is the meaning of the procedural signal "DE"?
A. "From" or "this is," as in "W0AIH DE KA9FOX"
B. "Directional Emissions" from your antenna
C. "Received all correctly"
D. "Calling any station"

N2A08
(A)
Page 2-3

N2A09
What is the meaning of the procedural signal "K"?
A. "Any station transmit"
B. "All received correctly"
C. "End of message"
D. "Called station only transmit"

N2A09
(A)
Page 2-3

N2A10
What is meant by the term "DX"?
A. Best regards
B. Distant station
C. Calling any station
D. Go ahead

N2A10
(B)
Page 2-4

N2A11
(B)
Page 2-4

N2A11
What is the meaning of the term "73"?
A. Long distance
B. Best regards
C. Love and kisses
D. Go ahead

N2A12
(C)
Page 2-5

N2A12
What are RST signal reports?
A. A short way to describe ionospheric conditions
B. A short way to describe transmitter power
C. A short way to describe signal reception
D. A short way to describe sunspot activity

N2A13
(D)
Page 2-5

N2A13
What does RST mean in a signal report?
A. Recovery, signal strength, tempo
B. Recovery, signal speed, tone
C. Readability, signal speed, tempo
D. Readability, signal strength, tone

N2A14
(B)
Page 2-2

N2A14
What is one meaning of the Q signal "QRS"?
A. "Interference from static"
B. "Send more slowly"
C. "Send RST report"
D. "Radio station location is"

N2A15
(D)
Page 2-2

N2A15
What is one meaning of the Q signal "QTH"?
A. "Time here is"
B. "My name is"
C. "Stop sending"
D. "My location is"

N2A16
(C)
Page 2-4

N2A16
What is a QSL card in the amateur service?
A. A letter or postcard from an amateur pen pal
B. A Notice of Violation from the FCC
C. A written acknowledgment of communications between two amateurs
D. A postcard reminding you when your license will expire

N2A17
(C)
Page 2-8

N2A17
What is the correct way to call CQ when using voice?
A. Say "CQ" once, followed by "this is," followed by your call sign spoken three times
B. Say "CQ" at least five times, followed by "this is," followed by your call sign spoken once
C. Say "CQ" three times, followed by "this is," followed by your call sign spoken three times
D. Say "CQ" at least ten times, followed by "this is," followed by your call sign spoken once

N2A18
How should you answer a voice CQ call?
A. Say the other station's call sign at least ten times, followed by "this is," then your call sign at least twice
B. Say the other station's call sign at least five times phonetically, followed by "this is," then your call sign at least once
C. Say the other station's call sign at least three times, followed by "this is," then your call sign at least five times phonetically
D. Say the other station's call sign once, followed by "this is," then your call sign given phonetically

N2A18
(D)
Page 2-9

N2A19
What is the proper Q signal to use to see if a frequency is in use before transmitting on CW?
A. QRV?
B. QRU?
C. QRL?
D. QRZ?

N2A19
(C)
Page 2-3

N2A20
What is one meaning of the Q signal "QSY"?
A. "Change frequency"
B. "Send more slowly"
C. "Send faster"
D. "Use more power"

N2A20
(A)
Page 2-2

N2A21
What is one meaning of the Q signal "QSO"?
A. A contact is confirmed
B. A conversation is in progress
C. I can communicate with
D. A conversation is desired

N2A21
(B)
Page 2-2

N2A22
What is the proper Q signal to use to ask if someone is calling you on CW?
A. QSL?
B. QRZ?
C. QRL?
D. QRT?

N2A22
(B)
Page 2-5

N2A23
To make your call sign better understood when using voice transmissions, what should you do?
A. Use Standard International Phonetics for each letter of your call
B. Use any words that start with the same letters as your call sign for each letter of your call
C. Talk louder
D. Turn up your microphone gain

N2A23
(A)
Page 2-9

N2B Radio teleprinting; packet; repeater operating procedures; special operations

N2B01
(B)
Page 2-17

N2B01
What is the correct way to call CQ when using RTTY?
A. Send the letters "CQ" three times, followed by "DE," followed by your call sign sent once
B. Send the letters "CQ" three to six times, followed by "DE," followed by your call sign sent three times
C. Send the letters "CQ" ten times, followed by the procedural signal "DE," followed by your call sign sent twice
D. Send the letters "CQ" over and over

N2B02
(B)
Page 2-15

N2B02
What speed should you use when answering a CQ call using RTTY?
A. Half the speed of the received signal
B. The same speed as the received signal
C. Twice the speed of the received signal
D. Any speed, since RTTY systems adjust to any signal speed

N2B03
(B)
Page 2-15

N2B03
What does the abbreviation "RTTY" stand for?
A. "Returning to you," meaning "your turn to transmit"
B. Radioteletype
C. A general call to all digital stations
D. Morse code practice over the air

N2B04
(C)
Page 2-20

N2B04
What does "connected" mean in a packet-radio link?
A. A telephone link is working between two stations
B. A message has reached an amateur station for local delivery
C. A transmitting station is sending data to only one receiving station; it replies that the data is being received correctly
D. A transmitting and receiving station are using a digipeater, so no other contacts can take place until they are finished

N2B05
(D)
Page 2-19

N2B05
What does "monitoring" mean on a packet-radio frequency?
A. The FCC is copying all messages
B. A member of the Amateur Auxiliary to the FCC's Compliance and Information Bureau is copying all messages
C. A receiving station is displaying all messages sent to it, and replying that the messages are being received correctly
D. A receiving station is displaying all messages on the frequency, and is not replying to any messages

N2B06
(A)
Page 2-20

N2B06
What is a digipeater?
A. A packet-radio station that retransmits only data that is marked to be retransmitted
B. A packet-radio station that retransmits any data that it receives
C. A repeater that changes audio signals to digital data
D. A repeater built using only digital electronics parts

N2B07
What does "network" mean in packet radio?
A. A way of connecting terminal-node controllers by telephone so data can be sent over long distances
B. A way of connecting packet-radio stations so data can be sent over long distances
C. The wiring connections on a terminal-node controller board
D. The programming in a terminal-node controller that rejects other callers if a station is already connected

N2B07
(B)
Page 2-21

N2B08
What is a good way to make contact on a repeater?
A. Say the call sign of the station you want to contact three times
B. Say the other operator's name, then your call sign three times
C. Say the call sign of the station you want to contact, then your call sign
D. Say, "Breaker, breaker," then your call sign

N2B08
(C)
Page 2-12

N2B09
When using a repeater to communicate, which of the following do you need to know about the repeater?
A. Its input frequency and offset
B. Its call sign
C. Its power level
D. Whether or not it has an autopatch

N2B09
(A)
Page 2-10

N2B10
What does it mean to say that a repeater has an input and an output frequency?
A. The repeater receives on one frequency and transmits on another
B. The repeater offers a choice of operating frequency, in case one is busy
C. One frequency is used to control the repeater and another is used to retransmit received signals
D. The repeater must receive an access code on one frequency before retransmitting received signals

N2B10
(A)
Page 2-10

N2B11
What is an autopatch?
A. Something that automatically selects the strongest signal to be repeated
B. A device that connects a mobile station to the next repeater if it moves out of range of the first
C. A device that allows repeater users to make telephone calls from their stations
D. A device that locks other stations out of a repeater when there is an important conversation in progress

N2B11
(C)
Page 2-14

N2B12
What is the purpose of a repeater time-out timer?
A. It lets a repeater have a rest period after heavy use
B. It logs repeater transmit time to predict when a repeater will fail
C. It tells how long someone has been using a repeater
D. It limits the amount of time someone can transmit on a repeater

N2B12
(D)
Page 2-13

N2B13
What is a CTCSS (or PL) tone?
A. A special signal used for telecommand control of model craft
B. A sub-audible tone, added to a carrier, which may cause a receiver to accept a signal
C. A tone used by repeaters to mark the end of a transmission
D. A special signal used for telemetry between amateur space stations and Earth stations

N2B13
(B)
Page 2-11

N2B14
(A)
Page 2-13

N2B14
What is simplex operation?
A. Transmitting and receiving on the same frequency
B. Transmitting and receiving over a wide area
C. Transmitting on one frequency and receiving on another
D. Transmitting one-way communications

N2B15
(B)
Page 2-13

N2B15
When should you use simplex operation instead of a repeater?
A. When the most reliable communications are needed
B. When a contact is possible without using a repeater
C. When an emergency telephone call is needed
D. When you are traveling and need some local information

Subelement N3

Subelement N3 - Radio-Wave Propagation [1 Exam Question — 1 Group]

N3A Line of sight, ground wave, HF propagation characteristics; sunspots and the sunspot cycle; and reflection of VHF/UHF signals

N3A01
(A)
Page 3-2

N3A01
When a signal travels in a straight line from one antenna to another, what is this called?
A. Line-of-sight propagation
B. Straight line propagation
C. Knife-edge diffraction
D. Tunnel ducting

N3A02
(B)
Page 3-2

N3A02
How do VHF and UHF radio waves usually travel from a transmitting antenna to a receiving antenna?
A. They bend through the ionosphere
B. They go in a straight line
C. They wander in any direction
D. They move in a circle going either east or west from the transmitter

N3A03
(D)
Page 3-2

N3A03
When a signal travels along the surface of the Earth, what is this called?
A. Sky-wave propagation
B. Knife-edge diffraction
C. E-region propagation
D. Ground-wave propagation

N3A04
(B)
Page 3-3

N3A04
How does the range of sky-wave propagation compare to ground-wave propagation?
A. It is much shorter
B. It is much longer
C. It is about the same
D. It depends on the weather

N3A05
When a signal is returned to Earth by the ionosphere, what is this called?
A. Sky-wave propagation
B. Earth-Moon-Earth propagation
C. Ground-wave propagation
D. Tropospheric propagation

N3A05
(A)
Page 3-3

N3A06
What is the usual cause of sky-wave propagation?
A. Signals are reflected by a mountain
B. Signals are reflected by the Moon
C. Signals are bent back to Earth by the ionosphere
D. Signals are retransmitted by a repeater

N3A06
(C)
Page 3-3

N3A07
What is a skip zone?
A. An area covered by ground-wave propagation
B. An area covered by sky-wave propagation
C. An area that is too far away for ground-wave propagation, but too close for sky-wave propagation
D. An area that is too far away for ground-wave or sky-wave propagation

N3A07
(C)
Page 3-4

N3A08
What are the regions of ionized gases high above the Earth called?
A. The ionosphere
B. The troposphere
C. The gas region
D. The ion zone

N3A08
(A)
Page 3-3

N3A09
What is the name of the area of the atmosphere that makes long-distance radio communications possible by bending radio waves?
A. Troposphere
B. Stratosphere
C. Magnetosphere
D. Ionosphere

N3A09
(D)
Page 3-3

N3A10
What causes the ionosphere to form?
A. Solar radiation ionizing the outer atmosphere
B. Temperature changes ionizing the outer atmosphere
C. Lightning ionizing the outer atmosphere
D. Release of fluorocarbons into the atmosphere

N3A10
(A)
Page 3-3

N3A11
What type of solar radiation is most responsible for ionization in the outer atmosphere?
A. Thermal
B. Non-ionized particle
C. Ultraviolet
D. Microwave

N3A11
(C)
Page 3-3

N3A12
Which ionospheric region is closest to the Earth?
A. The A region
B. The D region
C. The E region
D. The F region

N3A12
(B)
Page 3-4

N3A13
(D)
Page 3-5

N3A13
Which region of the ionosphere is mainly responsible for long-distance sky-wave radio communications?
A. D region
B. E region
C. F1 region
D. F2 region

N3A14
(B)
Page 3-5

N3A14
Which of the ionospheric regions may split into two regions only during the daytime?
A. Troposphere
B. F
C. Electrostatic
D. D

N3A15
(C)
Page 3-5

N3A15
Which two daytime ionospheric regions combine into one region at night?
A. E and F1
B. D and E
C. F1 and F2
D. E1 and E2

N3A16
(A)
Page 3-3

N3A16
How does the number of sunspots relate to the amount of ionization in the ionosphere?
A. The more sunspots there are, the greater the ionization
B. The more sunspots there are, the less the ionization
C. Unless there are sunspots, the ionization is zero
D. Sunspots do not affect the ionosphere

N3A17
(C)
Page 3-3

N3A17
How long is an average sunspot cycle?
A. 2 years
B. 5 years
C. 11 years
D. 17 years

N3A18
(C)
Page 3-2

N3A18
What can happen to VHF or UHF signals going towards a metal-framed building?
A. They will go around the building
B. They can be bent by the ionosphere
C. They can be easily reflected by the building
D. They are sometimes scattered in the ecosphere

Subelement N4 — Amateur Radio Practices
[4 Exam Questions — 4 Groups]

Subelement N4

N4A Preventing unauthorized use; lightning protection and station grounding

N4A01
How could you best keep unauthorized persons from using your amateur station at home?
A. Use a carrier-operated relay in the main power line
B. Use a key-operated on/off switch in the main power line
C. Put a "Danger - High Voltage" sign in the station
D. Put fuses in the main power line

N4A01
(B)
Page 4-3

N4A02
How could you best keep unauthorized persons from using a mobile amateur station in your car?
A. Disconnect the microphone when you are not using it
B. Put a "do not touch" sign on the radio
C. Turn the radio off when you are not using it
D. Tune the radio to an unused frequency when you are done using it

N4A02
(A)
Page 4-6

N4A03
Why would you use a key-operated on/off switch in the main power line of your station?
A. To keep unauthorized persons from using your station
B. For safety, in case the main fuses fail
C. To keep the power company from turning off your electricity during an emergency
D. For safety, to turn off the station in the event of an emergency

N4A03
(A)
Page 4-3

N4A04
How can an antenna system best be protected from lightning damage?
A. Install a balun at the antenna feed point
B. Install an RF choke in the antenna feed line
C. Ground all antennas when they are not in use
D. Install a fuse in the antenna feed line

N4A04
(C)
Page 4-7

N4A05
How can amateur station equipment best be protected from lightning damage?
A. Use heavy insulation on the wiring
B. Never turn off the equipment
C. Disconnect the ground system from all radios
D. Disconnect all equipment from the power lines and antenna cables

N4A05
(D)
Page 4-7

N4A06
For best protection from electrical shock, what should be grounded in an amateur station?
A. The power supply primary
B. All station equipment
C. The antenna feed line
D. The AC power mains

N4A06
(B)
Page 4-4

N4A07
(D)
Page 4-7

N4A07
Why should you ground all antenna and rotator cables when your amateur station is not in use?
A. To lock the antenna system in one position
B. To avoid radio frequency interference
C. To save electricity
D. To protect the station and building from lightning damage

N4A08
(A)
Page 4-4

N4A08
What document describes safe grounding practices for electrical wiring, antennas and other electrical equipment, such as would be used in an amateur station?
A. The National Electrical Code
B. FCC Rules, Part 97
C. The National Construction Trades Association Manual
D. The National Association of Broadcaster's Safety Manual

N4A09
(C)
Page 4-4

N4A09
Where should you connect the chassis of each piece of your station equipment to best protect against electrical shock?
A. To insulated shock mounts
B. To the antenna
C. To a good ground connection
D. To a circuit breaker

N4A10
(B)
Page 4-3

N4A10
Which of these materials is best for a ground rod driven into the earth?
A. Hard plastic
B. Copper or copper-clad steel
C. Iron or steel
D. Fiberglass

N4A11
(C)
Page 4-3

N4A11
If you ground your station equipment to a ground rod driven into the earth, what is the shortest length the rod should be?
A. 4 feet
B. 6 feet
C. 8 feet
D. 10 feet

N4B Safety interlocks, antenna installation safety procedures

N4B01
(D)
Page 4-6

N4B01
Why would there be an interlock switch in a high-voltage power supply to turn off the power if its cabinet is opened?
A. To keep dangerous RF radiation from leaking out through an open cabinet
B. To keep dangerous RF radiation from coming in through an open cabinet
C. To turn the power supply off when it is not being used
D. To keep anyone opening the cabinet from getting shocked by dangerous high voltages

N4B02
(A)
Page 4-6

N4B02
What is the name used for a safety switch inside a power-supply cabinet that turns off power when the cabinet door is opened?
A. An interlock switch
B. A circuit breaker
C. A deadman switch
D. The main switch

N4B03
What kind of safety equipment should you wear if you are working on an antenna tower?
A. A grounding chain and rubber-sole shoes
B. A reflective vest of approved color
C. Electrical-insulating safety gloves and a static discharge line
D. A carefully inspected safety belt, hard hat and safety glasses

N4B03
(D)
Page 9-27

N4B04
Why should you wear a safety belt if you are working on an antenna tower?
A. To safely hold your tools so they don't fall and injure someone on the ground
B. To keep the tower from becoming unbalanced while you are working
C. To safely bring any tools you might use up and down the tower
D. To prevent you from accidentally falling

N4B04
(D)
Page 9-27

N4B05
Why should you wear a hard hat and safety glasses if you are on the ground helping someone work on an antenna tower?
A. So you won't be hurt if the tower should accidentally fall
B. To keep RF energy away from your head during antenna testing
C. To protect your head from something dropped from the tower
D. So someone passing by will know that work is being done on the tower and will stay away

N4B05
(C)
Page 9-27

N4B06
What is an advantage to using copper-clad steel wire for an HF wire antenna?
A. It will stretch rather than break under strain
B. It is very flexible and easy to handle
C. It will not rust with age
D. It is much stronger than the same gauge drawn-copper wire

N4B06
(D)
Page 9-9

N4B07
What is one disadvantage to using small gauge wire when constructing an HF antenna?
A. It is not strong, and will stretch or break easily
B. It is difficult to see from the ground
C. It can only be fed with coaxial cable
D. It can only be fed with parallel-conductor feed line

N4B07
(A)
Page 9-9

N4B08
What safety factors must you consider when using a bow and arrow or slingshot and weight to shoot an antenna-support line over a tree?
A. You must ensure that the line is strong enough to withstand the shock of shooting the weight
B. You must ensure that the arrow or weight has a safe flight path if the line breaks
C. You must ensure that the bow and arrow or slingshot is in good working condition
D. All of these choices are correct

N4B08
(D)
Page 9-13

N4B09
Which of the following is the best way to install your antenna in relation to overhead electric power lines?
A. Always be sure your antenna wire is higher than the power line, and crosses it at a 90-degree angle
B. Always be sure your antenna and feed line are well clear of any power lines
C. Always be sure your antenna is lower than the power line, and crosses it at a small angle
D. Only use vertical antennas within 100 feet of a power line

N4B09
(B)
Page 9-11

N4B10
(A)
Page 4-6

N4B10
What circuit should be controlled by a safety interlock switch in an amateur transceiver or power amplifier?
A. The power supply
B. The IF amplifier
C. The audio amplifier
D. The cathode bypass circuit

N4B11
(B)
Page 4-6

N4B11
What electrical rating should a safety interlock switch have?
A. Sufficient capacitance to prevent any leakage current
B. Sufficient voltage rating and current capacity for the protected circuit
C. Sufficient inductance to ensure a strong magnetic field to hold the cover when power is on
D. Sufficient resistance to ensure proper current limiting

N4B12
(D)
Page 4-6

N4B12
In which of the following devices should there be a safety interlock switch to protect anyone from dangerous voltages?
A. A vacuum-tube power amplifier
B. A high-voltage power supply
C. A station-monitor oscilloscope
D. All of these choices are correct

N4B13
(C)
Page 9-13

N4B13
What is one disadvantage of using inexpensive polypropylene rope to support your HF dipole antenna?
A. Birds like to pick at the brightly colored rope as a source of nest-building material
B. The texture grabs rough tree limbs and does not slide easily
C. It disintegrates rapidly when exposed to sunlight and weather
D. It is a good conductor of electricity

N4C SWR meaning and measurements

N4C01
(B)
Page 9-6

N4C01
What does an SWR reading of 1:1 mean?
A. An antenna for another frequency band is probably connected
B. The best impedance match has been attained
C. No power is going to the antenna
D. The SWR meter is broken

N4C02
(C)
Page 9-7

N4C02
What does an SWR reading of less than 1.5:1 mean?
A. An impedance match that is too low
B. An impedance mismatch; something may be wrong with the antenna system
C. A fairly good impedance match
D. An antenna gain of 1.5

N4C03
(D)
Page 9-7

N4C03
What does an SWR reading of 4:1 mean?
A. An impedance match that is too low
B. An impedance match that is good, but not the best
C. An antenna gain of 4
D. An impedance mismatch; something may be wrong with the antenna system

N4C04
What kind of SWR reading may mean poor electrical contact between parts of an antenna system?
A. A jumpy reading
B. A very low reading
C. No reading at all
D. A negative reading

N4C04
(A)
Page 9-8

N4C05
What does a very high SWR reading mean?
A. The antenna is the wrong length, or there may be an open or shorted connection somewhere in the feed line
B. The signals coming from the antenna are unusually strong, which means very good radio conditions
C. The transmitter is putting out more power than normal, showing that it is about to go bad
D. There is a large amount of solar radiation, which means very poor radio conditions

N4C05
(A)
Page 9-8

N4C06
If an SWR reading at the low frequency end of an amateur band is 2.5:1, increasing to 5:1 at the high frequency end of the same band, what does this tell you about your 1/2-wavelength dipole antenna?
A. The antenna is broadbanded
B. The antenna is too long for operation on the band
C. The antenna is too short for operation on the band
D. The antenna is just right for operation on the band

N4C06
(B)
Page 9-8

N4C07
If an SWR reading at the low frequency end of an amateur band is 5:1, decreasing to 2.5:1 at the high frequency end of the same band, what does this tell you about your 1/2-wavelength dipole antenna?
A. The antenna is broadbanded
B. The antenna is too long for operation on the band
C. The antenna is too short for operation on the band
D. The antenna is just right for operation on the band

N4C07
(C)
Page 9-8

N4C08
If you use a 3-30 MHz RF-power meter at UHF frequencies, how accurate will its readings be?
A. They may not be accurate at all
B. They will be accurate enough to get by
C. They will be accurate but the readings must be divided by two
D. They will be accurate but the readings must be multiplied by two

N4C08
(A)
Page 9-8

N4C09
What instrument is used to measure standing wave ratio?
A. An ohmmeter
B. An ammeter
C. An SWR meter
D. A current bridge

N4C09
(C)
Page 9-6

N4C10
What instrument is used to measure the relative impedance match between an antenna and its feed line?
A. An ammeter
B. An ohmmeter
C. A voltmeter
D. An SWR meter

N4C10
(D)
Page 9-6

N4C11
(A)
Page 9-7

N4C11
Where would you connect an SWR meter to measure standing wave ratio?
A. Between the feed line and the antenna
B. Between the transmitter and the power supply
C. Between the transmitter and the receiver
D. Between the transmitter and the ground

N4D RFI and its complications, resolution and responsibility

N4D01
(C)
Page 8-4

N4D01
What is meant by receiver overload?
A. Too much voltage from the power supply
B. Too much current from the power supply
C. Interference caused by strong signals from a nearby source
D. Interference caused by turning the volume up too high

N4D02
(A)
Page 8-2

N4D02
What is meant by harmonic radiation?
A. Unwanted signals at frequencies that are multiples of the fundamental (chosen) frequency
B. Unwanted signals that are combined with a 60-Hz hum
C. Unwanted signals caused by sympathetic vibrations from a nearby transmitter
D. Signals that cause skip propagation to occur

N4D03
(A)
Page 8-6

N4D03
Why is harmonic radiation from an amateur station not wanted?
A. It may cause interference to other stations and may result in out-of-band signals
B. It uses large amounts of electric power
C. It may cause sympathetic vibrations in nearby transmitters
D. It may cause auroras in the air

N4D04
(A)
Page 8-7

N4D04
What type of interference may come from a multi-band antenna connected to a poorly tuned transmitter?
A. Harmonic radiation
B. Auroral distortion
C. Parasitic excitation
D. Intermodulation

N4D05
(C)
Page 8-4

N4D05
What is the main purpose of shielding in a transmitter?
A. It gives the low-pass filter a solid support
B. It helps the sound quality of transmitters
C. It prevents unwanted RF radiation
D. It helps keep electronic parts warmer and more stable

N4D06
(B)
Page 8-6

N4D06
What type of filter might be connected to an amateur HF transmitter to cut down on harmonic radiation?
A. A key-click filter
B. A low-pass filter
C. A high-pass filter
D. A CW filter

N4D07
What is one way to tell if radio-frequency interference to a receiver is caused by front-end overload?
A. If connecting a low-pass filter to the transmitter greatly cuts down the interference
B. If the interference is about the same no matter what frequency is transmitted
C. If connecting a low-pass filter to the receiver greatly cuts down the interference
D. If grounding the receiver makes the problem worse

N4D07
(B)
Page 8-5

N4D08
If your neighbor reports television interference whenever you are transmitting from your amateur station, no matter what frequency band you use, what is probably the cause of the interference?
A. Too little transmitter harmonic suppression
B. Receiver VR tube discharge
C. Receiver overload
D. Incorrect antenna length

N4D08
(C)
Page 8-5

N4D09
If your neighbor reports television interference on one or two channels only when you are transmitting on the 15-meter band, what is probably the cause of the interference?
A. Too much low-pass filtering on the transmitter
B. De-ionization of the ionosphere near your neighbor's TV antenna
C. TV receiver front-end overload
D. Harmonic radiation from your transmitter

N4D09
(D)
Page 8-6

N4D10
What type of filter should be connected to a TV receiver as the first step in trying to prevent RF overload from an amateur HF station transmission?
A. Low-pass
B. High-pass
C. Band pass
D. Notch

N4D10
(B)
Page 8-5

N4D11
What first step should be taken at a cable TV receiver when trying to prevent RF overload from an amateur HF station transmission?
A. Install a low-pass filter in the cable system transmission line
B. Tighten all connectors and inspect the cable system transmission line
C. Make sure the center conductor of the cable system transmission line is well grounded
D. Install a ceramic filter in the cable system transmission line

N4D11
(B)
Page 8-5

N4D12
What effect might a break in a cable television transmission line have on amateur communications?
A. Cable lines are shielded and a break cannot affect amateur communications
B. Harmonic radiation from the TV receiver may cause the amateur transmitter to transmit off-frequency
C. TV interference may result when the amateur station is transmitting, or interference may occur to the amateur receiver
D. The broken cable may pick up very high voltages when the amateur station is transmitting

N4D12
(C)
Page 8-5

N4D13
(A)
Page 8-4

N4D13
If you are told that your amateur station is causing television interference, what should you do?
A. First make sure that your station is operating properly, and that it does not cause interference to your own television
B. Immediately turn off your transmitter and contact the nearest FCC office for assistance
C. Connect a high-pass filter to the transmitter output and a low-pass filter to the antenna-input terminals of the television
D. Continue operating normally, because you have no reason to worry about the interference

N4D14
(C)
Page 8-6

N4D14
If harmonic radiation from your transmitter is causing interference to television receivers in your neighborhood, who is responsible for taking care of the interference?
A. The owners of the television receivers are responsible
B. Both you and the owners of the television receivers share the responsibility
C. You alone are responsible, since your transmitter is causing the problem
D. The FCC must decide if you or the owners of the television receivers are responsible

N4D15
(D)
Page 8-5

N4D15
If signals from your transmitter are causing front-end overload in your neighbor's television receiver, who is responsible for taking care of the interference?
A. You alone are responsible, since your transmitter is causing the problem
B. Both you and the owner of the television receiver share the responsibility
C. The FCC must decide if you or the owner of the television receiver are responsible
D. The owner of the television receiver is responsible

Subelement N5

Subelement N5 — Electrical Principles [4 Exam Questions — 4 Groups]

N5A Metric prefixes, i.e. pico, nano, micro, milli, centi, kilo, mega, giga

N5A01
(B)
Page 5-2

N5A01
If a dial marked in kilohertz shows a reading of 7125 kHz, what would it show if it were marked in megahertz?
A. 0.007125 MHz
B. 7.125 MHz
C. 71.25 MHz
D. 7,125,000 MHz

N5A02
(C)
Page 5-2

N5A02
If a dial marked in megahertz shows a reading of 3.525 MHz, what would it show if it were marked in kilohertz?
A. 0.003525 kHz
B. 35.25 kHz
C. 3525 kHz
D. 3,525,000 kHz

N5A03
If a dial marked in kilohertz shows a reading of 3725 kHz, what would it show if it were marked in hertz?
A. 3.725 Hz
B. 37.25 Hz
C. 3725 Hz
D. 3,725,000 Hz

N5A03
(D)
Page 5-2

N5A04
If an antenna is 400 centimeters long, what is its length in meters?
A. 0.0004 meters
B. 4 meters
C. 40 meters
D. 40,000 meters

N5A04
(B)
Page 5-2

N5A05
If an ammeter marked in amperes is used to measure a 3000-milliampere current, what reading would it show?
A. 0.003 amperes
B. 0.3 amperes
C. 3 amperes
D. 3,000,000 amperes

N5A05
(C)
Page 5-2

N5A06
If a voltmeter marked in volts is used to measure a 3500-millivolt potential, what reading would it show?
A. 0.35 volts
B. 3.5 volts
C. 35 volts
D. 350 volts

N5A06
(B)
Page 5-2

N5A07
How many farads is 500,000 microfarads?
A. 0.0005 farads
B. 0.5 farads
C. 500 farads
D. 500,000,000 farads

N5A07
(B)
Page 5-2

N5A08
How many microfarads is 1,000,000 picofarads?
A. 0.001 microfarads
B. 1 microfarad
C. 1000 microfarads
D. 1,000,000,000 microfarads

N5A08
(B)
Page 5-2

N5A09
How many hertz are in a kilohertz?
A. 10
B. 100
C. 1000
D. 1,000,000

N5A09
(C)
Page 5-2

N5A10
(C)
Page 5-2

N5A10
How many kilohertz are in a megahertz?
A. 10
B. 100
C. 1000
D. 1,000,000

N5A11
(B)
Page 5-2

N5A11
If you have a hand-held transceiver with an output of 500 milliwatts, how many watts would this be?
A. 0.02
B. 0.5
C. 5
D. 50

N5A12
(C)
Page 5-2

N5A12
If you have a hand-held transceiver with an output of 250 milliwatts, how many watts would this be?
A. 0.01
B. 25
C. 0.25
D. 0.125

N5A13
(C)
Page 5-2

N5A13
If your station is transmitting on a frequency of 1.265 GHz, what would the frequency be if it were given in MHz?
A. 12.65 MHz
B. 126.5 MHz
C. 1265 MHz
D. 12,650 MHz

N5B Concepts and measurement of current, voltage, resistance; concept of conductor and insulator

N5B01
(D)
Page 5-6

N5B01
What is the name for the flow of electrons in an electric circuit?
A. Voltage
B. Resistance
C. Capacitance
D. Current

N5B02
(C)
Page 5-7

N5B02
What is the basic unit of electric current?
A. The volt
B. The watt
C. The ampere
D. The ohm

N5B03
(D)
Page 5-7

N5B03
Which instrument would you use to measure electric current?
A. An ohmmeter
B. A wavemeter
C. A voltmeter
D. An ammeter

N5B04
What is the name of the pressure that forces electrons to flow through a circuit?
A. Magnetomotive force, or inductance
B. Electromotive force, or voltage
C. Farad force, or capacitance
D. Thermal force, or heat

N5B04
(B)
Page 5-6

N5B05
What is the basic unit of electromotive force (EMF)?
A. The volt
B. The watt
C. The ampere
D. The ohm

N5B05
(A)
Page 5-6

N5B06
How much voltage does an automobile battery usually supply?
A. About 12 volts
B. About 30 volts
C. About 120 volts
D. About 240 volts

N5B06
(A)
Page 5-6

N5B07
How much voltage does a wall outlet usually supply (in the US)?
A. About 12 volts
B. About 30 volts
C. About 120 volts
D. About 480 volts

N5B07
(C)
Page 5-6

N5B08
Which instrument would you use to measure electric potential or electromotive force?
A. An ammeter
B. A voltmeter
C. A wavemeter
D. An ohmmeter

N5B08
(B)
Page 5-6

N5B09
What limits the current that flows through a circuit for a particular applied DC voltage?
A. Reliance
B. Reactance
C. Saturation
D. Resistance

N5B09
(D)
Page 5-9

N5B10
What is the basic unit of resistance?
A. The volt
B. The watt
C. The ampere
D. The ohm

N5B10
(D)
Page 5-9

N5B11
Which instrument would you use to measure resistance?
A. An ammeter
B. A voltmeter
C. An ohmmeter
D. A wavemeter

N5B11
(C)
Page 5-9

N5B12
(C)
Page 5-7

N5B12
What are three good electrical conductors?
A. Copper, gold, mica
B. Gold, silver, wood
C. Gold, silver, aluminum
D. Copper, aluminum, paper

N5B13
(A)
Page 5-7

N5B13
What are four good electrical insulators?
A. Glass, air, plastic, porcelain
B. Glass, wood, copper, porcelain
C. Paper, glass, air, aluminum
D. Plastic, rubber, wood, carbon

N5B14
(B)
Page 5-7

N5B14
What does an electrical insulator do?
A. It lets electricity flow through it in one direction
B. It does not let electricity flow through it
C. It lets electricity flow through it when light shines on it
D. It lets electricity flow through it

N5C Ohm's Law (any calculations will be kept to a very low level—no fractions or decimals) and the concepts of energy and power, and open and short circuits

N5C01
(A)
Page 5-9

N5C01
What formula shows how voltage, current and resistance relate to each other in an electric circuit?
A. Ohm's Law
B. Kirchhoff's Law
C. Ampere's Law
D. Tesla's Law

N5C02
(C)
Page 5-9

N5C02
Which of the following principles is used when working with almost any electronic circuit?
A. Ampere's Law
B. Coulomb's Law
C. Ohm's Law
D. Tesla's Law

N5C03
(C)
Page 5-10

N5C03
If a current of 2 amperes flows through a 50-ohm resistor, what is the voltage across the resistor?
A. 25 volts
B. 52 volts
C. 100 volts
D. 200 volts

N5C04
(B)
Page 5-10

N5C04
If a 100-ohm resistor is connected to 200 volts, what is the current through the resistor?
A. 1 ampere
B. 2 amperes
C. 300 amperes
D. 20,000 amperes

N5C05
If a current of 3 amperes flows through a resistor connected to 90 volts, what is the resistance?
A. 3 ohms
B. 30 ohms
C. 93 ohms
D. 270 ohms

N5C05
(B)
Page 5-10

N5C06
What term describes how fast electrical energy is used?
A. Resistance
B. Current
C. Power
D. Voltage

N5C06
(C)
Page 5-12

N5C07
If you have light bulbs marked 60 watts, 75 watts and 100 watts, which one will use more electrical energy in one hour?
A. The 60 watt bulb
B. The 75 watt bulb
C. The 100 watt bulb
D. They will all be the same

N5C07
(C)
Page 5-12

N5C08
What is the basic unit of electrical power?
A. The ohm
B. The watt
C. The volt
D. The ampere

N5C08
(B)
Page 5-12

N5C09
Which electrical circuit can have no current?
A. A closed circuit
B. A short circuit
C. An open circuit
D. A complete circuit

N5C09
(C)
Page 5-11

N5C10
What type of electrical circuit is created when a fuse blows?
A. A closed circuit
B. A bypass circuit
C. An open circuit
D. A short circuit

N5C10
(C)
Page 5-11

N5C11
Which electrical circuit draws too much current?
A. An open circuit
B. A dead circuit
C. A closed circuit
D. A short circuit

N5C11
(D)
Page 5-11

N5D Concepts of frequency, including AC vs. DC, frequency units, AF vs. RF and wavelength

N5D01
(B)
Page 5-13

N5D01
What is the name of a current that flows only in one direction?
A. An alternating current
B. A direct current
C. A normal current
D. A smooth current

N5D02
(D)
Page 5-13

N5D02
Which of the following will produce an alternating current (AC)?
A. A lead-acid automotive battery
B. A solar array
C. A fuel cell
D. A commercial generating station

N5D03
(A)
Page 5-13

N5D03
What is the name of a current that flows back and forth, first in one direction, then in the opposite direction?
A. An alternating current
B. A direct current
C. A rough current
D. A steady state current

N5D04
(A)
Page 5-13

N5D04
Which of the following will produce a direct current (DC)?
A. A NiCd battery
B. An RF signal generator
C. A commercial generating station
D. A crystal calibrator

N5D05
(D)
Page 5-13

N5D05
What term means the number of times per second that an alternating current flows back and forth?
A. Pulse rate
B. Speed
C. Wavelength
D. Frequency

N5D06
(A)
Page 5-13

N5D06
What is the basic unit of frequency?
A. The hertz
B. The watt
C. The ampere
D. The ohm

N5D07
(B)
Page 5-14

N5D07
Most humans can hear sounds in what frequency range?
A. 0 - 20 Hz
B. 20 - 20,000 Hz
C. 200 - 200,000 Hz
D. 10,000 - 30,000 Hz

N5D08
Why do we call electrical signals in the frequency range of 20 Hz to 20,000 Hz audio frequencies?
A. Because the human ear cannot sense anything in this range
B. Because the human ear can sense sounds in this range
C. Because this range is too low for radio energy
D. Because the human ear can sense radio waves in this range

N5D08
(B)
Page 5-14

N5D09
What is the lowest frequency of electrical energy that is usually known as a radio frequency?
A. 20 Hz
B. 2,000 Hz
C. 20,000 Hz
D. 1,000,000 Hz

N5D09
(C)
Page 5-14

N5D10
Electrical energy at a frequency of 7125 kHz is in what frequency range?
A. Audio
B. Radio
C. Hyper
D. Super-high

N5D10
(B)
Page 5-14

N5D11
If a radio wave makes 3,725,000 cycles in one second, what does this mean?
A. The radio wave's voltage is 3725 kilovolts
B. The radio wave's wavelength is 3725 kilometers
C. The radio wave's frequency is 3725 kilohertz
D. The radio wave's speed is 3725 kilometers per second

N5D11
(C)
Page 5-13

N5D12
What does 60 hertz (Hz) mean?
A. 6000 cycles per second
B. 60 cycles per second
C. 6000 meters per second
D. 60 meters per second

N5D12
(B)
Page 5-13

N5D13
What is the name for the distance an AC signal travels during one complete cycle?
A. Wave speed
B. Waveform
C. Wavelength
D. Wave spread

N5D13
(C)
Page 5-13

N5D14
What happens to a signal's wavelength as its frequency increases?
A. It gets shorter
B. It gets longer
C. It stays the same
D. It disappears

N5D14
(A)
Page 5-15

N5D15
What happens to a signal's frequency as its wavelength gets longer?
A. It goes down
B. It goes up
C. It stays the same
D. It disappears

N5D15
(A)
Page 5-15

Subelement N6

Subelement N6 — Circuit Components [2 Exam Questions — 2 Groups]

N6A Electrical function and/or schematic representation of resistor, switch, fuse, or battery

N6A01
(B)
Page 6-3

N6A01
Why would you use a single-pole, double-throw switch?
A. To switch one input to one output
B. To switch one input to either of two outputs
C. To switch two inputs at the same time, one input to either of two outputs, and the other input to either of two outputs
D. To switch two inputs at the same time, one input to one output, and the other input to another output

N6A02
(D)
Page 6-3

N6A02
Why would you use a double-pole, single-throw switch?
A. To switch one input to one output
B. To switch one input to either of two outputs
C. To switch two inputs at the same time, one input to either of two outputs, and the other input to either of two outputs
D. To switch two inputs at the same time, one input to one output, and the other input to the other output

N6A03
(D)
Page 6-4

N6A03
Why would you use a fuse?
A. To create a short circuit when there is too much current in a circuit
B. To change direct current into alternating current
C. To change alternating current into direct current
D. To create an open circuit when there is too much current in a circuit

N6A04
(A)
Page 6-4

N6A04
Which of these components has a positive and a negative side?
A. A battery
B. A potentiometer
C. A fuse
D. A resistor

N6A05
(B)
Page 6-1

N6A05
Which of these components has a value that can be varied?
A. A single-cell battery
B. A potentiometer
C. A fuse
D. A resistor

N6A06
(B)
Page 6-2

N6A06
In Figure N6-1, which symbol represents a variable resistor or potentiometer?
A. Symbol 1
B. Symbol 2
C. Symbol 3
D. Symbol 6

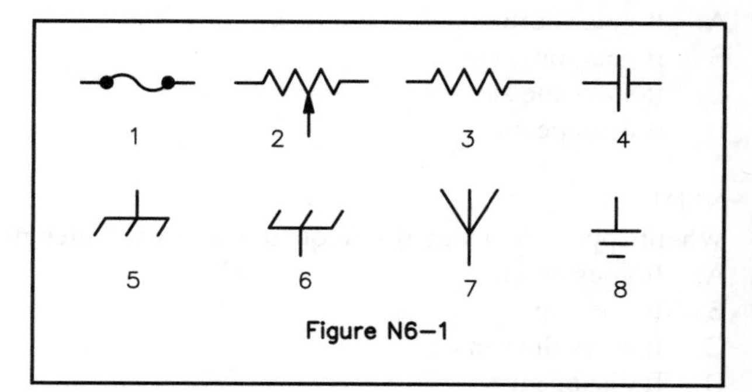

Figure N6—1

N6A07
In Figure N6-1, which symbol represents a fixed resistor?
A. Symbol 1
B. Symbol 2
C. Symbol 3
D. Symbol 4

N6A07
(C)
Page 6-2

N6A08
In Figure N6-1, which symbol represents a fuse?
A. Symbol 1
B. Symbol 3
C. Symbol 5
D. Symbol 7

N6A08
(A)
Page 6-3

N6A09
In Figure N6-1, which symbol represents a single-cell battery?
A. Symbol 7
B. Symbol 5
C. Symbol 1
D. Symbol 4

N6A09
(D)
Page 6-4

N6A10
In Figure N6-2, which symbol represents a single-pole, single-throw switch?
A. Symbol 1
B. Symbol 2
C. Symbol 3
D. Symbol 4

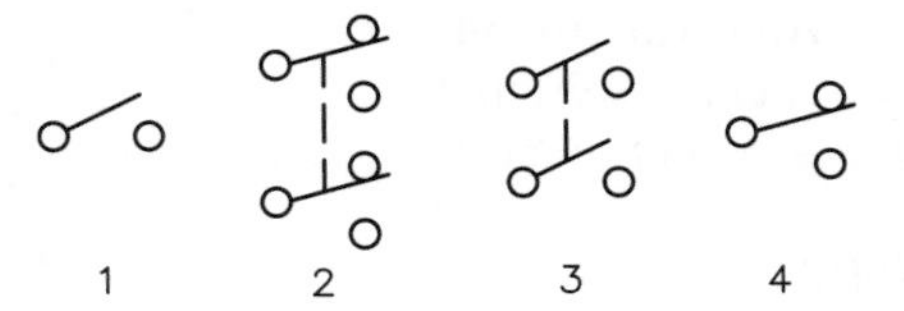

Figure N6–2

N6A10
(A)
Page 6-2

N6A11
In Figure N6-2, which symbol represents a single-pole, double-throw switch?
A. Symbol 1
B. Symbol 2
C. Symbol 3
D. Symbol 4

N6A11
(D)
Page 6-2

N6A12
In Figure N6-2, which symbol represents a double-pole, single-throw switch?
A. Symbol 1
B. Symbol 2
C. Symbol 3
D. Symbol 4

N6A12
(C)
Page 6-2

N6A13
In Figure N6-2, which symbol represents a double-pole, double-throw switch?
A. Symbol 1
B. Symbol 2
C. Symbol 3
D. Symbol 4

N6A13
(B)
Page 6-2

N6B Electrical function and/or schematic representation of a ground, antenna, transistor, integrated circuit or vacuum tube

N6B01
(A)
Page 6-6

N6B01
Which component can amplify a small signal using low voltages?
A. A PNP transistor
B. A variable resistor
C. An electrolytic capacitor
D. A multiple-cell battery

N6B02
(B)
Page 6-6

N6B02
Which component conducts electricity from a negative emitter to a positive collector when its base voltage is made positive?
A. A variable resistor
B. An NPN transistor
C. A triode vacuum tube
D. A multiple-cell battery

N6B03
(A)
Page 6-5

N6B03
Which component is used to radiate radio energy?
A. An antenna
B. An earth ground
C. A chassis ground
D. A potentiometer

N6B04
(D)
Page 6-5

N6B04
In Figure N6-1, which symbol represents an earth ground?
A. Symbol 2
B. Symbol 5
C. Symbol 6
D. Symbol 8

N6B05
(B)
Page 6-5

N6B05
In Figure N6-1, which symbol represents a chassis ground?
A. Symbol 2
B. Symbol 5
C. Symbol 6
D. Symbol 8

N6B06
(D)
Page 6-5

N6B06
In Figure N6-1, which symbol represents an antenna?
A. Symbol 2
B. Symbol 3
C. Symbol 6
D. Symbol 7

N6B07
(D)
Page 6-6

N6B07
In Figure N6-3, which symbol represents an NPN transistor?
A. Symbol 1
B. Symbol 2
C. Symbol 3
D. Symbol 4

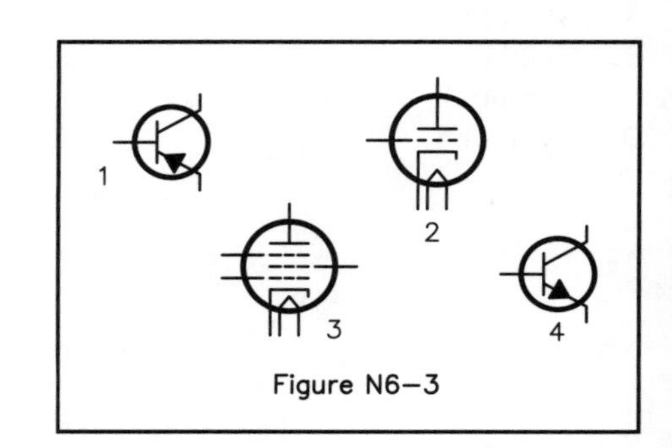

Figure N6–3

N6B08
In Figure N6-3, which symbol represents a PNP transistor?
A. Symbol 1
B. Symbol 2
C. Symbol 3
D. Symbol 4

N6B08
(A)
Page 6-6

N6B09
In Figure N6-3, which symbol represents a triode vacuum tube?
A. Symbol 1
B. Symbol 2
C. Symbol 3
D. Symbol 4

N6B09
(B)
Page 6-7

N6B10
In Figure N6-3, which symbol represents a pentode vacuum tube?
A. Symbol 1
B. Symbol 2
C. Symbol 3
D. Symbol 4

N6B10
(C)
Page 6-7

N6B11
What is one reason a triode vacuum tube might be used instead of a transistor in a circuit?
A. It handles higher power
B. It uses lower voltages
C. It operates more efficiently
D. It is much smaller

N6B11
(A)
Page 6-7

N6B12
Which component can amplify a small signal but must use high voltages?
A. A transistor
B. An electrolytic capacitor
C. A vacuum tube
D. A multiple-cell battery

N6B12
(C)
Page 6-7

N6B13
What is one advantage of using ICs (integrated circuits) instead of vacuum tubes in a circuit?
A. ICs usually combine several functions into one package
B. ICs can handle high-power input signals
C. ICs can handle much higher voltages
D. ICs can handle much higher temperatures

N6B13
(A)
Page 6-7

Subelement N7

Subelement N7 — Practical Circuits
[2 Exam Questions — 2 Groups]

N7A Functional layout of station components including transmitter, transceiver, receiver, power supply, antenna, antenna switch, antenna feed line, impedance-matching device and SWR meter

N7A01
(B)
Page 7-3

N7A01
What would you connect to your transceiver if you wanted to switch it between several antennas?
A. A terminal-node switch
B. An antenna switch
C. A telegraph key switch
D. A high-pass filter

N7A02
(C)
Page 7-3

N7A02
What device might allow use of an antenna on a band it was not designed for?
A. An SWR meter
B. A low-pass filter
C. An antenna tuner
D. A high-pass filter

N7A03
(D)
Page 7-3

N7A03
What connects your transceiver to your antenna?
A. A dummy load
B. A ground wire
C. The power cord
D. A feed line

N7A04
(B)
Page 7-3

N7A04
What might you connect between your transceiver and an antenna switch connected to several antennas?
A. A high-pass filter
B. An SWR meter
C. A key-click filter
D. A mixer

N7A05
(D)
Page 7-4

N7A05
If your SWR meter is connected to an antenna tuner on one side, what would you connect to the other side of it?
A. A power supply
B. An antenna
C. An antenna switch
D. A transceiver

N7A06
(D)
Page 7-3

N7A06
Which of the following should never be connected to a transceiver output?
A. An antenna switch
B. An SWR meter
C. An antenna
D. A receiver

N7A07
If your mobile transceiver works in your car but not in your home, what should you check first?
A. The power supply
B. The speaker
C. The microphone
D. The SWR meter

N7A07
(A)
Page 7-2

N7A08
What does an antenna tuner do?
A. It matches a transceiver output impedance to the antenna system impedance
B. It helps a receiver automatically tune in stations that are far away
C. It switches an antenna system to a transceiver when sending, and to a receiver when listening
D. It switches a transceiver between different kinds of antennas connected to one feed line

N7A08
(A)
Page 7-3

N7A09
In Figure N7-1, if block 1 is a transceiver and block 3 is a dummy antenna, what is block 2?
A. A terminal-node switch
B. An antenna switch
C. A telegraph key switch
D. A high-pass filter

1 2 3

Figure N7—1

N7A09
(B)
Page 7-3

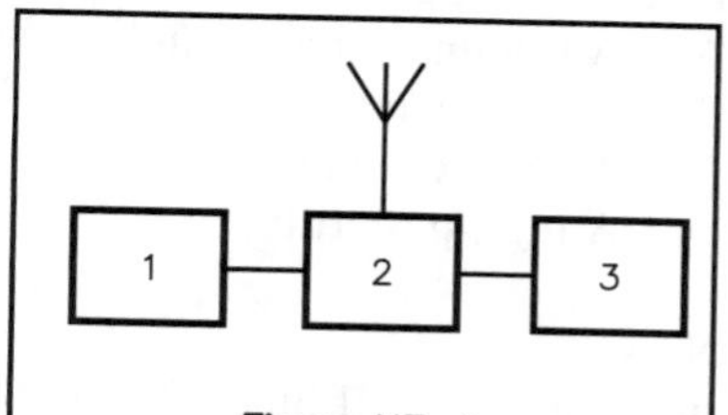

N7A10
In Figure N7-1, if block 2 is an antenna switch and block 3 is a dummy antenna, what is block 1?
A. A terminal-node switch
B. A dipole antenna
C. A transceiver
D. A high-pass filter

N7A10
(C)
Page 7-3

N7A11
In Figure N7-1, if block 1 is a transceiver and block 2 is an antenna switch, what is block 3?
A. A terminal-node switch
B. An SWR meter
C. A telegraph key switch
D. A dummy antenna

N7A11
(D)
Page 7-3

N7A12
In Figure N7-2, if block 2 is an SWR meter and block 3 is an antenna switch, what is block 1?
A. A transceiver
B. A high-pass filter
C. An antenna tuner
D. A modem

N7A12
(A)
Page 7-3

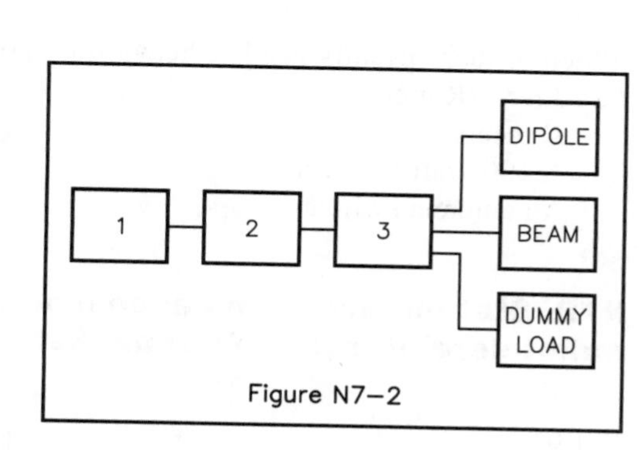

Figure N7—2

N7A13
In Figure N7-2, if block 1 is a transceiver and block 3 is an antenna switch, what is block 2?
A. A terminal-node switch
B. A dipole antenna
C. An SWR meter
D. A high-pass filter

N7A13
(C)
Page 7-3

N7A14
(D)
Page 7-2

N7A14
In Figure N7-2, if block 1 is a transceiver and block 2 is an SWR meter, what is block 3?
A. A terminal-node switch
B. A power supply
C. A telegraph key switch
D. An antenna switch

N7A15
(B)
Page 7-4

N7A15
In Figure N7-3, if block 1 is a transceiver and block 2 is an SWR meter, what is block 3?
A. An antenna switch
B. An antenna tuner
C. A key-click filter
D. A terminal-node controller

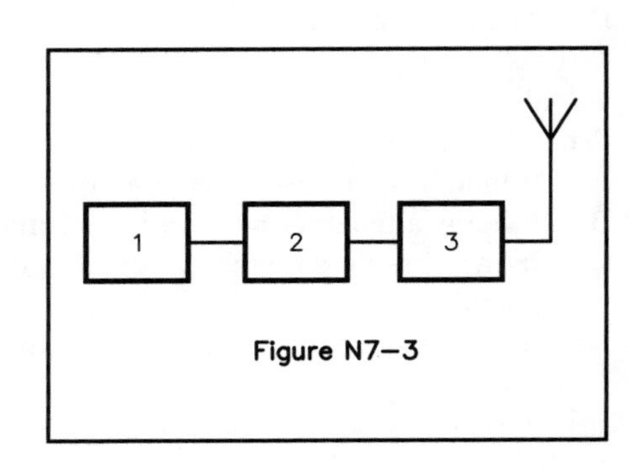

Figure N7—3

N7A16
(C)
Page 7-4

N7A16
In Figure N7-3, if block 1 is a transceiver and block 3 is an antenna tuner, what is block 2?
A. A terminal-node switch
B. A dipole antenna
C. An SWR meter
D. A high-pass filter

N7A17
(D)
Page 7-4

N7A17
In Figure N7-3, if block 2 is an SWR meter and block 3 is an antenna tuner, what is block 1?
A. A terminal-node switch
B. A power supply
C. A telegraph key switch
D. A transceiver

N7A18
(C)
Page 7-2

N7A18
What device converts household current to 12 VDC?
A. A catalytic converter
B. A low-pass filter
C. A power supply
D. An RS-232 interface

N7A19
(C)
Page 7-2

N7A19
Which of these usually needs a heavy-duty power supply?
A. An SWR meter
B. A receiver
C. A 100-watt transceiver
D. An antenna switch

N7B Station layout and accessories for telegraphy, radiotelephone, radioteleprinter (RTTY) or packet

N7B01
(B)
Page 7-4

N7B01
What would you connect to a transceiver to send Morse code?
A. A key-click filter
B. A telegraph key
C. An SWR meter
D. An antenna switch

N7B02
Where would you connect a telegraph key to send Morse code?
A. To a modem
B. To an antenna switch
C. To a transceiver
D. To an antenna

N7B02
(C)
Page 7-4

N7B03
What do many amateurs use to help form good Morse code characters?
A. A key-operated on/off switch
B. An electronic keyer
C. A key-click filter
D. A DTMF keypad

N7B03
(B)
Page 7-4

N7B04
Where would you connect a microphone for voice operation?
A. To a power supply
B. To an antenna switch
C. To a transceiver
D. To an antenna

N7B04
(C)
Page 7-5

N7B05
What would you connect to a transceiver for voice operation?
A. A splatter filter
B. A terminal-voice controller
C. A receiver audio filter
D. A microphone

N7B05
(D)
Page 7-5

N7B06
What would you connect to a transceiver for RTTY operation?
A. A modem and a teleprinter or computer system
B. A computer, a printer and a RTTY refresh unit
C. A data-inverter controller
D. A modem, a monitor and a DTMF keypad

N7B06
(A)
Page 7-5

N7B07
What would you connect between a transceiver and a computer system or teleprinter for RTTY operation?
A. An RS-432 interface
B. A DTMF keypad
C. A modem
D. A terminal-network controller

N7B07
(C)
Page 7-5

N7B08
What would you connect between a computer system and a transceiver for packet-radio operation?
A. A terminal-node controller
B. A DTMF keypad
C. An SWR bridge
D. An antenna tuner

N7B08
(A)
Page 7-5

N7B09
Where would you connect a terminal-node controller for packet-radio operation?
A. Between your antenna and transceiver
B. Between your computer and monitor
C. Between your computer and transceiver
D. Between your keyboard and computer

N7B09
(C)
Page 7-5

N7B10
(D)
Page 7-5

N7B10
In RTTY operation, what equipment connects to a modem?
A. A DTMF keypad, a monitor and a transceiver
B. A DTMF microphone, a monitor and a transceiver
C. A transceiver and a terminal-network controller
D. A transceiver and a teleprinter or computer system

N7B11
(B)
Page 7-5

N7B11
In packet-radio operation, what equipment connects to a terminal-node controller?
A. A transceiver and a modem
B. A transceiver and a terminal or computer system
C. A DTMF keypad, a monitor and a transceiver
D. A DTMF microphone, a monitor and a transceiver

N7B12
(D)
Page 7-6

N7B12
What important feature must an HF transceiver have for digital operation?
A. A digital readout
B. Loud audio
C. A fully solid-state receiver
D. A fast T/R switching time

N7B13
(C)
Page 7-6

N7B13
What circuit can improve CW reception during crowded band conditions?
A. A high-pass filter
B. A crystal oscillator
C. A digital signal processor
D. A signal generator

Subelement N8

Subelement N8 — Signals and Emissions [2 Exam Questions — 2 Groups]

N8A CW, phone, RTTY and data emission types

N8A01
(B)
Page 2-3

N8A01
How is a CW signal usually transmitted?
A. By frequency-shift keying an RF signal
B. By on/off keying an RF signal
C. By audio-frequency-shift keying an oscillator tone
D. By on/off keying an audio-frequency signal

N8A02
(C)
Page 2-3

N8A02
What is another name for international Morse code emissions?
A. RTTY
B. Data
C. CW
D. Phone

N8A03
(D)
Page 2-3

N8A03
What type of emission is transmitted by turning an RF signal on and off?
A. Frequency-shift-keyed RTTY
B. Phase-shift-keyed RTTY
C. Frequency modulated CW
D. CW

N8A04
What is the name for voice emissions?
A. RTTY
B. Data
C. CW
D. Phone

N8A04
(D)
Page 2-7

N8A05
Which sideband is commonly used for 10-meter phone operation?
A. Upper sideband
B. Lower sideband
C. Amplitude-compandored sideband
D. Double sideband

N8A05
(A)
Page 2-8

N8A06
What does the term "phone transmissions" usually mean?
A. The use of telephones to set up an amateur contact
B. A phone patch between Amateur Radio and the telephone system
C. AM, FM or SSB voice transmissions by radiotelephony
D. Placing the telephone handset near a transceiver's microphone and speaker to relay a telephone call

N8A06
(C)
Page 2-7

N8A07
How is an HF RTTY signal usually produced?
A. By frequency-shift keying an RF signal
B. By on/off keying an RF signal
C. By digital pulse-code keying of an unmodulated carrier
D. By on/off keying an audio-frequency signal

N8A07
(A)
Page 2-14

N8A08
What is the name for narrow-band direct-printing telegraphy emissions?
A. RTTY
B. Data
C. CW
D. Phone

N8A08
(A)
Page 2-15

N8A09
What is another name for packet-radio emissions?
A. RTTY
B. Data
C. CW
D. Phone

N8A09
(B)
Page 2-18

N8A10
Which of the following devices would you need to conduct Amateur Radio communications using a data emission?
A. A telegraph key
B. A computer
C. A transducer
D. A telemetry sensor

N8A10
(B)
Page 2-16

N8A11
(D)
Page 2-14

N8A11
What are two advantages to using modern data-transmission techniques for communications?
A. Very simple and low-cost equipment
B. No parity-checking required and high transmission speed
C. Easy for mobile stations to use and no additional cabling required
D. High transmission speed and communications reliability

N8B Harmonics and unwanted signals; chirp; superimposed hum; equipment and adjustments to help reduce interference to others

N8B01
(C)
Page 8-2

N8B01
How does the frequency of a harmonic compare to the desired transmitting frequency?
A. It is slightly more than the desired frequency
B. It is slightly less than the desired frequency
C. It is exactly two, or three, or more times the desired frequency
D. It is much less than the desired frequency

N8B02
(A)
Page 8-2

N8B02
What is the fourth harmonic of a 7160-kHz signal?
A. 28,640 kHz
B. 35,800 kHz
C. 28,160 kHz
D. 1790 kHz

N8B03
(C)
Page 8-2

N8B03
If you are told your station was heard on 21,375 kHz, but at the time you were operating on 7125 kHz, what is one reason this could happen?
A. Your transmitter's power-supply filter capacitor was bad
B. You were sending CW too fast
C. Your transmitter was radiating harmonic signals
D. Your transmitter's power-supply filter choke was bad

N8B04
(D)
Page 8-2

N8B04
If someone tells you that signals from your hand-held transceiver are interfering with other signals on a frequency near yours, what may be the cause?
A. You may need a power amplifier for your hand-held
B. Your hand-held may have chirp from weak batteries
C. You may need to turn the volume up on your hand-held
D. Your hand-held may be transmitting spurious emissions

N8B05
(D)
Page 8-1

N8B05
If your transmitter sends signals outside the band where it is transmitting, what is this called?
A. Off-frequency emissions
B. Transmitter chirping
C. Side tones
D. Spurious emissions

N8B06
(A)
Page 8-4

N8B06
What problem may occur if your transmitter is operated without the cover and other shielding in place?
A. It may transmit spurious emissions
B. It may transmit a chirpy signal
C. It may transmit a weak signal
D. It may transmit a phase-inverted signal

N8B07
What may happen if an SSB transmitter is operated with the microphone gain set too high?
A. It may cause digital interference to computer equipment
B. It may cause splatter interference to other stations operating near its frequency
C. It may cause atmospheric interference in the air around the antenna
D. It may cause interference to other stations operating on a higher frequency band

N8B07
(B)
Page 8-3

N8B08
What may happen if an SSB transmitter is operated with too much speech processing?
A. It may cause digital interference to computer equipment
B. It may cause splatter interference to other stations operating near its frequency
C. It may cause atmospheric interference in the air around the antenna
D. It may cause interference to other stations operating on a higher frequency band

N8B08
(B)
Page 8-4

N8B09
What may happen if an FM transmitter is operated with the microphone gain or deviation control set too high?
A. It may cause digital interference to computer equipment
B. It may cause interference to other stations operating near its frequency
C. It may cause atmospheric interference in the air around the antenna
D. It may cause interference to other stations operating on a higher frequency band

N8B09
(B)
Page 8-4

N8B10
What may your FM hand-held or mobile transceiver do if you shout into its microphone?
A. It may cause digital interference to computer equipment
B. It may cause interference to other stations operating near its frequency
C. It may cause atmospheric interference in the air around the antenna
D. It may cause interference to other stations operating on a higher frequency band

N8B10
(B)
Page 8-4

N8B11
What can you do if you are told your FM hand-held or mobile transceiver is over-deviating?
A. Talk louder into the microphone
B. Let the transceiver cool off
C. Change to a higher power level
D. Talk farther away from the microphone

N8B11
(D)
Page 8-4

N8B12
What does chirp mean?
A. An overload in a receiver's audio circuit whenever CW is received
B. A high-pitched tone that is received along with a CW signal
C. A small change in a transmitter's frequency each time it is keyed
D. A slow change in transmitter frequency as the circuit warms up

N8B12
(C)
Page 8-9

N8B13
What can be done to keep a CW transmitter from chirping?
A. Add a low-pass filter
B. Use an RF amplifier
C. Keep the power supply current very steady
D. Keep the power supply voltages very steady

N8B13
(D)
Page 8-9

N8B14
What may cause a buzzing or hum in the signal of an HF transmitter?
A. Using an antenna that is the wrong length
B. Energy from another transmitter
C. Bad design of the transmitter's RF power output circuit
D. A bad filter capacitor in the transmitter's power supply

N8B14
(D)
Page 8-9

N8B15
(B)
Page 8-4

N8B15
What should you check if you change your transceiver's microphone from a mobile type to a base station type?
A. Check the CTCSS levels on the oscilloscope
B. Make an on-the-air radio check to ensure the quality of your signal
C. Check the amount of current the transceiver is now using
D. Check to make sure the frequency readout is now correct

N8B16
(C)
Page 8-7

N8B16
Why is good station grounding needed when connecting your computer to your transceiver to receive high-frequency data signals?
A. Good grounding raises the receiver's noise floor
B. Good grounding protects the computer from nearby lightning strikes
C. Good grounding will minimize stray noise on the receiver
D. FCC rules require all equipment to be grounded

Subelement N9

Subelement N9 — Antennas and Feed Lines [3 Exam Questions — 3 Groups]

N9A Wavelength vs. antenna length; multiband antenna advantages and disadvantages

N9A01
(D)
Page 9-9

N9A01
How do you calculate the length (in feet) of a half-wavelength dipole antenna?
A. Divide 150 by the antenna's operating frequency (in MHz) [150/f (in MHz)]
B. Divide 234 by the antenna's operating frequency (in MHz) [234/f (in MHz)]
C. Divide 300 by the antenna's operating frequency (in MHz) [300/f (in MHz)]
D. Divide 468 by the antenna's operating frequency (in MHz) [468/f (in MHz)]

N9A02
(B)
Page 9-16

N9A02
How do you calculate the length (in feet) of a quarter-wavelength vertical antenna?
A. Divide 150 by the antenna's operating frequency (in MHz) [150/f (in MHz)]
B. Divide 234 by the antenna's operating frequency (in MHz) [234/f (in MHz)]
C. Divide 300 by the antenna's operating frequency (in MHz) [300/f (in MHz)]
D. Divide 468 by the antenna's operating frequency (in MHz) [468/f (in MHz)]

N9A03
(A)
Page 9-9

N9A03
How long should you make a half-wavelength dipole antenna for 3725 kHz (measured to the nearest foot)?
A. 126 ft
B. 81 ft
C. 63 ft
D. 40 ft

N9A04
(C)
Page 9-9

N9A04
How long should you make a half-wavelength dipole antenna for 28.150 MHz (measured to the nearest foot)?
A. 22 ft
B. 11 ft
C. 17 ft
D. 34 ft

N9A05
How long should you make a quarter-wavelength vertical antenna for 7125 kHz (measured to the nearest foot)?
A. 11 ft
B. 16 ft
C. 21 ft
D. 33 ft

N9A05
(D)
Page 9-17

N9A06
How long should you make a quarter-wavelength vertical antenna for 21.125 MHz (measured to the nearest foot)?
A. 7 ft
B. 11 ft
C. 4 ft
D. 22 ft

N9A06
(B)
Page 9-17

N9A07
How long should you make a half-wavelength vertical antenna for 223 MHz (measured to the nearest inch)?
A. 112 inches
B. 50 inches
C. 25 inches
D. 12 inches

N9A07
(C)
Page 9-9

N9A08
If an antenna is made longer, what happens to its resonant frequency?
A. It decreases
B. It increases
C. It stays the same
D. It disappears

N9A08
(A)
Page 9-14

N9A09
If an antenna is made shorter, what happens to its resonant frequency?
A. It decreases
B. It increases
C. It stays the same
D. It disappears

N9A09
(B)
Page 9-14

N9A10
How could you decrease the resonant frequency of a dipole antenna?
A. Lengthen the antenna
B. Shorten the antenna
C. Use less feed line
D. Use a smaller size feed line

N9A10
(A)
Page 9-14

N9A11
How could you increase the resonant frequency of a dipole antenna?
A. Lengthen the antenna
B. Shorten the antenna
C. Use more feed line
D. Use a larger size feed line

N9A11
(B)
Page 9-14

N9A12
(A)
Page 9-15

N9A12
What is one advantage to using a multiband antenna?
A. You can operate on several bands with a single feed line
B. Multiband antennas always have high gain
C. You can transmit on several frequencies simultaneously
D. Multiband antennas offer poor harmonic suppression

N9A13
(D)
Page 9-16

N9A13
What is one disadvantage to using a multiband antenna?
A. It must always be used with a balun
B. It will always have low gain
C. It cannot handle high power
D. It can radiate unwanted harmonics

N9B Yagi parts, concept of directional antennas

N9B01
(B)
Page 9-24

N9B01
In what direction does a Yagi antenna send out radio energy?
A. It goes out equally in all directions
B. Most of it goes in one direction
C. Most of it goes equally in two opposite directions
D. Most of it is aimed high into the air

N9B02
(C)
Page 9-24

N9B02
Approximately how long is the driven element of a Yagi antenna?
A. 1/4 wavelength
B. 1/3 wavelength
C. 1/2 wavelength
D. 1 wavelength

N9B03
(D)
Page 9-25

N9B03
In Figure N9-1, what is the name of element 2 of the Yagi antenna?
A. Director
B. Reflector
C. Boom
D. Driven element

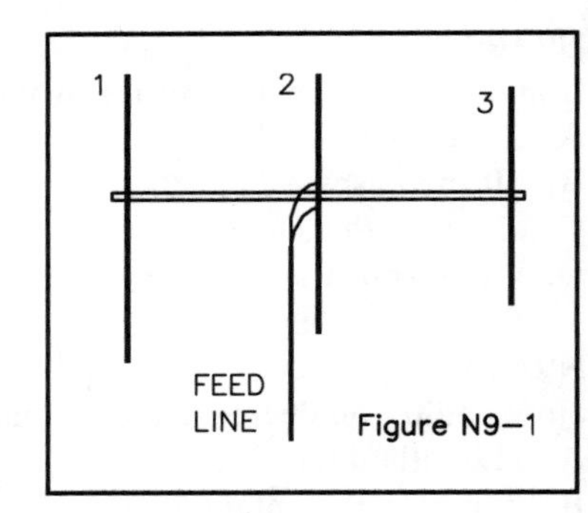

Figure N9—1

N9B04
(A)
Page 9-25

N9B04
In Figure N9-1, what is the name of element 3 of the Yagi antenna?
A. Director
B. Reflector
C. Boom
D. Driven element

N9B05
(B)
Page 9-25

N9B05
In Figure N9-1, what is the name of element 1 of the Yagi antenna?
A. Director
B. Reflector
C. Boom
D. Driven element

N9B06
Looking at the Yagi antenna in Diagram N9-1, in which direction on the page would it send most of its radio energy?
A. Left
B. Right
C. Top
D. Bottom

N9B06
(B)
Page 9-25

N9B07
Which of the following factors has the greatest effect on the gain of a properly designed Yagi antenna?
A. The number of elements
B. Boom length
C. Element spacing
D. Element diameter

N9B07
(B)
Page 9-24

N9B08
What is one advantage of a 5/8-wavelength vertical antenna as compared to a 1/4-wavelength vertical antenna for VHF or UHF mobile operations?
A. A 5/8-wavelength antenna can handle more power
B. A 5/8-wavelength antenna has more gain
C. A 5/8-wavelength antenna has less corona loss
D. A 5/8-wavelength antenna is easier to install on a car

N9B08
(B)
Page 9-18

N9B09
In what direction does a vertical antenna send out radio energy?
A. Most of it goes in two opposite directions
B. Most of it goes high into the air
C. Most of it goes equally in all horizontal directions
D. Most of it goes in one direction

N9B09
(C)
Page 9-16

N9B10
If the ends of a half-wave dipole antenna (mounted at least a half-wavelength high) point east and west, which way would the antenna send out radio energy?
A. Equally in all directions
B. Mostly up and down
C. Mostly north and south
D. Mostly east and west

N9B10
(C)
Page 9-11

N9B11
Which is true of "rubber duck" antennas for hand-held transceivers?
A. The shorter they are, the better they perform
B. They are much less efficient than a quarter-wavelength telescopic antenna
C. They offer the highest amount of gain possible for any hand-held transceiver antenna
D. They have a good long-distance communications range

N9B11
(B)
Page 9-18

N9C Feed lines, baluns and polarization via element orientation

N9C01
What is a coaxial cable?
A. Two wires side-by-side in a plastic ribbon
B. Two wires side-by-side held apart by insulating rods
C. Two wires twisted around each other in a spiral
D. A center wire inside an insulating material covered by a metal sleeve or shield

N9C01
(D)
Page 9-3

N9C02
(B)
Page 9-3

N9C02
Why does coaxial cable make a good antenna feed line?
A. You can make it at home, and its impedance matches most amateur antennas
B. It is weatherproof, and it can be used near metal objects
C. It is weatherproof, and its impedance is higher than that of most amateur antennas
D. It can be used near metal objects, and its impedance is higher than that of most amateur antennas

N9C03
(B)
Page 9-3

N9C03
Which kind of antenna feed line can carry radio energy very well even if it is buried in the ground?
A. Twin lead
B. Coaxial cable
C. Parallel conductor
D. Twisted pair

N9C04
(A)
Page 9-3

N9C04
Why should you use only good quality coaxial cable and connectors for a UHF antenna system?
A. To keep RF loss low
B. To keep television interference high
C. To keep the power going to your antenna system from getting too high
D. To keep the standing-wave ratio of your antenna system high

N9C05
(A)
Page 9-3

N9C05
What is the best antenna feed line to use if it must be put near grounded metal objects?
A. Coaxial cable
B. Twin lead
C. Twisted pair
D. Ladder-line

N9C06
(B)
Page 9-4

N9C06
What is parallel-conductor feed line?
A. Two wires twisted around each other in a spiral
B. Two wires side-by-side held apart by insulating material
C. A center wire inside an insulating material that is covered by a metal sleeve or shield
D. A metal pipe that is as wide or slightly wider than a wavelength of the signal it carries

N9C07
(D)
Page 9-4

N9C07
What are some reasons to use parallel-conductor, open-wire feed line?
A. It has low impedance and will operate with a high SWR
B. It will operate with a high SWR and it works well when tied down to metal objects
C. It has a low impedance and has less loss than coaxial cable
D. It will operate with a high SWR and has less loss than coaxial cable

N9C08
(A)
Page 9-4

N9C08
What are some reasons not to use ladder line to feed your antenna?
A. It does not work well when tied down to metal objects and you must use an impedance-matching device with your transceiver
B. It is difficult to make at home and it does not work very well with a high SWR
C. It does not work well when tied down to metal objects and it cannot operate under high power
D. You must use an impedance-matching device with your transceiver and it does not work very well with a high SWR

N9C09
What kind of antenna feed line is made of two conductors held apart by insulated rods?
A. Coaxial cable
B. Open-conductor ladder line
C. Twin lead in a plastic ribbon
D. Twisted pair

N9C09
(B)
Page 9-4

N9C10
What would you use to connect a coaxial cable of 50-ohms impedance to an antenna of 35-ohms impedance?
A. A terminating resistor
B. An SWR meter
C. An impedance-matching device
D. A low-pass filter

N9C10
(C)
Page 9-2

N9C11
What does balun mean?
A. Balanced antenna network
B. Balanced unloader
C. Balanced unmodulator
D. Balanced to unbalanced

N9C11
(D)
Page 9-6

N9C12
Where would you install a balun to feed a dipole antenna with 50-ohm coaxial cable?
A. Between the coaxial cable and the antenna
B. Between the transmitter and the coaxial cable
C. Between the antenna and the ground
D. Between the coaxial cable and the ground

N9C12
(A)
Page 9-6

Subelement N0 — RF Safety
[5 Exam Questions — 5 Groups]

Subelement N0

N0A RF safety fundamentals

N0A01
What factors affect the resulting RF fields radiated into the environment by an amateur transceiver?
A. Frequency and power level of the RF field
B. Antenna height and distance from the antenna to a person
C. Radiation pattern of the antenna
D. All of these answers are correct

N0A01
(D)
Page 10-8

N0A02
Which of the following effects on the human body are a result of exposure to high levels of RF energy?
A. Very rapid hair growth
B. Very rapid growth of fingernails and toenails
C. Possible heating of body tissue
D. High levels of RF energy have no known effect on the human body

N0A02
(C)
Page 10-2

N0A03
(D)
Page 10-7

N0A03
Why should you not stand within reach of any transmitting antenna when it is being fed with 1500 watts of RF energy?
A. It could result in the loss of the ability to move muscles
B. Your body would reflect the RF energy back to its source
C. It could cause cooling of body tissue
D. You could accidentally touch the antenna and be injured

N0A04
(A)
Page 10-2

N0A04
What impact does a high concentration of RF energy have on the human body?
A. It can heat tissue below the body's surface
B. There are no known adverse biological effects
C. It can cause rapid, uncontrolled weight gain
D. It can produce arthritis-like joint pains

N0A05
(B)
Page 10-2

N0A05
What is one effect of nonionizing radiation on the human body?
A. Cooling of body tissue
B. Heating of body tissue
C. Rapid dehydration
D. Sudden hair loss

N0A06
(A)
Page 10-6

N0A06
What factors determine the location of the boundary between the near and far fields of an antenna?
A. Wavelength and the physical size of the antenna
B. Antenna height and element length
C. Boom length and element diameter
D. Transmitter power and antenna gain

N0A07
(C)
Page 10-7

N0A07
Why should you not stand within reach of a high-gain 3-element "triband" Yagi transmitting antenna when it is being fed with 1500 watts of RF energy?
A. It could result in the loss of the ability to move muscles
B. Your body would reflect the RF energy back to its source
C. You could accidentally touch the antenna and be injured
D. It could cause cooling of body tissue

N0A08
(C)
Page 10-7

N0A08
Why should you not stand within reach of a transmitting antenna when it is being fed with 1000 watts of RF energy?
A. It could result in the loss of the ability to move muscles
B. Your body would reflect the RF energy back to its source
C. You could accidentally touch the antenna and be injured
D. It could cause cooling of body tissue

N0A09
(A)
Page 10-7

N0A09
Why should you not stand within reach of a high-gain parabolic-dish transmitting antenna when it is being fed with 1500 watts of RF energy?
A. You could accidentally touch the antenna and be injured
B. It could result in the loss of the ability to move muscles
C. Your body would reflect the RF energy back to its source
D. It could cause cooling of body tissue

N0A10
Why should you not stand within reach of a high-gain 17-element Yagi transmitting antenna transmitting on 146 MHz when it is being fed with 1000 watts of RF energy?
A. You could accidentally touch the antenna and be injured
B. It could result in the loss of the ability to move muscles
C. Your body would reflect the RF energy back to its source
D. It could cause cooling of body tissue

N0A10
(A)
Page 10-7

N0A11
Why should you not stand within reach of a high-gain multiple-antenna transmitting array when it is being fed with 1500 watts of RF energy?
A. It could cause heating of body tissue
B. It could result in the loss of the ability to move muscles
C. Your body would reflect the RF energy back to its source
D. It could cause cooling of body tissue

N0A11
(A)
Page 10-7

N0B RF safety terms and definitions

N0B01
In what type of RF radiation exposure environment are amateurs and their households considered to be located?
A. An excluded RF radiation exposure environment
B. A "controlled" RF environment
C. An "uncontrolled" RF environment
D. Both a "controlled" and "uncontrolled" environment

N0B01
(B)
Page 10-8

N0B02
What does the term "uncontrolled RF environment" mean when it is applied to RF radiation exposure?
A. A radio operator is not exercising proper antenna radiation safety
B. A location where there is RF radiation exposure to persons who have no knowledge or control of their exposure
C. A location where there is RF radiation exposure to persons who are aware of the potential for exposure
D. A transmitting station lacks the proper RF safety certification

N0B02
(B)
Page 10-3

N0B03
What does the term "controlled RF environment" mean when it is applied to RF radiation exposure?
A. A location where there is RF radiation exposure to persons who are aware of the potential for exposure
B. A location that has been made inaccessible by a security fence
C. A location where there is RF radiation exposure to persons who have no knowledge or control of their exposure
D. A transmitter has been certified by the FCC to be safe for use at all frequencies

N0B03
(A)
Page 10-3

N0B04
What unit of measurement specifies RF electric field strength?
A. Coulombs (C) at one wavelength from the antenna
B. Volts per meter (V/m)
C. Microfarads (μF) at the transmitter output
D. Microhenrys (μH) per square centimeter

N0B04
(B)
Page 10-3

N0B05
(D)
Page 10-3

N0B05
What unit of measurement specifies RF magnetic field strength?
A. Coulombs (C) at one wavelength from the antenna
B. microfarads (μF) at the transmitter output
C. In polar units (Pu) at the antenna terminals
D. Amperes per meter (A/m)

N0B06
(D)
Page 10-1

N0B06
Which of the following is considered to be nonionizing radiation?
A. X-radiation
B. Gamma radiation
C. Ultra violet radiation
D. Radio frequency radiation

N0B07
(A)
Page 10-8

N0B07
In what type of RF environment are amateurs and their immediate families considered to be located?
A. They are in a "controlled" RF environment
B. They are excluded from the RF radiation exposure guidelines
C. They are in an "uncontrolled" or "general public" environment
D. None of these choices are correct

N0B08
(A)
Page 10-7

N0B08
What is radiofrequency radiation?
A. Waves of electric and magnetic energy between 3 kHz and 300 GHz
B. Ultra-violet rays emitted by the sun between 20 Hz and 300 GHz
C. Sound energy given off by a radio receiver
D. Beams of X-Rays and Gamma rays emitted by a radio transmitter

N0B09
(D)
Page 10-8

N0B09
Why are residential neighbors of an amateur station considered to be in an "uncontrolled" environment?
A. Because they are not under the jurisdiction of the Federal Communications Commission
B. Because they are generally aware of the potential for RF radiation exposure
C. Because the RF environment is primarily controlled by the sun
D. Because they cannot exercise control over their RF radiation exposure

N0B10
(A)
Page 10-11

N0B10
Which of the following units of measurement are used to specify the power density of a radiated RF signal?
A. Milliwatts per square centimeter
B. Volts per meter
C. Amperes per meter
D. All of these choices are correct

N0B11
(A)
Page 10-11

N0B11
Referring to Figure NT0-1, which of the following equations should you use to calculate the maximum permissible exposure (MPE) on the Novice HF bands for a controlled RF radiation exposure environment?
A. Maximum permissible power density in mW per square cm equals 900 divided by the square of the operating frequency, in MHz
B. Maximum permissible power density in mW per square cm equals 180 divided by the square of the operating frequency, in MHz
C. Maximum permissible power density in mW per square cm equals 900 divided by the operating frequency, in MHz
D. Maximum permissible power density in mW per square cm equals 180 divided by the operating frequency, in MHz

(A) Limits for Occupational/Controlled Exposure				
Frequency Range (MHz)	Electric Field Strength (V/m)	Magnetic Field (A/m)	Power Density (mW/cm²)	Averaging Time (minutes)
0.3-3.0	614	1.63	(100)*	6
3.0-30	1842/f	4.89/f	$(900/f^2)$*	6
30-300	61.4	0.163	1.0	6
300-1500	—	—	f/300	6
1500-100,000	—	—	5	6
f = frequency in MHz		* = Plane-wave equivalent power density		

(B) Limits for General Population/Uncontrolled Exposure				
Frequency Range (MHz)	Electric Field Strength (V/m)	Magnetic Field (A/m)	Power Density (mW/cm²)	Averaging Time (minutes)
0.3-1.34	614	1.63	(100)*	30
1.34-30	824/f	2.19/f	$(180/f^2)$*	30
30-300	27.5	0.073	0.2	30
300-1500	—	—	f/1500	30
1500-100,000	—	—	1.0	30
f = frequency in MHz		* = Plane-wave equivalent power density		

Figure NT0-1—RF Exposure Limits

N0B12
Referring to Figure NT0-1, which of the following equations should you use to calculate the maximum permissible exposure (MPE) on the Novice HF bands for a controlled RF radiation exposure environment?

A. Maximum permissible electric field strength in volts per meter equals 824 divided by the operating frequency, in MHz
B. Maximum permissible electric field strength in volts per meter equals 1842 divided by the operating frequency, in MHz
C. Maximum permissible electric field strength in volts per meter equals 1842 divided by the square of the operating frequency, in MHz
D. Maximum permissible electric field strength in volts per meter equals 824 divided by the operating frequency, in MHz

N0B12
(B)
Page 10-11

N0B13
Referring to Figure NT0-1, which of the following equations should you use to calculate the maximum permissible exposure (MPE) on the Novice HF bands for an uncontrolled RF radiation exposure environment?

A. Maximum permissible power density in mW per square cm equals 900 divided by the square of the operating frequency, in MHz
B. Maximum permissible power density in mW per square cm equals 180 divided by the square of the operating frequency, in MHz
C. Maximum permissible power density in mW per square cm equals 900 divided by the operating frequency, in MHz
D. Maximum permissible power density in mW per square cm equals 180 divided by the operating frequency, in MHz

N0B13
(B)
Page 10-11

N0B14
(C)
Page 10-11

N0B14
Referring to Figure NT0-1, which of the following equations should you use to calculate the maximum permissible exposure (MPE) on the Novice HF bands for an uncontrolled RF radiation exposure environment?
A. Maximum permissible magnetic field strength in amperes per meter equals 2.19 divided by the square of the operating frequency, in MHz
B. Maximum permissible magnetic field strength in amperes per meter equals 4.89 divided by the square of the operating frequency, in MHz
C. Maximum permissible magnetic field strength in amperes per meter equals 2.19 divided by the operating frequency, in MHz
D. Maximum permissible magnetic field strength in amperes per meter equals 4.89 divided by the operating frequency, in MHz

N0C RF safety rules and guidelines

N0C01
(D)
Page 10-8

N0C01
What amateur stations must comply with the requirements for RF radiation exposure spelled out in Part 97?
A. Stations with antennas that exceed 10 dBi of gain.
B. Stations that have a duty cycle greater than 50 percent.
C. Stations that run more than 50 watts peak envelope power (PEP)
D. All amateur stations regardless of power

N0C02
(C)
Page 10-8

N0C02
Who is responsible for ensuring that an amateur station complies with FCC Rules about RF radiation exposure?
A. The Federal Communications Commission
B. The Environmental Protection Agency
C. The licensee of the amateur station
D. The Food and Drug Administration

N0C03
(C)
Page 10-8

N0C03
At what frequencies do the FCC's RF radiation exposure guidelines incorporate limits for Maximum Permissible Exposure (MPE)?
A. All frequencies below 30 MHz
B. All frequencies between 20,000 Hz and 10 MHz
C. All frequencies between 300 kHz and 100 GHz
D. All frequencies above 300 GHz

N0C04

N0C04
This question has been withdrawn.

N0C05
(D)
Page 10-8

N0C05
To determine compliance with the maximum permitted exposure (MPE) levels, safe exposure levels for RF energy are averaged for an "uncontrolled" RF environment over what time period?
A. 6 minutes
B. 10 minutes
C. 15 minutes
D. 30 minutes

N0C06
To determine compliance with the maximum permitted exposure (MPE) levels, safe exposure levels for RF energy are averaged for a "controlled" RF environment over what time period?
A. 6 minutes
B. 10 minutes
C. 15 minutes
D. 30 minutes

N0C06
(A)
Page 10-8

N0C07
What do the FCC RF radiation exposure regulations establish?
A. Maximum radiated field strength
B. Minimum permissible HF antenna height
C. Maximum permissible exposure limits
D. All of these choices are correct

N0C07
(C)
Page 10-8

N0C08
What do the FCC Rules specify for the maximum RF radiation field strength?
A. Amateur stations may not exceed an RF radiated field strength of 5 volts per meter
B. No station may transmit a signal that produces an RF radiated field strength greater than 10 amperes per meter
C. The maximum permissible power density from an amateur station antenna is 50 watts per square meter
D. The FCC Rules do not specify maximum RF radiation field strengths

N0C08
(D)
Page 10-8

N0C09
What is the averaging time to be considered for maximum permissible exposure (MPE) in controlled RF exposure environments?
A. 3 minutes
B. 6 minutes
C. 30 minutes
D. 60 minutes

N0C09
(B)
Page 10-8

N0C10
What is the averaging time to be considered for maximum permissible exposure (MPE) in uncontrolled RF exposure environments?
A. 3 minutes
B. 6 minutes
C. 30 minutes
D. 60 minutes

N0C10
(C)
Page 10-8

N0C11
Referring to Figure NT0-1, what is the maximum permissible exposure (MPE) limit for controlled environments on 3.7 MHz?
A. Equivalent far-field power density of 1.32 milliwatts per square centimeter
B. Equivalent far-field power density of 13.1 milliwatts per square centimeter
C. Equivalent far-field power density of 65.7 milliwatts per square centimeter
D. Equivalent far-field power density of 500 milliwatts per square centimeter

N0C11
(C)
Page 10-11

N0C12
Referring to Figure NT0-1, what is the maximum permissible exposure (MPE) limit for uncontrolled environments on 28.4 MHz?
A. Equivalent far-field power density of 0.077 milliwatts per square centimeter
B. Equivalent far-field power density of 0.22 milliwatts per square centimeter
C. Equivalent far-field power density of 1.1 milliwatts per square centimeter
D. Equivalent far-field power density of 29.1 milliwatts per square centimeter

N0C12
(B)
Page 10-11

N0C13
(B)
Page 10-11

N0C13
Referring to Figure NT0-1, what is the maximum permissible exposure (MPE) limit for uncontrolled environments on the 222-MHz Novice band?
A. 0.073 milliwatts per square centimeter
B. 0.2 milliwatts per square centimeter
C. 1 milliwatts per square centimeter
D. 27.5 milliwatts per square centimeter

N0C14
(C)
Page 10-11

N0C14
Referring to Figure NT0-1, what is the maximum permissible exposure (MPE) limit for controlled environments on the 1270-MHz Novice band?
A. 0.011 milliwatts per square centimeter
B. 0.85 milliwatts per square centimeter
C. 4.2 milliwatts per square centimeter
D. 100 milliwatts per square centimeter

N0C15
(A)
Page 10-13

N0C15
How does an Amateur Radio operator demonstrate that he or she has read and understood the FCC rules about RF-radiation exposure?
A. By indicating his or her understanding of this requirement on the Form 610 at the time of application
B. By posting a copy of Part 97 at the station
C. By completing an FCC Environmental Assessment Form
D. By completing an FCC Environmental Impact Statement

N0C16
(C)
Page 10-14

N0C16
What is the minimum safe distance for an uncontrolled RF radiation environment from a station using a half-wavelength dipole antenna on 3.5 MHz at 100 watts PEP, as specified in Table NT0-1?
A. 6 feet
B. 0.7 foot
C. 1.5 feet
D. 3 feet

N0C17
(A)
Page 10-14

N0C17
What is the minimum safe distance for a controlled RF radiation environment from a station using a quarter-wave vertical antenna on 28 MHz at 100 watts PEP, as specified in Table NT0-1?
A. 4.9 feet
B. 3.5 feet
C. 7 feet
D. 11 feet

N0C18
(A)
Page 10-14

N0C18
What is the minimum safe distance for a controlled RF radiation environment from a station using a half-wavelength dipole antenna on 7 MHz at 100 watts PEP, as specified in Table NT0-1?
A. 1.4 foot
B. 2 feet
C. 3.1 feet
D. 6.5 feet

Table NT0-1

RF Exposure Limites

Section B.

Estimated distances to meet RF power density guidelines with an omni-directional HF quarter-wave vertical or ground-plane antenna (estimated gain, 1 dBi). Calculations include the EPA ground reflection factor of 2.56.

Frequency: 7 MHz
Estimated antenna gain: 1 dBi
Controlled limit: 18.37 mW/cm²
Uncontrolled limit: 3.67 mW/cm²

Transmitter power (watts)	Distance to controlled limit	Distance to uncontrolled limit
100	1.2′	2.7′
500	2.7′	6.1′
1000	3.9′	8.7′
1500	4.7′	10.6′

Frequency: 21 MHz
Estimated antenna gain: 1 dBi
Controlled limit: 2.04 mW/cm²
Uncontrolled limit: 0.408 mW/cm²

Transmitter power (watts)	Distance to controlled limit	Distance to uncontrolled limit
100	3.7′	8.2′
500	8.2′	18.4′
1000	11.6′	26′
1500	14.2′	31.9′

Frequency: 28 MHz
Estimated antenna gain: 1 dBi
Controlled limit: 1.15 mW/cm²
Uncontrolled limit: 0.23 mW/cm²

Transmitter power (watts)	Distance to controlled limit	Distance to uncontrolled limit
100	4.9′	11′
500	11′	24.5′
1000	15.5′	34.7′
1500	19′	42.5′

Section C.

Estimated distances to meet RF power density guidelines with a horizontal half-wave dipole antenna (estimated gain, 2 dBi). Calculations include the EPA ground reflection factor of 2.56.

Frequency: 3.5 MHz
Estimated antenna gain: 2 dBi
Controlled limit: 73.5 mW/cm²
Uncontrolled limit: 14.7 mW/cm²

Transmitter power (watts)	Distance to controlled limit	Distance to uncontrolled limit
100	0.7′	1.5′
500	1.5′	3.4′
1000	2.2′	4.9′
1500	2.7′	6′

Frequency: 7 MHz
Estimated antenna gain: 2 dBi
Controlled limit: 18.37 mW/cm²
Uncontrolled limit: 3.67 mW/cm²

Transmitter power (watts)	Distance to controlled limit	Distance to uncontrolled limit
100	1.4′	3.1′
500	3.1′	6.9′
1000	4.3′	9.7′
1500	5.3′	11.9′

Frequency: 21 MHz
Estimated antenna gain: 2 dBi
Controlled limit: 2.04 mW/cm²
Uncontrolled limit: 0.408 mW/cm²

Transmitter power (watts)	Distance to controlled limit	Distance to uncontrolled limit
100	4.1′	9.2′
500	9.2′	20.6′
1000	13′	29.2′
1500	16′	35.7′

Frequency: 28 MHz
Estimated antenna gain: 2 dBi
Controlled limit: 1.15 mW/cm²
Uncontrolled limit: 0.23 mW/cm²

Transmitter power (watts)	Distance to controlled limit	Distance to uncontrolled limit
100	5.5′	12.3′
500	12.3′	27.5′
1000	17.4′	38.9′
1500	21.3′	47.7′

N0C19
(C)
Page 10-14

N0C19
What is the minimum safe distance for a controlled RF radiation environment from a station using a half-wavelength dipole antenna on 21 MHz at 100 watts PEP, as specified in Table NT0-1?
A. 1.5 feet
B. 2 feet
C. 4.1 feet
D. 9.2 feet

N0C20
(B)
Page 10-14

N0C20
What is the minimum safe distance for an uncontrolled RF radiation environment from a station using a half-wavelength dipole antenna on 21 MHz at 100 watts PEP, as specified in Table NT0-1?
A. 4.1 feet
B. 9.2 feet
C. 8 feet
D. 13 feet

N0C21
(C)
Page 10-14

N0C21
Using Table NT0-1 what is the minimum safe distance for a controlled RF radiation environment from a station using a half-wavelength dipole antenna on 3.5 MHz at 100 watts?
A. 6 feet
B. 1.5 feet
C. 0.7 foot
D. 3 feet

N0C22
(B)
Page 10-14

N0C22
Using Table NT0-1 what is the minimum safe distance for an uncontrolled RF radiation environment from a station using a quarter-wave vertical antenna on 7 MHz at 100 watts?
A. 4.0 feet
B. 2.7 feet
C. 1.2 feet
D. 7.5 feet

N0C23
(A)
Page 10-14

N0C23
What is the minimum safe distance for a controlled RF radiation environment from a station using a half-wavelength dipole antenna on 7 MHz at 100 watts PEP, as specified in Table NT0-1?
A. 1.4 foot
B. 3.1 feet
C. 4.5 feet
D. 6.5 feet

N0C24
(A)
Page 10-14

N0C24
Using Table NT0-1 what is the controlled limit for a station using a 21 MHz quarter-wave vertical at 100 watts?
A. 3.7 feet
B. 6 feet
C. 8.2 feet
D. 20 feet

N0C25
Using Table NT0-1 what is the uncontrolled limit for a station using a 21 MHz quarter-wave vertical at 100 watts?
A. 3.7 feet
B. 8.2 feet
C. 11.5 feet
D. 20 feet

N0C25
(B)
Page 10-14

N0C26
What is the minimum safe distance for an uncontrolled RF radiation environment from a station using a half-wavelength dipole antenna on 21 MHz at 100 watts PEP, as specified in Table NT0-1?
A. 2.5 feet
B. 9.2 feet
C. 4.1 feet
D. 13 feet

N0C26
(B)
Page 10-14

N0C27
Using Table NT0-1 what is the minimum safe distance for an uncontrolled RF radiation environment from a station using a 28 MHz half-wavelength dipole antenna at 100 watts?
A. 12.3 feet
B. 5.5 feet
C. 26.5 feet
D. 30 feet

N0C27
(A)
Page 10-14

N0D Routine station evaluation

N0D01
Which of the following antennas would (generally) create a stronger RF field on the ground beneath the antenna?
A. A horizontal loop at 30 meters above ground
B. A 3-element Yagi at 30 meters above ground
C. A 1/2 wave dipole antenna 5 meters above ground
D. A 3-element Quad at 30 meters above ground

N0D01
(C)
Page 10-9

N0D02
How does an amateur determine if his or her transmitted signal is within the RF radiation exposure guidelines?
A. By calling the FCC for a station inspection
B. By determining or analyzing transmitted field strength and power density
C. Compliance is determined by the transmitter manufacturer
D. By the use of a reflectometer and standing wave ratio (SWR) readings

N0D02
(B)
Page 10-11

N0D03
How may an amateur determine that his or her station complies with FCC RF-exposure regulations?
A. By calculation, based on FCC OET Bulletin No. 65
B. By calculation, based on computer modeling
C. By measurement, measuring the field strength using calibrated equipment
D. Any of these choices

N0D03
(D)
Page 10-8

N0D04

N0D04
This question has been withdrawn.

N0D05

N0D05
This question has been withdrawn.

N0D06
(D)
Page 10-13

N0D06
What must you do with the records of a routine RF radiation exposure evaluation?
A. They must be sent to the nearest FCC field office
B. They must be sent to the Environmental Protection Agency
C. They must be attached to each Form 610 when it is sent to the FCC for processing
D. Though not required, records may prove useful if the FCC asks for documentation to substantiate that an evaluation has been performed

N0D07
(A)
Page 10-12

N0D07
Which of the following instruments might you use to measure the RF radiation exposure levels in the vicinity of your station?
A. A calibrated field strength meter with a calibrated field strength sensor
B. A calibrated in-line wattmeter with a calibrated length of feed line
C. A calibrated RF impedance bridge
D. An amateur receiver with an S meter calibrated to National Bureau of Standards and Technology station WWV

N0D08
(D)
Page 10-12

N0D08
What factors can affect the accuracy of field strength measurements?
A. Interaction of the probe and measurement personnel with the near field
B. Frequency response of the test equipment and probes
C. Orientation of the probe with respect to the antenna polarity
D. All of these choices are correct

N0D09
(A)
Page 10-13

N0D09
What effect does the antenna gain and directivity have on a routine RF exposure evaluation?
A. Gain and directivity are part of the formulas used to perform calculations
B. The maximum permissible exposure (MPE) limits are directly proportional to antenna gain
C. The maximum permissible exposure (MPE) limits are inversely proportional to antenna directivity
D. All of these choices are correct

N0D10
What effect will nearby conductors such as telephone wiring or aluminum siding have on the field strength at any point near an antenna?
A. Conductors that are not part of the actual antenna will have no effect on the field strength
B. Conductors in the near field will interact with the field to add or subtract intensity, resulting in areas of varying field strength
C. Conductors in the near field will always interact with the field to increase the strength of the signal radiated from the antenna
D. Conductors in the near field will always interact with the field to decrease the strength of the signal radiated from the antenna

N0D10
(B)
Page 10-12

N0D11
As a general rule, what effect does antenna height above ground have on the RF exposure environment?
A. Power density is not related to antenna height or distance from the RF exposure environment
B. Antennas that are farther above ground produce higher maximum permissible exposures (MPE)
C. The higher the antenna the less the RF radiation exposure at ground level
D. RF radiation exposure is increased when the antenna is higher above ground

N0D11
(C)
Page 10-9

N0E Practical applications

N0E01
Which of the following steps is not helpful in reducing RF radiation exposure?
A. Reduce power
B. Adjust operating times or mode to produce a lower duty cycle
C. Locate the antenna more distant from areas of controlled and uncontrolled exposure
D. Install a low-pass filter in the antenna feed line

N0E01
(D)
Page 10-8

N0E02
Which of the following steps would help you to comply with RF-radiation exposure guidelines for uncontrolled RF environments?
A. Reduce transmitting times within a 6-minute period to reduce the station duty cycle
B. Operate only during periods of high solar absorption
C. Reduce transmitting times within a 30-minute period to reduce the station duty cycle
D. Operate only on high duty cycle modes

N0E02
(C)
Page 10-10

N0E03
Which of the following steps would help you to comply with RF-exposure guidelines for controlled RF environments?
A. Reduce transmitting times within a 30-minute period to reduce the station duty cycle
B. Operate only during periods of high solar absorption
C. Reduce transmitting times within a 6-minute period to reduce the station duty cycle
D. Operate only on high duty cycle modes

N0E03
(C)
Page 10-10

N0E04
Why should you make sure the antenna of a hand-held transceiver is not too close to your head when transmitting?
A. To help the antenna radiate energy equally in all directions
B. To reduce your exposure to the radio-frequency energy
C. To use your body to reflect the signal in one direction
D. To keep electrostatic charges from harming the operator

N0E04
(B)
Page 10-6

N0E05
(A)
Page 10-6

N0E05
What should you do for safety if you put up a UHF transmitting antenna?
A. Make sure the antenna will be in a place where no one can get near it when you are transmitting
B. Make sure that RF field screens are in place
C. Make sure the antenna is near the ground to keep its RF energy pointing in the correct direction
D. Make sure you connect an RF leakage filter at the antenna feed point

N0E06
(A)
Page 10-6

N0E06
How should you position the antenna of a hand-held transceiver while you are transmitting?
A. Away from your head and away from others
B. Towards the station you are contacting
C. Away from the station you are contacting
D. Down to bounce the signal off the ground

N0E07
(B)
Page 10-2

N0E07
Why should your antennas be located so that no one can touch them while you are transmitting?
A. Touching the antenna might cause television interference
B. Touching the antenna might cause RF burns
C. Touching the antenna might cause it to radiate harmonics
D. Touching the antenna might cause it to go into self-oscillation

N0E08
(D)
Page 10-2

N0E08
Why should you make sure that no one can touch an open-wire feed line while you are transmitting with it?
A. Because open-wire feed lines radiate large electric fields
B. Because the radiation from open-wire feed lines can cause body tissue cooling
C. Because contact might cause spurious emissions
D. Because high-voltage radio energy might burn the person

N0E09
(C)
Page 10-6

N0E09
For the least RF exposure, what is the best thing to do with your transmitting antennas?
A. Use vertical polarization
B. Use horizontal polarization
C. Mount the antennas where no one can come near them
D. Mount the antenna close to the ground

N0E10
(B)
Page 10-6

N0E10
To avoid excessively high human exposure to RF fields, how should amateur antennas generally be mounted?
A. With a high current point near ground
B. As far away from accessible areas as possible
C. On a nonmetallic mast
D. With the elements in a horizontal polarization

N0E11
(A)
Page 10-7

N0E11
For the least RF radiation exposure, what is the minimum height at which you should place your horizontal wire antenna?
A. High enough to ensure compliance with the FCC RF radiation exposure guidelines
B. As close to the ground as possible
C. Just high enough so you can easily reach it for adjustments or repairs
D. Above high-voltage electrical lines

N0E12
What action can amateur operators take to prevent exposure to RF radiation in excess of the FCC-specified limits?
A. Alter antenna patterns
B. Relocate antennas
C. Revise station technical parameters, such as frequency, power, or emission type
D. All of these choices are correct

N0E12
(D)
Page 10-8

N0E13
Which of the following radio frequency emissions will result in the least RF radiation exposure if they all have the same peak envelope power (PEP)?
A. Two-way exchanges of phase-modulated (PM) telephony
B. Two-way exchanges of frequency-modulated (FM) telephony
C. Two-way exchanges of single-sideband (SSB) telephony
D. Two-way exchanges of Morse code (CW) communication

N0E13
(C)
Page 10-10

N0E14
What is the minimum safe distance for an uncontrolled RF radiation environment from a station using a 3-element "triband" Yagi antenna on 21 MHz at 100 watts PEP, as specified in Table NT0-1?
A. 16.4 feet
B. 14.5 feet
C. 7.3 feet
D. 23 feet

N0E14
(A)
Page 10-14

Table NT0-1
RF Exposure Limits

Section A.

Estimated distances to meet RF power density guidelines in the main beam of a typical 3-element "triband" Yagi for the 14, 21 and 28 MHz Amateur Radio bands. Calculations include the EPA ground reflection factor of 2.56.

Frequency: 21 MHz
Antenna gain: 7 dBi
Controlled limit: 2.04 mW/cm^2
Uncontrolled limit: 0.408 mW/cm^2

Transmitter power (watts)	Distance to controlled limit	Distance to uncontrolled limit
100	7.3′	16.4′
500	16.4′	36.7′
1000	23.2′	51.9′
1500	28.4′	63.6′

Frequency: 28 MHz
Antenna gain: 8 dBi
Controlled limit: 1.15 mW/cm^2
Uncontrolled limit: 0.23 mW/cm^2

Transmitter power (watts)	Distance to controlled limit	Distance to uncontrolled limit
100	11′	24.5′
500	24.5′	54.9′
1000	34.7′	77.6′
1500	42.5′	95.1′

N0E15
(A)
Page 10-14

N0E15
What is the minimum safe distance for a controlled RF radiation environment from a station using a 3-element "triband" Yagi antenna on 21 MHz at 100 watts PEP, as specified in Table NT0-1?
A. 7.3 feet
B. 10 feet
C. 16.4 feet
D. 23 feet

N0E16
(B)
Page 10-14

N0E16
What is the minimum safe distance for a controlled RF radiation environment from a station using a 3-element "triband" Yagi antenna on 28 MHz at 100 watts PEP, as specified in Table NT0-1?
A. 15 feet
B. 11 feet
C. 24.5 feet
D. 18 feet

N0E17
(C)
Page 10-14

N0E17
What is the minimum safe distance for an uncontrolled RF radiation environment from a station using a 3-element "triband" Yagi antenna on 28 MHz at 100 watts PEP, as specified in Table NT0-1?
A. 7 feet
B. 11 feet
C. 24.5 feet
D. 34 feet

14 Technician (Element 3A) Question Pool —With Answers

The questions in this chapter will help you review your understanding of the material in the first 10 chapters of this book as you prepare for your exam.

Don't Start Here!

This chapter contains the complete question pool for the Element 3A exam. Element 3A is part of the Technician license exam. To earn a Technician license, you must also pass the Element 2 exam. The Element 2 question pool is printed in Chapter 13. The Technician license does not require a Morse code test. If you do pass the 5-wpm Morse code exam, however, you can earn a Technician Plus license.

Before you read the questions and answers printed in this chapter, be sure to read the text in Chapters 1 through 10. Use these questions as review exercises, when the text tells you to study them. Don't try to memorize all the

Table 14-1
Technician Exam Content

Subelement	*Topic*	*Number of Questions*
T1	Commission's Rules	5
T2	Operating Procedures	3
T3	Radio-Wave Propagation	3
T4	Amateur Radio Practices	4
T5	Electrical Principles	2
T6	Circuit Components	2
T7	Practical Circuits	1
T8	Signals and Emissions	2
T9	Antennas and Feed Lines	3
T0	RF Safety	5

questions and answers.

This question pool, released by the Volunteer Examiner Coordinators' Question Pool Committee, will be used on exams beginning July 1, 1997. The pool is scheduled to be used until June 30, 2001. Changes to FCC Rules and other factors may result in earlier revisions. Such changes will be announced in *QST* and other Amateur Radio publications. Normally, the Question Pool Committee will simply withdraw specific questions from the question pool when such Rules changes occur.

How Many Questions?

The FCC specifies that an Element 3A exam must include 30 questions. This question pool is divided into 10 sections, called subelements. (A subelement is a portion of the exam element, in this case Element 3A.) The FCC also specifies the number of questions from each subelement that must appear on your test. For example, there must be three questions from the Operating Procedures section, Subelement T2. There must also be three questions from the Antennas and Feed Lines section, Subelement T9. Five of the questions on your exam must come from the RF Safety section, Subelement T0. Table 14-1 summarizes the number of questions from each subelement that make up an Element 3A exam. The number of questions to be used from each subelement appears at the beginning of that subelement in the question pool, too.

The Volunteer Examiner Coordinators' Question Pool Committee has broken the subelements into smaller groups. There are the same number of groups as there are questions from each subelement, and the committee intends for one question to come from each of the smaller groups. This is not an FCC requirement, however, so your examiners may not select one question from each smaller group when they select questions for your exam.

There is a list of topics printed in bold type at the beginning of each small group. This list represents the syllabus, or study guide, topics for that section. The entire Technician syllabus is printed at the end of the Introduction chapter.

The small groups are listed alphabetically within each subelement. For example, since there are five exam questions from Subelement T1, these sections are labeled T1 through T1E. Two exam questions come from the Circuit Components subelement, so that subelement has sections labeled T6A and T6B.

The question numbers used in the question pool relate to the syllabus or study guide printed at the end of the Introduction chapter. The syllabus is an outline of topics covered by the exam. Each question number begins with a T. This indicates the question is from the Technician question pool. Next is a number to indicate which subelement the question is from. These numbers will range from 1 to 0. Following this number is a letter to indicate which group the question is from in that subelement. Each question number ends with a two-digit number to specify its position in the set. So question number T2A01 is the first question in the A group of the second subelement. Question T9C08 is the eighth question in the C group of the ninth subelement.

Who Picks the Questions?

The FCC allows Volunteer Examiner Teams to select the questions that will be used on amateur exams. If your test is coordinated by the ARRL/VEC, your test will be prepared by the VEC, or using a computer program supplied by the VEC. All VECs and examiners must use the questions, answers and distracters (incorrect answers) printed here.

This question pool contains more than 10 times the number of questions necessary to make up an exam. This ensures that the examiners have sufficient questions to choose from when they make up an exam.

Who Gives the Test?

All Amateur Radio license exams are given by teams of three or more Volunteer Examiners (VEs). Each of the examiners is accredited by a Volunteer Examiner Coordinator (VEC) to give exams under their program. A VEC is an organization that has entered into an agreement with the FCC to coordinate the efforts of VEs. The VEC reviews the paperwork for each exam session, and then forwards the information to the FCC.

Question Pool Format

The rest of this chapter contains the entire Element 3A question pool. We have printed the answer key to these questions along the edge of the page. There is a line to indicate where you should fold the page under to hide the answer key while you study. After making your best effort to answer the questions, you can look at the answers to check your understanding. We also have included page references along with the answers. These page numbers indicate where you will find the text discussion related to each question. If you have any problems with a question, refer to the page listed for that question. You may have to study beyond the listed page number to review all the related material. We also have included references to sections of Part 97 for the Commission's Rules Subelement, T1. This is to help you identify the particular rule citations for these questions. The complete text of the FCC Rules in Part 97 is included in the ARRL publication, *The FCC Rule Book*.

The Question Pool Committee included all the drawings for the question pool on one page, with two additional pages for tables related to the RF radiation safety questions in subelement T0. We placed a copy of these pages at the beginning of the question pool, and also placed the individual figures with the questions that reference them.

Good luck with your studies.

Element 3A (Technician)

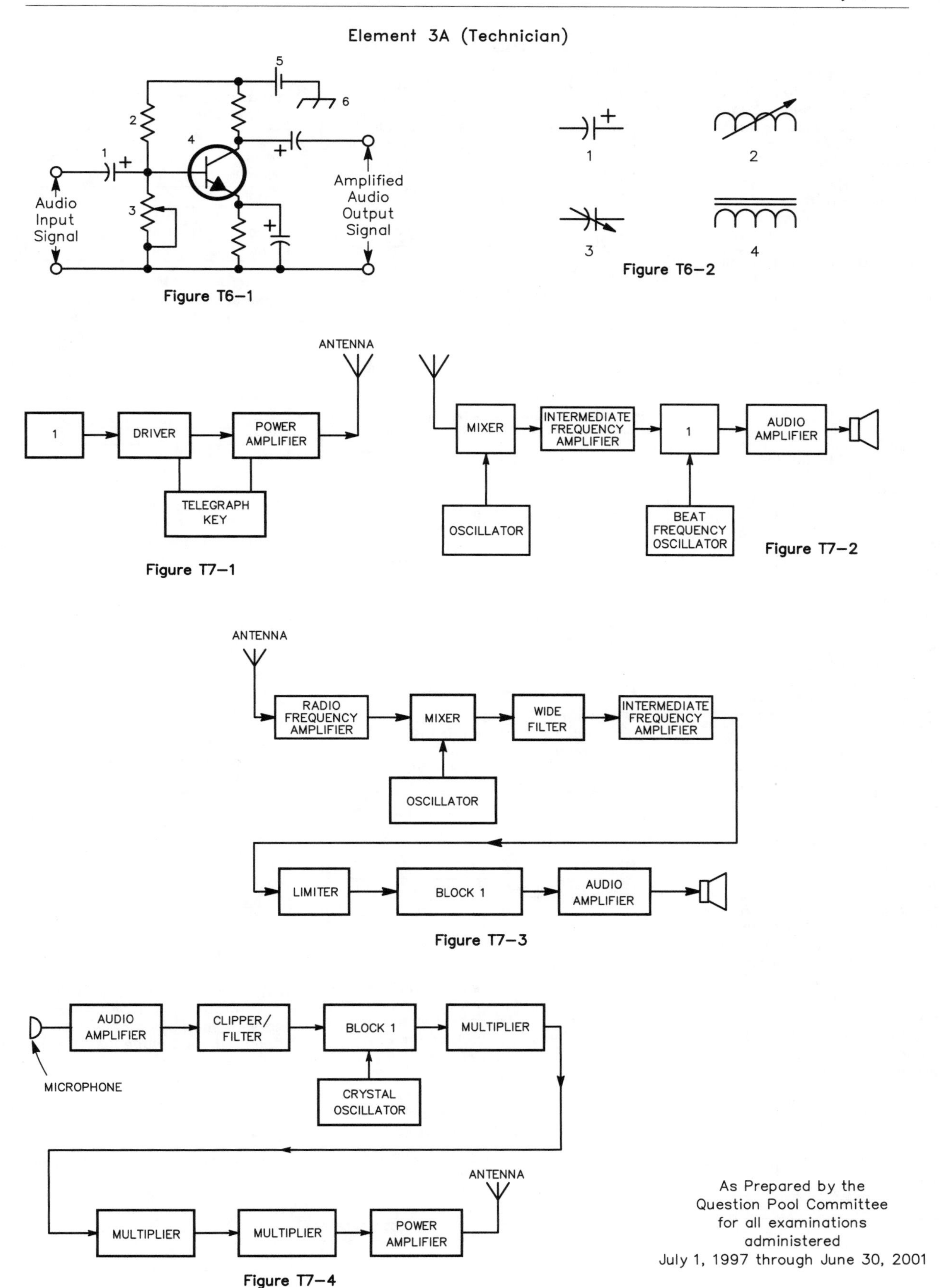

Figure T6–1

Figure T6–2

Figure T7–1

Figure T7–2

Figure T7–3

Figure T7–4

As Prepared by the
Question Pool Committee
for all examinations
administered
July 1, 1997 through June 30, 2001

Table NT0-1

RF Exposure Limits

Section A.

Estimated distances to meet RF power density guidelines in the main beam of a typical 3-element "triband" Yagi for the 14, 21 and 28 MHz Amateur Radio bands. Calculations include the EPA ground reflection factor of 2.56.

Frequency: 14 MHz
Antenna gain: 6.5 dBi
Controlled limit: 4.59 mW/cm²
Uncontrolled limit: 0.92 mW/cm²

Transmitter power (watts)	Distance to controlled limit	Distance to uncontrolled limit
100	4.6′	10.3′
500	10.3′	23.1′
1000	14.6′	32.7′
1500	17.9′	40′

Frequency: 21 MHz
Antenna gain: 7 dBi
Controlled limit: 2.04 mW/cm²
Uncontrolled limit: 0.408 mW/cm²

Transmitter power (watts)	Distance to controlled limit	Distance to uncontrolled limit
100	7.3′	16.4′
500	16.4′	36.7′
1000	23.2′	51.9′
1500	28.4′	63.6′

Frequency: 28 MHz
Antenna gain: 8 dBi
Controlled limit: 1.15 mW/cm²
Uncontrolled limit: 0.23 mW/cm²

Transmitter power (watts)	Distance to controlled limit	Distance to uncontrolled limit
100	11′	24.5′
500	24.5′	54.9′
1000	34.7′	77.6′
1500	42.5′	95.1′

Section B.

Estimated distances to meet RF power density guidelines with an omni-directional HF quarter-wave vertical or ground-plane antenna (estimated gain, 1 dBi). Calculations include the EPA ground reflection factor of 2.56.

Frequency: 3.5 MHz
Estimated antenna gain: 1 dBi
Controlled limit: 73.5 mW/cm²
Uncontrolled limit: 14.7 mW/cm²

Transmitter power (watts)	Distance to controlled limit	Distance to uncontrolled limit
100	0.6′	1.4′
500	1.4′	3.1′
1000	1.9′	4.3′
1500	2.4′	5.3′

Frequency: 7 MHz
Estimated antenna gain: 1 dBi
Controlled limit: 18.37 mW/cm²
Uncontrolled limit: 3.67 mW/cm²

Transmitter power (watts)	Distance to controlled limit	Distance to uncontrolled limit
100	1.2′	2.7′
500	2.7′	6.1′
1000	3.9′	8.7′
1500	4.7′	10.6′

Frequency: 14 MHz
Estimated antenna gain: 1 dBi
Controlled limit: 4.59 mW/cm²
Uncontrolled limit: 0.918 mW/cm²

Transmitter power (watts)	Distance to controlled limit	Distance to uncontrolled limit
100	2.5′	5.5′
500	5.5′	12.3′
1000	7.8′	17.3′
1500	9.5′	21.2′

Frequency: 21 MHz
Estimated antenna gain: 1 dBi
Controlled limit: 2.04 mW/cm²
Uncontrolled limit: 0.408 mW/cm²

Transmitter power (watts)	Distance to controlled limit	Distance to uncontrolled limit
100	3.7′	8.2′
500	8.2′	18.4′
1000	11.6′	26′
1500	14.2′	31.9′

Frequency: 28 MHz
Estimated antenna gain: 1 dBi
Controlled limit: 1.15 mW/cm²
Uncontrolled limit: 0.23 mW/cm²

Transmitter power (watts)	Distance to controlled limit	Distance to uncontrolled limit
100	4.9′	11′
500	11′	24.5′
1000	15.5′	34.7′
1500	19′	42.5′

Section C.

Estimated distances to meet RF power density guidelines with a horizontal half-wave dipole antenna (estimated gain, 2 dBi). Calculations include the EPA ground reflection factor of 2.56.

Frequency: 3.5 MHz
Estimated antenna gain: 2 dBi
Controlled limit: 73.5 mW/cm²
Uncontrolled limit: 14.7 mW/cm²

Transmitter power (watts)	Distance to controlled limit	Distance to uncontrolled limit
100	0.7′	1.5′
500	1.5′	3.4′
1000	2.2′	4.9′
1500	2.7′	6′

Frequency: 7 MHz
Estimated antenna gain: 2 dBi
Controlled limit: 18.37 mW/cm²
Uncontrolled limit: 3.67 mW/cm²

Transmitter power (watts)	Distance to controlled limit	Distance to uncontrolled limit
100	1.4′	3.1′
500	3.1′	6.9′
1000	4.3′	9.7′
1500	5.3′	11.9′

Frequency: 14 MHz
Estimated antenna gain: 2 dBi
Controlled limit: 4.59 mW/cm²
Uncontrolled limit: 0.918 mW/cm²

Transmitter power (watts)	Distance to controlled limit	Distance to uncontrolled limit
100	2.8′	6.2′
500	6.2′	13.8′
1000	8.7′	19.5′
1500	10.7′	23.8′

Frequency: 21 MHz
Estimated antenna gain: 2 dBi
Controlled limit: 2.04 mW/cm²
Uncontrolled limit: 0.408 mW/cm²

Transmitter power (watts)	Distance to controlled limit	Distance to uncontrolled limit
100	4.1′	9.2′
500	9.2′	20.6′
1000	13′	29.2′
1500	16′	35.7′

Frequency: 28 MHz
Estimated antenna gain: 2 dBi
Controlled limit: 1.15 mW/cm²
Uncontrolled limit: 0.23 mW/cm²

Transmitter power (watts)	Distance to controlled limit	Distance to uncontrolled limit
100	5.5′	12.3′
500	12.3′	27.5′
1000	17.4′	38.9′
1500	21.3′	47.7′

Section D.

Estimated distances to meet RF power density guidelines with a VHF quarter-wave ground-plane or mobile whip antenna (estimated gain, 1 dBi). Calculations include the EPA ground reflection factor of 2.56.

Frequency: 146 MHz
Estimated antenna gain: 1 dBi
Controlled limit: 1 mW/cm²
Uncontrolled limit: 0.2 mW/cm²

Transmitter power (watts)	Distance to controlled limit	Distance to uncontrolled limit
10	1.7′	3.7′
50	3.7′	8.3′
150	6.4′	14.4′

Section E.

Estimated distances to meet RF power density guidelines in the main beam of a UHF 5/8-wavelength ground-plane or mobile whip antenna (estimated gain, 4 dBi). Calculations include the EPA ground reflection factor of 2.56.

Frequency: 446 MHz
Estimated antenna gain: 4 dBi
Controlled limit: 1.49 mW/cm²
Uncontrolled limit: 0.3 mW/cm²

Transmitter power (watts)	Distance to controlled limit	Distance to uncontrolled limit
10	1.9′	4.3′
50	4.3′	9.6′
150	7.5′	16.7′

Section F.

Estimated distances to meet RF power density guidelines in the main beam of a 17-element Yagi on a five-wavelength boom designed for weak signal communications on the 144 MHz Amateur Radio band (estimated gain, 16.8 dBi). Calculations include the EPA ground reflection factor of 2.56.

Frequency: 144 MHz
Estimated antenna gain: 16.8 dBi
Controlled limit: 1 mW/cm²
Uncontrolled limit: 0.2 mW/cm²

Transmitter power (watts)	Distance to controlled limit	Distance to uncontrolled limit
10	10.2′	22.9′
100	32.4′	72.4′
500	72.4′	162′
1500	125.5′	280.6′

Section G.

Estimated distances to meet RF power density guidelines in the main beam of an array of eight 17-element Yagis with five-wavelength booms designed for earth-moon-earth ("moonbounce") communications on the 144 MHz Amateur Radio band (estimated gain, 24 dBi). Calculations include the EPA ground reflection factor of 2.56.

Frequency: 144 MHz
Estimated antenna gain: 24 dBi
Controlled limit: 1 mW/cm²
Uncontrolled limit: 0.2 mW/cm²

Transmitter power (watts)	Distance to controlled limit	Distance to uncontrolled limit
150	90.9′	203.3′
500	166′	371.1′
1500	287.4′	642.7′

RF Exposure Limits

(A) Limits for Occupational/Controlled Exposure

Frequency Range (MHz)	Electric Field Strength (V/m)	Magnetic Field (A/m)	Power Density (mW/cm²)	Averaging Time (minutes)
0.3-3.0	614	1.63	(100)*	6
3.0-30	1842/f	4.89/f	(900/f²)*	6
30-300	61.4	0.163	1.0	6
300-1500	——	——	f/300	6
1500-100,000	——	——	5	6
f = frequency in MHz		* = Plane-wave equivalent power density		

(B) Limits for General Population/Uncontrolled Exposure

Frequency Range (MHz)	Electric Field Strength (V/m)	Magnetic Field (A/m)	Power Density (mW/cm²)	Averaging Time (minutes)
0.3-1.34	614	1.63	(100)*	30
1.34-30	824/f	2.19/f	(180/f²)*	30
30-300	27.5	0.073	0.2	30
300-1500	——	——	f/1500	30
1500-100,000	——	——	1.0	30
f = frequency in MHz		* = Plane-wave equivalent power density		

Figure NT0-1

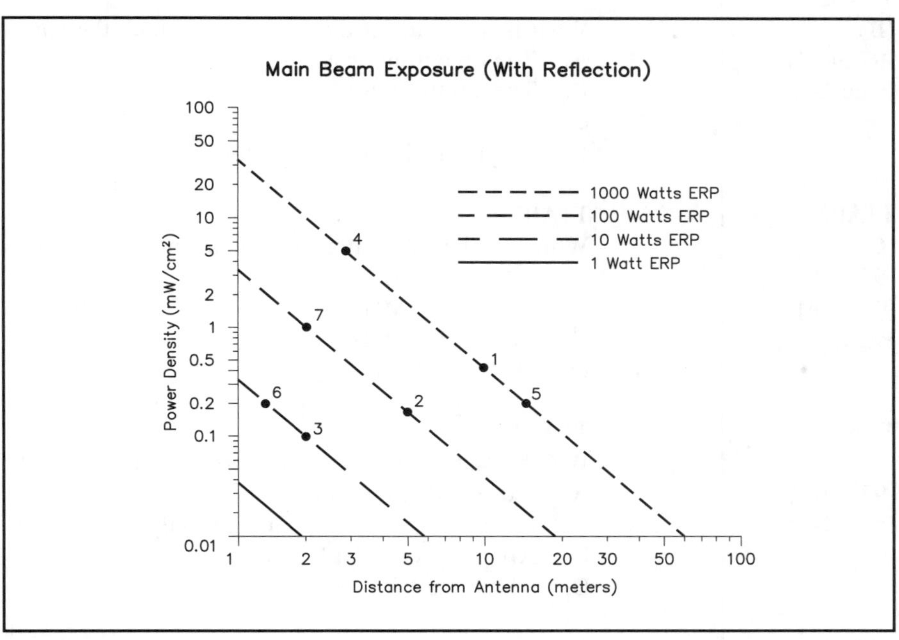

Figure NT0-2

As prepared by the Question Pool Committee for all examinations administered July 1, 1997 through June 30, 2001

Element 3A (Technician Class) Question Pool

Answer Key

Page Numbers tell you where to look in this book for more information.

Subelement T1 — Commission's Rules

Subelement T1

Numbers in [square brackets] indicate sections in FCC Part 97, the Amateur Radio Rules.

T1A Station control; frequency privileges authorized to the Technician and Technician Plus class control operator; term of licenses, grace periods and modifications of licenses

T1A01
(D)
[97.3a13]
Page 1-7

T1A01
What is the control point of an amateur station?
A. The on/off switch of the transmitter
B. The input/output port of a packet controller
C. The variable frequency oscillator of a transmitter
D. The location at which the control operator function is performed

T1A02
(B)
[97.3a13]
Page 1-7

T1A02
What is the term for the location at which the control operator function is performed?
A. The operating desk
B. The control point
C. The station location
D. The manual control location

T1A03
(C)
[97.301, 97.305e]
Page 1-24

T1A03
Which of the following frequencies may a Technician Plus operator use?
A. 7.1 - 7.2 MHz
B. 14.1 - 14.2 MHz
C. 21.1 - 21.2 MHz
D. 28.1 - 29.2 MHz

T1A04
(C)
[97.301a]
Page 1-25

T1A04
Which amateur licenses authorize privileges on 52.525 MHz?
A. Extra and Advanced only
B. Extra, Advanced and General only
C. All classes except Novice
D. All classes

T1A05
(B)
[97.301a]
Page 1-25

T1A05
Which amateur licenses authorize privileges on 146.52 MHz?
A. All classes
B. All classes except Novice
C. Extra, Advanced and General only
D. Extra and Advanced only

T1A06
Which amateur licenses authorize privileges on 223.50 MHz?
A. All classes
B. Extra, Advanced, General and Technician only
C. Extra, Advanced and General only
D. Extra and Advanced only

T1A06
(A)
[97.301a,f]
Page 1-25

T1A07
Which amateur licenses authorize privileges on 446.0 MHz?
A. All classes
B. All classes except Novice
C. Extra, Advanced and General only
D. Extra and Advanced only

T1A07
(B)
[97.301a]
Page 1-25

T1A08
In addition to passing both the Novice and Technician written examinations [Elements 2 and 3A], what else must you do before you are allowed to use the amateur bands below 30 MHz?
A. Pass the General class theory test
B. Notify the FCC that you intend to operate on the HF bands
C. Attend a class to learn about HF communications
D. Pass a Morse code test at a minimum speed of 5 WPM

T1A08
(D)
[97.301e]
Page 1-23

T1A09
If you are a Technician licensee awaiting the grant of your Technician Plus license, what must you have to prove that you are authorized to use the Novice amateur frequencies below 30 MHz?
A. A certificate from the FCC showing that you have notified them that you will be using the HF bands
B. A certificate showing that you have attended a class in HF communications
C. A Certificate of Successful Completion of Examination showing that you have passed a Morse code test
D. No special proof is required

T1A09
(C)
[97.9b]
Page 1-24

T1A10
What is the normal term for which a new amateur station license is granted?
A. 5 years
B. 7 years
C. 10 years
D. For the lifetime of the licensee

T1A10
(C)
[97.25a]
Page 1-6

T1A11
What is the "grace period" during which the FCC will renew an expired 10-year license?
A. 2 years
B. 5 years
C. 10 years
D. There is no grace period

T1A11
(A)
[97.21b]
Page 1-6

T1A12
What can you do to renew or change your operator/primary station license?
A. Properly fill out FCC Form 610 and send it to the FCC in Gettysburg, PA or a VEC who will file it electronically
B. Properly fill out FCC Form 610 and mail or fax it to the nearest FCC field office
C. Properly fill out FCC Form 610 and send it to the FCC in Washington, DC or e-mail the information to that office
D. Nothing; an amateur license never needs changing or renewing

T1A12
(A)
[97.21a3]
Page 1-6

T1A13
(C)
[97.27a1]
Page 1-26

T1A13
Under what conditions, if any, may the FCC modify an amateur license?
A. None; only the US Congress has this authority
B. Whenever it so desires
C. Whenever such action will promote the public interest, convenience, and necessity
D. Only when a state of emergency exists

T1B Emission privileges for Technician and Technician Plus class control operator; frequency selection and sharing; transmitter power

T1B01
(D)
[97.305c]
Page 1-24

T1B01
On what HF band may a Technician Plus licensee use FM phone emission?
A. 10 meters
B. 15 meters
C. 75 meters
D. None

T1B02
(C)
[97.301e]
Page 1-24

T1B02
What additional privileges are available to a Technician who upgrades to Technician Plus?
A. Only CW in the 3.675 - 3.725 MHz frequency band
B. All privileges in the 1.8 - 2.0 MHz frequency band
C. All HF privileges available to Novice operators
D. All privileges in the 28.0 - 29.7 MHz frequency band

T1B03
(B)
[97.305c]
Page 1-25

T1B03
On what frequencies within the 6-meter band may phone emissions be transmitted?
A. 50.0 - 54.0 MHz only
B. 50.1 - 54.0 MHz only
C. 51.0 - 54.0 MHz only
D. 52.0 - 54.0 MHz only

T1B04
(A)
[97.305c]
Page 1-25

T1B04
On what frequencies within the 2-meter band may image emissions be transmitted?
A. 144.1 - 148.0 MHz only
B. 146.0 - 148.0 MHz only
C. 144.0 - 148.0 MHz only
D. 146.0 - 147.0 MHz only

T1B05
(D)
[97.301a,
97.305c]
Page 1-25

T1B05
What frequencies within the 2-meter band are reserved exclusively for CW operations?
A. 146 - 147 MHz
B. 146.0 - 146.1 MHz
C. 145 - 148 MHz
D. 144.0 - 144.1 MHz

T1B06
(C)
[97.303]
Page 1-26

T1B06
If the FCC rules say that the amateur service is a secondary user of a frequency band, and another service is a primary user, what does this mean?
A. Nothing special; all users of a frequency band have equal rights to operate
B. Amateurs are only allowed to use the frequency band during emergencies
C. Amateurs are allowed to use the frequency band only if they do not cause harmful interference to primary users
D. Amateurs must increase transmitter power to overcome any interference caused by primary users

T1B07
If you are using a frequency within a band assigned to the amateur service on a secondary basis, and a station assigned to the primary service on that band causes interference, what action should you take?
A. Notify the FCC's regional Engineer in Charge of the interference
B. Increase your transmitter's power to overcome the interference
C. Attempt to contact the station and request that it stop the interference
D. Change frequencies; you may be causing harmful interference to the other station, in violation of FCC rules

T1B07
(D)
[97.303]
Page 1-26

T1B08
What rule applies if two amateur stations want to use the same frequency?
A. The station operator with a lesser class of license must yield the frequency to a higher-class licensee
B. The station operator with a lower power output must yield the frequency to the station with a higher power output
C. Both station operators have an equal right to operate on the frequency
D. Station operators in ITU Regions 1 and 3 must yield the frequency to stations in ITU Region 2

T1B08
(C)
[97.101b]
Page 1-26

T1B09
If a repeater is causing harmful interference to another repeater and a frequency coordinator has recommended the operation of one station only, who is responsible for resolving the interference?
A. The licensee of the unrecommended repeater
B. Both repeater licensees
C. The licensee of the recommended repeater
D. The frequency coordinator

T1B09
(A)
[97.205c]
Page 1-27

T1B10
If a repeater is causing harmful interference to another amateur repeater and a frequency coordinator has recommended the operation of both stations, who is responsible for resolving the interference?
A. The licensee of the repeater that has been recommended for the longest period of time
B. The licensee of the repeater that has been recommended the most recently
C. The frequency coordinator
D. Both repeater licensees

T1B10
(D)
[97.205c]
Page 1-27

T1B11
If a repeater is causing harmful interference to another repeater and a frequency coordinator has NOT recommended either station, who is primarily responsible for resolving the interference?
A. Both repeater licensees
B. The licensee of the repeater that has been in operation for the longest period of time
C. The licensee of the repeater that has been in operation for the shortest period of time
D. The frequency coordinator

T1B11
(A)
[97.205c]
Page 1-26

T1B12
What is the term for the average power supplied to an antenna transmission line during one RF cycle at the crest of the modulation envelope?
A. Peak transmitter power
B. Peak output power
C. Average radio-frequency power
D. Peak envelope power

T1B12
(D)
[97.3b6]
Page 1-25

T1B13
(D)
[97.313b]
Page 1-25

T1B13
What is the maximum transmitting power permitted an amateur station on 146.52 MHz?
A. 200 watts PEP output
B. 500 watts ERP
C. 1000 watts DC input
D. 1500 watts PEP output

T1C Digital communications, station identification, ID with authorization of Certificate of Successful Completion of Examination

T1C01
(C)
[97.307f3,4]
Page 1-26

T1C01
What is the maximum frequency shift permitted for RTTY or data transmissions below 50 MHz?
A. 0.1 kHz
B. 0.5 kHz
C. 1 kHz
D. 5 kHz

T1C02
(D)
[97.307]
Page 1-26

T1C02
What is the maximum frequency shift permitted for RTTY or data transmissions above 50 MHz?
A. 0.1 kHz or the sending speed in bauds, whichever is greater
B. 0.5 kHz or the sending speed in bauds, whichever is greater
C. 5 kHz or the sending speed in bauds, whichever is greater
D. The FCC rules do not specify a maximum frequency shift above 50 MHz

T1C03
(B)
[97.307f4]
Page 1-26

T1C03
What is the maximum symbol rate permitted for packet transmissions on the 10-meter band?
A. 300 bauds
B. 1200 bauds
C. 19.6 kilobauds
D. 56 kilobauds

T1C04
(C)
[97.307f5]
Page 1-26

T1C04
What is the maximum symbol rate permitted for packet transmissions on the 2-meter band?
A. 300 bauds
B. 1200 bauds
C. 19.6 kilobauds
D. 56 kilobauds

T1C05
(C)
[97.307f4]
Page 1-26

T1C05
What is the maximum symbol rate permitted for RTTY or data transmissions on the 10-meter band?
A. 56 kilobauds
B. 19.6 kilobauds
C. 1200 bauds
D. 300 bauds

T1C06
(B)
[97.307f5]
Page 1-26

T1C06
What is the maximum symbol rate permitted for RTTY or data transmissions on the 6- and 2-meter bands?
A. 56 kilobauds
B. 19.6 kilobauds
C. 1200 bauds
D. 300 bauds

T1C07
What is the maximum authorized bandwidth of RTTY, data or multiplexed emissions using an unspecified digital code on the 6- and 2-meter bands?
A. 20 kHz
B. 50 kHz
C. The total bandwidth shall not exceed that of a single-sideband phone emission
D. The total bandwidth shall not exceed 10 times that of a CW emission

T1C07
(A)
[97.307f5]
Page 1-26

T1C08
What is the maximum symbol rate permitted for RTTY or data transmissions above 222 MHz?
A. 300 bauds
B. 1200 bauds
C. 19.6 kilobauds
D. 56 kilobauds

T1C08
(D)
[97.307f6]
Page 1-26

T1C09
On what exclusive frequency band may packet network relays operate on a secondary basis (with specific permission)?
A. 50 - 51 MHz
B. 146 - 147 MHz
C. 219 - 220 MHz
D. 440 - 450 MHz

T1C09
(C)
[97.301a, 97.303e3]
Page 1-26

T1C10
What is the maximum output power permitted for digital network relays on 219-220 MHz?
A. 25 W PEP
B. 50 W PEP
C. 100 W PEP
D. 1500 W PEP

T1C10
(B)
[97.313h]
Page 1-26

T1C11
What license class must be held by the control operator of a station communicating through an amateur satellite?
A. Extra or Advanced
B. Any class except Novice
C. Any class
D. Technician with satellite endorsement

T1C11
(C)
[97.209a]
Page 1-22

T1C12
What emission type may always be used for station identification, regardless of the transmitting frequency?
A. CW
B. RTTY
C. MCW
D. Phone

T1C12
(A)
[97.305a]
Page 1-29

T1C13
What is the fastest code speed a repeater may use for automatic identification?
A. 13 words per minute
B. 20 words per minute
C. 30 words per minute
D. There is no limitation

T1C13
(B)
[97.119b1]
Page 1-29

T1C14
(B)
[97.119a]
Page 1-29

T1C14
How often must a Technician class operator identify his or her station when operating simplex FM phone from an automobile?
A. Once every 15 minutes
B. At least every ten minutes, and at the end of each communication
C. At the beginning and end of each transmission
D. Once every 30 minutes

T1C15
(A)
[97.119e1]
Page 1-29

T1C15
If you are a Novice licensee with a Certificate of Successful Completion of Examination (CSCE) for Technician Plus privileges, how should you identify your station when transmitting on 146.34 MHz?
A. You must give your call sign, followed by any suitable word that denotes the slant mark and the identifier "KT"
B. You may not operate on 146.34 MHz until your new license arrives
C. No special form of identification is needed
D. You must give your call sign and the location of the VE examination where you obtained the CSCE

T1C16
(C)
[97.119e]
Page 1-29

T1C16
If you are a Technician licensee with a Certificate of Successful Completion of Examination (CSCE) for Technician Plus privileges, how should you identify your station when transmitting on 28.4 MHz?
A. You must give your call sign followed by the words "plus plus"
B. You must give your call sign followed by the words "temporary plus"
C. No special form of identification is needed
D. You must give your call sign and the location of the VE examination where you obtained the CSCE

T1D Correct language, phonetics, beacons and radio control of model craft and vehicles

T1D01
(C)
[97.119b2]
Page 1-29

T1D01
If you are using a language besides English to make a contact, what language must you use when identifying your station?
A. The language being used for the contact
B. The language being used for the contact, provided the US has a third-party communications agreement with that country
C. English
D. Any language of a country that is a member of the International Telecommunication Union

T1D02
(B)
[97.119b2]
Page 1-29

T1D02
Which language, besides English, may you use for amateur communications?
A. Any language, provided you identify your station in both English and French
B. Any language, provided you identify your station in English
C. Only German, Spanish, French or Japanese
D. Only languages common within your ITU region

T1D03
(C)
[97.119b2]
Page 1-29

T1D03
What do the FCC Rules suggest you use as an aid for correct station identification when using phone?
A. A speech compressor
B. Q signals
C. A phonetic alphabet
D. Unique words of your choice

T1D04
What is the advantage in using the International Telecommunication Union (ITU) phonetic alphabet when identifying your station?
A. The words are internationally recognized substitutes for letters
B. There is no advantage
C. The words have been chosen to be easily pronounced by Asian cultures
D. It preserves traditions begun in the early days of Amateur Radio

T1D04
(A)
[97.119b2]
Page 1-29

T1D05
What is one reason to avoid using "cute" phrases or word combinations to identify your station?
A. They are not easily understood by non-English-speaking amateurs
B. They might offend English-speaking amateurs
C. They do not meet FCC identification requirements
D. They might be interpreted as codes or ciphers intended to obscure the meaning of your identification

T1D05
(A)
[97.119b2]
Page 1-29

T1D06
What is an amateur station called that transmits communications for the purpose of observation of propagation and reception?
A. A beacon
B. A repeater
C. An auxiliary station
D. A radio control station

T1D06
(A)
[97.3a9]
Page 1-28

T1D07
What is the maximum transmitting power permitted an amateur station in beacon operation?
A. 10 watts PEP output
B. 100 watts PEP output
C. 500 watts PEP output
D. 1500 watts PEP output

T1D07
(B)
[97.203c]
Page 1-28

T1D08
What minimum class of amateur license must you hold to operate a beacon or a repeater station?
A. Novice
B. Technician
C. General
D. Amateur Extra

T1D08
(B)
[97.205a]
Page 1-28

T1D09
What minimum information must be on a label affixed to a transmitter used for telecommand (control) of model craft?
A. Station call sign
B. Station call sign and the station licensee's name
C. Station call sign and the station licensee's name and address
D. Station call sign and the station licensee's class of license

T1D09
(C)
[97.215a]
Page 1-18

T1D10
What are the station identification requirements for an amateur transmitter used for telecommand (control) of model craft?
A. Once every ten minutes
B. Once every ten minutes, and at the beginning and end of each transmission
C. At the beginning and end of each transmission
D. Station identification is not required if the transmitter is labeled with the station licensee's name, address and call sign

T1D10
(D)
[97.215a]
Page 1-18

T1D11
(B)
[97.215c]
Page 1-18

T1D11
What is the maximum transmitter power an amateur station is allowed when used for telecommand (control) of model craft?
A. One milliwatt
B. One watt
C. 25 watts
D. 100 watts

T1E Emergency communications; broadcasting; permissible one-way, satellite and third-party communication; indecent and obscene language

T1E01
(A)
[97.401a]
Page 1-27

T1E01
If a disaster disrupts normal communication systems in an area where the amateur service is regulated by the FCC, what kinds of transmissions may stations make?
A. Those that are necessary to meet essential communication needs and facilitate relief actions
B. Those that allow a commercial business to continue to operate in the affected area
C. Those for which material compensation has been paid to the amateur operator for delivery into the affected area
D. Those that are to be used for program production or news gathering for broadcasting purposes

T1E02
(C)
[97.401c]
Page 1-28

T1E02
What information is included in an FCC declaration of a temporary state of communication emergency?
A. A list of organizations authorized to use radio communications in the affected area
B. A list of amateur frequency bands to be used in the affected area
C. Any special conditions and special rules to be observed during the emergency
D. An operating schedule for authorized amateur emergency stations

T1E03
(A)
[97.3a10]
Page 1-18

T1E03
What is meant by the term broadcasting?
A. Transmissions intended for reception by the general public, either direct or relayed
B. Retransmission by automatic means of programs or signals from non-amateur stations
C. One-way radio communications, regardless of purpose or content
D. One-way or two-way radio communications between two or more stations

T1E04
(B)
[97.3a10, 97.113b]
Page 1-28

T1E04
Which of the following one-way communications may not be transmitted in the amateur service?
A. Telecommands to model craft
B. Broadcasts intended for the general public
C. Brief transmissions to make adjustments to the station
D. Morse code practice

T1E05
(A)
[97.209b2]
Page 1-22

T1E05
Which band may NOT be used by Earth stations for satellite communications?
A. 6 meters
B. 2 meters
C. 70 centimeters
D. 23 centimeters

T1E06
If you wanted to use your amateur station to retransmit communications between a space shuttle and its associated Earth stations, what agency must first give its approval?
A. The FCC in Washington, DC
B. The office of your local FCC Engineer In Charge (EIC)
C. The National Aeronautics and Space Administration (NASA)
D. The Department of Defense (DOD)

T1E06
(C)
[97.113e]
Page 1-21

T1E07
What kind of payment is allowed for third-party messages sent by an amateur station?
A. Any amount agreed upon in advance
B. Donation of repairs to amateur equipment
C. Donation of amateur equipment
D. No payment of any kind is allowed

T1E07
(D)
[97.113a2]
Page 1-19

T1E08
When are third-party messages allowed to be sent to a foreign country?
A. When sent by agreement of both control operators
B. When the third party speaks to a relative
C. They are not allowed under any circumstances
D. When the US has a third-party agreement with the foreign country or the third party is qualified to be a control operator

T1E08
(D)
[97.115a2]
Page 1-19

T1E09
If you let an unlicensed third party use your amateur station, what must you do at your station's control point?
A. You must continuously monitor and supervise the third-party's participation
B. You must monitor and supervise the communication only if contacts are made in countries that have no third-party communications agreement with the US
C. You must monitor and supervise the communication only if contacts are made on frequencies below 30 MHz
D. You must key the transmitter and make the station identification

T1E09
(A)
[97.115b1]
Page 1-20

T1E10
When may you send obscene words from your amateur station?
A. Only when they do not cause interference to other communications
B. Never; obscene words are not allowed in amateur transmissions
C. Only when they are not retransmitted through a repeater
D. Any time, but there is an unwritten rule among amateurs that they should not be used on the air

T1E10
(B)
[97.113a4]
Page 1-21

T1E11
When may you send indecent words from your amateur station?
A. Only when they do not cause interference to other communications
B. Only when they are not retransmitted through a repeater
C. Any time, but there is an unwritten rule among amateurs that they should not be used on the air
D. Never; indecent words are not allowed in amateur transmissions

T1E11
(D)
[97.113a4]
Page 1-21

Subelement T2

Subelement T2 — Operating Procedures [3 Exam Questions — 3 Groups]

T2A Repeater operation; autopatch, definition and proper use; courteous operation; repeater frequency coordination

T2A01 (A) Page 2-24

T2A01
What is the usual input/output frequency separation for repeaters in the 2-meter band?
A. 600 kHz
B. 1.0 MHz
C. 1.6 MHz
D. 5.0 MHz

T2A02 (C) Page 2-24

T2A02
What is the usual input/output frequency separation for repeaters in the 1.25-meter band?
A. 600 kHz
B. 1.0 MHz
C. 1.6 MHz
D. 5.0 MHz

T2A03 (D) Page 2-24

T2A03
What is the usual input/output frequency separation for repeaters in the 70-centimeter band?
A. 600 kHz
B. 1.0 MHz
C. 1.6 MHz
D. 5.0 MHz

T2A04 (C) Page 2-14

T2A04
What is an autopatch?
A. An automatic digital connection between a US and a foreign amateur
B. A digital connection used to transfer data between a hand-held radio and a computer
C. A device that allows radio users to access the public telephone system
D. A video interface allowing images to be patched into a digital data stream

T2A05 (B) Page 2-13

T2A05
What is the purpose of repeater operation?
A. To cut your power bill by using someone else's higher power system
B. To help mobile and low-power stations extend their usable range
C. To transmit signals for observing propagation and reception
D. To communicate with stations in services other than amateur

T2A06 (B) Page 2-13

T2A06
What causes a repeater to "time out"?
A. The repeater's battery supply runs out
B. Someone's transmission goes on longer than the repeater allows
C. The repeater gets too hot and stops transmitting until its circuitry cools off
D. Something is wrong with the repeater

T2A07 (D) Page 2-13

T2A07
During commuting rush hours, which type of repeater operation should be discouraged?
A. Mobile stations
B. Low-power stations
C. Highway traffic information nets
D. Third-party communications nets

T2A08
What is a courtesy tone (used in repeater operations)?
A. A sound used to identify the repeater
B. A sound used to indicate when a transmission is complete
C. A sound used to indicate that a message is waiting for someone
D. A sound used to activate a receiver in case of severe weather

T2A08
(B)
Page 2-13

T2A09
What is the meaning of: "Your signal is full quieting..."?
A. Your signal is strong enough to overcome all receiver noise
B. Your signal has no spurious sounds
C. Your signal is not strong enough to be received
D. Your signal is being received, but no audio is being heard

T2A09
(A)
Page 2-9

T2A10
How do you call another station on a repeater if you know the station's call sign?
A. Say "break, break 79," then say the station's call sign
B. Say the station's call sign, then identify your own station
C. Say "CQ" three times, then say the station's call sign
D. Wait for the station to call "CQ," then answer it

T2A10
(B)
Page 2-12

T2A11
What is a repeater called that is available for anyone to use?
A. An open repeater
B. A closed repeater
C. An autopatch repeater
D. A private repeater

T2A11
(A)
Page 2-11

T2A12
Why should local amateur communications use VHF and UHF frequencies instead of HF frequencies?
A. To minimize interference on HF bands capable of long-distance communication
B. Because greater output power is permitted on VHF and UHF
C. Because HF transmissions are not propagated locally
D. Because signals are louder on VHF and UHF frequencies

T2A12
(A)
Page 2-10

T2A13
How might you join a closed repeater system?
A. Contact the control operator and ask to join
B. Use the repeater until told not to
C. Use simplex on the repeater input until told not to
D. Write the FCC and report the closed condition

T2A13
(A)
Page 2-11

T2A14
How can on-the-air interference be minimized during a lengthy transmitter testing or loading-up procedure?
A. Choose an unoccupied frequency
B. Use a dummy load
C. Use a non-resonant antenna
D. Use a resonant antenna that requires no loading-up procedure

T2A14
(B)
Page 2-7

T2A15
What is the proper way to ask someone their location when using a repeater?
A. Say, "What is your QTH?"
B. Say, "What is your 20?"
C. Say, "Where are you?"
D. Locations are not normally told by radio

T2A15
(C)
Page 2-12

T2A16
(C)
Page 2-13

T2A16
Why should you pause briefly between transmissions when using a repeater?
A. To check the SWR of the repeater
B. To reach for pencil and paper for third-party communications
C. To listen for anyone wanting to break in
D. To dial up the repeater's autopatch

T2A17
(A)
Page 2-13

T2A17
Why should you keep transmissions short when using a repeater?
A. A long transmission may prevent someone with an emergency from using the repeater
B. To see if the receiving station operator is still awake
C. To give any listening non-hams a chance to respond
D. To keep long-distance charges down

T2A18
(D)
Page 2-12

T2A18
What is the proper way to break into a conversation on a repeater?
A. Wait for the end of a transmission and start calling the desired party
B. Shout, "break, break!" to show that you're eager to join the conversation
C. Turn on an amplifier and override whoever is talking
D. Say your call sign during a break between transmissions

T2A19
(D)
Page 2-11

T2A19
What is a repeater frequency coordinator?
A. Someone who organizes the assembly of a repeater station
B. Someone who provides advice on what kind of repeater to buy
C. The person whose call sign is used for a repeater's identification
D. A person or group that recommends frequencies for repeater operation

T2A20
(D)
Page 2-11

T2A20
What is it called if the frequency coordinator recommends that you operate on a specific repeater frequency pair?
A. FCC type acceptance
B. FCC type approval
C. Frequency division multiplexing
D. Repeater frequency coordination

T2B Simplex operations; RST signal reporting; choice of equipment for desired communications; communications modes including amateur television (ATV), packet radio and SSB/CW weak signal operations

T2B01
(C)
Page 2-13

T2B01
Why should simplex be used where possible, instead of using a repeater?
A. Signal range will be increased
B. Long distance toll charges will be avoided
C. The repeater will not be tied up unnecessarily
D. Your antenna's effectiveness will be better tested

T2B02
(A)
Page 2-13

T2B02
If you are talking to a station using a repeater, how would you find out if you could communicate using simplex instead?
A. See if you can clearly receive the station on the repeater's input frequency
B. See if you can clearly receive the station on a lower frequency band
C. See if you can clearly receive a more distant repeater
D. See if a third station can clearly receive both of you

T2B03
If you are operating simplex on a repeater frequency, why would it be good amateur practice to change to another frequency?
A. The repeater's output power may ruin your station's receiver
B. There are more repeater operators than simplex operators
C. Changing the repeater's frequency is not practical
D. Changing the repeater's frequency requires the authorization of the FCC

T2B03
(C)
Page 2-13

T2B04
Which of the following is the best way to perform an on-the-air test of a pair of hand-held transceivers on your work bench?
A. Operate them through a local repeater
B. Operate them on an unoccupied simplex frequency
C. Operate them into separate inverting loads
D. Operate them into linear amplifiers

T2B04
(B)
Page 2-13

T2B05
What is the meaning of: "Your signal report is five seven..."?
A. Your signal is perfectly readable and moderately strong
B. Your signal is perfectly readable, but weak
C. Your signal is readable with considerable difficulty
D. Your signal is perfectly readable with near pure tone

T2B05
(A)
Page 2-9

T2B06
What is the meaning of: "Your signal report is three three..."?
A. The contact is serial number thirty-three
B. The station is located at latitude 33 degrees
C. Your signal is readable with considerable difficulty and weak in strength
D. Your signal is unreadable, very weak in strength

T2B06
(C)
Page 2-9

T2B07
What is the meaning of: "Your signal report is five nine plus 20 dB..."?
A. Your signal strength has increased by a factor of 100
B. Repeat your transmission on a frequency 20 kHz higher
C. The bandwidth of your signal is 20 decibels above linearity
D. A relative signal-strength meter reading is 20 decibels greater than strength 9

T2B07
(D)
Page 2-9

T2B08
Which of the following would be the most useful for an emergency search and rescue operation?
A. A high-gain antenna, such as a 6-foot dish
B. A hand-held VHF transceiver set up to access a local repeater
C. An HF multiband transceiver capable of world-wide communications
D. A portable 40-meter dipole that could be temporarily mounted on any available support

T2B08
(B)
Page 2-25

T2B09
Which of the following modes of communication are NOT available to a Technician class operator?
A. CW and SSB on HF bands
B. Amateur television (ATV)
C. EME (Moon bounce)
D. VHF packet, CW and SSB

T2B09
(A)
Page 2-25

T2B10
(B)
Page 2-23

T2B10
When should digital transmissions be used on 2-meter simplex voice frequencies?
A. In between voice syllables
B. Digital operations should be avoided on simplex voice frequencies
C. Only in the evening
D. At any time, so as to encourage the best use of the band

T2B11
(A)
Page 2-23

T2B11
What operating mode should your packet TNC include if you want to participate in the amateur TCP/IP network?
A. KISS mode
B. Command mode
C. Monitor mode
D. CW interface mode

T2B12
(B)
Page 2-25

T2B12
Which of the following will allow you to monitor Amateur Television (ATV) on the 70-cm band?
A. A portable video camera
B. A cable ready TV receiver
C. An SSTV converter
D. A TV flyback transformer

T2B13
(B)
Page 2-25

T2B13
Which of the following would be useful to create an effective weak signal VHF Amateur Radio station?
A. A hand-held VHF FM transceiver
B. A multi-mode VHF transceiver
C. An omni directional antenna
D. A mobile VHF FM transceiver

T2C Distress calling and emergency drills and communications - operations and equipment, Radio Amateur Civil Emergency Service (RACES)

T2C01
(A)
Page 2-25

T2C01
What is the proper distress call to use when operating phone?
A. Say "MAYDAY" several times
B. Say "HELP" several times
C. Say "EMERGENCY" several times
D. Say "SOS" several times

T2C02
(D)
Page 2-25

T2C02
What is the proper distress call to use when operating CW?
A. MAYDAY
B. QRRR
C. QRZ
D. SOS

T2C03
(A)
Page 2-12

T2C03
What is the proper way to interrupt a repeater conversation to signal a distress call?
A. Say "BREAK" twice, then your call sign
B. Say "HELP" as many times as it takes to get someone to answer
C. Say "SOS," then your call sign
D. Say "EMERGENCY" three times

T2C04
What is one reason for using tactical call signs such as "command post" or "weather center" during an emergency?
A. They keep the general public informed about what is going on
B. They are more efficient and help coordinate public-service communications
C. They are required by the FCC
D. They increase goodwill between amateurs

T2C04
(B)
Page 2-25

T2C05
What type of messages concerning a person's well-being are sent into or out of a disaster area?
A. Routine traffic
B. Tactical traffic
C. Formal message traffic
D. Health and Welfare traffic

T2C05
(D)
Page 2-26

T2C06
What are messages called that are sent into or out of a disaster area concerning the immediate safety of human life?
A. Tactical traffic
B. Emergency traffic
C. Formal message traffic
D. Health and Welfare traffic

T2C06
(B)
Page 2-26

T2C07
Why is it a good idea to have a way to operate your amateur station without using commercial AC power lines?
A. So you may use your station while mobile
B. So you may provide communications in an emergency
C. So you may operate in contests where AC power is not allowed
D. So you will comply with the FCC rules

T2C07
(B)
Page 2-25

T2C08
What is the most important accessory to have for a hand-held radio in an emergency?
A. An extra antenna
B. A portable amplifier
C. Several sets of charged batteries
D. A microphone headset for hands-free operation

T2C08
(C)
Page 2-25

T2C09
Which type of antenna would be a good choice as part of a portable HF amateur station that could be set up in case of an emergency?
A. A three-element quad
B. A three-element Yagi
C. A dipole
D. A parabolic dish

T2C09
(C)
Page 2-25

T2C10
With what organization must you register before you can participate in RACES drills?
A. A local Amateur Radio club
B. A local racing organization
C. The responsible civil defense organization
D. The Federal Communications Commission

T2C10
(C)
Page 2-26

T2C11
(A)
Page 2-26

T2C11
What is the maximum number of hours allowed per week for RACES drills?
A. One
B. Seven, but not more than one hour per day
C. Eight
D. As many hours as you want

T2C12
(D)
Page 2-26

T2C12
How must you identify messages sent during a RACES drill?
A. As emergency messages
B. As amateur traffic
C. As official government messages
D. As drill or test messages

Subelement T3

Subelement T3 — Radio-Wave Propagation [3 Exam Questions — 3 Groups]

T3A VHF/UHF/Microwave Propagation

T3A01
(B)
Page 3-9

T3A01
How are VHF signals propagated within the range of the visible horizon?
A. By sky wave
B. By line of sight
C. By plane wave
D. By geometric refraction

T3A02
(C)
Page 3-9

T3A02
Ducting occurs in which region of the atmosphere?
A. F2
B. Ecosphere
C. Troposphere
D. Stratosphere

T3A03
(A)
Page 3-9

T3A03
What effect does tropospheric bending have on 2-meter radio waves?
A. It lets you contact stations farther away
B. It causes them to travel shorter distances
C. It garbles the signal
D. It reverses the sideband of the signal

T3A04
(D)
Page 3-9

T3A04
What causes tropospheric ducting of radio waves?
A. A very low pressure area
B. An aurora to the north
C. Lightning between the transmitting and receiving stations
D. A temperature inversion

T3A05
(B)
Page 3-9

T3A05
What causes VHF radio waves to be propagated several hundred miles over oceans?
A. A polar air mass
B. A widespread temperature inversion
C. An overcast of cirriform clouds
D. A high-pressure zone

T3A06
In which of the following frequency ranges does tropospheric ducting most often occur?
A. UHF
B. MF
C. HF
D. VHF

T3A06
(A)
Page 3-9

T3A07
In which of the following frequency ranges does sky-wave propagation least often occur?
A. LF
B. UHF
C. HF
D. VHF

T3A07
(B)
Page 3-10

T3A08
What weather condition may cause tropospheric ducting?
A. A stable high-pressure system
B. An unstable low-pressure system
C. A series of low-pressure waves
D. Periods of heavy rainfall

T3A08
(A)
Page 3-9

T3A09
What band conditions might indicate long-range skip on the 6-meter and 2-meter bands?
A. Noise on the 80-meter band
B. The absence of signals on the 10-meter band
C. Very long-range skip on the 10-meter band
D. Strong signals on the 10-meter band from stations about 500 - 600 miles away

T3A09
(D)
Page 3-10

T3A10
Which ionospheric region most affects sky-wave propagation on the 6-meter band?
A. The D region
B. The E region
C. The F1 region
D. The F2 region

T3A10
(B)
Page 3-10

T3A11
How does the signal loss for a given path through the troposphere vary with frequency?
A. There is no relationship
B. The path loss decreases as the frequency increases
C. The path loss increases as the frequency increases
D. There is no path loss at all

T3A11
(C)
Page 3-9

T3A12
What type of propagation usually occurs from one hand-held VHF transceiver to another nearby?
A. Tunnel propagation
B. Sky-wave propagation
C. Line-of-sight propagation
D. Auroral propagation

T3A12
(C)
Page 3-2

T3A13
Which frequency band, open to Technician class amateurs, experiences summertime sporadic E propagation?
A. 23 centimeters
B. 6 meters
C. 70 centimeters
D. 1.25 meters

T3A13
(B)
Page 3-10

T3A14
(A)
Page 3-10

T3A14
Which of the following emission modes are considered to be weak-signal modes and have the greatest potential for DX contacts?
A. Single sideband and CW
B. Packet radio and RTTY
C. Frequency modulation
D. Amateur television

T3A15
(D)
Page 3-10

T3A15
Which Technician frequency band could offer you the best chance of sky-wave propagation?
A. 1.25 meters
B. 70 centimeters
C. 23 centimeters
D. 6 meters

T3B Ionospheric absorption, causes and variation, maximum usable frequency

T3B01
(D)
Page 3-4

T3B01
Which region of the ionosphere is mainly responsible for absorbing MF/HF radio signals during the daytime?
A. The F2 region
B. The F1 region
C. The E region
D. The D region

T3B02
(B)
Page 3-4

T3B02
When does ionospheric absorption of radio signals occur?
A. When tropospheric ducting occurs
B. When long-wavelength signals enter the D region
C. When signals travel to the F region at night
D. When a temperature inversion occurs

T3B03
(A)
Page 3-4

T3B03
What effect does the D region of the ionosphere have on lower-frequency HF signals in the daytime?
A. It absorbs the signals
B. It bends the radio waves out into space
C. It refracts the radio waves back to earth
D. It has little or no effect on 80-meter radio waves

T3B04
(B)
Page 3-4

T3B04
What causes the ionosphere to absorb radio waves?
A. The weather below the ionosphere
B. The ionization of the D region
C. The presence of ionized clouds in the E region
D. The splitting of the F region

T3B05
(C)
Page 3-7

T3B05
If you are receiving a weak and distorted signal from a distant station on a frequency close to the maximum usable frequency, what type of propagation is probably occurring?
A. Ducting
B. Line-of-sight
C. Scatter
D. Ground-wave

T3B06
Which ionospheric region limits daytime radio communications on the 80-meter band to short distances?
A. The D region
B. The E region
C. The F1 region
D. The F2 region

T3B06
(A)
Page 3-4

T3B07
Which region of the ionosphere is the least useful for long-distance radio-wave propagation?
A. The D region
B. The E region
C. The F1 region
D. The F2 region

T3B07
(A)
Page 3-4

T3B08
What is the condition of the ionosphere above a particular area of the Earth just before local sunrise?
A. Atmospheric attenuation is at a maximum
B. The D region is above the E region
C. The E region is above the F region
D. Ionization is at a minimum

T3B08
(D)
Page 3-4

T3B09
When is the ionosphere above a particular area of the Earth most ionized?
A. Dusk
B. Midnight
C. Midday
D. Dawn

T3B09
(C)
Page 3-4

T3B10
When is the ionosphere above a particular area of the Earth least ionized?
A. Shortly before dawn
B. Just after noon
C. Just after dusk
D. Shortly before midnight

T3B10
(A)
Page 3-4

T3B11
When is the E region above a particular area of the Earth most ionized?
A. Dawn
B. Midday
C. Dusk
D. Midnight

T3B11
(B)
Page 3-5

T3B12
What happens to signals that take off vertically from the antenna and are higher in frequency than the critical frequency?
A. They pass through the ionosphere
B. They are absorbed by the ionosphere
C. Their frequency is changed by the ionosphere to be below the maximum usable frequency
D. They are reflected back to their source

T3B12
(A)
Page 3-7

T3B13
(C)
Page 3-3

T3B13
What causes the maximum usable frequency to vary?
A. The temperature of the ionosphere
B. The speed of the winds in the upper atmosphere
C. The amount of radiation received from the sun, mainly ultraviolet
D. The type of weather just below the ionosphere

T3B14
(A)
Page 3-3

T3B14
In relation to sky-wave propagation, what does the term "maximum usable frequency" (MUF) mean?
A. The highest frequency signal that will reach its intended destination
B. The lowest frequency signal that will reach its intended destination
C. The highest frequency signal that is most absorbed by the ionosphere
D. The lowest frequency signal that is most absorbed by the ionosphere

T3C Amateur satellite and EME operations

T3C01
(A)
Page 3-11

T3C01
Why might you have to retune your receiver while listening to signals from an amateur satellite?
A. Because of the Doppler effect
B. Because of the Einstein effect
C. Because of the Edison effect
D. Because of the Faraday effect

T3C02
(B)
Page 3-11

T3C02
How does the Doppler effect change an amateur satellite's signal as the satellite passes overhead?
A. The signal's amplitude increases or decreases
B. The signal's frequency increases or decreases
C. The signal's polarization changes from horizontal to vertical
D. The signal's circular polarization rotates

T3C03
(C)
Page 3-11

T3C03
Why do many satellites and satellite operators use circularly polarized antennas?
A. To correct for Doppler shift on transmitted signals
B. To obtain a wider beamwidth and eliminate the need to track the satellite
C. To reduce the fading effects of non-spin-stabilized satellites
D. To reduce the effects of terrestrial interference

T3C04
(D)
Page 3-10

T3C04
Why do many amateur satellites operate on the VHF/UHF bands?
A. To take advantage of the skip zone
B. Because VHF/UHF equipment costs less than HF equipment
C. To give Technician class operators greater access to modern communications technology
D. Because VHF and UHF signals easily pass through the ionosphere

T3C05
(B)
Page 3-11

T3C05
Why are high-gain antennas normally used for EME (moonbounce) communications?
A. To reduce the scattering of the reflected signal as it returns to Earth
B. To overcome the extreme path losses of this mode
C. To reduce the effects of polarization changes in the received signal
D. To overcome the high levels of solar noise at the receiver

T3C06
Why is the Doppler effect not important when operating EME (moonbounce)?
A. The Doppler effect does not occur beyond the ionosphere
B. EME antennas are always circularly polarized to eliminate any Doppler effect
C. The distance between the earth and the moon does not change rapidly enough to produce the Doppler effect
D. The rough surface of the moon scatters signals enough to eliminate the Doppler effect

T3C06
(C)
Page 3-11

T3C07
Which of the following antenna systems would be the best choice for an EME (moonbounce) station?
A. A single dipole antenna
B. An isotropic antenna
C. A ground-plane antenna
D. A high-gain array of Yagi antennas

T3C07
(D)
Page 3-11

T3C08
Which antenna system would NOT be a good choice for an EME (moonbounce) station?
A. A parabolic-dish antenna
B. A multi-element array of collinear antennas
C. A ground-plane antenna
D. A high-gain array of Yagi antennas

T3C08
(C)
Page 3-11

T3C09
Why is it necessary to use high-gain antennas and high transmitter power for EME (moonbounce) operation?
A. To overcome path losses and poor reflectivity of the moon's surface
B. To overcome the effects of Faraday rotation
C. To reduce the effects of Doppler shift
D. To reduce the effects of the solar wind

T3C09
(A)
Page 3-11

T3C10
When is it necessary to use a higher transmitter power level when conducting satellite communications?
A. When the satellite is at its perigee
B. When the satellite is low to the horizon
C. When the satellite is fully illuminated by the sun
D. When the satellite is near directly overhead

T3C10
(B)
Page 3-11

T3C11
Which of the following conditions must be met before two stations can conduct real-time communications through a satellite?
A. Both stations must use circularly polarized antennas
B. The satellite must be illuminated by the sun during the communications
C. The satellite must be in view of both stations simultaneously
D. Both stations must use high-gain antenna systems

T3C11
(C)
Page 3-10

Subelement T4

Subelement T4 — Amateur Radio Practices
[4 Exam Questions — 4 Groups]

T4A Electrical wiring, including switch location, dangerous voltages and currents

T4A01
(C)
Page 4-9

T4A01
Where should the green wire in a three-wire AC line cord be connected in a power supply?
A. To the fuse
B. To the "hot" side of the power switch
C. To the chassis
D. To the white wire

T4A02
(D)
Page 4-9

T4A02
Where should the black (or red) wire in a three-wire AC line cord be connected in a power supply?
A. To the white wire, which connects to the "hot" side of the power switch
B. To the green wire, which connects to ground
C. To the chassis
D. To the fuse, which connects to the "hot" side of the power switch

T4A03
(B)
Page 4-9

T4A03
Where should the white wire in a three-wire AC line cord be connected in a power supply?
A. To the neutral side of the power transformer's primary winding, which has a fuse
B. To the neutral side of the power transformer's primary winding, which does not have a fuse
C. To the chassis
D. To the black wire

T4A04
(C)
Page 4-9

T4A04
What is the correct color code for a 120 VAC three-conductor power cord?
A. The green wire connects to the neutral terminal, white connects to the hot terminal, and black connects to the ground terminal
B. The black wire connects to the neutral terminal, green connects to the hot terminal, and the white wire connects to the ground terminal
C. The white wire connects to the neutral terminal, black connects to the hot terminal, and green connects to the ground terminal
D. The red wire connects to the neutral terminal, black connects to the ground terminal, and white connects to the hot terminal.

T4A05
(B)
Page 4-9

T4A05
Why is the retaining screw in one terminal of a wall outlet made of brass while the other one is silver colored?
A. To prevent corrosion
B. To indicate correct wiring polarity
C. To better conduct current
D. To reduce skin effect

T4A06
(C)
Page 4-9

T4A06
What is an important safety rule concerning the main electrical box in your home?
A. Make sure the door cannot be opened easily
B. Make sure something is placed in front of the door so no one will be able to get to it easily
C. Make sure others in your home know where it is and how to shut off the electricity
D. Warn others in your home never to touch the switches, even in an emergency

T4A07
Where should the main power switch for a high-voltage power supply be located?
A. Inside the cabinet, to kill the power if the cabinet is opened
B. On the back side of the cabinet, out of sight
C. Anywhere that can be seen and reached easily
D. A high-voltage power supply should not be switch-operated

T4A07
(C)
Page 4-10

T4A08
What document is used by almost every US city as the basis for electrical safety requirements for power wiring and antennas?
A. The Code of Federal Regulations
B. The Proceedings of the IEEE
C. The ITU Radio Regulations
D. The National Electrical Code

T4A08
(D)
Page 4-4

T4A09
What document would you use to see if you comply with standard electrical safety rules when building an amateur antenna?
A. The Code of Federal Regulations
B. The Proceedings of the IEEE
C. The National Electrical Code
D. The ITU Radio Regulations

T4A09
(C)
Page 4-4

T4A10
What is the minimum voltage that is usually dangerous to humans?
A. 30 volts
B. 100 volts
C. 1000 volts
D. 2000 volts

T4A10
(A)
Page 4-11

T4A11
What precaution should you take when leaning over a power amplifier?
A. Take your shoes off
B. Watch out for loose jewelry contacting high voltage
C. Shield your face from the heat produced by the power supply
D. Watch out for sharp edges that may snag your clothing

T4A11
(B)
Page 4-10

T4A12
What should you do if you discover someone who is being burned by high voltage?
A. Run from the area so you won't be burned too
B. Turn off the power, call for emergency help and give CPR if needed
C. Immediately drag the person away from the high voltage
D. Wait for a few minutes to see if the person can get away from the high voltage on their own, then try to help

T4A12
(B)
Page 4-11

T4A13
Where should fuses be connected on a mobile transceiver's DC power cable?
A. Between the red and black wires
B. In series with just the black wire
C. In series with just the red wire
D. In series with both the red and black wires

T4A13
(D)
Page 4-6

T4A14
(A)
Page 4-11

T4A14
How much electrical current flowing through the human body will probably be fatal?
A. As little as 1/10 of an ampere
B. Approximately 10 amperes
C. More than 20 amperes
D. Current through the human body is never fatal

T4A15
(A)
Page 4-11

T4A15
Which body organ can be fatally affected by a very small amount of electrical current?
A. The heart
B. The brain
C. The liver
D. The lungs

T4A16
(A)
Page 4-11

T4A16
How much electrical current flowing through the human body is usually painful?
A. As little as 1/500 of an ampere
B. Approximately 10 amperes
C. More than 20 amperes
D. Current flow through the human body is never painful

T4B Meters and their placement in circuits, including volt, amp, multi, peak-reading and RF watt; ratings of fuses and switches

T4B01
(B)
Page 4-12

T4B01
How is a voltmeter usually connected to a circuit under test?
A. In series with the circuit
B. In parallel with the circuit
C. In quadrature with the circuit
D. In phase with the circuit

T4B02
(A)
Page 4-12

T4B02
How is an ammeter usually connected to a circuit under test?
A. In series with the circuit
B. In parallel with the circuit
C. In quadrature with the circuit
D. In phase with the circuit

T4B03
(A)
Page 4-13

T4B03
Where should an RF wattmeter be connected for the most accurate readings of transmitter output power?
A. At the transmitter output connector
B. At the antenna feed point
C. One-half wavelength from the transmitter output
D. One-half wavelength from the antenna feed point

T4B04
(C)
Page 4-12

T4B04
How can the range of a voltmeter be increased?
A. By adding resistance in series with the circuit under test
B. By adding resistance in parallel with the circuit under test
C. By adding resistance in series with the meter, between the meter and the circuit under test
D. By adding resistance in parallel with the meter, between the meter and the circuit under test

T4B05
What happens inside a voltmeter when you switch it from a lower to a higher voltage range?
A. Resistance is added in series with the meter
B. Resistance is added in parallel with the meter
C. Resistance is reduced in series with the meter
D. Resistance is reduced in parallel with the meter

T4B05
(A)
Page 4-12

T4B06
How can the range of an ammeter be increased?
A. By adding resistance in series with the circuit under test
B. By adding resistance in parallel with the circuit under test
C. By adding resistance in series with the meter
D. By adding resistance in parallel with the meter

T4B06
(D)
Page 4-12

T4B07
For which measurements would you normally use a multimeter?
A. SWR and power
B. Resistance, capacitance and inductance
C. Resistance and reactance
D. Voltage, current and resistance

T4B07
(D)
Page 4-12

T4B08
What might happen if you switch a multimeter to measure resistance while you have it connected to measure voltage?
A. The multimeter would read half the actual voltage
B. It would probably destroy the meter circuitry
C. The multimeter would read twice the actual voltage
D. Nothing unusual would happen; the multimeter would measure the circuit's resistance

T4B08
(B)
Page 4-13

T4B09
If you switch a multimeter to read microamps and connect it into a circuit drawing 5 amps, what might happen?
A. The multimeter would read half the actual current
B. The multimeter would read twice the actual current
C. It would probably destroy the meter circuitry
D. The multimeter would read a very small value of current

T4B09
(C)
Page 4-12

T4B10
At what line impedance do most RF watt meters usually operate?
A. 25 ohms
B. 50 ohms
C. 100 ohms
D. 300 ohms

T4B10
(B)
Page 4-13

T4B11
What does a directional wattmeter measure?
A. Forward and reflected power
B. The directional pattern of an antenna
C. The energy used by a transmitter
D. Thermal heating in a load resistor

T4B11
(A)
Page 4-13

T4B12
(B)
Page 4-13

T4B12
If a directional RF wattmeter reads 90 watts forward power and 10 watts reflected power, what is the actual transmitter output power?
A. 10 watts
B. 80 watts
C. 90 watts
D. 100 watts

T4B13
(C)
Page 4-13

T4B13
If a directional RF wattmeter reads 96 watts forward power and 4 watts reflected power, what is the actual transmitter output power?
A. 80 watts
B. 88 watts
C. 92 watts
D. 100 watts

T4B14
(A)
Page 4-13

T4B14
Why might you use a peak-reading RF wattmeter at your station?
A. To make sure your transmitter's output power is not higher than that authorized by your license class
B. To make sure your transmitter is not drawing too much power from the AC line
C. To make sure all your transmitter's power is being radiated by your antenna
D. To measure transmitter input and output power at the same time

T4B15
(C)
Page 4-10

T4B15
What could happen to your transceiver if you replace its blown 5 amp AC line fuse with a 30 amp fuse?
A. The 30-amp fuse would better protect your transceiver from using too much current
B. The transceiver would run cooler
C. The transceiver could use more current than 5 amps and a fire could occur
D. The transceiver would not be able to produce as much RF output

T4B16
(D)
Page 4-10

T4B16
Why shouldn't you use a switch rated at 1 amp to switch power to a mobile transceiver that draws 8 amps?
A. This would be against FCC Rules
B. This would be against state motor vehicle laws
C. The transceiver would not be able to produce as much RF output
D. The switch could overheat and become a safety hazard

T4C Marker generator, crystal calibrator, signal generators and impedance-match indicator

T4C01
(A)
Page 4-14

T4C01
What is a marker generator?
A. A high-stability oscillator that generates reference signals at exact frequency intervals
B. A low-stability oscillator that "sweeps" through a range of frequencies
C. A low-stability oscillator used to inject a signal into a circuit under test
D. A high-stability oscillator that can produce a wide range of frequencies and amplitudes

T4C02
(A)
Page 4-14

T4C02
What is one use for a marker generator?
A. To calibrate the tuning dial on a receiver
B. To calibrate the volume control on a receiver
C. To test the amplitude linearity of a transmitter
D. To test the frequency integration of a transmitter

T4C03
What device is used to inject a frequency calibration signal into a receiver?
A. A calibrated voltmeter
B. A calibrated oscilloscope
C. A calibrated wavemeter
D. A crystal calibrator

T4C03
(D)
Page 4-14

T4C04
What device produces a stable, low-level signal that can be set to a desired frequency?
A. A wavemeter
B. A reflectometer
C. A signal generator
D. An oscilloscope

T4C04
(C)
Page 4-14

T4C05
What is one use for an RF signal generator?
A. Measuring AF signal amplitudes
B. Aligning tuned circuits
C. Adjusting transmitter impedance-neutralizing networks
D. Measuring transmission-line impedances

T4C05
(B)
Page 4-14

T4C06
What device can measure an impedance mismatch in your antenna system?
A. A field-strength meter
B. An ammeter
C. A wavemeter
D. A reflectometer

T4C06
(D)
Page 4-15

T4C07
Where should a reflectometer be connected for best accuracy when reading the impedance match between an antenna and its feed line?
A. At the antenna feed point
B. At the transmitter output connector
C. At the midpoint of the feed line
D. Anywhere along the feed line

T4C07
(A)
Page 4-15

T4C08
If you use an RF power meter designed to operate on 3-30 MHz for VHF measurements, how accurate will its readings be?
A. They are not likely to be accurate
B. They will be accurate enough to get by
C. If it properly calibrates to full scale in the set position, they may be accurate
D. They will be accurate providing the readings are multiplied by 4.5

T4C08
(A)
Page 4-13

T4C09
If you use an SWR meter designed to operate on 3-30 MHz for VHF measurements, how accurate will its readings be?
A. They will not be accurate
B. They will be accurate enough to get by
C. If it properly calibrates to full scale in the set position, they may be accurate
D. They will be accurate providing the readings are multiplied by 4.5

T4C09
(C)
Page 4-15

T4C10
(B)
Page 4-14

T4C10
What frequency standard may be used to calibrate the tuning dial of a receiver?
A. A calibrated voltmeter
B. Signals from WWV and WWVH
C. A deviation meter
D. A sweep generator

T4C11
(C)
Page 4-14

T4C11
What is the most accurate way to check the calibration of your receiver's tuning dial?
A. Monitor the BFO frequency of a second receiver
B. Tune to a popular amateur net frequency
C. Tune to one of the frequencies of station WWV or WWVH
D. Tune to another amateur station and ask what frequency the operator is using

T4D Dummy antennas and S-meters

T4D01
(D)
Page 4-15

T4D01
What device should be connected to a transmitter's output when you are making transmitter adjustments?
A. A multimeter
B. A reflectometer
C. A receiver
D. A dummy antenna

T4D02
(B)
Page 4-15

T4D02
What is a dummy antenna?
A. An nondirectional transmitting antenna
B. A nonradiating load for a transmitter
C. An antenna used as a reference for gain measurements
D. A flexible antenna usually used on hand-held transceivers

T4D03
(C)
Page 4-16

T4D03
What is the main component of a dummy antenna?
A. A wire-wound resistor
B. An iron-core coil
C. A noninductive resistor
D. An air-core coil

T4D04
(B)
Page 4-15

T4D04
What device is used in place of an antenna during transmitter tests so that no signal is radiated?
A. An antenna matcher
B. A dummy antenna
C. A low-pass filter
D. A decoupling resistor

T4D05
(A)
Page 4-15

T4D05
Why would you use a dummy antenna?
A. For off-the-air transmitter testing
B. To reduce output power
C. To give comparative signal reports
D. To allow antenna tuning without causing interference

T4D06
What minimum rating should a dummy antenna have for use with a 100 watt single-sideband phone transmitter?
A. 100 watts continuous
B. 141 watts continuous
C. 175 watts continuous
D. 200 watts continuous

T4D06
(A)
Page 4-16

T4D07
Why might a dummy antenna get warm when in use?
A. Because it stores electric current
B. Because it stores radio waves
C. Because it absorbs static electricity
D. Because it changes RF energy into heat

T4D07
(D)
Page 4-16

T4D08
Would a 100 watt light bulb make a good dummy load for tuning a transceiver?
A. Yes; a light bulb behaves exactly like a dummy load
B. No; the impedance of the light bulb changes as the filament gets hot
C. No; the light bulb would act like an open circuit
D. No; the light bulb would act like a short circuit

T4D08
(B)
Page 4-16

T4D09
What is used to measure relative signal strength in a receiver?
A. An S meter
B. An RST meter
C. A signal deviation meter
D. An SSB meter

T4D09
(A)
Page 4-16

T4D10
Why might two radios using the same antenna and receiving the same signal show two very different S-meter readings?
A. S meters are always referenced to the maximum RF output available from the transceiver
B. S meters are always referenced to the maximum discernible signal the receiver can hear
C. Receiver S meters give only a relative indication of received signal strength
D. Some S meters are calibrated to US standards while others are calibrated to foreign standards

T4D10
(C)
Page 4-16

T4D11
What does your transceiver "S meter" indicate?
A. The transmitted audio strength
B. The final RF transistor amplifier source voltage
C. The percentage of secondary modulation
D. The relative received signal strength

T4D11
(D)
Page 4-16

Subelement T5

Subelement T5 — Electrical Principles [2 Exam Questions — 2 Groups]

T5A Definition and unit of measurement of resistance, inductance and capacitance

T5A01 (D) Page 5-9

T5A01
What does resistance do in an electric circuit?
A. It stores energy in a magnetic field
B. It stores energy in an electric field
C. It provides electrons by a chemical reaction
D. It opposes the flow of electrons

T5A02 (B) Page 5-9

T5A02
What is the definition of 1 ohm?
A. The reactance of a circuit in which a 1-microfarad capacitor is resonant at 1 MHz
B. The resistance of a circuit in which a 1-amp current flows when 1 volt is applied
C. The resistance of a circuit in which a 1-milliamp current flows when 1 volt is applied
D. The reactance of a circuit in which a 1-millihenry inductor is resonant at 1 MHz

T5A03 (C) Page 5-9

T5A03
What is the basic unit of resistance?
A. The farad
B. The watt
C. The ohm
D. The resistor

T5A04 (D) Page 5-9

T5A04
What is one reason resistors are used in electronic circuits?
A. To block the flow of direct current while allowing alternating current to pass
B. To block the flow of alternating current while allowing direct current to pass
C. To increase the voltage of the circuit
D. To control the amount of current that flows for a particular applied voltage

T5A05 (D) Page 5-19

T5A05
What is the ability to store energy in a magnetic field called?
A. Admittance
B. Capacitance
C. Resistance
D. Inductance

T5A06 (C) Page 5-20

T5A06
What is the basic unit of inductance?
A. The coulomb
B. The farad
C. The henry
D. The ohm

T5A07 (C) Page 5-20

T5A07
What is a henry?
A. The basic unit of admittance
B. The basic unit of capacitance
C. The basic unit of inductance
D. The basic unit of resistance

T5A08
What is one reason inductors are used in electronic circuits?
A. To block the flow of direct current while allowing alternating current to pass
B. To reduce the flow of AC while allowing DC to pass freely
C. To change the time constant of the applied voltage
D. To change alternating current to direct current

T5A08
(B)
Page 5-19

T5A09
What is the ability to store energy in an electric field called?
A. Inductance
B. Resistance
C. Tolerance
D. Capacitance

T5A09
(D)
Page 5-20

T5A10
What is the basic unit of capacitance?
A. The farad
B. The ohm
C. The volt
D. The henry

T5A10
(A)
Page 5-21

T5A11
What is a farad?
A. The basic unit of resistance
B. The basic unit of capacitance
C. The basic unit of inductance
D. The basic unit of admittance

T5A11
(B)
Page 5-21

T5A12
What is one reason capacitors are used in electronic circuits?
A. To block the flow of direct current while allowing alternating current to pass
B. To block the flow of alternating current while allowing direct current to pass
C. To change the time constant of the applied voltage
D. To change alternating current to direct current

T5A12
(A)
Page 5-21

T5B Concepts and calculation of resistance, inductance and capacitance values in series and parallel circuits

T5B01
How is the current in a DC circuit directly calculated when the voltage and resistance are known?
A. I = R × E [current equals resistance multiplied by voltage]
B. I = R / E [current equals resistance divided by voltage]
C. I = E / R [current equals voltage divided by resistance]
D. I = E / P [current equals voltage divided by power]

T5B01
(C)
Page 5-9

T5B02
How is the resistance in a DC circuit calculated when the voltage and current are known?
A. R = I / E [resistance equals current divided by voltage]
B. R = E / I [resistance equals voltage divided by current]
C. R = I × E [resistance equals current multiplied by voltage]
D. R = P / E [resistance equals power divided by voltage]

T5B02
(B)
Page 5-9

T5B03
(C)
Page 5-10

T5B03
How is the voltage in a DC circuit directly calculated when the current and resistance are known?
A. E = I / R [voltage equals current divided by resistance]
B. E = R / I [voltage equals resistance divided by current]
C. E = I × R [voltage equals current multiplied by resistance]
D. E = I / P [voltage equals current divided by power]

T5B04
(D)
Page 5-10

T5B04
If a 12-volt battery supplies 0.25 ampere to a circuit, what is the circuit's resistance?
A. 0.25 ohm
B. 3 ohms
C. 12 ohms
D. 48 ohms

T5B05
(D)
Page 5-10

T5B05
If a 12-volt battery supplies 0.15 ampere to a circuit, what is the circuit's resistance?
A. 0.15 ohm
B. 1.8 ohms
C. 12 ohms
D. 80 ohms

T5B06
(B)
Page 5-10

T5B06
If a 4800-ohm resistor is connected to 120 volts, how much current will flow through it?
A. 4 A
B. 25 mA
C. 25 A
D. 40 mA

T5B07
(D)
Page 5-10

T5B07
If a 48,000-ohm resistor is connected to 120 volts, how much current will flow through it?
A. 400 A
B. 40 A
C. 25 mA
D. 2.5 mA

T5B08
(A)
Page 5-10

T5B08
If a 4800-ohm resistor is connected to 12 volts, how much current will flow through it?
A. 2.5 mA
B. 25 mA
C. 40 A
D. 400 A

T5B09
(A)
Page 5-10

T5B09
If a 48,000-ohm resistor is connected to 12 volts, how much current will flow through it?
A. 250 μA
B. 250 mA
C. 4000 mA
D. 4000 A

T5B10
(D)
Page 5-17

T5B10
If two resistors are connected in series, what is their total resistance?
A. The difference between the individual resistor values
B. Always less than the value of either resistor
C. The product of the individual resistor values
D. The sum of the individual resistor values

T5B11
If two resistors are connected in parallel, what is their total resistance?
A. The difference between the individual resistor values
B. Always less than the value of either resistor
C. The product of the two values
D. The sum of the individual resistors

T5B11
(B)
Page 5-18

T5B12
If two equal-value inductors are connected in series, what is their total inductance?
A. Half the value of one inductor
B. Twice the value of one inductor
C. The same as the value of either inductor
D. The value of one inductor times the value of the other

T5B12
(B)
Page 5-20

T5B13
If two equal-value inductors are connected in parallel, what is their total inductance?
A. Half the value of one inductor
B. Twice the value of one inductor
C. The same as the value of either inductor
D. The value of one inductor times the value of the other

T5B13
(A)
Page 5-20

T5B14
If two equal-value capacitors are connected in series, what is their total capacitance?
A. Twice the value of one capacitor
B. The same as the value of either capacitor
C. Half the value of either capacitor
D. The value of one capacitor times the value of the other

T5B14
(C)
Page 5-21

T5B15
If two equal-value capacitors are connected in parallel, what is their total capacitance?
A. Twice the value of one capacitor
B. Half the value of one capacitor
C. The same as the value of either capacitor
D. The value of one capacitor times the value of the other

T5B15
(A)
Page 5-21

Subelement T6 — Circuit Components [2 Exam Questions — 2 Groups]

Subelement T6

T6A Resistors, construction types, variable and fixed, color code, power ratings, schematic symbols

T6A01
Which of the following are common resistor types?
A. Plastic and porcelain
B. Film and wire-wound
C. Electrolytic and metal-film
D. Iron core and brass core

T6A01
(B)
Page 6-10

T6A02
What does a variable resistor or potentiometer do?
A. Its resistance changes when AC is applied to it
B. It transforms a variable voltage into a constant voltage
C. Its resistance changes when its slide or contact is moved
D. Its resistance changes when it is heated

T6A02
(C)
Page 6-11

T6A03
(B)
Page 6-12

T6A03
How do you find a resistor's value?
A. By using a voltmeter
B. By using the resistor's color code
C. By using Thevenin's theorem for resistors
D. By using the Baudot code

T6A04
(B)
Page 6-12

T6A04
How do you find a resistor's tolerance rating?
A. By using a voltmeter
B. By reading the resistor's color code
C. By using Thevenin's theorem for resistors
D. By reading its Baudot code

T6A05
(A)
Page 6-12

T6A05
What do the first three color bands on a resistor indicate?
A. The value of the resistor in ohms
B. The resistance tolerance in percent
C. The power rating in watts
D. The resistance material

T6A06
(B)
Page 6-12

T6A06
What does the fourth color band on a resistor indicate?
A. The value of the resistor in ohms
B. The resistance tolerance in percent
C. The power rating in watts
D. The resistance material

T6A07
(A)
Page 6-12

T6A07
Why do resistors sometimes get hot when in use?
A. Some electrical energy passing through them is lost as heat
B. Their reactance makes them heat up
C. Hotter circuit components nearby heat them up
D. They absorb magnetic energy, which makes them hot

T6A08
(C)
Page 6-12

T6A08
Why would a large size resistor be used instead of a smaller one of the same resistance value?
A. For better response time
B. For a higher current gain
C. For greater power dissipation
D. For less impedance in the circuit

T6A09
(C)
Page 6-12

T6A09
What range of resistance values are possible with a 100-ohm resistor that has a 10% tolerance?
A. 90 to 100 ohms
B. 10 to 100 ohms
C. 90 to 110 ohms
D. 80 to 120 ohms

T6A10
(A)
Page 6-12

T6A10
Which tolerance rating would indicate a high-precision resistor?
A. 0.1%
B. 5%
C. 10%
D. 20%

T6A11
Which tolerance rating would indicate a low-precision resistor?
A. 0.1%
B. 5%
C. 10%
D. 20%

T6A11
(D)
Page 6-12

T6A12
Which symbol of Figure T6-1 represents a fixed resistor?
A. Symbol 2
B. Symbol 3
C. Symbol 4
D. Symbol 5

T6A12
(A)
Pge 6-9

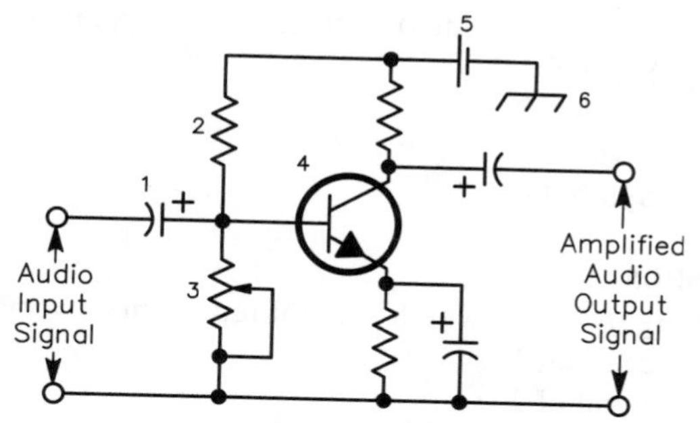

Figure T6–1

T6A13
Which symbol of Figure T6-1 represents a variable resistor?
A. Symbol 1
B. Symbol 2
C. Symbol 3
D. Symbol 6

T6A13
(C)
Page 6-11

T6A14
What type of resistor does symbol 2 represent in Figure T6-1?
A. A wire-wound resistor
B. A carbon-film resistor
C. A carbon composition resistor
D. Symbol 2 gives no information about the resistor's type

T6A14
(D)
Page 6-10

T6B Inductor and capacitor schematic symbols; construction of variable and fixed inductors and capacitors; factors affecting inductance and capacitance

T6B01
Which symbol of Figure T6-2 represents a fixed-value capacitor?
A. Symbol 1
B. Symbol 2
C. Symbol 3
D. Symbol 4

T6B01
(A)
Page 6-16

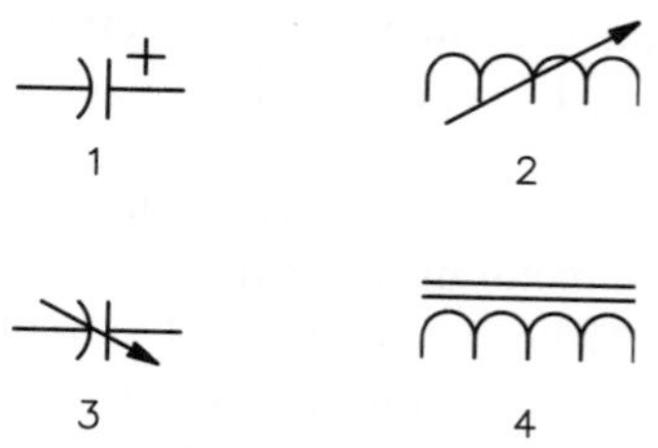

Figure T6–2

T6B02
In Figure T6-2, which symbol represents an adjustable inductor?
A. Symbol 1
B. Symbol 2
C. Symbol 3
D. Symbol 4

T6B02
(B)
Page 6-13

T6B03
In Figure T6-2, which symbol represents a fixed-value iron-core inductor?
A. Symbol 1
B. Symbol 2
C. Symbol 3
D. Symbol 4

T6B03
(D)
Page 6-13

T6B04
(D)
Page 6-14

T6B04
In Figure T6-2, which symbol represents an inductor wound over a toroidal core?
A. Symbol 1
B. Symbol 2
C. Symbol 3
D. Symbol 4

T6B05
(A)
Page 6-16

T6B05
In Figure T6-2, which symbol represents an electrolytic capacitor?
A. Symbol 1
B. Symbol 2
C. Symbol 3
D. Symbol 4

T6B06
(C)
Page 6-16

T6B06
In Figure T6-2, which symbol represents a variable capacitor?
A. Symbol 1
B. Symbol 2
C. Symbol 3
D. Symbol 4

T6B07
(D)
Page 6-13

T6B07
What is an inductor core?
A. The place where a coil is tapped for resonance
B. A tight coil of wire used in a transformer
C. Insulating material placed between the wires of a transformer
D. The place inside an inductor where its magnetic field is concentrated

T6B08
(C)
Pge 6-13

T6B08
What does an inductor do?
A. It stores energy electrostatically and opposes a change in voltage
B. It stores energy electrochemically and opposes a change in current
C. It stores energy electromagnetically and opposes a change in current
D. It stores energy electromechanically and opposes a change in voltage

T6B09
(D)
Page 6-13

T6B09
What determines the inductance of a coil?
A. The core material, the core diameter, the length of the coil and whether the coil is mounted horizontally or vertically
B. The core diameter, the number of turns of wire used to wind the coil and the type of metal used for the wire
C. The core material, the number of turns used to wind the core and the frequency of the current through the coil
D. The core material, the core diameter, the length of the coil and the number of turns of wire used to wind the coil

T6B10
(A)
Page 6-13

T6B10
As an iron core is inserted in a coil, what happens to the coil's inductance?
A. It increases
B. It decreases
C. It stays the same
D. It disappears

T6B11
What can happen if you tune a ferrite-core coil with a metal tool?
A. The metal tool can change the coil's inductance and cause you to tune the coil incorrectly
B. The metal tool can become magnetized so much that you might not be able to remove it from the coil
C. The metal tool can pick up enough magnetic energy to become very hot
D. The metal tool can pick up enough magnetic energy to become a shock hazard

T6B11
(A)
Page 6-13

T6B12
What describes a capacitor?
A. Two or more layers of silicon material with an insulating material between them
B. Two or more turns of wire wound around a core material
C. Two or more conductive plates with an insulating material between them
D. Two or more insulating plates with a conductive material between them

T6B12
(C)
Page 6-14

T6B13
What does a capacitor do?
A. It stores energy electrochemically and opposes a change in current
B. It stores energy electrostatically and opposes a change in voltage
C. It stores energy electromagnetically and opposes a change in current
D. It stores energy electromechanically and opposes a change in voltage

T6B13
(B)
Page 6-14

T6B14
What determines the capacitance of a capacitor?
A. The material between the plates, the area of one side of one plate, the number of plates and the spacing between the plates
B. The material between the plates, the number of plates and the size of the wires connected to the plates
C. The number of plates, the spacing between the plates and whether the dielectric material is N type or P type
D. The material between the plates, the area of one plate, the number of plates and the material used for the protective coating

T6B14
(A)
Page 6-15

T6B15
As the plate area of a capacitor is increased, what happens to its capacitance?
A. It decreases
B. It increases
C. It stays the same
D. It disappears

T6B15
(B)
Page 6-15

T6B16
Which of the following best describes a variable capacitor?
A. A set of fixed capacitors whose connections can be varied
B. Two sets of insulating plates separated by a conductor, which can be varied in distance from each other
C. A set of capacitors connected in a series-parallel circuit
D. Two sets of rotating conducting plates separated by an insulator, which can be varied in surface area exposed to each other

T6B16
(D)
Page 6-16

Subelement T7

Subelement T7 — Practical Circuits
[1 Exam Question — 1 Group]

T7A Transmitter and receiver block diagrams; purpose and operation of low-pass, high-pass and band-pass filters

T7A01
(D)
Page 7-11

T7A01
What circuit has a variable-frequency oscillator connected to a driver and a power amplifier?
A. A packet-radio transmitter
B. A crystal-controlled transmitter
C. A single-sideband transmitter
D. A VFO-controlled transmitter

T7A02
(B)
Pge 7-14

T7A02
What circuit combines signals from an IF amplifier stage and a beat-frequency oscillator (BFO), to produce an audio signal?
A. An AGC circuit
B. A detector circuit
C. A power supply circuit
D. A VFO circuit

T7A03
(D)
Page 7-14

T7A03
What circuit uses a limiter and a frequency discriminator to produce an audio signal?
A. A double-conversion receiver
B. A variable-frequency oscillator
C. A superheterodyne receiver
D. An FM receiver

T7A04
(D)
Page 7-11

T7A04
What circuit is pictured in Figure T7-1 if block 1 is a variable-frequency oscillator?
A. A packet-radio transmitter
B. A crystal-controlled transmitter
C. A single-sideband transmitter
D. A VFO-controlled transmitter

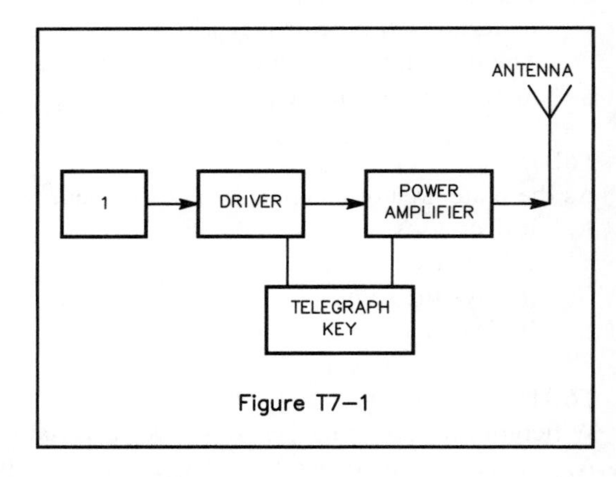

Figure T7—1

T7A05
(A)
Page 7-11

T7A05
What circuit is pictured in Figure T7-1 if block 1 is a crystal oscillator?
A. A crystal-controlled transmitter
B. A VFO-controlled transmitter
C. A single-sideband transmitter
D. A CW transceiver

T7A06
(B)
Page 7-11

T7A06
What purpose does block 1 serve in the simple CW transmitter pictured in Figure T7-1?
A. It detects the CW signal
B. It controls the transmitter frequency
C. It controls the transmitter output power
D. It filters out spurious emissions from the transmitter

T7A07
What is block 1 in Figure T7-2?
A. An AGC circuit
B. A detector
C. A power supply
D. A VFO circuit

T7A07
(B)
Page 7-14

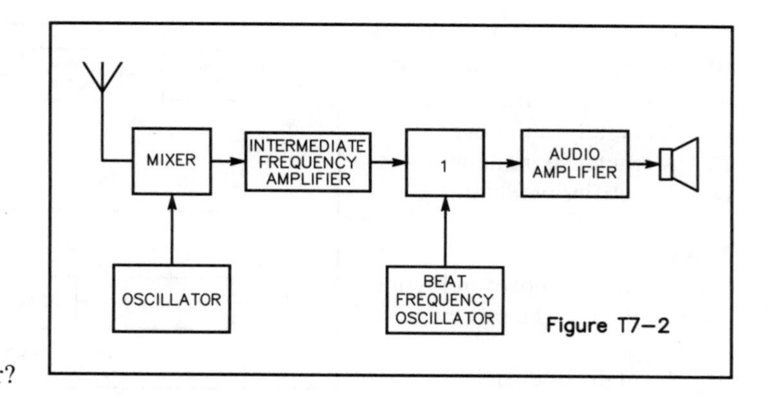

MIXER
INTERMEDIATE FREQUENCY AMPLIFIER
1
AUDIO AMPLIFIER
OSCILLATOR
BEAT FREQUENCY OSCILLATOR
Figure T7–2

T7A08
What type of circuit does Figure T7-2 represent if block 1 is a product detector?
A. A simple phase modulation receiver
B. A simple FM receiver
C. A simple CW and SSB receiver
D. A double-conversion multiplier

T7A08
(C)
Page 7-14

T7A09
If Figure T7-2 is a diagram of a simple single-sideband receiver, what type of circuit should be shown in block 1?
A. A high pass filter
B. A ratio detector
C. A low pass filter
D. A product detector

T7A09
(D)
Page 7-14

T7A10
What circuit is pictured in Figure T7-3, if block 1 is a frequency discriminator?
A. A double-conversion receiver
B. A variable-frequency oscillator
C. A superheterodyne receiver
D. An FM receiver

T7A10
(D)
Page 7-14

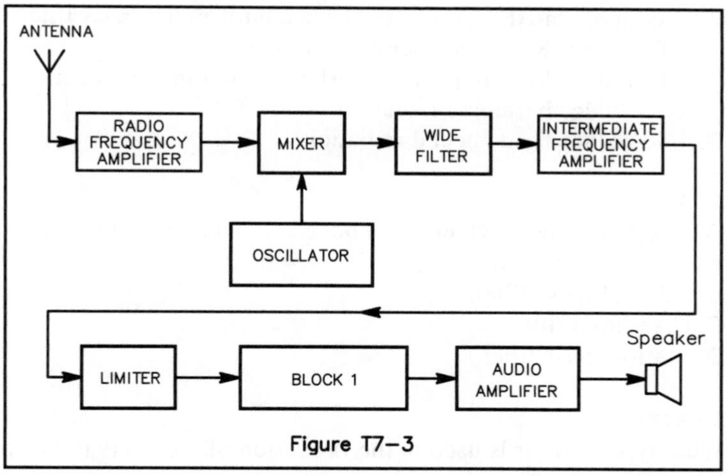

ANTENNA
RADIO FREQUENCY AMPLIFIER
MIXER
WIDE FILTER
INTERMEDIATE FREQUENCY AMPLIFIER
OSCILLATOR
Speaker
LIMITER
BLOCK 1
AUDIO AMPLIFIER
Figure T7–3

T7A11
What is block 1 in the FM receiver shown in Figure T7-3?
A. A frequency discriminator
B. A product detector
C. A frequency-shift modulator
D. A phase inverter

T7A11
(A)
Page 7-14

T7A12
What would happen if block 1 failed to function in the FM receiver diagram shown in Figure T7-3?
A. The audio output would sound loud and distorted
B. There would be no audio output
C. There would be no effect
D. The receiver's power supply would be short-circuited

T7A12
(B)
Pae 7-14

T7A13
(C)
Page 7-12

T7A13
What is block 1 in Figure T7-4?
A. A band-pass filter
B. A crystal oscillator
C. A reactance modulator
D. A rectifier modulator

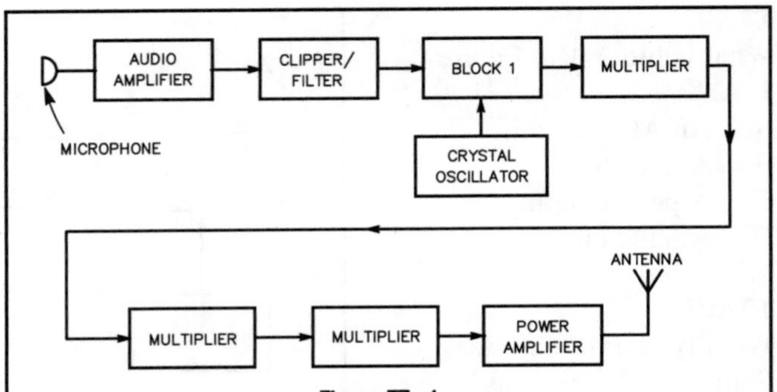

Figure T7–4

T7A14
(C)
Page 7-12

T7A14
What circuit is shown in Figure T7-4 if block 1 is a reactance modulator?
A. A single-sideband transmitter
B. A double-sideband AM transmitter
C. An FM transmitter
D. A product transmitter

T7A15
(D)
Page 7-11

T7A15
How would the output of the FM transmitter shown in Figure T7-4 be affected if the audio amplifier failed to operate (assuming block 1 is a reactance modulator)?
A. There would be no output from the transmitter
B. The output would be 6-dB below the normal output power
C. The transmitted audio would be distorted but understandable
D. The output would be an unmodulated carrier

T7A16
(C)
Page 7-8

T7A16
Why do modern HF transmitters have a built-in low-pass filter in their RF output circuits?
A. To reduce RF energy below a cutoff point
B. To reduce low-frequency interference to other amateurs
C. To reduce harmonic radiation
D. To reduce fundamental radiation

T7A17
(A)
Page 7-10

T7A17
What circuit blocks RF energy above and below certain limits?
A. A band-pass filter
B. A high-pass filter
C. An input filter
D. A low-pass filter

T7A18
(A)
Page 7-10

T7A18
What type of filter is used in the IF section of receivers to block energy outside a certain frequency range?
A. A band-pass filter
B. A high-pass filter
C. An input filter
D. A low-pass filter

T7A19
(C)
Page 7-13

T7A19
What circuit function is found in all types of receivers?
A. An audio filter
B. A beat-frequency oscillator
C. A detector
D. An RF amplifier

T7A20
What would you use to connect a dual-band antenna to a mobile transceiver which has separate VHF and UHF outputs?
A. A dual-needle SWR meter
B. A full-duplex phone patch
C. Twin high-pass filters
D. A duplexer

T7A20
(D)
Page 7-15

Subelement T8 — Signals and Emissions [2 Exam Questions — 2 Groups]

Subelement T8

T8A Concepts and types of modulation

T8A01
What is the name for unmodulated carrier wave emissions?
A. Phone
B. Test
C. MCW
D. RTTY

T8A01
(B)
Page 2-3

T8A02
What is the name for emissions produced by switching a transmitter's output on and off?
A. Phone
B. Test
C. CW
D. RTTY

T8A02
(C)
Page 2-3

T8A03
What term describes the process of combining an information signal with a radio signal?
A. Superposition
B. Modulation
C. Demodulation
D. Phase-inversion

T8A03
(B)
Page 2-7

T8A04
What is the name for packet-radio emissions?
A. CW
B. Data
C. Phone
D. RTTY

T8A04
(B)
Page 2-18

T8A05
How is tone-modulated Morse code produced?
A. By feeding a microphone's audio signal into an FM transmitter
B. By feeding an on/off keyed audio tone into a CW transmitter
C. By on/off keying of a carrier
D. By feeding an on/off keyed audio tone into a transmitter

T8A05
(D)
Page 2-15

T8A06
What is the name of the voice emission most used on VHF/UHF repeaters?
A. Single-sideband phone
B. Pulse-modulated phone
C. Slow-scan phone
D. Frequency-modulated phone

T8A06
(D)
Page 2-10

T8A07
(A)
Page 2-7

T8A07
Which of the following voice emission modes begins by amplitude modulating an RF carrier?
A. Single-sideband phone
B. Pulse-modulated phone
C. Phase-modulated phone
D. Width-modulated phone

T8A08
(A)
Page 2-7

T8A08
What is meant by the upper-sideband (USB)?
A. The part of a single-sideband signal that is above the carrier frequency
B. The part of a single-sideband signal that is below the carrier frequency
C. Any frequency above 10 MHz
D. The carrier frequency of a single-sideband signal

T8A09
(D)
Page 7-12

T8A09
What emissions are produced by a transmitter using a reactance modulator?
A. CW
B. Test
C. Single-sideband, suppressed-carrier phone
D. Phase-modulated phone

T8A10
(C)
Page 7-12

T8A10
What other emission does phase modulation most resemble?
A. Amplitude modulation
B. Pulse modulation
C. Frequency modulation
D. Single-sideband modulation

T8A11
(B)
Page 2-15

T8A11
What is the name for emissions produced by an on/off keyed audio tone?
A. RTTY
B. MCW
C. CW
D. Phone

T8A12
(D)
Page 7-12

T8A12
If you receive a phase-modulated voice signal and a frequency-modulated voice signal, what difference will you notice?
A. Phase-modulated signals cannot be detected with most amateur equipment
B. Phase-modulated signals do not sound as clear as frequency-modulated signals
C. Phase-modulated signals are more difficult to tune in than frequency-modulated signals
D. The signals will sound the same

T8B RF carrier, definition and typical bandwidths and FM deviation

T8B01
(A)
Page 2-7

T8B01
What is another name for a constant-amplitude radio-frequency signal?
A. An RF carrier
B. An AF carrier
C. A sideband carrier
D. A subcarrier

T8B02
What is an RF carrier?
A. The part of a transmitter that carries the signal to the transmitter antenna
B. The part of a receiver that carries the signal from the antenna to the detector
C. A radio frequency signal that is modulated to produce a radiotelephone signal
D. A modulation that changes a radio frequency signal to produce a radiotelephone signal

T8B02
(C)
Page 2-7

T8B03
What kind of emission would your FM transmitter produce if its microphone failed to work?
A. An unmodulated carrier
B. A phase-modulated carrier
C. An amplitude-modulated carrier
D. A frequency-modulated carrier

T8B03
(A)
Page 7-11

T8B04
How would you modulate a 2-meter FM transceiver to produce packet-radio emissions?
A. Connect a terminal-node controller to interrupt the transceiver's carrier wave
B. Connect a terminal-node controller to the transceiver's microphone input
C. Connect a keyboard to the transceiver's microphone input
D. Connect a DTMF key pad to the transceiver's microphone input

T8B04
(B)
Page 2-19

T8B05
Why is FM voice best for local VHF/UHF radio communications?
A. The carrier is not detectable
B. It is more resistant to distortion caused by reflected signals
C. It has audio that is less affected by interference from static-type electrical noise
D. Its RF carrier stays on frequency better than the AM modes

T8B05
(C)
Page 2-10

T8B06
Why do many radio receivers have several IF filters of different bandwidths that can be selected by the operator?
A. Because some frequency bands are wider than others
B. Because different bandwidths help increase the receiver sensitivity
C. Because different bandwidths improve S-meter readings
D. Because some emission types need a wider bandwidth than others to be received properly

T8B06
(D)
Page 8-10

T8B07
Which list of emission types is in order from the narrowest bandwidth to the widest bandwidth?
A. RTTY, CW, SSB voice, FM voice
B. CW, FM voice, RTTY, SSB voice
C. CW, RTTY, SSB voice, FM voice
D. CW, SSB voice, RTTY, FM voice

T8B07
(C)
Page 8-10

T8B08
What is the usual bandwidth of a single-sideband amateur signal?
A. 1 kHz
B. 2 kHz
C. Between 3 and 6 kHz
D. Between 2 and 3 kHz

T8B08
(D)
Page 8-10

T8B09
What is the usual bandwidth of a frequency-modulated amateur signal?
A. Less than 5 kHz
B. Between 5 and 10 kHz
C. Between 10 and 20 kHz
D. Greater than 20 kHz

T8B09
(C)
Page 8-10

T8B10
(B)
Page 8-11

T8B10
What is the usual bandwidth of UHF amateur fast-scan television?
A. More than 6 MHz
B. About 6 MHz
C. About 3 MHz
D. About 1 MHz

T8B11
(B)
Page 8-4

T8B11
What is the result of over deviation in an FM transmitter?
A. Increased transmitter power
B. Out-of-channel emissions
C. Increased transmitter range
D. Poor carrier suppression

T8B12
(C)
Page 8-3

T8B12
What causes splatter interference?
A. Keying a transmitter too fast
B. Signals from a transmitter's output circuit are being sent back to its input circuit
C. Overmodulation of a transmitter
D. The transmitting antenna is the wrong length

Subelement T9

Subelement T9 — Antennas and Feed Lines [3 Exam Questions — 3 Groups]

T9A Parasitic beam and non-directional antennas

T9A01
(C)
Page 9-31

T9A01
What is a directional antenna?
A. An antenna that sends and receives radio energy equally well in all directions
B. An antenna that cannot send and receive radio energy by skywave or skip propagation
C. An antenna that sends and receives radio energy mainly in one direction
D. An antenna that uses a directional coupler to measure power transmitted

T9A02
(A)
Page 9-31

T9A02
How is a Yagi antenna constructed?
A. Two or more straight, parallel elements are fixed in line with each other
B. Two or more square or circular loops are fixed in line with each other
C. Two or more square or circular loops are stacked inside each other
D. A straight element is fixed in the center of three or more elements that angle toward the ground

T9A03
(C)
Page 9-31

T9A03
What type of beam antenna uses two or more parallel straight elements arranged in line with each other?
A. A delta loop antenna
B. A quad antenna
C. A Yagi antenna
D. A Zepp antenna

T9A04
How many directly driven elements do most parasitic beam antennas have?
A. None
B. One
C. Two
D. Three

T9A04
(B)
Page 9-31

T9A05
What is a parasitic beam antenna?
A. An antenna in which some elements obtain their radio energy by induction or radiation from a driven element
B. An antenna in which wave traps are used to magnetically couple the elements
C. An antenna in which all elements are driven by direct connection to the feed line
D. An antenna in which the driven element obtains its radio energy by induction or radiation from director elements

T9A05
(A)
Pge 9-31

T9A06
What are the parasitic elements of a Yagi antenna?
A. The driven element and any reflectors
B. The director and the driven element
C. Only the reflectors (if any)
D. Any directors or any reflectors

T9A06
(D)
Page 9-31

T9A07
What is a cubical quad antenna?
A. Four straight, parallel elements in line with each other, each approximately 1/2-electrical wavelength long
B. Two or more parallel four-sided wire loops, each approximately one-electrical wavelength long
C. A vertical conductor 1/4-electrical wavelength high, fed at the bottom
D. A center-fed wire 1/2-electrical wavelength long

T9A07
(B)
Page 9-32

T9A08
What is a delta loop antenna?
A. An antenna similar to a cubical quad antenna, except with triangular elements rather than square
B. A large copper ring or wire loop, used in direction finding
C. An antenna system made of three vertical antennas, arranged in a triangular shape
D. An antenna made from several triangular coils of wire on an insulating form

T9A08
(A)
Page 9-33

T9A09
Which of the following antennas is NOT an example of a parasitic beam?
A. A quad
B. A Yagi
C. A collinear array
D. A delta loop

T9A09
(C)
Page 9-31

T9A10
What type of non-directional antenna is easy to make at home and works well outdoors?
A. A Yagi
B. A delta loop
C. A cubical quad
D. A ground plane

T9A10
(D)
Page 9-16

T9A11
(D)
Page 9-18

T9A11
What type of antenna is made when a magnetic-base whip antenna is placed on the roof of a car?
A. A Yagi
B. A delta loop
C. A cubical quad
D. A ground plane

T9A12
(A)
Page 9-18

T9A12
If a magnetic-base whip antenna is placed on the roof of a car, in what direction does it send out radio energy?
A. It goes out equally well in all horizontal directions
B. Most of it goes in one direction
C. Most of it goes equally in two opposite directions
D. Most of it is aimed high into the air

T9B Polarization, impedance matching and SWR, feed lines, balanced vs. unbalanced (including baluns)

T9B01
(B)
Page 9-30

T9B01
What does horizontal wave polarization mean?
A. The magnetic lines of force of a radio wave are parallel to the Earth's surface
B. The electric lines of force of a radio wave are parallel to the Earth's surface
C. The electric lines of force of a radio wave are perpendicular to the Earth's surface
D. The electric and magnetic lines of force of a radio wave are perpendicular to the Earth's surface

T9B02
(C)
Page 9-30

T9B02
What does vertical wave polarization mean?
A. The electric lines of force of a radio wave are parallel to the Earth's surface
B. The magnetic lines of force of a radio wave are perpendicular to the Earth's surface
C. The electric lines of force of a radio wave are perpendicular to the Earth's surface
D. The electric and magnetic lines of force of a radio wave are parallel to the Earth's surface

T9B03
(C)
Page 9-16

T9B03
What is one advantage of using a single element vertical antenna?
A. It usually has a high angle of radiation
B. It is always a ground-independent antenna
C. It usually has a low-angle radiation pattern
D. It usually creates a high SWR on the transmission line

T9B04
(C)
Page 9-30

T9B04
What electromagnetic-wave polarization does a Yagi antenna have when its elements are parallel to the Earth's surface?
A. Circular
B. Helical
C. Horizontal
D. Vertical

T9B05
(D)
Page 9-30

T9B05
What electromagnetic-wave polarization does a half-wavelength antenna have when it is perpendicular to the Earth's surface?
A. Circular
B. Horizontal
C. Parabolical
D. Vertical

T9B06
What electromagnetic-wave polarization does most man-made electrical noise have in the HF and VHF spectrum?
A. Horizontal
B. Left-hand circular
C. Right-hand circular
D. Vertical

T9B06
(D)
Page 9-30

T9B07
What electromagnetic-wave polarization do most repeaters have in the VHF and UHF spectrum?
A. Horizontal
B. Vertical
C. Right-hand circular
D. Left-hand circular

T9B07
(B)
Page 9-30

T9B08
What electromagnetic-wave polarization is used for most satellite operation?
A. Only horizontal
B. Only vertical
C. Circular
D. No polarization

T9B08
(C)
Page 9-31

T9B09
What does standing-wave ratio mean?
A. The ratio of maximum to minimum inductances on a feed line
B. The ratio of maximum to minimum capacitances on a feed line
C. The ratio of maximum to minimum impedances on a feed line
D. The ratio of maximum to minimum voltages on a feed line

T9B09
(D)
Page 9-6

T9B10
Why should you try to maintain a low SWR when a VHF parasitic beam is fed with coaxial cable?
A. A low SWR reduces spurious emissions
B. A low SWR allows the transmission line to warm up in cold weather
C. A low SWR results in a more efficient transfer of energy from the transmission line to the antenna
D. A low SWR reduces front-end overload in neighboring television receivers

T9B10
(C)
Page 9-6

T9B11
What does forward power mean?
A. The power traveling from the transmitter to the antenna
B. The power radiated from the top of an antenna system
C. The power produced during the positive half of an RF cycle
D. The power used to drive a linear amplifier

T9B11
(A)
Page 9-6

T9B12
What does reflected power mean?
A. The power radiated down to the ground from an antenna
B. The power returned towards the source on a transmission line
C. The power produced during the negative half of an RF cycle
D. The power returned to an antenna by buildings and trees

T9B12
(B)
Page 9-6

T9B13
(C)
Page 9-3

T9B13
What happens to radio energy when it is sent through a poor quality coaxial cable?
A. It causes spurious emissions
B. It is returned to the transmitter's chassis ground
C. It is converted to heat in the cable
D. It causes interference to other stations near the transmitting frequency

T9B14
(A)
Page 9-4

T9B14
What is one disadvantage of using parallel-conductor open-wire transmission line?
A. It is more difficult to properly install
B. It is more expensive than coax
C. Its balanced characteristics cannot be matched to the 50-ohm output impedance of modern transceivers
D. It cannot be operated efficiently with a high SWR

T9B15
(C)
Page 9-5

T9B15
What is an unbalanced line?
A. A feed line with neither conductor connected to ground
B. A feed line with both conductors connected to ground
C. A feed line with one conductor connected to ground
D. All of these answers are correct

T9B16
(D)
Page 9-5

T9B16
What is a balanced line?
A. A feed line that has its inner conductor balanced with the outer shield
B. A feed line that is always operated at a low SWR to preserve its balance
C. A feed line with an impedance that is balanced at 450 ohms or more
D. A feed line made of 2 parallel conductors with a uniform space between them

T9B17
(A)
Page 9-6

T9B17
What device can be installed to feed a balanced antenna with an unbalanced feed line?
A. A balun
B. A loading coil
C. A triaxial transformer
D. A wavetrap

T9B18
(C)
Page 9-6

T9B18
Which of the following would you NOT use to make a balun?
A. A toroid
B. A length of transmission line
C. A pair of tantalum capacitors
D. A pair of air-wound coils

T9C Line losses by line type, length and frequency

T9C01
(B)
Page 9-29

T9C01
What common connector usually joins RG-213 coaxial cable to an HF transceiver?
A. An F-type cable connector
B. A PL-259 connector
C. A banana plug connector
D. A binding post connector

T9C02
What common connector usually joins a hand-held transceiver to its antenna?
A. A BNC connector
B. A PL-259 connector
C. An F-type cable connector
D. A binding post connector

T9C02
(A)
Page 9-29

T9C03
Which of these common connectors has the lowest loss at UHF?
A. An F-type cable connector
B. A type-N connector
C. A BNC connector
D. A PL-259 connector

T9C03
(B)
Page 9-30

T9C04
If you install a 6-meter Yagi antenna on a tower 150 feet from your transmitter, which of the following feed lines is best?
A. RG-213
B. RG-58
C. RG-59
D. RG-174

T9C04
(A)
Page 9-29

T9C05
If you have a transmitter and an antenna that are 50 feet apart, but are connected by 200 feet of RG-58 coaxial cable, what should you do to reduce antenna system loss?
A. Cut off the excess cable so the feed line is an even number of wavelengths long
B. Cut off the excess cable so the feed line is an odd number of wavelengths long
C. Cut off the excess cable
D. Roll the excess cable into a coil that is as small as possible

T9C05
(C)
Page 9-29

T9C06
As the length of a feed line is changed, what happens to signal loss?
A. Signal loss is the same for any length of feed line
B. Signal loss increases as length increases
C. Signal loss decreases as length increases
D. Signal loss is the least when the length is the same as the signal's wavelength

T9C06
(B)
Page 9-29

T9C07
As the frequency of a signal is changed, what happens to signal loss in a feed line?
A. Signal loss is the same for any frequency
B. Signal loss increases with increasing frequency
C. Signal loss increases with decreasing frequency
D. Signal loss is the least when the signal's wavelength is the same as the feed line's length

T9C07
(B)
Page 9-29

T9C08
If your antenna feed line gets hot when you are transmitting, what might this mean?
A. You should transmit using less power
B. The conductors in the feed line are not insulated very well
C. The feed line is too long
D. The SWR may be too high, or the feed line loss may be high

T9C08
(D)
Page 9-28

T9C09
Why should you regularly clean, tighten and re-solder all antenna connectors?
A. To help keep their resistance at a minimum
B. To keep them looking nice
C. To keep them from getting stuck in place
D. To increase their capacitance

T9C09
(A)
Page 9-30

T9C10
(C)
Page 9-29

T9C10
Which of the following is a reason to use good-quality, large-diameter coax in your VHF installations?
A. To allow operation with a high SWR
B. To keep the signal confined to the center conductor
C. To keep losses to a minimum
D. To allow operation on harmonically related bands

T9C11
(B)
Page 9-29

T9C11
Why is household lamp cord (zip-cord) not a good feed line to use for a 6 meter antenna installation?
A. The line would not warm up properly at this frequency
B. Line losses would be great at this frequency
C. Line impedance would be too great at this frequency
D. Line impedance would be too low at this frequency

Subelement T0

Subelement T0 — RF Safety [5 Exam Questions — 5 Groups]

T0A RF safety fundamentals

T0A01
(B)
Page 10-6

T0A01
Why is it a good idea to adhere to the FCC's Rules for using the minimum power needed when you are transmitting with your hand-held radio?
A. Large fines are always imposed on operators violating this rule
B. To reduce the level of RF radiation exposure to the operator's head
C. To reduce calcification of the NiCd battery pack
D. To eliminate self oscillation in the receiver RF amplifier

T0A02
(D)
Page 10-8

T0A02
Over what frequency range are the FCC Regulations most stringent for RF radiation exposure?
A. Frequencies below 300 kHz
B. Frequencies between 300 kHz and 3 MHz
C. Frequencies between 3 MHz and 30 MHz
D. Frequencies between 30 MHz and 300 MHz

T0A03
(C)
Page 10-2

T0A03
What is one biological effect to the eye that can result from RF exposure?
A. The strong magnetic fields can cause blurred vision
B. The strong magnetic fields can cause polarization lens
C. It can cause heating, which can result in the formation of cataracts
D. It can cause heating, which can result in astigmatism

T0A04
(B)
Page 10-12

T0A04
How do you calculate the boundary between the near field and the far field of a full sized dipole or Yagi antenna?
A. Multiply the square root of the antenna length by 2 and divide by the frequency of the signal
B. Multiply the square of the antenna length by 2 and divide by the wavelength of the signal
C. Divide the antenna length by 2 and multiply by the frequency
D. Divide the square of the antenna length by 2 and multiply by the wavelength

T0A05
In the far field, as the distance from the source increases, how does power density vary?
A. The power density is proportional to the square of the distance
B. The power density is proportional to the square root of the distance
C. The power density is proportional to the inverse square of the distance
D. The power density is proportional to the inverse cube of the distance

T0A05
(C)
Page 10-12

T0A06
In the near field, how does the field strength vary with distance from the source?
A. It always increases with the cube of the distance
B. It always decreases with the cube of the distance
C. It varies as a sine wave with distance
D. It depends on the type of antenna being used

T0A06
(D)
Page 10-12

T0A07
In the far field, what is the relationship between the electric (E) field and magnetic (H) field?
A. In the formula 50 ohms equals E divided by H; it is a fixed relationship
B. In the formula 72 ohms equals H divided by E; it is a fixed relationship
C. In the formula 377 ohms equals E divided by H; it is a fixed relationship
D. In the formula 450 ohms equals H divided by E; it is a fixed relationship

T0A07
(C)
Page 10-12

T0A08
Why should you never look into the open end of a waveguide while the transmitter is operating?
A. You may be exposing your eyes to more than the maximum permissible exposure level of RF radiation
B. You may be exposing your eyes to more than the maximum permissible exposure level of infrared radiation
C. You may be exposing your eyes to more than the maximum permissible exposure level of ultraviolet radiation
D. All of these choices are correct

T0A08
(A)
Page 10-6

T0A09
Why should you never look into the open end of a microwave feed horn antenna while the transmitter is operating?
A. You may be exposing your eyes to more than the maximum permissible exposure of RF radiation
B. You may be exposing your eyes to more than the maximum permissible exposure level of infrared radiation
C. You may be exposing your eyes to more than the maximum permissible exposure level of ultraviolet radiation
D. All of these choices are correct

T0A09
(A)
Page 10-6

T0A10
Why are Amateur Radio operators required to meet the FCC RF radiation exposure limits?
A. The standards are applied equally to all radio services
B. To ensure that RF radiation occurs only in a desired direction
C. Because amateur station operations are more easily adjusted than those of commercial radio services
D. To ensure a safe operating environment for amateurs, their families and neighbors

T0A10
(D)
Page 10-8

T0A11
(A)
Page 10-2

T0A11
Why are the maximum permissible exposure (MPE) levels not uniform throughout the radio spectrum?
A.. The human body absorbs energy differently at various frequencies
B. Some frequency ranges have a cooling effect while others have a heating effect on the body
C. Some frequency ranges have no effect on the body
D. Radiation at some frequencies can have a catalytic effect on the body

T0B RF safety terms and definitions

T0B01
(C)
Page 10-2

T0B01
What does the term "specific absorption rate" or SAR mean?
A. The degree of RF energy consumed by the ionosphere
B. The rate at which transmitter energy is lost because of a poor feed line
C. The rate at which RF energy is absorbed into the human body
D. The amount of signal weakening caused by atmospheric phenomena

T0B02
(C)
Page 10-1

T0B02
Which of the following terms best describe RF radiation?
A. Cohesive radiation
B. Ionizing radiation
C. Nonionizing radiation
D. Impulse radiation

T0B03
(B)
Page 10-1

T0B03
Why is RF energy classified as nonionizing radiation?
A. Because the frequency is too high for there to be enough photon energy to ionize atoms
B. Because the frequency is too low for there to be enough photon energy to ionize atoms
C. Because it has no polar component
D. Because it has no power factor

T0B04
(D)
Page 10-2

T0B04
On what value are the maximum permissible exposure (MPE) limits based?
A. The square of the mass of the exposed body
B. The square root of the mass of the exposed body
C. The whole-body specific gravity (WBSG)
D. The whole-body specific absorption rate (SAR)

T0B05
(C)
Page 10-2

T0B05
Why do exposure limits vary with frequency?
A. Lower-frequency RF fields have more energy than higher-frequency fields
B. Lower-frequency RF fields penetrate deeper into the body than higher-frequency fields
C. The body's ability to absorb RF energy varies with frequency
D. It is impossible to measure specific absorption rates at some frequencies

T0B06
(A)
Page 10-10

T0B06
Why is the concept of "duty cycle" one factor used to determine safe RF radiation exposure levels?
A. It takes into account the amount of time the transmitter is operating at full power during a single transmission
B. It takes into account the transmitter power supply rating
C. It takes into account the antenna feed line loss
D. It takes into account the thermal effects of the final amplifier

T0B07
Why is the concept of "time averaging" one factor used to determine safe RF radiation exposure levels?
A. It takes into account the operating frequency
B. It takes into account the transmit/receive time ratio during normal amateur communication
C. It takes into account the overall efficiency of the final amplifier
D. It takes into account the antenna feed line loss

T0B07
(B)
Page 10-8

T0B08
Why is the concept of "specific absorption rate (SAR)" one factor used to determine safe RF radiation exposure levels?
A. It takes into account the overall efficiency of the final amplifier
B. It takes into account the transmit/receive time ratio during normal amateur communication
C. It takes into account the rate at which the human body absorbs RF energy at a particular frequency
D. It takes into account the antenna feed line loss

T0B08
(C)
Page 10-2

T0B09
Why must the frequency of an RF source be considered when evaluating RF radiation exposure?
A. Lower-frequency RF fields have more energy than higher-frequency fields
B. Lower-frequency RF fields penetrate deeper into the body than higher-frequency fields
C. Higher-frequency RF fields are transient in nature, and do not affect the human body
D. The human body absorbs more RF energy at some frequencies than at others

T0B09
(D)
Page 10-2

T0B10
Which radio frequency emission has the shortest duty cycle?
A. Two-way exchanges of phase modulated signals
B. Two-way exchanges of FM telephony
C. Two-way exchanges of SSB, single-sideband signals
D. Two-way exchanges of CW, Morse code signals

T0B10
(C)
Page 10-10

T0B11
From an RF safety standpoint, what impact does the duty cycle have on the minimum safe distance separating an antenna and the neighboring environment?
A. The lower the duty cycle, the shorter the compliance distance
B. The compliance distance is increased with an increase in the duty cycle
C. Lower duty cycles subject the environment to lower radio-frequency radiation
D. All of these answers are correct

T0B11
(D)
Page 10-10

T0B12
What effect does a 50% duty cycle have on the calculated "key down" RF safety distance from an amateur antenna to a neighboring residence?
A. The compliance distance is reduced
B. You must also multiply the distance by 50%
C. Duty cycle is not a consideration in the RF safety calculations
D. You divide the duty cycle into the inverse square of the distance

T0B12
(A)
Page 10-10

T0C RF safety rules and guidelines

T0C01
(A)
Page 10-11

T0C01
Referring to Figure NT0-1, what is the formula for calculating the maximum permissible exposure (MPE) limit for controlled environments on the 1.25-meter (222 MHz) band?

A. There is no formula, MPE is a fixed power density of 1.0 milliwatt per square centimeter averaged over any 6 minutes
B. There is no formula, MPE is a fixed power density of 0.2 milliwatt per square centimeter averaged over any 30 minutes
C. The MPE in milliwatts per square centimeter equals the frequency in megahertz divided by 300 averaged over any 6 minutes
D. The MPE in milliwatts per square centimeter equals the frequency in megahertz divided by 1500 averaged over any 30 minutes

T0C02
(B)
Page 10-11

T0C02
Referring to Figure NT0-1, what is the formula for calculating the maximum permissible exposure (MPE) limit for uncontrolled environments on the 2-meter (146 MHz) band?

A. There is no formula, MPE is a fixed power density of 1.0 milliwatt per square centimeter averaged over any 6 minutes
B. There is no formula, MPE is a fixed power density of 0.2 milliwatt per square centimeter averaged over any 30 minutes
C. The MPE in milliwatts per square centimeter equals the frequency in megahertz divided by 300 averaged over any 6 minutes
D. The MPE in milliwatts per square centimeter equals the frequency in megahertz divided by 1500 averaged over any 30 minutes

RF Exposure Limits

(A) Limits for Occupational/Controlled Exposure				
Frequency Range (MHz)	Electric Field Strength (V/m)	Magnetic Field (A/m)	Power Density (mW/cm^2)	Averaging Time (minutes)
0.3-3.0	614	1.63	(100)*	6
3.0-30	1842/f	4.89/f	($900/f^2$)*	6
30-300	61.4	0.163	1.0	6
300-1500	——	——	f/300	6
1500-100,000	——	——	5	6
f = frequency in MHz		* = Plane-wave equivalent power density		

(B) Limits for General Population/Uncontrolled Exposure				
Frequency Range (MHz)	Electric Field Strength (V/m)	Magnetic Field (A/m)	Power Density (mW/cm^2)	Averaging Time (minutes)
0.3-1.34	614	1.63	(100)*	30
1.34-30	824/f	2.19/f	($180/f^2$)*	30
30-300	27.5	0.073	0.2	30
300-1500	——	——	f/1500	30
1500-100,000	——	——	1.0	30
f = frequency in MHz		* = Plane-wave equivalent power density		

Figure NT0-1

T0C03
Referring to Figure NT0-1, what is the formula for calculating the maximum permissible exposure (MPE) limit for controlled environments on the 70-centimeter (440 MHz) band?

A. There is no formula, MPE is a fixed power density of 1.0 milliwatt per square centimeter averaged over any 6 minutes
B. There is no formula, MPE is a fixed power density of 0.2 milliwatt per square centimeter averaged over any 30 minutes
C. The MPE in milliwatts per square centimeter equals the frequency in megahertz divided by 300 averaged over any 6 minutes
D. The MPE in milliwatts per square centimeter equals the frequency in megahertz divided by 1500 averaged over any 30 minutes

T0C03
(C)
Page 10-11

T0C04
Referring to Figure NT0-1, what is the formula for calculating the maximum permissible exposure (MPE) limit for uncontrolled environments on the 1240 to 1300-MHz band?

A. There is no formula, MPE is a fixed power density of 1.0 milliwatt per square centimeter averaged over any 6 minutes
B. There is no formula, MPE is a fixed power density of 0.2 milliwatt per square centimeter averaged over any 30 minutes
C. The MPE in milliwatts per square centimeter equals the frequency in megahertz divided by 300 averaged over any 6 minutes
D. The MPE in milliwatts per square centimeter equals the frequency in megahertz divided by 1500 averaged over any 30 minutes

T0C04
(D)
Page 10-11

T0C05
Referring to Figure NT0-1, what is the electric field strength of the maximum permissible exposure (MPE) limit for controlled environments on the 2-meter (144 MHz) band?

A. 61.4 volts per meter
B. 27.5 volts per meter
C. 0.163 volts per meter
D. 0.073 volts per meter

T0C05
(A)
Page 10-11

T0C06
Referring to Figure NT0-1, what is the electric field strength of the maximum permissible exposure (MPE) limit for uncontrolled environments on the 1.25-meter (222 MHz) band?

A. 61.4 volts per meter
B. 27.5 volts per meter
C. 0.163 volts per meter
D. 0.073 volts per meter

T0C06
(B)
Page 10-11

T0C07
On which of the following amateur bands will the maximum permissible exposure (MPE) limits be a constant value for controlled RF radiation exposure environments?

A. 1240 to 1300 MHz
B. 902 to 928 MHz
C. 420 to 450 MHz
D. 222 to 225 MHz

T0C07
(D)
Page 10-11

T0C08
On which of the following amateur bands will the maximum permissible exposure (MPE) limits be a constant value for uncontrolled RF radiation exposure environments?

A. 1240 to 1300 MHz
B. 902 to 928 MHz
C. 420 to 450 MHz
D. 144 to 148 MHz

T0C08
(D)
Page 10-11

T0C09
(C)
Page 10-10

T0C09
Where will you find the applicable FCC RF radiation maximum permissible exposure (MPE) limits defined?
A. FCC Part 97 Amateur Service Rules and Regulations
B. FCC Part 15 Radiation Exposure Rules and Regulations
C. FCC Part 1 and Office of Engineering and Technology (OET) Bulletin 65
D. Environmental Protection Agency Regulation 65

T0C10
(A)
Page 10-17

T0C10
What factors must you consider if your repeater station antenna will be located at a site that is occupied by antennas for transmitters in other services?
A. Your radiated signal must be considered as part of the total RF radiation from the site when determining RF radiation exposure levels
B. Each individual transmitting station at a multiple-transmitter site must meet the RF radiation exposure levels
C. Each station at a multiple-transmitter site may add no more than 1% of the maximum permissible exposure (MPE) for that site
D. Amateur stations are categorically excluded from RF radiation exposure evaluation at multiple-transmitter sites

T0C11
(B)
Page 10-16

T0C11
Which of the following categories describes most common amateur use of a hand-held transceiver?
A. Mobile devices
B. Portable devices
C. Fixed devices
D. None of these choices is correct

T0C12
(C)
Page 10-16

T0C12
Why does the FCC consider a hand-held transceiver to be a portable device when evaluating for RF radiation exposure?
A. Because it is generally a low-power device
B. Because it is designed to be carried close to your body
C. Because it's transmitting antenna is generally within 20 centimeters of the human body
D. All of these choices are correct

T0C13
(C)
Page 10-13

T0C13
Using Table NT0-1 what is the minimum safe distance for an uncontrolled RF radiation environment from a station using a half-wavelength dipole antenna on 3.5 MHz at 100 watts?
A. 6 feet
B. 3.4 feet
C. 1.5 feet
D. 3 feet

T0C14
(B)
Page 10-13

T0C14
Using Table NT0-1 what is the minimum safe distance for an uncontrolled RF radiation environment from a station using a quarter-wave vertical antenna on 7 MHz at 100 watts?
A. 4.0 feet
B. 2.7 feet
C. 1.2 feet
D. 7.5 feet

Table NT0-1

RF Exposure Limits

Section B.

Estimated distances to meet RF power density guidelines with an omni-directional HF quarter-wave vertical or ground-plane antenna (estimated gain, 1 dBi). Calculations include the EPA ground reflection factor of 2.56.

Frequency: 7 MHz
Estimated antenna gain: 1 dBi
Controlled limit: 18.37 mW/cm²
Uncontrolled limit: 3.67 mW/cm²

Transmitter power (watts)	Distance to controlled limit	Distance to uncontrolled limit
100	1.2′	2.7′
500	2.7′	6.1′
1000	3.9′	8.7′
1500	4.7′	10.6′

Frequency: 21 MHz
Estimated antenna gain: 1 dBi
Controlled limit: 2.04 mW/cm²
Uncontrolled limit: 0.408 mW/cm²

Transmitter power (watts)	Distance to controlled limit	Distance to uncontrolled limit
100	3.7′	8.2′
500	8.2′	18.4′
1000	11.6′	26′
1500	14.2′	31.9′

Frequency: 28 MHz
Estimated antenna gain: 1 dBi
Controlled limit: 1.15 mW/cm²
Uncontrolled limit: 0.23 mW/cm²

Transmitter power (watts)	Distance to controlled limit	Distance to uncontrolled limit
100	4.9′	11′
500	11′	24.5′
1000	15.5′	34.7′
1500	19′	42.5′

Section C.

Estimated distances to meet RF power density guidelines with a horizontal half-wave dipole antenna (estimated gain, 2 dBi). Calculations include the EPA ground reflection factor of 2.56.

Frequency: 3.5 MHz
Estimated antenna gain: 2 dBi
Controlled limit: 73.5 mW/cm²
Uncontrolled limit: 14.7 mW/cm²

Transmitter power (watts)	Distance to controlled limit	Distance to uncontrolled limit
100	0.7′	1.5′
500	1.5′	3.4′
1000	2.2′	4.9′
1500	2.7′	6′

Frequency: 7 MHz
Estimated antenna gain: 2 dBi
Controlled limit: 18.37 mW/cm²
Uncontrolled limit: 3.67 mW/cm²

Transmitter power (watts)	Distance to controlled limit	Distance to uncontrolled limit
100	1.4′	3.1′
500	3.1′	6.9′
1000	4.3′	9.7′
1500	5.3′	11.9′

Frequency: 21 MHz
Estimated antenna gain: 2 dBi
Controlled limit: 2.04 mW/cm²
Uncontrolled limit: 0.408 mW/cm²

Transmitter power (watts)	Distance to controlled limit	Distance to uncontrolled limit
100	4.1′	9.2′
500	9.2′	20.6′
1000	13′	29.2′
1500	16′	35.7′

Frequency: 28 MHz
Estimated antenna gain: 2 dBi
Controlled limit: 1.15 mW/cm²
Uncontrolled limit: 0.23 mW/cm²

Transmitter power (watts)	Distance to controlled limit	Distance to uncontrolled limit
100	5.5′	12.3′
500	12.3′	27.5′
1000	17.4′	38.9′
1500	21.3′	47.7′

T0C15
(A)
Page 10-13

T0C15
Using Table NT0-1 what is the minimum safe distance for a controlled RF radiation environment from a station using a quarter-wave vertical on 28 MHz at 100 watts?
A. 4.9 feet
B. 3.5 feet
C. 7 feet
D. 11 feet

T0C16
(A)
Page 10-13

T0C16
What is the minimum safe distance for a controlled RF radiation environment from a station using a half-wavelength dipole antenna on 7 MHz at 100 watts PEP, as specified in Table NT0-1?
A. 1.4 feet
B. 2 feet
C. 3.1 feet
D. 6.5 feet

T0C17
(A)
Page 10-13

T0C17
Using Table NT0-1 what is the uncontrolled limit for a station using a 3.5 MHz half-wavelength dipole antenna at 100 watts?
A. 1.5 feet
B. 2 feet
C. 3 feet
D. 3.4 feet

T0C18
(A)
Page 10-13

T0C18
Using Table NT0-1 what is the controlled limit for a station using a 21 MHz quarter-wave vertical at 100 watts?
A. 3.7 feet
B. 6 feet
C. 8.2 feet
D. 20 feet

T0C19
(B)
Page 10-13

T0C19
Using Table NT0-1 what is the uncontrolled limit for a station using a 21 MHz quarter-wave vertical at 100 watts?
A. 3.7 feet
B. 8.2 feet
C. 14.5 feet
D. 26.5 feet

T0C20
(C)
Page 10-13

T0C20
What is the minimum safe distance for a controlled RF radiation environment from a station using a half-wavelength dipole antenna on 21 MHz at 100 watts PEP, as specified in Table NT0-1?
A. 1.5 feet
B. 2 feet
C. 4.1 feet
D. 9.2 feet

T0C21
What is the minimum safe distance for an uncontrolled RF radiation environment from a station using a half-wavelength dipole antenna on 21 MHz at 100 watts PEP, as specified in Table NT0-1?
A. 2.5 feet
B. 9.2 feet
C. 8 feet
D. 20.6 feet

T0C21
(B)
Page 10-13

T0C22
Using Table NT0-1 what is the minimum safe distance for an uncontrolled RF radiation environment from a station using a 28 MHz half-wavelength dipole antenna at 100 watts?
A. 12.3 feet
B. 14.5 feet
C. 27.5 feet
D. 30 feet

T0C22
(A)
Page 10-13

T0D Routine station evaluation

T0D01
This question has been withdrawn.

T0D01

T0D02
Which of the following factors must be taken into account when using a computer program to model RF fields at your station?
A. Height above sea level at your station
B. Ionization level in the F2 region of the ionosphere
C. Ground interactions
D. The latitude and longitude of your station location

T0D02
(C)
Page 10-7

T0D03
In which of the following areas is it most difficult to accurately evaluate the effects of RF radiation exposure?
A. In the far field
B. In the cybersphere
C. In the near field
D. In the low-power field

T0D03
(C)
Page 10-6

T0D04
Is it necessary for you to perform mathematical calculations of the RF radiation exposure if your station transmits with more than 50 watts peak envelope power (PEP)?
A. Yes, calculations are always required to ensure greatest accuracy
B. Calculations are required if your station is located in a densely populated neighborhood
C. No, calculations may not give accurate results, so measurements are always required
D. No, there are alternate means to determine if your station meets the RF radiation exposure limits

T0D04
(D)
Page 10-8

T0D05
(A)
Page 10-16

T0D05
Which point on Figure NT0-2 represents the power density in the main beam of an antenna transmitting 1000 watts effective radiated power (ERP) at a location 10 meters from the antenna?
A. Point 1
B. Point 2
C. Point 3
D. Point 4

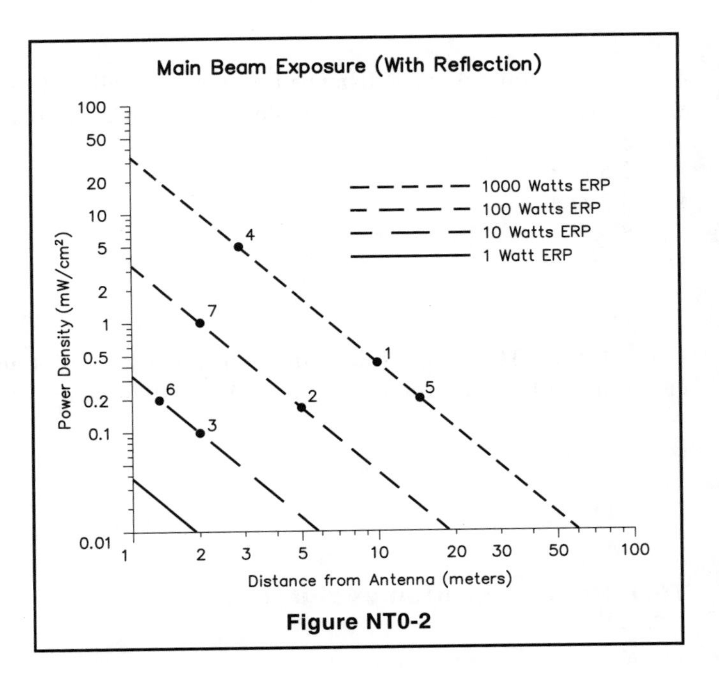

Figure NT0-2

T0D06
(B)
Page 10-16

T0D06
Which point on Figure NT0-2 represents the power density in the main beam of an antenna transmitting 100 watts effective radiated power (ERP) at a location 5 meters from the antenna?
A. Point 1
B. Point 2
C. Point 3
D. Point 6

T0D07
(C)
Page 10-16

T0D07
Which point on Figure NT0-2 represents the power density in the main beam of an antenna transmitting 10 watts effective radiated power (ERP) at a location 2 meters from the antenna?
A. Point 1
B. Point 2
C. Point 3
D. Point 6

T0D08
(C)
Page 10-16

T0D08
Which point on Figure NT0-2 represents the power density in the main beam of an antenna transmitting 1000 watts effective radiated power (ERP) at a location 3 meters from the antenna?
A. Point 1
B. Point 3
C. Point 4
D. Point 5

T0D09
(A)
Page 10-16

T0D09
Which point on Figure NT0-2 represents a power density of 0.2 milliwatts per square centimeter in the main beam of an antenna transmitting 1000 watts effective radiated power (ERP)?
A. Point 5
B. Point 2
C. Point 3
D. Point 4

T0D10
For what purpose might you use the graph shown in Figure NT0-2?
A. To determine the maximum permissible radiated power for your license class
B. To help evaluate the worst case RF radiation exposure from your station
C. To help evaluate the efficiency of your antenna system
D. All of these choices are correct

T0D10
(B)
Page 10-16

T0D11
Which point on Figure NT0-2 represents the power density at a location 10 meters from the rear of an antenna transmitting 1000 watts effective radiated power (ERP)?
A. Point 2
B. Point 3
C. Point 6
D. None of these choices is correct

T0D11
(D)
Page 10-16

T0D12
Using Table NT0-1 what is the minimum safe distance for a controlled RF radiation environment from a station using a 146 MHz quarter-wave vertical antenna at 10 watts?
A. 1.7 feet
B. 2.5 feet
C. 1.2 feet
D. 2 feet

T0D12
(A)
Page 10-13

Table NT0-1

RF Exposure

Section D.

Estimated distances to meet RF power density guidelines with a VHF quarter-wave ground-plane or mobile whip antenna (estimated gain, 1 dBi). Calculations include the EPA ground reflection factor of 2.56.

Frequency: 146 MHz
Estimated antenna gain: 1 dBi
Controlled limit: 1 mW/cm²
Uncontrolled limit: 0.2 mW/cm²

Transmitter power (watts)	Distance to controlled limit	Distance to uncontrolled limit
10	1.7′	3.7′
50	3.7′	8.3′
150	6.4′	14.4′

T0D13
Using Table NT0-1 what is the minimum safe distance for a controlled RF radiation environment from a station using a 146 MHz quarter-wave vertical antenna at 50 watts?
A. 3.7 feet
B. 3 feet
C. 4 feet
D. 8.3 feet

T0D13
(A)
Page 10-13

T0D14
Using Table NT0-1 what is the minimum safe distance for a controlled RF radiation environment from a station using a 146 MHz quarter-wave vertical antenna at 150 watts?
A. 5 feet
B. 6.4 feet
C. 14.4 feet
D. 9 feet

T0D14
(B)
Page 10-13

T0D15
Using Table NT0-1 what is the minimum safe distance for an uncontrolled RF radiation environment from a station using a 146 MHz quarter-wave vertical antenna at 150 watts?
A. 6 feet
B. 14.4 feet
C. 8.3 feet
D. 10.5 feet

T0D15
(B)
Page 10-13

T0D16
(C)
Page 10-13

T0D16
Using Table NT0-1 what is the minimum safe distance for an uncontrolled RF radiation environment from a station using a 146 MHz quarter-wave vertical antenna at 50 watts?
A. 4 feet
B. 3.7 feet
C. 8.3 feet
D. 9 feet

T0D17
(B)
Page 10-13

T0D17
Using Table NT0-1 what is the minimum safe distance for an uncontrolled RF radiation environment from a station using a 146 MHz quarter-wave vertical antenna at 10 watts?
A. 1.7 feet
B. 3.7 feet
C. 3 feet
D. 4 feet

T0D18
(B)
Page 10-13

T0D18
Using Table NT0-1 what is the minimum safe distance for an uncontrolled RF radiation environment from a station using a 446 MHz 5/8-wave vertical antenna at 10 watts?
A. 1 foot
B. 4.3 feet
C. 9.6 feet
D. 6 feet

Table NT0-1

RF Exposure Limits

Section E.

Estimated distances to meet RF power density guidelines in the main beam of a UHF 5/8-wavelength ground-plane or mobile whip antenna (estimated gain, 4 dBi). Calculations include the EPA ground reflection factor of 2.56.

Frequency: 446 MHz
Estimated antenna gain: 4 dBi
Controlled limit: 1.49 mW/cm²
Uncontrolled limit: 0.3 mW/cm²

Transmitter power (watts)	Distance to controlled limit	Distance to uncontrolled limit
10	1.9′	4.3′
50	4.3′	9.6′
150	7.5′	16.7′

T0D19
(C)
Page 10-13

T0D19
Using Table NT0-1 what is the minimum safe distance for an uncontrolled RF radiation environment from a station using a 446 MHz 5/8-wave vertical antenna at 50 watts?
A. 2.5 foot
B. 4.3 feet
C. 9.6 feet
D. 9 feet

T0D20
(A)
Page 10-13

T0D20
Using Table NT0-1 what is the minimum safe distance for an uncontrolled RF radiation environment from a station using a 446 MHz 5/8-wave vertical antenna at 150 watts?
A. 16.7 feet
B. 7.5 feet
C. 6 feet
D. 10.5 feet

T0D21
(B)
Page 10-13

T0D21
Using Table NT0-1 what is the minimum safe distance for a controlled RF radiation environment from a station using a 446 MHz 5/8-wave vertical antenna at 150 watts?
A. 16.7 feet
B. 7.5 feet
C. 2.5 feet
D. 1 foot

T0D22
Using Table NT0-1 what is the minimum safe distance for a controlled RF radiation environment from a station using a 446 MHz 5/8-wave vertical antenna at 50 watts?
A. 1 foot
B. 4.3 feet
C. 1.9 feet
D. 6 feet

T0D22
(B)
Page 10-13

T0D23
Using Table NT0-1 what is the minimum safe distance for a controlled RF radiation environment from a station using a 446 MHz 5/8-wave vertical antenna at 10 watts?
A. 1.9 feet
B. 2.5 feet
C. 4 feet
D. 4.3 feet

T0D23
(A)
Page 10-13

T0E Practical applications for VHF/UHF and above operations

T0E01
For the lowest RF radiation exposure to passengers, where would you mount your mobile antenna?
A. On the trunk lid
B. On the roof
C. On a front fender opposite the broadcast radio antenna
D. On one side of the rear bumper

T0E01
(B)
Page 10-16

T0E02
What should you do for safety before removing the shielding on a UHF power amplifier?
A. Make sure all RF screens are in place at the antenna feed line
B. Make sure the antenna feed line is properly grounded
C. Make sure the amplifier cannot accidentally be turned on
D. Make sure that RF leakage filters are connected

T0E02
(C)
Page 10-17

T0E03
Why might mobile transceivers produce less RF radiation exposure than hand-held transceivers in mobile operations?
A. They do not produce less exposure because they usually have higher power levels.
B. They have a higher duty cycle
C. When mounted on a metal vehicle roof, mobile antennas are generally well shielded from vehicle occupants
D. Larger transmitters dissipate heat and energy more readily

T0E03
(C)
Page 10-17

T0E04
What are some reasons you should never operate a power amplifier unless its covers are in place?
A. To maintain the required high operating temperatures of the equipment and reduce RF radiation exposure
B. To reduce the risk of shock from high voltages and reduce RF radiation exposure
C. To ensure that the amplifier will go into self oscillation and to minimize the effects of stray capacitance
D. To minimize the effects of stray inductance and to reduce the risk of shock from high voltages

T0E04
(B)
Page 10-17

T0E05
(C)
Page 10-17

T0E05
Considering RF radiation exposure, which of the following conditions may be a reason to modify your station's antenna system?
A. An SWR of 1:1
B. High feed line losses
C. Feed line radiation
D. Nonresonant parasitic elements

T0E06
(D)
Page 10-16

T0E06
Which of the following RF radiation exposure precautions might you use to ensure a safe operating environment at your amateur station?
A. Avoid conditions leading to "RF in the shack"
B. Use roof-mounted antennas for mobile operation whenever possible
C. Avoid conditions leading to feed line radiation
D. All of these choices are correct

T0E07
(D)
Page 10-12

T0E07
Which of the following statements are true about a broadband instrument used to measure RF fields?
A. It is calibrated over a wide frequency range
B. It responds instantaneously over a wide frequency range
C. It requires no tuning
D. All of these choices are correct

T0E08
(D)
Page 10-12

T0E08
Which of the following statements are true about a narrow bandwidth instrument used to measure RF fields?
A. It may operate over a wide frequency range
B. It's instantaneous bandwidth may be only a few kilohertz
C. It must be tuned to the frequency of interest
D. All of these choices are correct

T0E09
(A)
Page 10-6

T0E09
Why is it dangerous to look into the open end of a microwave feed horn antenna with power applied?
A. Fields are concentrated at the open end of a microwave feed horn
B. The feed horn antenna disperses the radiated energy over a wide area, to increase radiation exposure
C. The feed horn antenna inverts the phase of the radiated energy, resulting in a strong cooling effect on nearby tissue
D. The feed horn antenna converts RF radiation into powerful audio signals

T0E10
(B)
Page 10-15

T0E10
What is one way you can demonstrate compliance with the FCC RF radiation exposure limits?
A. Ensure a good RF ground connection for all transmitting antennas
B. Restrict accessibility to areas of high RF radiation levels
C. Use open-wire feed line for all transmitting antennas
D. Use only BNC and N-type connectors in your transmission lines

T0E11
(C)
Page 10-8

T0E11
What is the maximum emission power density permitted from an amateur station under the FCC RF radiation exposure limits?
A. The FCC Rules specify a maximum emission of 1.0 milliwatt per square centimeter
B. The FCC Rules specify a maximum emission of 5.0 milliwatts per square centimeter
C. The FCC Rules specify exposure limits, not emission limits
D. The FCC Rules specify maximum emission limits that vary with frequency

T0E12
Using Table NT0-1 what is the minimum safe distance for an uncontrolled RF radiation environment from a 3-element "triband" Yagi on 21 MHz at 100 watts?
A. 16.4 feet
B. 7.3 feet
C. 4.5 feet
D. 23 feet

T0E12
(A)
Page 10-13

T0E13
Using Table NT0-1 what is the minimum safe distance for a controlled RF radiation environment from a station using a 3-element "triband" Yagi on 28 MHz at 100 watts?
A. 15 feet
B. 11 feet
C. 22 feet
D. 24.5 feet

T0E13
(B)
Page 10-13

T0E14
Using Table NT0-1 what is the minimum safe distance for an uncontrolled RF radiation environment from a station using a 3-element "triband" Yagi on 28 MHz at 100 watts?
A. 7 feet
B. 24.5 feet
C. 15 feet
D. 34.7 feet

T0E14
(B)
Page 10-13

T0E15
Using Table NT0-1 what is the minimum safe distance for an uncontrolled RF radiation environment from a station using a 17-element Yagi on a five-wavelength boom on 144 MHz at 10 watts?
A. 32.4 feet
B. 22.9 feet
C. 2.5 feet
D. 20 feet

T0E15
(B)
Page 10-13

T0E16
Using Table NT0-1 what is the minimum safe distance for an uncontrolled RF radiation environment from a station using a 17-element Yagi on a five-wavelength boom on 144 MHz at 100 watts?
A. 14.5 feet
B. 20 feet
C. 72.4 feet
D. 32.4 feet

T0E16
(C)
Page 10-13

T0E17
Using Table NT0-1 what is the minimum safe distance for an uncontrolled RF radiation environment from a station using a 17-element Yagi on a five-wavelength boom on 144 MHz at 500 watts?
A. 20 feet
B. 72.4 feet
C. 162 feet
D. 175.5 feet

T0E17
(C)
Page 10-13

Table NT0-1

RF Exposure Limits

Section A.

Estimated distances to meet RF power density guidelines in the main beam of a typical 3-element "triband" Yagi for the 14, 21 and 28 MHz Amateur Radio bands. Calculations include the EPA ground reflection factor of 2.56.

Frequency: 21 MHz
Antenna gain: 7 dBi
Controlled limit: 2.04 mW/cm²
Uncontrolled limit: 0.408 mW/cm²

Transmitter power (watts)	Distance to controlled limit	Distance to uncontrolled limit
100	7.3′	16.4′
500	16.4′	36.7′
1000	23.2′	51.9′
1500	28.4′	63.6′

Frequency: 28 MHz
Antenna gain: 8 dBi
Controlled limit: 1.15 mW/cm²
Uncontrolled limit: 0.23 mW/cm²

Transmitter power (watts)	Distance to controlled limit	Distance to uncontrolled limit
100	11′	24.5′
500	24.5′	54.9′
1000	34.7′	77.6′
1500	42.5′	95.1′

Section F.

Estimated distances to meet RF power density guidelines in the main beam of a 17-element Yagi on a five-wavelength boom designed for weak signal communications on the 144 MHz Amateur Radio band (estimated gain, 16.8 dBi). Calculations include the EPA ground reflection factor of 2.56.

Frequency: 144 MHz
Estimated antenna gain: 16.8 dBi
Controlled limit: 1 mW/cm²
Uncontrolled limit: 0.2 mW/cm²

Transmitter power (watts)	Distance to controlled limit	Distance to uncontrolled limit
10	10.2′	22.9′
100	32.4′	72.4′
500	72.4′	162′
1500	125.5′	280.6′

T0E18
(D)
Page 10-13

T0E18
Using Table NT0-1 what is the minimum safe distance for an uncontrolled RF radiation environment from a station using a 17-element Yagi on a five-wavelength boom on 144 MHz at 1500 watts?
A. 45.5 feet
B. 78.5 feet
C. 125.5 feet
D. 280.6 feet

T0E19
(B)
Page 10-13

T0E19
Using Table NT0-1 what is the minimum safe distance for a controlled RF radiation environment from a station using a 17-element Yagi on a five-wavelength boom on 144 MHz at 1500 watts?
A. 45.5 feet
B. 125.5 feet
C. 162 feet
D. 175.5 feet

T0E20
(A)
Page 10-13

T0E20
Using Table NT0-1 what is the minimum safe distance for a controlled RF radiation environment from a station using a 17-element Yagi on a five-wavelength boom on 144 MHz at 500 watts?
A. 72.4 feet
B. 78.5 feet
C. 101 feet
D. 125.5 feet

T0E21
(D)
Page 10-13

T0E21
Using Table NT0-1 what is the minimum safe distance for a controlled RF radiation environment from a station using a 17-element Yagi on a five-wavelength boom on 144 MHz at 100 watts?
A. 45.5 feet
B. 78.5 feet
C. 10.2 feet
D. 32.4 feet

T0E22
(C)
Page 10-13

T0E22
Using Table NT0-1 what is the minimum safe distance for a controlled RF radiation environment from a station using a 17-element Yagi on a five-wavelength boom on 144 MHz at 10 watts?
A. 32.4 feet
B. 78.5 feet
C. 10.2 feet
D. 20 feet

T0E23
(A)
Page 10-13

T0E23
Using Table NT0-1 what is the minimum safe distance for a controlled RF radiation environment from a station using eight 17-element Yagis on five-wavelength booms for moonbounce (EME) on 144 MHz at 150 watts?
A. 90.9 feet
B. 57 feet
C. 78.5 feet
D. 181.8 feet

Table NT0-1

RF Exposure Limits

Section G.

Estimated distances to meet RF power density guidelines in the main beam of an array of eight 17-element Yagis with five-wavelength booms designed for earth-moon-earth ("moonbounce") communications on the 144 MHz Amateur Radio band (estimated gain, 24 dBi). Calculations include the EPA ground reflection factor of 2.56.

Frequency: 144 MHz
Estimated antenna gain: 24 dBi
Controlled limit: 1 mW/cm²
Uncontrolled limit: 0.2 mW/cm²

Transmitter power (watts)	Distance to controlled limit	Distance to uncontrolled limit
150	90.9′	203.3′
500	166′	371.1′
1500	287.4′	642.7′

T0E24
Using Table NT0-1 what is the minimum safe distance for a controlled RF radiation environment from a station using eight 17-element Yagis on five-wavelength booms for moonbounce (EME) on 144 MHz at 500 watts?
A. 90.9 feet
B. 175.5 feet
C. 127 feet
D. 166 feet

T0E24
(D)
Page 10-13

T0E25
Using Table NT0-1 what is the minimum safe distance for a controlled RF radiation environment from a station using eight 17-element Yagis on five-wavelength booms for moonbounce (EME) on 144 MHz at 1500 watts?
A. 287.4 feet
B. 166 feet
C. 127 feet
D. 232 feet

T0E25
(A)
Page 10-13

T0E26
Using Table NT0-1 what is the uncontrolled limit for an RF radiation environment from a station using eight 17-element Yagis on five-wavelength booms for moonbounce (EME) on 144 MHz at 1500 watts?
A. 371.1 feet
B. 175.5 feet
C. 642.7 feet
D. 232 feet

T0E26
(C)
Page 10-13

T0E27
Using Table NT0-1 what is the uncontrolled limit for an RF radiation environment from a station using eight 17-element Yagis on five-wavelength booms for moonbounce (EME) on 144 MHz at 500 watts?
A. 203.3 feet
B. 127 feet
C. 401.5 feet
D. 371.1 feet

T0E27
(D)
Page 10-13

T0E28
Using Table NT0-1 what is the uncontrolled limit for an RF radiation environment from a station using eight 17-element Yagis on five-wavelength booms for moonbounce (EME) on 144 MHz at 150 watts?
A. 203.3 feet
B. 127 feet
C. 371.1 feet
D. 232 feet

T0E28
(A)
Page 10-13

APPENDIX

In this Appendix you'll find an array of data tables, charts and other information to aid your review before the exam — and assist you after you've passed.

HELPFUL DATA TABLES

Standard Resistance Values

Numbers in **bold** type are ± 10% values. Others are 5% values.

Ohms										*Megohms*				
1.0	3.6	**12**	43	**150**	510	**1800**	6200	**22000**	75000	0.24	0.62	1.6	4.3	11.0
1.1	**3.9**	13	**47**	160	**560**	2000	**6800**	24000	**82000**	**0.27**	**0.68**	**1.8**	**4.7**	**12.0**
1.2	4.3	**15**	51	**180**	620	**2200**	7500	**27000**	91000	0.30	0.75	2.0	5.1	13.0
1.3	**4.7**	16	**56**	200	**680**	2400	**8200**	30000	**100000**	**0.33**	**0.82**	**2.2**	**5.6**	**15.0**
1.5	5.1	**18**	62	**220**	750	**2700**	9100	**33000**	110000	0.36	0.91	2.4	6.2	16.0
1.6	**5.6**	20	**68**	240	**820**	3000	**10000**	36000	**120000**	**0.39**	**1.0**	**2.7**	**6.8**	**18.0**
1.8	6.2	**22**	75	**270**	910	**3300**	11000	**39000**	130000	0.43	1.1	3.0	7.5	20.0
2.0	**6.8**	24	**82**	300	**1000**	3600	**12000**	43000	**150000**	**0.47**	**1.2**	**3.3**	**8.2**	**22.0**
2.2	7.5	**27**	91	**330**	1100	**3900**	13000	**47000**	160000	0.51	1.3	3.6	9.1	
2.4	**8.2**	30	**100**	360	**1200**	4300	**15000**	51000	**180000**	**0.56**	**1.5**	**3.9**	**10.0**	
2.7	9.1	**33**	110	**390**	1300	**4700**	16000	**56000**	200000					
3.0	**10.0**	36	**120**	430	**1500**	5100	**18000**	62000	**220000**					
3.3	11.0	**39**	130	**470**	1600	**5600**	20000	**68000**						

Standard Values for 1000-V Disc-Ceramic Capacitors

pF	*pF*	*pF*	*pF*
3.3	39	250	1000
5	47	270	1200
6	50	300	1500
6.8	51	330	1800
8	56	360	2000
10	68	390	2500
12	75	400	2700
15	82	470	3000
18	100	500	3300
20	120	510	3900
22	130	560	4700
24	150	600	5000
25	180	680	5600
27	200	750	6800
30	220	820	8200
33	240	910	10000

Resistor Color Code

Color	*Sig. Figure*	*Decimal Multiplier*	*Tolerance (%)*	*Color*	*Sig. Figure*	*Decimal Multiplier*	*Tolerance (%)*
Black	0	1		Violet	7	10,000,000	
Brown	1	10		Gray	8	100,000,000	
Red	2	100		White	9	1,000,000,000	
Orange	3	1,000		Gold	—	0.1	5
Yellow	4	10,000		Silver	—	0.01	10
Green	5	100,000		No color	—		20
Blue	6	1,000,000					

Common Values for Small Electrolytic Capacitors

µF	V*	µF	V*
33	6.3	10	35
33	10	22	35
100	10	33	35
220	10	47	35
330	10	100	35
470	10	220	35
10	16	330	35
22	16	470	35
33	16	1000	35
47	16	1	50
100	16	2.2	50
220	16	3.3	50
470	16	4.7	50
1000	16	10	50
2200	16	33	50
4.7	25	47	50
22	25	100	50
33	25	220	50
47	25	330	50
100	25	470	50
220	25	10	63
330	25	22	63
470	25	47	63
1000	25	1	100
2200	25	10	100
4.7	35	33	100

*Working voltage

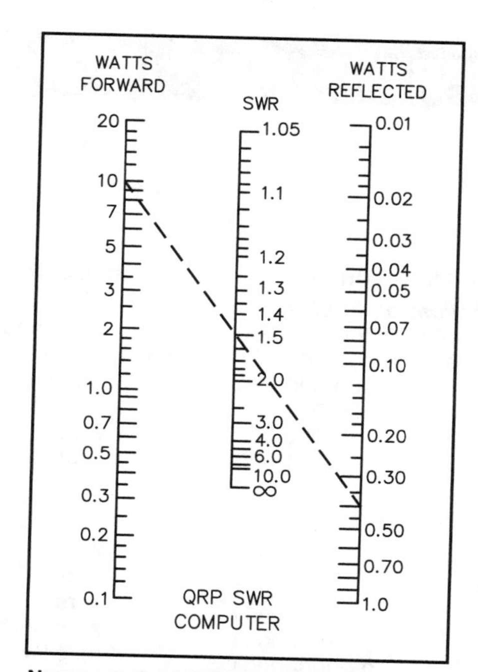

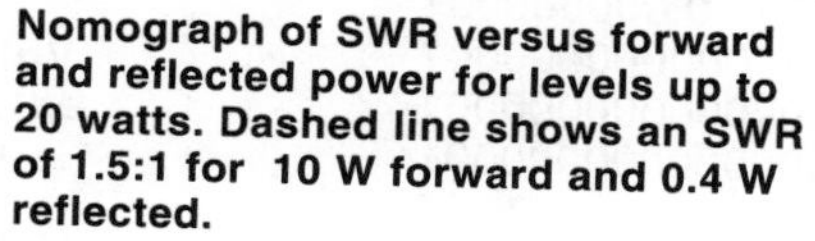

Nomograph of SWR versus forward and reflected power for levels up to 20 watts. Dashed line shows an SWR of 1.5:1 for 10 W forward and 0.4 W reflected.

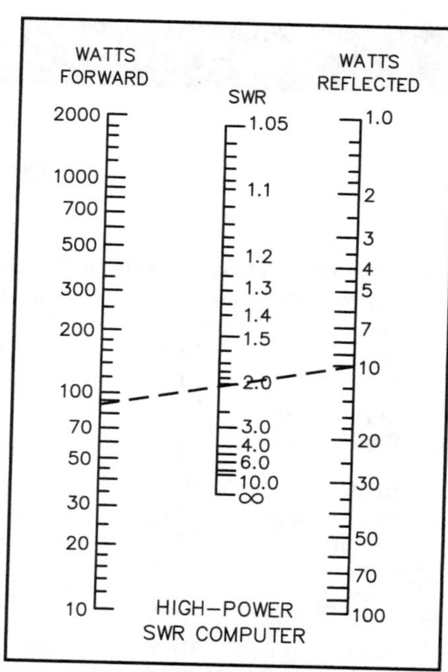

Nomograph of SWR versus forward and reflected power for levels up to 2000 watts. Dashed line shows an SWR of 2:1 for 90 W forward and 10 W reflected.

Fractions of an Inch with Metric Equivalents

Fractions Of An Inch		*Decimals Of An Inch*	*Millimeters*
	1/64	0.0156	0.397
1/32		0.0313	0.794
	3/64	0.0469	1.191
1/16		0.0625	1.588
	5/64	0.0781	1.984
3/32		0.0938	2.381
	7/64	0.1094	2.778
1/8		0.1250	3.175
	9/64	0.1406	3.572
5/32		0.1563	3.969
	11/64	0.1719	4.366
3/16		0.1875	4.763
	13/64	0.2031	5.159
7/32		0.2188	5.556
	15/64	0.2344	5.953
1/4		0.2500	6.350
	17/64	0.2656	6.747
9/32		0.2813	7.144
	19/64	0.2969	7.541
5/16		0.3125	7.938
	21/64	0.3281	8.334
11/32		0.3438	8.731
	23/64	0.3594	9.128
3/8		0.3750	9.525
	25/64	0.3906	9.922
13/32		0.4063	10.319
	27/64	0.4219	10.716
7/16		0.4375	11.113
	29/64	0.4531	11.509
15/32		0.4688	11.906
	31/64	0.4844	12.303
1/2		0.5000	12.700
	33/64	0.5156	13.097
17/32		0.5313	13.494
	35/64	0.5469	13.891
9/16		0.5625	14.288
	37/64	0.5781	14.684
19/32		0.5938	15.081
	39/64	0.6094	15.478
5/8		0.6250	15.875
	41/64	0.6406	16.272
21/32		0.6563	16.669
	43/64	0.6719	17.066
11/16		0.6875	17.463
	45/64	0.7031	17.859
23/32		0.7188	18.256
	47/64	0.7344	18.653
3/4		0.7500	19.050
	49/64	0.7656	19.447
25/32		0.7813	19.844
	51/64	0.7969	20.241
13/16		0.8125	20.638
	53/64	0.8281	21.034
27/32		0.8438	21.431
	55/64	0.8594	21.828
7/8		0.8750	22.225
	57/64	0.8906	22.622
29/32		0.9063	23.019
	59/64	0.9219	23.416
15/16		0.9375	23.813
	61/64	0.9531	24.209
31/32		0.9688	24.606
	63/64	0.9844	25.003
1		1.0000	25.400

US CUSTOMARY TO METRIC CONVERSIONS

International System of Units (SI)—Metric Units

Prefix	*Symbol*			*Multiplication Factor*
exa	E	10^{18}	=	1 000 000 000 000 000 000
peta	P	10^{15}	=	1 000 000 000 000 000
tera	T	10^{12}	=	1 000 000 000 000
giga	G	10^{9}	=	1 000 000 000
mega	M	10^{6}	=	1 000 000
kilo	k	10^{3}	=	1 000
hecto	h	10^{2}	=	100
deca	da	10^{1}	=	10
(unit)		10^{0}	=	1
deci	d	10^{-1}	=	0.1
centi	c	10^{-2}	=	0.01
milli	m	10^{-3}	=	0.001
micro	µ	10^{-6}	=	0.000001
nano	n	10^{-9}	=	0.000000001
pico	p	10^{-12}	=	0.000000000001
femto	f	10^{-15}	=	0.000000000000001
atto	a	10^{-18}	=	0.000000000000000001

Linear

1 metre (m) = 100 centrimetres (cm) = 1000 millimetres (mm)

Area

1 $m^2 = 1 \times 10^4$ $cm^2 = 1 \times 10^6$ mm^2

Volume

1 $m^3 = 1 \times 10^6$ $cm^3 = 1 \times 10^9$ mm^3
1 litre (l) = 1000 $cm^3 = 1 \times 10^6$ mm^3

Mass

1 kilogram (kg) = 1 000 grams (g)
(Approximately the mass of 1 litre of water)
1 metric ton (or tonne) = 1 000 kg

US Customary Units

Linear Units

12 inches (in) = 1 foot (ft)
36 inches = 3 feet = 1 yard (yd)
1 rod = 5½ yards = 16½ feet
1 statute mile = 1 760 yards = 5 280 feet
1 nautical mile = 6 076.11549 feet

Area

1 ft^2 = 144 in^2
1 yd^2 = 9 ft^2 = 1 296 in^2
1 rod^2 = 30¼ yd^2
1 acre = 4840 yd^2 = 43 560 ft^2
1 acre = 160 rod^2
1 $mile^2$ = 640 acres

Volume

1 ft^3 = 1 728 in^3
1 yd^3 = 27 ft^3

Liquid Volume Measure

1 fluid ounce (fl oz) = 8 fluidrams = 1.804 in^3
1 pint (pt) = 16 fl oz
1 quart (qt) = 2 pt = 32 fl oz = 57¾ in^3
1 gallon (gal) = 4 qt = 231 in^3
1 barrel = 31½ gal

Dry Volume Measure

1 quart (qt) = 2 pints (pt) = 67.2 in^3
1 peck = 8 qt
1 bushel = 4 pecks = 2 150.42 in^3

Avoirdupois Weight

1 dram (dr) = 27.343 grains (gr) or (gr a)
1 ounce (oz) = 437.5 gr
1 pound (lb) = 16 oz = 7 000 gr
1 short ton = 2 000 lb, 1 long ton = 2 240 lb

Troy Weight

1 grain troy (gr t) = 1 grain avoirdupois
1 pennyweight (dwt) or (pwt) = 24 gr t
1 ounce troy (oz t) = 480 grains
1 lb t = 12 oz t = 5 760 grains

Apothecaries' Weight

1 grain apothecaries' (gr ap) = 1 gr t = 1 gr a
1 dram ap (dr ap) = 60 gr
1 oz ap = 1 oz t = 8 dr ap = 480 fr
1 lb ap = 1 lb t = 12 oz ap = 5 760 gr

Multiply →

Metric Unit = Conversion Factor × US Customary Unit

← **Divide**

Metric Unit ÷ Conversion Factor = US Customary Unit

Metric Unit =	Conversion Factor ×	US Unit
(Length)		
mm	25.4	inch
cm	2.54	inch
cm	30.48	foot
m	0.3048	foot
m	0.9144	yard
km	1.609	mile
km	1.852	nautical mile
(Area)		
mm^2	645.16	$inch^2$
cm^2	6.4516	in^2
cm^2	929.03	ft^2
m^2	0.0929	ft^2
cm^2	8361.3	yd^2
m^2	0.83613	yd^2
m^2	4047	acre
km^2	2.59	mi^2
(Mass)	(Avoirdupois Weight)	
grams	0.0648	grains
g	28.349	oz
g	453.59	lb
kg	0.45359	lb
tonne	0.907	short ton
tonne	1.016	long ton

Metric Unit =	Conversion Factor ×	US Unit
(Volume)		
mm^3	16387.064	in^3
cm^3	16.387	in^3
m^3	0.028316	ft^3
m^3	0.764555	yd^3
ml	16.387	in^3
ml	29.57	fl oz
ml	473	pint
ml	946.333	quart
l	28.32	ft^3
l	0.9463	quart
l	3.785	gallon
l	1.101	dry quart
l	8.809	peck
l	35.238	bushel
(Mass)	(Troy Weight)	
g	31.103	oz t
g	373.248	lb t
(Mass)	(Apothecaries' Weight)	
g	3.387	dr ap
g	31.103	oz ap
g	373.248	lb ap

US AMATEUR FREQUENCY AND MODE ALLOCATIONS

160 METERS

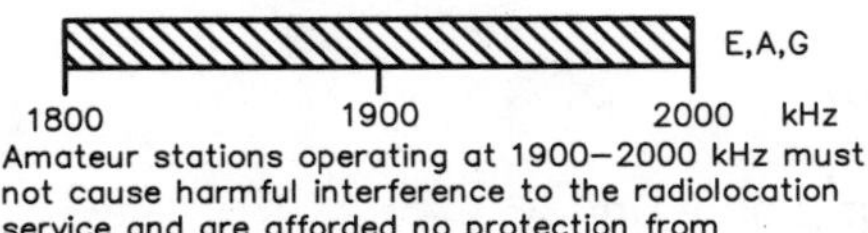

Amateur stations operating at 1900–2000 kHz must not cause harmful interference to the radiolocation service and are afforded no protection from radiolocation operations.

80 METERS

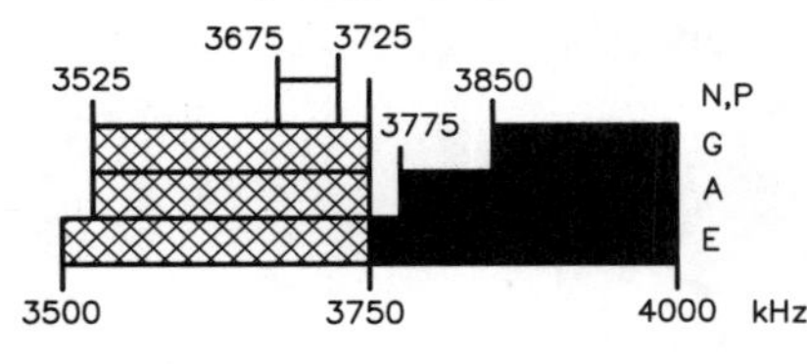

5167.5 kHz (SSB only): Alaska emergency use only.

40 METERS

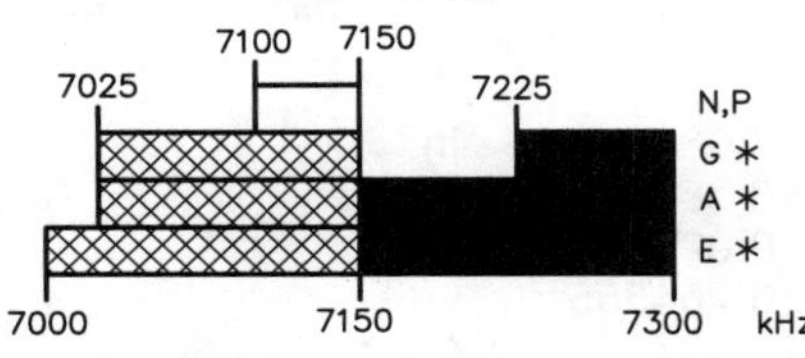

∗ Phone operation is allowed on 7075–7100 kHz in Puerto Rico, US Virgin islands and areas of the Caribbean south of 20 degrees north latitude; and in Hawaii and areas near ITU Region 3, including Alaska.

30 METERS

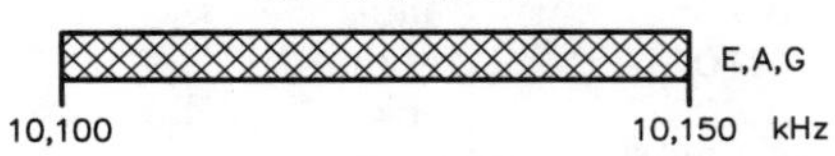

Maximum power on 30 meters is 200 watts PEP output. Amateurs must avoid interference to the fixed service outside the US.

20 METERS

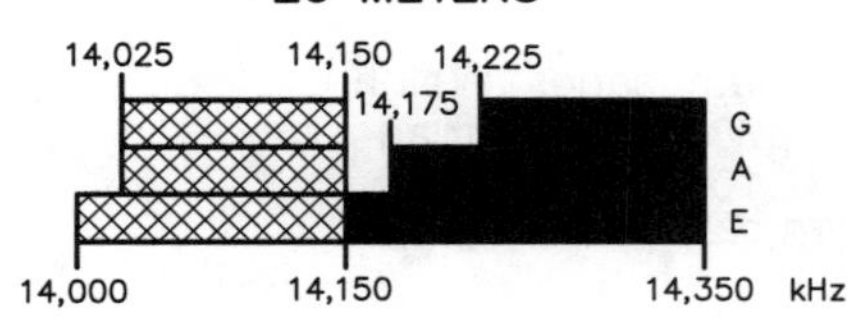

17 METERS

15 METERS

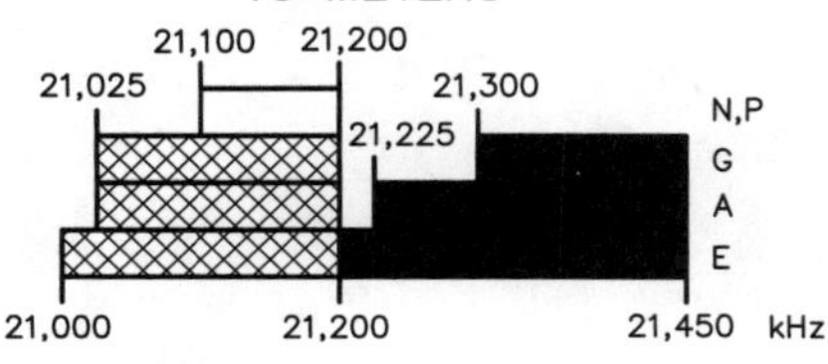

12 METERS

10 METERS

Novices and Technician Plus Licensees are limited to 200 watts PEP output on 10 meters.

6 METERS

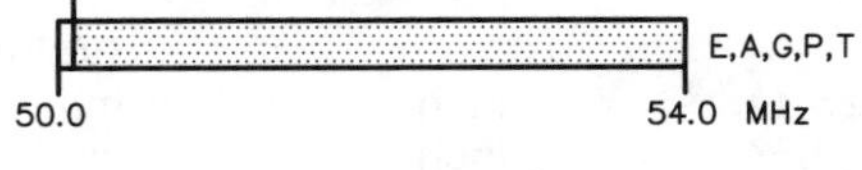

2 METERS

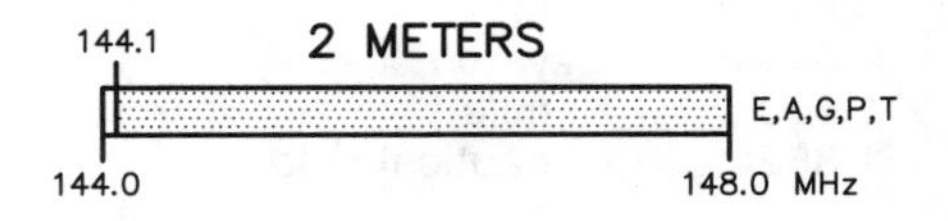

1.25 METERS

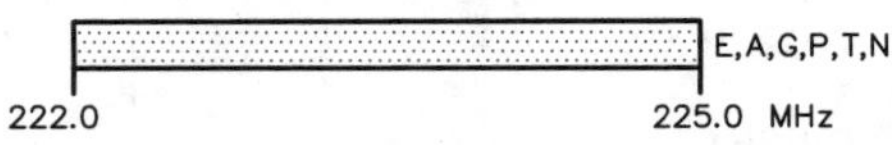

Novices are limited to 25 watts PEP output from 222 to 225 MHz.

70 CENTIMETERS∗∗

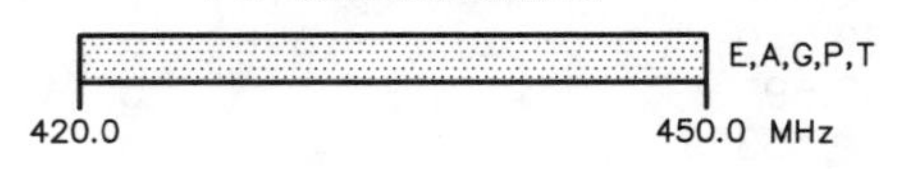

33 CENTIMETERS∗∗

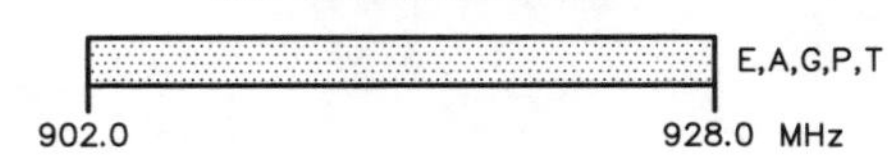

23 CENTIMETERS∗∗

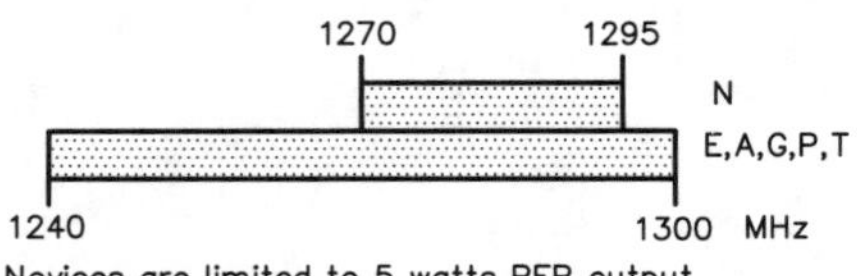

Novices are limited to 5 watts PEP output from 1270 to 1295 MHz.

US AMATEUR BANDS

December 20, 1994

US AMATEUR POWER LIMITS

At all times, transmitter power should be kept down to that necessary to carry out the desired communications. Power is rated in watts PEP output. Unless otherwise stated, the maximum power output is 1500 W. Power for all license classes is limited to 200 W in the 10,100–10,150 kHz band and in all Novice subbands below 28,100 kHz. Novices and Technicians are restricted to 200 W in the 28,100–28,500 kHz subbands. In addition, Novices are restricted to 25 W in the 222–225 MHz band and 5 W in the 1270–1295 MHz subband.

Operators with Technician class licenses and above may operate on all bands above 50 MHz. For more detailed information see The FCC Rule Book.

KEY

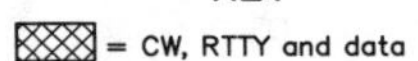

= CW, RTTY and data
= CW, RTTY, data, MCW, test, phone and image
= CW, phone and image
= CW and phone
= CW, RTTY, data, phone, and image
= CW only

E = EXTRA CLASS
A = ADVANCED
G = GENERAL
P = TECHNICIAN PLUS
T = TECHNICIAN
N = NOVICE

∗∗ Geographical and power restrictions apply to these bands. See The FCC Rule Book for more information about your area.

Above 23 Centimeters:

All licensees except Novices are authorized all modes on the following frequencies:
2300–2310 MHz
2390–2450 MHz
3300–3500 MHz
5650–5925 MHz
10.0–10.5 GHz
24.0–24.25 GHz
47.0–47.2 GHz
75.5–81.0 GHz
119.98–120.02 GHz
142–149 GHz
241–250 GHz
All above 300 GHz

For band plans and sharing arrangements, see The ARRL Operating Manual or The FCC Rule Book.

SCHEMATIC SYMBOLS

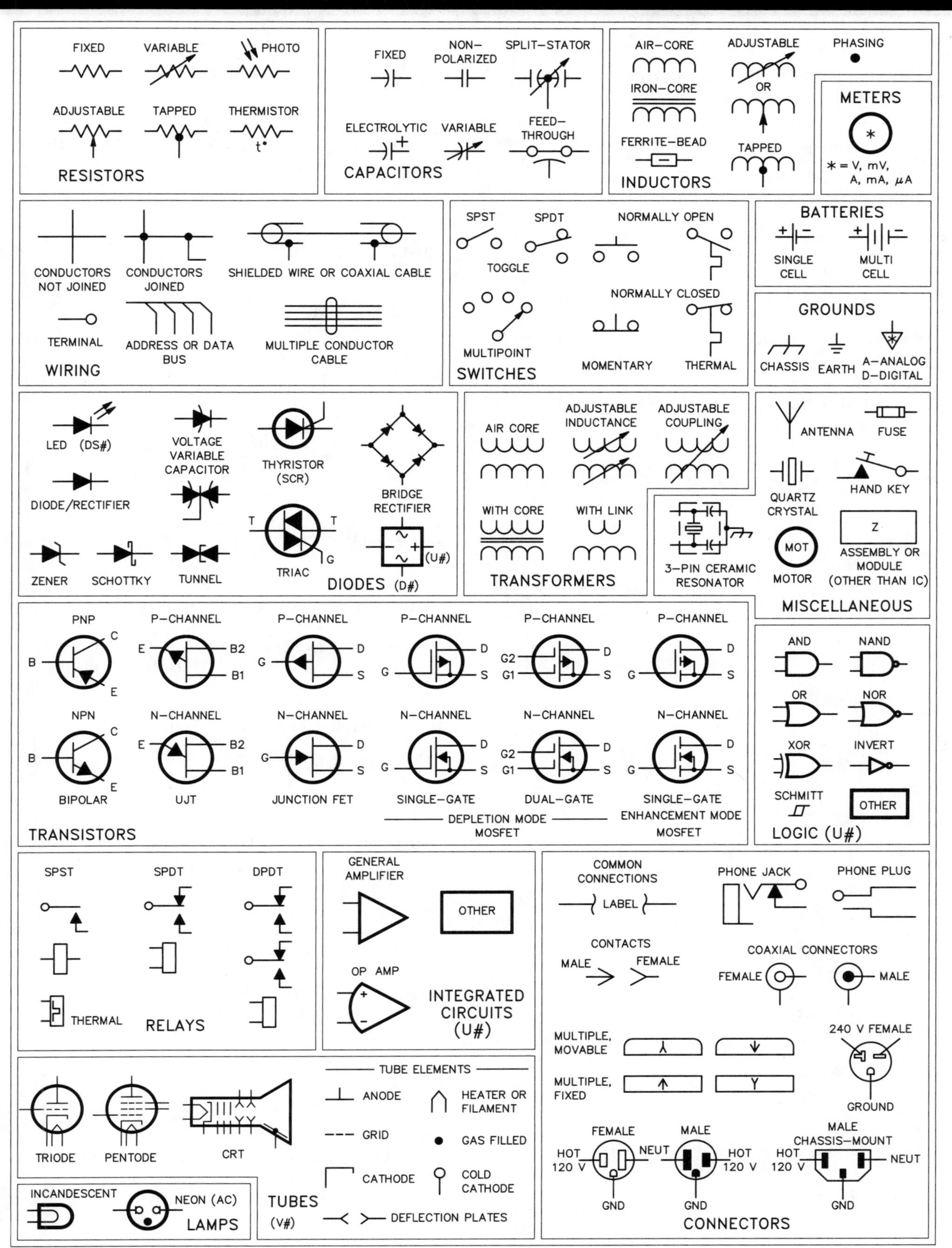

EQUATIONS USED IN THIS BOOK

True forward power = Forward power reading– Reflected power reading (Eq 4-1)

$$\text{Current} = \frac{\text{Voltage}}{\text{Resistance}} \quad \text{(Eq 5-1)}$$

$$\text{Resistance} = \frac{\text{Voltage}}{\text{Current}} \quad \text{(Eq 5-2)}$$

$$\text{Voltage} = \text{Current} \times \text{Resistance} \quad \text{(Eq 5-3)}$$

$$E = IR \text{ (volts = amperes} \times \text{ohms} \quad \text{(Eq 5-4)}$$

$$I = \frac{E}{R} \text{ (amperes = volts divided by ohms)} \quad \text{(Eq 5-5)}$$

$$R = \frac{E}{I} = \text{(ohms = volts divided by amperes)} \quad \text{(Eq 5-6)}$$

$$c = f\lambda \quad \text{(Eq 5-7)}$$

$$f = \frac{c}{\lambda} \quad \text{(Eq 5-8)}$$

$$\lambda = \frac{c}{f} \quad \text{(Eq 5-9)}$$

$$R_{TOTAL} = R_1 + R_2 + R_3 + \ldots + R_n \quad \text{(Eq 5-10)}$$

$$R_{TOTAL} = \frac{1}{\frac{1}{R_1} + \frac{1}{R_2} + \frac{1}{R_3} + \ldots + \frac{1}{R_n}} \quad \text{(Eq 5-11)}$$

$$R_{TOTAL} = \frac{R_1 \times R_2}{R_1 + R_2} \quad \text{(Eq 5-12)}$$

$$L_{TOTAL} = L_1 + L_2 + L_3 + \ldots + L_n \quad \text{(Eq 5-13)}$$

$$L_{TOTAL} = \frac{1}{\frac{1}{L_1} + \frac{1}{L_2} + \frac{1}{L_3} + \ldots + \frac{1}{L_n}} \quad \text{(Eq 5-14)}$$

$$L_{TOTAL} = \frac{L_1 \times L_2}{L_1 + L_2} \quad \text{(Eq 5-15)}$$

$$C_{TOTAL} = C_1 + C_2 + C_3 + \ldots + C_n \quad \text{(Eq 5-16)}$$

$$C_{TOTAL} = \frac{1}{\frac{1}{C_1} + \frac{1}{C_2} + \frac{1}{C_3} + \ldots + \frac{1}{C_n}} \quad \text{(Eq 5-17)}$$

$$C_{TOTAL} = \frac{C_1 \times C_2}{C_1 + C_2} \quad \text{(Eq 5-18)}$$

$$Bw = 2 \times (D + M) \quad \text{(Eq 8-1)}$$

$$\lambda \text{ (in feet)} = \frac{984}{f \text{ (MHz)}} \quad \text{(Eq 9-1)}$$

$$\tfrac{1}{2}\lambda \text{ (in feet)} = \frac{468}{f \text{ (MHz)}} \quad \text{(Eq 9-2)}$$

$$\tfrac{1}{4}\lambda \text{ (in feet)} = \frac{234}{f \text{ (MHz)}} \quad \text{(Eq 9-3)}$$

HF Controlled Plane - Wave Equivalent (Eq 10-1)

$$\text{Power Density MPE} = \frac{900}{(f \text{ (MHz)})^2}$$

HF Uncontrolled Plane - Wave Equivalent (Eq 10-2)

$$\text{Power Density MPE} = \frac{180}{(f \text{ (MHz)})^2}$$

UHF Controlled Plane - Wave Equivalent (Eq 10-3)

$$\text{Power Density MPE} = \frac{f \text{ (MHz)}}{300}$$

UHF Uncontrolled Plane - Wave Equivalent (Eq 10-4)

$$\text{Power Density MPE} = \frac{f \text{ (MHz)}}{1500}$$

$$D \approx \frac{2L^2}{\lambda} \quad \text{(Eq 10-5)}$$

$$\frac{E}{H} = \text{Intrinsic Impedance of Free Space} = 377\ \Omega \quad \text{(Eq 10-6)}$$

GLOSSARY OF KEY WORDS

A1A emission — The FCC emission type designator that describes international Morse code telegraphy (**CW**) communications without audio modulation of the carrier.

Alternating current (ac) — Electrical current that flows first in one direction in a wire and then in the other. The applied voltage is also changing polarity. This direction reversal continues at a rate that depends on the frequency of the ac.

Amateur operator — A person holding a written authorization to be the control operator of an amateur station.

Amateur service — A radiocommunication service for the purpose of self-training, intercommunication and technical investigations carried out by amateurs, that is, duly authorized persons interested in radio technique solely with a personal aim and without **pecuniary** interest.

Amateur station — A station licensed in the amateur service, including necessary equipment, used for amateur communication.

Ampere (A) — The basic unit of electrical current. Current is a measure of the electron flow through a circuit. If we could count electrons, we would find that if there are 6.24×10^{18} electrons moving past a point in one second, we have a current of one ampere.[1] We abbreviate amperes as amps.

Amplitude modulation (AM) — A method of combining an information signal and an RF (radio-frequency) carrier. In double-sideband voice AM transmission, we use the voice information to vary (modulate) the amplitude of an RF carrier. Shortwave broadcast stations use this type of AM, as do stations in the Standard Broadcast Band (540-1600 kHz). Few amateurs use double-sideband voice AM, but a variation, known as single sideband, is very popular.

Antenna — A device that picks up or sends out radio waves.

Antenna switch — A switch used to connect one transmitter, receiver or transceiver to several different antennas.

Antenna tuner — See **antenna-matching network**.

Antenna-matching network — A device that matches the antenna system input impedance to the transmitter, receiver or transceiver output impedance. Also called an **antenna tuner**, **impedance-matching network** or *Transmatch*.

Audio frequency (AF) — The range of frequencies that the human ear can detect. Audio frequencies are usually listed as 20 Hz to 20,000 Hz.

Audio-frequency shift keying (AFSK) — A method of transmitting radioteletype information by switching between two audio tones fed into an FM transmitter. Most often used on VHF. Also see **frequency-shift keying (FSK)**.

Autopatch — A device that allows repeater users to make telephone calls through a repeater.

Balun — Contraction for ***bal***anced to ***un***balanced. A device to couple a balanced load to an unbalanced source, or vice versa.

Band spread — A receiver quality used to describe how far apart stations on different nearby frequencies will seem to be. We usually express band spread as the number of kilohertz that the frequency changes per tuning-knob rotation. Band spread and frequency resolution are related. The amount of band spread determines how easily signals can be tuned.

[1]Numbers written as a multiple of some power are expressed in exponential notation. This notation is explained in detail on page 5-2.

Bandwidth — A range of associated frequencies (measured in hertz). For example, bandwidth is used to describe the range of frequencies in the radio spectrum that a radio transmission occupies or the range of frequencies that will pass through a given filter.

Battery — A device that converts chemical energy into electrical energy. It provides excess electrons to produce a current and the voltage or EMF to push those electrons through a circuit.

Baudot — A code used in radioteletype communications. Each character is represented with a string of five bits of digital information. Each character has a different combination of bits.

Beam antenna — A directional antenna. A beam antenna must be rotated to provide coverage in different directions.

Beat-frequency oscillator (BFO) — A receiver circuit that provides a signal to the detector. The BFO signal mixes with the incoming signal to produce an audio tone for CW reception. A BFO is needed to copy CW and SSB signals.

Block diagram — A drawing using boxes to represent sections of a complicated device or process. The block diagram shows the connections between sections.

Broadcasting — Transmissions intended to be received by the general public, either direct or relayed.

Centi — The metric prefix for 10^{-2}, or divide by 100.

Chassis ground — The common connection for all parts of a circuit that connect to the negative side of the power supply.

Chirp — A slight shift in transmitter frequency each time you key the transmitter.

Closed repeater — A repeater that restricts access to those who know a special code. See **CTCSS**.

Closed, or **complete circuit** — An electrical circuit with an uninterrupted path for the current to follow. Turning a switch on, for example, closes or completes the circuit, allowing current to flow.

Coaxial cable — Coax (pronounced kó-aks). A type of feed line with one conductor inside the other.

Code key — A device used as a switch to generate Morse code. Also called a **hand key**, **straight key** or **telegraph key**.

Code-practice oscillator — A device that produces an audio tone, used for learning the code.

Conductor — A material that has a loose grip on its electrons, so an electrical current can pass through it.

Connected — The condition in which two packet-radio stations are sending information to each other. Each is acknowledging when the data has been received correctly.

Continuous wave (CW) — Morse code telegraphy. See **CW (Continuous wave)**.

Control operator — An amateur operator designated by the licensee of a station to be responsible for the transmissions of an amateur station.

Control point — The locations at which the control operator function is performed.

Controlled environment — Any area in which an RF signal may cause radiation exposure to people who are aware of the radiated electric and magnetic fields and who can exercise some control over their exposure to these fields. The FCC generally considers amateur operators and their families to be in a controlled RF exposure environment to determine the maximum permissible exposure levels.

CQ — "Calling any station": the general call when requesting a conversation with anyone.

CTCSS — A sub-audible tone system used on some repeaters. When added to a carrier, a CTCSS tone allows a receiver to accept a signal. Also called **PL**.

Current — A flow of electrons in an electrical circuit.

CW (continuous wave) — The FCC emission type that describes international Morse code telegraphy communication without audio modulation of the carrier. Hams usually produce Morse code signals by interrupting the continuous-wave signal from a transmitter to form the dots and dashes. A communications mode transmitted by on/off keying of a radio-frequency signal. Another name for international Morse code.

D region — The lowest region of the ionosphere. The D region contributes very little to short-wave radio propagation. It acts mainly to absorb energy from radio waves as they pass through it. This absorption has a significant effect on signals below about 7.5 MHz during daylight.

Dash — The long sound used in Morse code. Pronounce this as "dah" when verbally sounding Morse code characters. Dashes are three times longer than dots.

Data — Computer-based modes, such as RTTY and packet.

DE — The Morse code abbreviation for "from" or "this is."

Deci — The metric prefix for 10^{-1}, or divide by 10.

Digipeater — A packet-radio station used to retransmit signals that are specifically addressed to be retransmitted by that station.

Digital communications — Computer-based communications modes. Also see **Data**.

Dipole antenna — See **Half-wave dipole**. A dipole need not be 1/2 wavelength long.

Direct current (dc) — Electrical current that flows in one direction only.

Director — An element in front of the driven element in a Yagi and some other directional antennas.

Dot — The short sound used in Morse code. Pronounce this as "dit" when verbally sounding Morse code characters if the dot comes at the end of the character. If the dot comes at the beginning or in the middle of the character, pronounce it as "di."

Double-pole, double-throw (DPDT) switch — A switch that has six contacts. The DPDT switch has two center contacts. The two center contacts can each be connected to one of two other contacts.

Double-pole, single-throw (DPST) switch — A switch that connects two contacts to another set of contacts. A DPST switch turns two circuits on or off at the same time.

Driven element — The part of an antenna that connects directly to the feed line.

Dummy load — A station accessory that allows you to test or adjust transmitting equipment without sending a signal out over the air. Also called dummy antenna.

Duplex operation — Receiving and transmitting on two different frequencies. Also see **simplex operation**.

Duty cycle — The ratio of average power to peak envelope power (PEP) expressed as a percentage. A lower duty cycle means less RF radiation exposure for the same PEP output.

DX — Distance, foreign countries.

E region — The second lowest ionospheric region, the E region exists only during the day. Under certain conditions, it may refract radio waves enough to return them to Earth.

Earth ground — A circuit connection to a ground rod driven into the Earth or to a cold-water pipe made of copper that goes into the ground.

Earth station — An amateur station located on, or within 50 km of, the Earth's surface intended for communications with space stations or with other Earth stations by means of one or more other objects in space.

Electromotive force (EMF) — The force or pressure that pushes a current through a circuit.

Electron — A tiny, negatively charged particle, normally found in an area surrounding the nucleus of an atom. Moving electrons make up an electrical current.

Electronic keyer — A device that generates Morse code dots and dashes electronically.

Emergency — A situation where there is a danger to lives or property.

Emission — RF signals transmitted from a radio station.

Emission privilege — Permission to use a particular emission type (such as Morse code or voice).

Emission types — Term for the different modes authorized for use on the Amateur Radio bands. Examples are CW, SSB and FM.

Energy — The ability to do work; the ability to exert a force to move some object.

F region — A combination of the two highest ionospheric regions, the F1 and F2 regions. The F region refracts radio waves and returns them to Earth. Its height varies greatly depending on the time of day, season of the year and amount of sunspot activity.

False or deceptive signals — Transmissions that are intended to mislead or confuse those who may receive the transmissions. For example, distress calls transmitted when there is no actual emergency are false or deceptive signals.

Feed line — The wires or cable used to connect a transmitter, receiver or transceiver to an antenna. Also see **Transmission line**.

Fist — The unique rhythm of an individual amateur's Morse code sending.

Frequency — The number of complete cycles of an alternating current that occur per second.

Frequency bands — A group of frequencies where amateur communications are authorized.

Frequency coordinators — Individuals or groups that recommend repeater frequencies.

Frequency modulation (FM) — A method of combining an information signal with a radio signal (see **Amplitude modulation**). As you might suspect, we use voice or data to vary the frequency of the transmitted signal. FM broadcast stations and most professional communications (police, fire, taxi) use FM. VHF/UHF FM voice is the most popular amateur mode. It is the mode of voice (phone) communications used on repeaters.

Frequency privilege — Permission to use a particular group of frequencies.

Frequency-shift keying (FSK) — A method of transmitting radioteletype information by switching an RF carrier between two separate frequencies. FSK RTTY is most often used on HF. Also see **audio-frequency shift keying**.

Front-end overload — Interference to a receiver caused by a strong signal that overpowers the receiver RF amplifier ("front end"). See also **receiver overload**.

Fuse — A thin metal strip mounted in a holder. When too much current passes through the fuse, the metal strip melts and opens the circuit.

Gain — A measure of the directivity of an antenna, as compared with another antenna such as a dipole.

General-coverage receiver — A receiver used to listen to a wide range of frequencies. Most general-coverage receivers tune from frequencies below the standard-broadcast band to at least 30 MHz. These frequencies include the shortwave-broadcast bands and the amateur bands from 160 to 10 meters.

Giga — The metric prefix for 10^9, or times 1,000,000,000.

Grace period — The time the FCC allows following the expiration of an amateur license to renew that license without having to retake an examination. Those who hold an expired license may not operate an amateur station until the license is reinstated.

Ground connection — A connection made to the earth for electrical safety. This connection can be made inside (to a metal cold-water pipe) or outside (to a **ground rod**).

Ground rod — A copper or copper-clad steel rod that is driven into the earth. A heavy copper wire from the ham shack connects all station equipment to the ground rod.

Ground-wave propagation — The method by which radio waves travel along the Earth's surface.

Half-wave dipole — A basic antenna used by radio amateurs. It consists of a length of wire or tubing, opened and fed at the center. The entire antenna is $^1/_2$ wavelength long at the desired operating frequency.

Ham-bands-only receiver — A receiver designed to cover only the bands used by amateurs. Usually refers to the bands from 80 to 10 meters, sometimes including 160 meters.

Hand key — A simple switch used to send Morse code. Also called a **code key**, **straight key** or **telegraph key**.

Harmonics — Signals from a transmitter or oscillator occurring on whole-number multiples of the desired operating frequency.

Hertz (Hz) — An alternating-current frequency of one cycle per second. The basic unit of frequency.

High-pass filter — A filter designed to pass high-frequency signals, while blocking lower-frequency signals.

Impedance — The opposition to electric current that an antenna feed line presents. Impedance includes factors other than resistance, and applies to alternating currents. Ideally, the characteristic impedance of a feed line is the same as the transmitter output impedance and the antenna input impedance.

Impedance-matching device — A device that matches one impedance level to another. For example, it may match the impedance of an antenna system to the impedance of a transmitter or receiver. Amateurs also call such devices a Transmatch, impedance-matching network or antenna tuner.

Impedance-matching network — A device that matches the impedance of an antenna system to the impedance of a transmitter or receiver. Also called an **antenna-matching network**, **antenna tuner** or *Transmatch*.

Input frequency — A repeater's receiving frequency. To use a repeater, transmit on the input frequency and receive on the output frequency.

Insulator — A material that maintains a tight grip on its electrons, so that an electric current cannot pass through it (within voltage limits).

Integrated circuit (IC) — A modern electronics component that consists of many transistor elements on a single wafer of silicon.

Ionizing radiation — Electromagnetic radiation that has sufficient energy to knock electrons free from their atoms, producing positive and negative ions. X-rays, gamma rays and ultraviolet radiation are examples of ionizing radiation.

Ionosphere — A region of electrically charged (ionized) gases high in the atmosphere. The ionosphere bends radio waves as they travel through it, returning them to Earth. Also see **sky-wave propagation**.

K — The Morse code abbreviation for "any station respond."

Key click — A click or thump at the beginning or end of a CW signal.

Key-click filter — A circuit in a transmitter that reduces or eliminates key clicks.

Key-operated on-off switch — A good way to prevent unauthorized persons from using your station is to install a key-operated switch that controls station power.

Kilo — The metric prefix for 10^3, or times 1000.

Ladder line — Another name for **open-wire feed line**.

Lightning protection — There are several ways to help prevent lightning damage to your equipment (and your house), among them unplugging equipment, disconnecting antenna feed lines and using a lightning arrestor.

Line-of-sight propagation — The term used to describe VHF and UHF propagation in a straight line directly from one station to another.

Low-pass filter — A filter designed to pass low-frequency signals, while blocking higher-frequency signals.

Lower sideband (LSB) — The common single-sideband operating mode on the 160, 80 and 40-meter amateur bands.

Malicious (harmful) interference — Intentional, deliberate obstruction of radio transmissions.

Maximum usable frequency (MUF) — The highest-frequency radio signal that will reach a particular destination using **sky-wave propagation**, or *skip*. The MUF may vary for radio signals sent to different destinations.

MAYDAY — From the French *m'aidez* (help me), MAYDAY is used when calling for emergency assistance in voice modes.

MCW — The FCC emission type that describes international Morse code telegraphy communication with audio modulation of the carrier.

Mega — The metric prefix for 10^6, or times 1,000,000.

Metric prefixes — A series of terms used in the metric system of measurement. We use metric prefixes to describe a quantity as compared to a basic unit. The metric prefixes indicate multiples of 10.

Metric system — A system of measurement developed by scientists and used in most countries of the world. This system uses a set of prefixes that are multiples of 10 to indicate quantities larger or smaller than the basic unit.

Micro — The metric prefix for 10^{-6}, or divide by 1,000,000.

Microphone — A device that converts sound waves into electrical energy.

Milli — The metric prefix for 10^{-3}, or divide by 1000.

Mobile device — A radio transmitting device designed to be mounted in a vehicle. In Amateur Radio operation, the transmitter is usually activated by a push-to-talk (PTT) switch.

Modem — Short for *mo*dulator/*dem*odulator. A modem modulates a radio signal to transmit data and demodulates a received signal to recover transmitted data.

Monitor mode — One type of packet radio receiving mode. In monitor mode, everything transmitted on a packet frequency is displayed by the monitoring TNC. This occurs whether or not the transmissions are addressed to the monitoring station.

Morse code — (See **CW**).

Multimode transceiver — Transceiver capable of SSB, CW and FM operation.

Narrow-band direct-printing telegraphy — The technical term for **radioteletype (RTTY)**.

National Electrical Code — A set of guidelines governing electrical safety, including antennas.

Network — A term used to describe several packet stations linked together to transmit data over long distances.

Nonionizing radiation — Electromagnetic radiation that does not have sufficient energy to knock electrons free from their atoms. Radio frequency (RF) radiation is nonionizing.

NPN Transistor — A transistor that has a layer of P-type semiconductor material sandwiched between layers of N-type semiconductor material.

Offset — The 300 to 1000-Hz difference in CW transmitting and receiving frequencies in a transceiver. For a repeater, offset refers to the difference between its transmitting and receiving frequencies.

Ohm — The basic unit of electrical resistance, used to describe the amount of opposition to current.

Ohm's Law — A basic law of electronics. Ohm's Law gives a relationship between voltage (E), current (I) and resistance (R). The voltage applied to a circuit is equal to the current through the circuit times the resistance of the circuit (E = IR).

Open circuit — An electrical circuit that does not have a complete path, so current can't flow through the circuit.

Open repeater — A repeater that can be used by all hams who have a license that authorizes operation on the repeater frequencies.

Open-wire feed line — **Parallel-conductor feed line** that has air as its primary insulation material. (Also see **ladder line**.)

Operator/primary station license — An amateur license actually includes two licenses in one. The operator license is that portion of an Amateur Radio license that gives permission to operate an amateur station. The primary station license is that portion of an Amateur Radio license that authorizes an amateur station at a specific location. The station license also lists the call sign of that station.

Output frequency — A repeater's transmitting frequency. To use a repeater, transmit on the input frequency and receive on the output frequency.

Packet radio — A digital communications system in which information is broken into short bursts. The bursts (packets) also contain addressing and error-detection information.

Parallel-conductor feed line — Feed line with two conductors held a constant distance apart. (Also called **ladder line** or **open-wire feed line**.)

Peak envelope power (PEP) — The average power of a signal at its largest amplitude peak.

Pecuniary — Payment of any type, whether money or other goods. Amateurs may not operate their stations in return for any type of payment.

Pentode — A vacuum tube with five active elements: cathode, plate, control grid, screen grid and suppressor grid.

Phone — Another name for **voice communications**.

Phonetic alphabet — Standard words used on voice modes to make it easier to understand letters of the alphabet, such as those in call signs. The call sign KA6LMN stated phonetically is Kilo Alfa Six Lima November.

Pico — The metric prefix for 10^{-12}, or divide by 1,000,000,000,000.

PL (see **CTCSS**)

PNP Transistor — A transistor that has a layer of N-type semiconductor material sandwiched between layers of P-type semiconductor material.

Polarization — The electrical-field characteristic of a radio wave. An antenna that is parallel to the surface of the earth, such as a dipole, produces horizontally polarized waves. One that is perpendicular to the earth's surface, such as a quarter-wave vertical, produces vertically polarized waves. An antenna that has both horizontal and vertical polarization is said to be circularly polarized.

Portable device — A radio transmitting device designed to have a transmitting antenna that is generally within 20 centimeters of a human body.

Potentiometer — Another name for a **variable resistor**. The value of a potentiometer can be changed without removing it from a circuit.

Power — The rate of energy consumption. We calculate power in an electrical circuit by multiplying the voltage applied to the circuit times the current through the circuit (P = IE).

Power supply — An electrical circuit that provides excess electrons to flow into another circuit. The power supply also supplies the voltage or EMF to push the electrons along. Power supplies convert a power source (such as the ac mains) to a form useful for various circuits. A power supply usually provides a direct-current output at some desired voltage from an ac input voltage.

Procedural signal (prosign) — One or two letters sent as a single character. Amateurs use prosigns in CW contacts as a short way to indicate the operator's intention. Some examples are K for "Go Ahead," or AR for "End of Message." (The bar over the letters indicates that we send the prosign as one character.)

Propagation — The study of how radio waves travel.

Q signals — Three-letter symbols beginning with Q. Used on CW to save time and to improve communication. Some examples are QRS (send slower), QTH (location), QSO (ham conversation) and QSL (acknowledgment of receipt).

QRL? — Ham radio Q signal meaning "Is this frequency in use?"

QSL card — A postcard that serves as a confirmation of communication between two hams.

QSO — A conversation between two radio amateurs.

Quarter-wavelength vertical antenna — An antenna constructed of a quarter-wavelength long radiating element placed perpendicular to the earth.

Radio frequency (RF) — The range of frequencies that can be radiated through space in the form of electromagnetic radiation. We usually consider RF to be those frequencies higher than the audio frequencies, or above 20 kilohertz.

Radio-frequency interference (RFI) — Disturbance to electronic equipment caused by radio-frequency signals.

Radioteletype (RTTY) — Radio signals sent from one teleprinter machine to another machine. Anything that one operator types on his teleprinter will be printed on the other machine. Also known as **narrow-band direct-printing telegraphy**.

Ragchew — A lengthy conversation between two radio amateurs.

Receiver — A device that converts radio waves into signals we can hear or see.

Receiver incremental tuning (RIT) — A transceiver control that allows for a slight change in the receiver frequency without changing the transmitter frequency. Some manufacturers call this a clarifier (CLAR) control.

Receiver overload — Interference to a receiver caused by a strong RF signal that forces its way into the equipment. A signal that overloads the receiver RF amplifier (front end) causes **front-end overload**. Receiver overload is sometimes called **RF overload**.

Reflection — Signals that travel by **line-of-sight propagation** are reflected by large objects like buildings.

Reflector — An element behind the driven element in a Yagi and some other directional antennas.

Repeater — An amateur station that receives a signal and retransmits it for greater range.

Resistance — The ability to oppose an electric current.

Resistor — Any material that opposes a current in an electrical circuit. An electronic circuit component especially designed to oppose current, and used to control the current through a circuit.

Resonant frequency — The desired operating frequency of a tuned circuit. In an antenna, the resonant frequency is one where the feed-point impedance contains only resistance.

RF burn — A burn produced by coming in contact with RF voltages.

RF overload — Another term for **receiver overload**.

RF radiation — Waves of electric and magnetic energy. Such electromagnetic radiation with frequencies as low as 3 kHz and as high as 300 GHz are considered to be part of the RF region.

RF safety — Preventing injury or illness to humans from the effects of radio-frequency energy.

Rig — The radio amateur's term for a transmitter, receiver or transceiver.

RST — A system of numbers used for signal reports: R is readability, S is strength and T is tone. (On single-sideband phone, only R and S reports are used.)

Safety interlock — A switch that automatically turns off ac power to a piece of equipment when the top cover is removed.

Schematic symbol — A drawing used to represent a circuit component on a wiring diagram.

Selectivity — The ability of a receiver to separate two closely spaced signals.

Sensitivity — The ability of a receiver to detect weak signals. Shack — The room where an Amateur Radio operator keeps his or her station equipment.

Shack — The room where an Amateur Radio operator keeps his or her station equipment.

Short circuit — An electrical circuit in which the current does not take the desired path, but finds a shortcut instead. Often the current goes directly from the negative power-supply terminal to the positive one, bypassing the rest of the circuit.

Sidebands — The sum or difference frequencies generated when an RF carrier is mixed with an audio signal. Single-sideband phone (SSB) signals have an upper sideband (USB — that part of the signal above the carrier) and a lower sideband (LSB — the part of the signal below the carrier). SSB transceivers allow operation on either USB or LSB.

Simplex operation — Receiving and transmitting on the same frequency. See **duplex operation**.

Single sideband (SSB) — A common mode of voice operation on the amateur bands. SSB is a form of **amplitude modulation**.

Single-pole, double-throw (SPDT) switch — A switch that connects one center contact to one of two other contacts.

Single-pole, single-throw (SPST) switch — A switch that only connects one center contact to another contact.

Skip zone — An area of poor radio communication, too distant for ground waves and too close for sky waves.

Sky-wave propagation — The method by which radio waves travel through the ionosphere and back to Earth. Sometimes called *skip*, sky-wave propagation has a far greater range than **line-of-sight** and **ground-wave propagation**.

SOS — A Morse code call for emergency assistance.

Space station — An amateur station located more than 50 km above the Earth's surface.

Splatter — A type of interference to stations on nearby frequencies. Splatter occurs when a transmitter is overmodulated.

Spurious emissions — Signals from a transmitter on frequencies other than the operating frequency.

SSB — Emission mode that describes the type of voice emission used on the HF bands. Abbreviation for **single sideband**.

Standing-wave ratio (SWR) — Sometimes called voltage standing-wave ratio (VSWR). A measure of the impedance match between the feed line and the antenna. Also, with a Transmatch in use, a measure of the match between the feed line from the transmitter and the antenna system. The system includes the Transmatch and the line to the antenna. VSWR is the ratio of maximum voltage to minimum voltage along the feed line. Also the ratio of antenna impedance to feed-line impedance when the antenna is a purely resistive load.

Station grounding — Connecting all station equipment to a good earth ground improves both safety and station performance.

Straight key — Another name for a **code key, hand key** or **telegraph key**.

Sunspot cycle — The number of **sunspots** increases and decreases in a predictable cycle that lasts about 11 years.

Sunspots — Dark spots on the surface of the sun. When there are few sunspots, long-distance radio propagation is poor on the higher-frequency bands. When there are many sunspots, long-distance HF propagation improves.

Superimposed hum — A low-pitched buzz or hum on a radio signal.

Switch — A device used to connect or disconnect electrical contacts.

SWR meter — A device used to measure SWR.

SWR meter — A measuring instrument that can indicate when an antenna system is working well.

Telegraph key — Another name for a **code key, hand key** or **straight key**.

Teleprinter — A machine that can convert keystrokes (typing) into electrical impulses. The teleprinter can also convert the proper electrical impulses back into text. Computers have largely replaced teleprinters for amateur radioteletype work.

Television interference (TVI) — Interruption of television reception caused by another signal.

Terminal node controller (TNC) — A TNC accepts information from a computer and converts the information into packets. The TNC also receives packets and extracts information to be displayed by a computer.

Third-party communications — Messages passed from one amateur to another on behalf of a third person.

Third-party communications agreement — An official understanding between the United States and another country that allows amateurs in both countries to participate in third-party communications.

Third-party participation — The way an unlicensed person can participate in amateur communications. A control operator must ensure compliance with FCC Rules.

Ticket — A common name for an Amateur Radio license.

Time-out timer — A device that limits the amount of time any one person can talk through a repeater.

Traffic net — An on-the-air meeting of amateurs, for the purpose of relaying messages.

Transceiver — A radio transmitter and receiver combined in one unit.

Transistor — A solid-state device made of three layers of semiconductor material. See **NPN transistor** and **PNP transistor**.

Transmission line — The wires or cable used to connect a transmitter or receiver to an antenna. Also called a **feed line**.

Transmitter — A device that produces radio-frequency signals.

Triode — A vacuum tube with three active elements: cathode, plate and control grid.

Uncontrolled environment — Any area in which an RF signal may cause radiation exposure to people who may not be aware of the radiated electric and magnetic fields. The FCC generally considers members of the general public and an amateur's neighbors to be in an uncontrolled RF radiation exposure environment to determine the maximum permissible exposure levels.

Unidentified communications or signals — Signals or radio communications in which the transmitting station's call sign is not transmitted.

Upper sideband (USB) — The common single-sideband operating mode on the 20, 17, 15, 12 and 10-meter HF amateur bands, and all the VHF and UHF bands.

Variable resistor — A resistor whose value you can change without removing it from a circuit.

Vertical antenna — A common amateur antenna, usually made of metal tubing. The radiating element is vertical. There are usually four or more radial elements parallel to or on the ground.

Voice communications — Hams can use several voice modes, including FM and SSB.

Volt (V) — The basic unit of electrical pressure or EMF.

Voltage — The EMF or pressure that causes electrons to move through an electrical circuit.

Voltage source — Any source of excess electrons. A voltage source produces a current and the force to push the electrons through an electrical circuit.

Volunteer Examiner (VE) — A licensed amateur who is accredited by a **Volunteer Examiner Coordinator (VEC)** to administer amateur license exams.

Volunteer Examiner Coordinator (VEC) — An organization that has entered into an agreement with the FCC to coordinate amateur license exams.

W1AW — The headquarters station of the American Radio Relay League. This station is a memorial to the League's cofounder, Hiram Percy Maxim. The station provides daily on-the-air code practice and bulletins of interest to hams.

Watt (W) — The unit of power in the metric system. The watt describes how fast a circuit uses electrical energy.

Wavelength — Often abbreviated λ. The distance an ac signal (such as a radio wave) will travel during the time it takes the signal to go through one complete cycle. The wavelength relates to frequency. Higher frequencies have shorter wavelengths.

Yagi antenna — The most popular type of amateur directional (beam) antenna. It has one driven element and one or more additional elements.

73 — Ham lingo for "best regards." Used on both phone and CW toward the end of a contact.

INDEX

F

G

H

I

K

L

M

N

O

P

Q

R

S

ABOUT THE AMERICAN RADIO RELAY LEAGUE

The seed for Amateur Radio was planted in the 1890s, when Guglielmo Marconi began his experiments in wireless telegraphy. Soon he was joined by dozens, then hundreds, of others who were enthusiastic about sending and receiving messages through the air—some with a commercial interest, but others solely out of a love for this new communications medium. The United States government began licensing Amateur Radio operators in 1912.

By 1914, there were thousands of Amateur Radio operators—hams—in the United States. Hiram Percy Maxim, a leading Hartford, Connecticut, inventor and industrialist saw the need for an organization to band together this fledgling group of radio experimenters. In May 1914 he founded the American Radio Relay League (ARRL) to meet that need.

Today ARRL, with more than 170,000 members, is the largest organization of radio amateurs in the United States. The League is a not-for-profit organization that:

- promotes interest in Amateur Radio communications and experimentation
- represents US radio amateurs in legislative matters, and
- maintains fraternalism and a high standard of conduct among Amateur Radio operators.

At League headquarters in the Hartford suburb of Newington, the staff helps serve the needs of members. ARRL is also International Secretariat for the International Amateur Radio Union, which is made up of similar societies in more than 150 countries around the world.

ARRL publishes the monthly journal *QST*, as well as newsletters and many publications covering all aspects of Amateur Radio. Its headquarters station, W1AW, transmits Morse-code practice sessions and bulletins of interest to radio amateurs. The League also coordinates an extensive field organization, which provides technical and other support for radio amateurs as well as communications for public service activities. ARRL also represents US amateurs with the Federal Communications Commission and other government agencies in the US and abroad.

Membership in ARRL means much more than receiving *QST* each month. In addition to the services already described, ARRL offers membership services on a personal level, such as the ARRL Volunteer Examiner Coordinator Program and a QSL bureau.

Full ARRL membership (available only to licensed radio amateurs) gives you a voice in how the affairs of the organization are governed. League policy is set by a Board of Directors (one from each of 15 Divisions). Each year, half of the ARRL Board of Directors stands for election by the full members they represent. The day-to-day operation of ARRL HQ is managed by an Executive Vice President and a Chief Financial Officer.

No matter what aspect of Amateur Radio attracts you, ARRL membership is relevant and important. There would be no Amateur Radio as we know it today were it not for the ARRL. We would be happy to welcome you as a member! (An Amateur Radio license is not required for Associate Membership.) For more information about ARRL and answers to any questions you may have about Amateur Radio, write or call:

ARRL
225 MAIN STREET
NEWINGTON CT 06111-1494
(860) 594-0200

Prospective new amateurs call:
800-32-NEW HAM (800-326-3942)

You can also contact us via e-mail:
newham@arrl.org

or check out our World Wide Web site:
http://www.arrl.org/

NOW YOU'RE TALKING!

PROOF OF PURCHASE

Welcome to Amateur Radio!

Congratulations! You've taken your first step into a hobby—and a service—that knows no limits. Amateur Radio is a worldwide network of people from various cultures, united by a common love of wireless communication. Amateur Radio is as old as radio itself, and its future is no less fantastic than its past.

For most people, Amateur Radio is a lifelong pursuit. We want to make sure you get a good start, which is why we've published this book. But first, who are "we" ?

THE AMERICAN RADIO RELAY LEAGUE: WHAT'S IN IT FOR YOU?

◆ **Help for New Hams**: Are you a beginning ham looking for help in getting started in your new hobby? The hams at ARRL HQ in Newington, Connecticut, will be glad to assist you. Call 800-32-NEWHAM. ARRL maintains a computer data base of ham clubs and ham radio "helpers" from across the country who've told us they're interested in helping beginning hams. There are probably several clubs in your area! Contact us for more information.

◆ **Licensing Classes**: If you're going to become a ham, you'll need to find a local license exam opportunity sooner or later. ARRL Registered Instructors teach licensing classes all around the country, and ARRL-sponsored Volunteer Examiners are right there to administer your exams. To find the locations and dates of Amateur Radio Licensing classes and test sessions in your area, call the New Ham Desk at 800-32-NEWHAM.

◆ **Clubs**: As a beginning ham, one of the best moves you can make is to join a local ham club. Whether you join an all-around group or a special-interest club (repeaters, DXing, and so on), you'll make new friends, have a lot of fun, and you can tap into a ready reserve of ham radio knowledge and experience. To find the ham clubs in your area, call HQ's New Ham Desk at 800-32-NEWHAM.

◆ **Technical Information Service**: Do you have a question of a technical nature? (What new ham doesn't?) Contact the Technical Information Service (TIS) at HQ. Our resident technical experts will help you over the phone, send you specific information on your question (antennas, interference and so on) or refer you to your local ARRL Technical Coordinator or Technical Specialist. It's expert information—and it doesn't cost Members an extra cent!

◆ **Regulatory Information**: Need help with a thorny antenna zoning problem? Having trouble understanding an FCC regulation? Vacationing in a faraway place and want to know how to get permission to operate your ham radio there? HQ's Regulatory Information Branch has the answers you need!

◆ **Operating Awards**: Like to collect "wallpaper"? The ARRL sponsors a wide variety of certificates and Amateur Radio achievement awards. For information on awards you can qualify for, contact the Membership Services Department at HQ.

◆ **Equipment Insurance**: When it comes to protecting their Amateur Radio equipment investments, ARRL Members travel First Class. A. H. Wohlers Company provides the League's "all-risk" equipment insurance plan. (It can protect your ham radio computer, too.) It's comprehensive and cost effective, and it's available only to ARRL Members. Why worry about losing your valuable radio equipment when you can protect it for only a few dollars a year?

◆ **Amateur Radio Emergency Service**: If you're interested in providing public service and emergency communications for your community, you can join more than 25,000 other hams who have registered their communications capabilities with local Emergency Coordinators. Your EC will call on you and other ARES members for vital assistance if disaster should strike your community. Contact the Field Services Department at HQ for information.

◆ **Audio-Visual Programs**: Need a program for your next ham club meeting, informal get-together or public display? ARRL Affililated clubs can buy tapes from the League's video programs. There are many to choose from, including popular titles on Amateur Radio's role in Operation Desert Storm and space shuttle activities. Contact the Educational Activities Department at HQ for a complete list or to order a tape.

◆ **Blind, Disabled Ham Help**: For a list of available resources and information on the Courage HANDI-HAM System, contact the ARRL Program for the Disabled at HQ.

With your membership you also receive the monthly journal *QST*. Each 200-page issue is packed with valuable information you can use—including a special *New Ham Companion* section! And *QST* Product Reviews are the most respected source of information to help you get the most for your Amateur Radio equipment dollar. (For many hams, *QST* alone is worth far more than the cost of ARRL membership.)

The ARRL also publishes newsletters and dozens of books covering all aspects of Amateur Radio. Our Headquarters station, W1AW, transmits bulletins of interest to radio amateurs and Morse Code practice sessions.

When it comes to representing Amateur Radio's best interests in our nation's capital, the League's team in Washington, DC, is constantly working with the FCC, Congress and industry to protect and foster your privileges as a ham operator.

Regardless of your Amateur Radio interests, ARRL Membership is relevant and important. We will be happy to welcome you as a Member. Use the Invitation to Membership on the next page to **join today**. And don't hesitate to contact us if you have any questions!

Invitation to Membership

Class of License	Call Sign	Date of Application

Name

Address

City, State, ZIP

Membership Class	❑ Regular		❑ Family ❑ Blind	❑ 65 or older with Proof of Age
	US	ELSEWHERE		US
❑ 1 year	$34	$47	$5	$28
❑ 2 years	65	89	10	53
❑ 3 years	92	127	15	76

Currently unlicensed (check here) ❑

A member of the immediate family of a League member, living at the same address, may become a League member without *QST* at the special rate of $5 per year. Family membership must run con-current with that of the member receiving *QST*. Blind amateurs may join without *QST* for $5 per year.

If you are 21 or younger, a special rate may apply. Write HQ for details.

Payment enclosed ❑

Charge to: ❑ VISA (13 or 16 digits) ❑ MasterCard (16 digits) ❑ AMEX (15 digits) ❑ Discover (16 digits)

Card Number ____________

Card good from ________ to ________ (EXPIRATION DATE)

Signature ____________

If you do not wish your name and address made available for non-ARRL related mailings, please check this box. ❑

Dues rates subject to change without notice.

THE AMERICAN RADIO RELAY LEAGUE, INC

225 MAIN STREET NEWINGTON, CONNECTICUT 06111 USA 860-594-0200 FAX: 860-594-0303

YOU'VE GOT QUESTIONS? WE'VE GOT ANSWERS!

Now you've got what you need to go after your own Amateur Radio license and call sign! But you've probably still got a question or tow. Does anyone in my area teach classes? Where can I find a person who can give me my exam? Where in my area can I buy equipment? Which is better from my location, a dipole antenna or a vertical antenna? If I decide to learn Morse code, where can I find someone to help me practice? Can someone check my station to see if I've set up everything correctly? Who do I turn to for answers? The American Radio Relay League's New Ham Desk can send you a list of Amateur Radio clubs, instructors, examiners and Elmers who live in your area and enjoy helping newcomers. Here's your first question and answer: What's an Elmer? An Elmer is a person who helps you with whatever you need; an Elmer is another Amateur Radio tradition of hams helping others. For your list, send a postcard to ARRL NewHam Desk, 225 Main St, Newington, CT 06111, or call us at 1-800-32NEWHAM.

FEEDBACK

Please use this form to give us your comments on this book and what you'd like to see in future editions, or e-mail us at **pubsfdbk@arrl.org** (publications feedback). If you use e-mail, please include your name, call, e-mail address and the book title, edition and printing in the body of your message. Also indicate whether or not you are an ARRL member.

Please check the box that best answers these questions:

How well did this book prepare you for your exam? ☐ Very well ☐ Fairly well ☐ Not very well

How well did this book teach you about ham radio? ☐ Very well ☐ Fairly well ☐ Not very well

Which exam did you take (or will you be taking)? ☐ Novice ☐ Technician Did you pass? ☐ Yes ☐ No

If you checked Technician: Do you expect to learn Morse code at some point? ☐ Yes ☐ No

Which operating modes do you plan to use first?

☐ SSB ☐ FM ☐ Packet ☐ RTTY ☐ Image ☐ Morse code ☐ Other ________________

Where did you purchase this book? ☐ From ARRL directly ☐ From an ARRL dealer

Is there a dealer who carries ARRL publications within:

☐ 5 miles ☐ 15 miles ☐ 30 miles of your location? ☐ Not sure.

Name ______________________________ ARRL member? ☐ Yes ☐ No

______________________________ Call Sign ________________

Address ______________________________

City, State/Province, ZIP/Postal Code ______________________________

Daytime Phone ( ) ______________________________ Age ________________

If licensed, how long? ______________________________

Other hobbies ______________________________

____

Occupation ______________________________

For ARRL use only	NYT
Edition	3 4 5 6 7 8 9 10 11 12
Printing	2 3 4 5 6 7 8 9 10 11 12

From ______________________________

____

____

Please affix postage. Post Office will not deliver without postage.

EDITOR, NOW YOU'RE TALKING!
AMERICAN RADIO RELAY LEAGUE
225 MAIN STREET
NEWINGTON CT 06111-1494

please fold and tape

GAP: THE PERFECT ANTENNA

We at GAP realize there isn't a perfect antenna. No singular antenna will scream DX on 80 and be the best for local nets on 10. If anyone tells you there is, beware! The perfect antenna does not exist, but the right one for you may. If you want something to bust the pile on the low bands, then consider the Voyager. Just starting out in ham radio and need a great general coverage antenna, the Challenger is easy to assemble and for little effort will yield superior performance, especially on DX. Maybe you knowingly or unknowingly moved into one of those "restricted areas" where the Eagle's limited visibility, but unlimited ability is desired.

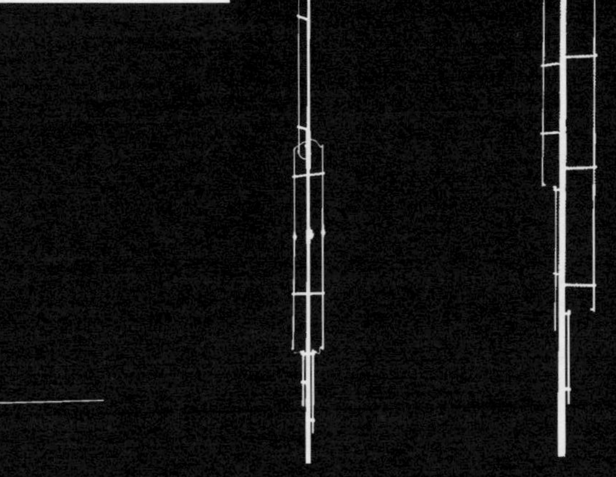

Eagle DX   Challenger DX   Voyager DX

This chart helps you select the right GAP antenna. W hen comparing GAPs, bandwidth is not a concern. With few exceptions, a GAP yields continuous coverage under 2:1 for the **ENTIRE BAND**.

All antennas utilize a GAP elevated asymmetric feed. A major benefit is the virtual elimination of the earth loss, so more RF radiates into the air instead of the ground. This feed is why a GAP requires **NO RADIALS**. Just as elevating a GAP offers no significant improvement to its performance, adding radials won't either, making set up a breeze.

A GAP antenna has no traps, coils or transformers. This is important. The greatest sources of failure in multiband antennas are these devices. Perhaps you heard someone discuss a trap that had melted, arced or became full of water. Improvements to these inherent problems are the focus of the antenna manufacturer, while the basic design of the antenna remains unchanged. **GAP improved the trap by eliminating it!** Removing these devices means they don't have to be tuned and, more importantly, won't be detuned by the first ice or rain. The absence of these devices improves antenna reliability, stability and increases bandwidth.

Another major advantage to a GAP antenna is its NO tune feature. Screws are simply inserted into predrilled holes with a supplied nutdriver.

The secret is out and people in the know say:

CQ–"The GAP consistently outperformed base-fed antennas...and was quieter."
73–"This is a real DX antenna, much quieter than other verticals."
RF–"To say this antenna is effective would be a real understatement. Switching back and forth on 40m between another multiband HF vertical and the GAP, there was no comparison. Signals were always stronger on the GAP, sometimes by S units, not just DB's."
Worldradio – "These guys have solved the problem associated with verticals. That is, an awful lot of RF is wallowing around and dropping into the dirt instead of going outward bound. A half-wave vertical does need radials if it is end fed (at the bottom). But the same half-wave vertical does not (as much, hardly at all) if is fed in the center."
IEEE–"Near field and power density analyses show another advantage of this antenna (asymmetric vertical dipole): it decreases the power density close to the ground, and so avoids power dissipation in the soil below it. The input impedance is very stable and almost independent of ground conductivity. This antenna can operate with high radiation efficiency in the MF AM standard broadcast band, without the classical buried ground plane, so as to yield easier installation and maintenance."

Latest Release: **TITAN DX**

This all purpose antenna is designed to operate 10m-80m, WARC bands included. It sits on a 1-1/4" pipe and can be mounted close to the ground or up on a roof. Its bandwidth and no tune feature make it an ideal antenna for the limited space environment as well as a terrific addition to the antenna farm.

MODEL	BANDS OF OPERATION											HT	WT	MOUNT	COUNTER-POISE	COST
	2m	6m	10m	12m	15m	17m	20m	30m	40m	80m	160m					
Challenger DX	■	■	■	■	■		■		■	■		31.5'	21 lbs	Drop In Ground Mount	3 Wires @ 25'	$259
Eagle DX			■	■	■	■	■		■			21.5'	19 lbs	1-1/4" pipe	80" Rigid	$269
Titan DX			■	■	■	■	■	■	■	■		25'	25 lbs	1-1/4" pipe	80" Rigid	$299
Voyager DX							■		■	■	■	45'	39 lbs	Hinged Base	3 Wires @ 57'	$399

GAP ANTENNA PRODUCTS INC.
6010 N. Old Dixie Hwy.
Vero Beach, FL 32967

TO ORDER, CALL
(561) 778-3728

Come Visit Us At gapantenna.com